Hydrometallurgy

Special Issue Editors

Suresh Bhargava
Mark Pownceby
Rahul Ram

MDPI • Basel • Beijing • Wuhan • Barcelona • Belgrade

MDPI

Special Issue Editors
Suresh Bhargava
RMIT University
Australia

Mark Pownceby
CSIRO Mineral Resources
Australia

Rahul Ram
RMIT University
Australia

Editorial Office
MDPI AG
St. Alban-Anlage 66
Basel, Switzerland

This edition is a reprint of the Special Issue published online in the open access journal *Metals* (ISSN 2075-4701) from 2015–2016 (available at: http://www.mdpi.com/journal/metals/special_issues/hydrometallurgy).

For citation purposes, cite each article independently as indicated on the article page online and as indicated below:

Author 1; Author 2. Article title. *Journal Name* **Year**, *Article number*, page range.

First Edition 2017

ISBN 978-3-03842-464-2 (Pbk)
ISBN 978-3-03842-465-9 (PDF)

Table of Contents

About the Special Issue Editors

Suresh Bhargava is a world-renowned interdisciplinary scientist who has achieved excellence in five disciplines and is recognized for delivering research that underpins significant industrial applications. He has published over 400 journal articles and 200 industrial reports. His research has been cited over 9,000 times. Out of his seven patents, five have gone to industries or licensed to commercialization. He has been quoted as being among the top 1% world scientists in the resource sector. As a passionate supporter of technological science and engineering for innovation, he provides consultancy and advisory services to many government and industrial bodies around the world, including BHP Billiton, Alcoa World Alumina, Rio Tinto and Mobil Exxon. Fellow of six Academies around the world, Professor Bhargava was awarded many prestigious national and international awards including the 2016 Khwarizmi International Award (KIA), the 2015 CHEMECA Medal, Indian National Science Academy's P. C. Ray Chair (distinguished lecture series 2014), the RMIT University Vice Chancellor's Research Excellence Award (2006 and 2014), the Applied Research award (2013), and the R. K. Murphy Medal (2008) by the Royal Australian Chemical Institute. Most recently, he was decorated with the title of Distinguished Professor at RMIT University.

Mark Pownceby is a Principal Research Scientist at CSIRO Mineral Resources, Melbourne Australia. An Earth Scientist by training, he spent 3 years as a visiting scientist at the Bayerisches Geoinstitut (Bayreuth, Germany) developing experimental techniques for measuring fundamental thermodynamic properties of alloys and oxides before joining CSIRO in 1992 as a process mineralogist. He has >25 years research experience in process mineralogy where his activities span a range of disciplines including: uranium ore characterization and hydrometallurgy, solid state chemistry and mineralogy, experimental phase equilibria, advanced resource characterization and processing of iron ore and heavy mineral sands. He has considerable expertise in minerals and materials characterization specializing in the application of electron probe microanalysis to ores and processed products and in using in-situ x-ray diffraction techniques for monitoring and quantifying mineralogical changes during processing. Mark is currently an Adjunct Professor at both RMIT and Swinburne Universities, Australia.

Rahul Ram is currently a Post-doctoral research fellow at RMIT University, Melbourne Australia. He received an APAI scholarship from BHP Billiton to conduct his PhD in 2013 on processing of uranium ores at RMIT University for which he received the Dr. Megan Clark Excellence Award. Following this, he worked as a Process Advisor with Rio Tinto G&I before returning to RMIT University as a key member of the ARC Linkage between Rio Tinto, Murdoch Uni, CSIRO and RMIT. He then received the URC fellowship award as a visiting research fellow at the University of Cape Town, South Africa before returning to RMIT. His research expertise includes hydrometallurgy, electrochemistry, geochemistry, materials synthesis and characterization, process modeling and sustainability design; with significant experience in both fundamental and applied research across various industries. He is currently the assistant editor of Hydrometallurgy, the premier journal in the field.

metals

MDPI

Editorial

Hydrometallurgy

Suresh K. Bhargava [1,*], Mark I. Pownceby [2,*] and Rahul Ram [1,*]

[1]   Centre of Advanced Materials & Industrial Chemistry, School of Applied Sciences, RMIT University,
GPO Box 2476, Melbourne, VIC 3000, Australia

[2]   CSIRO Mineral Resources, Private Bag 10, Clayton South, VIC 3169, Australia

*   Correspondence: suresh.bhargava@rmit.edu.au (S.K.B.); mark.pownceby@csiro.au (M.I.P.);
rahul.ram@rmit.edu.au (R.R.)

Received: 18 May 2016; Accepted: 18 May 2016; Published: 23 May 2016

Hydrometallurgy, which involves the use of aqueous solutions for the recovery of metals from ores, concentrates, and recycled or residual material, plays an integral role in the multi-billion dollar minerals processing industry. It involves either the selective separation of various metals in solution on the basis of thermodynamic preferences, or the recovery of metals from solution through electro-chemical reductive processes or through crystalisation of salts. There are numerous hydrometallurgical process technologies used for recovering metals, such as: agglomeration; leaching; solvent extraction/ion exchange; metal recovery; and remediation of tailings/waste. Hydrometallurgical processes are integral across various stages in a typical mining recovery and mineral processing circuits be it *in situ* leaching (where solution is pumped through rock matrices); heap leaching (of the ROM or crushed ore); tank/autoclave leaching (of the concentrate/matte obtained from floatation); electro-refining (of the blister product from smelting routes); and the treatment of waste tailings/slags from the aforementioned processes. Modern hydrometallurgical routes to extract metals from their ores are faced with a number of issues related to both the chemistry, geology and engineering aspects of the processes involved. These issues include declining ore grade, variations in mineralogy across the deposits and geo-metallurgical locations of the ore site; which would influence the hydrometallurgical route chosen. The development of technologies to improve energy efficiency, water/resources consumption and waste remediation (particularly acid-rock drainage) across the circuit is also an important factor to be considered. Therefore, there is an ongoing development of novel solutions to these existing problems at both fundamental scales and pilot plant scales in order to implement environmentally sustainable practices in the recovery of valuable metals.

The Present Issue

We are delighted to be the Guest Editors for this Special Issue of Hydrometallurgy published in the journal *Metals*. With a total of 22 papers covering both fundamental and applied research, this issue covers all aspects of hydrometallurgy from comprehensive review articles [1,2], theoretical modelling [3] and experimental simulations [4], surface studies of dissolution mechanisms and kinetics [5], pre-treatment by roasting [6] or carbonation [7] to enhance recovery, aqueous carbonation as a means of CO_2 sequestration [8], biological systems [9–11], solvent and liquid-liquid extraction [12–14], nanoparticle preparation [15,16] and the development of novel and/or environmentally sustainable methods for the treatment of wastes and effluents for the recovery of valuable metals and products [17–22]. The number and of quality of submissions makes this Special Issue of *Metals* the most successful to date. As Guest Editors, we would especially like to thank Dr. Jane Zhang, Managing Editor for her support and active role in the publication. We are also extremely grateful to the entire staff of the *Metals* Editorial Office, who productively collaborated on this endeavour. Furthermore, we would like to thank all of the contributing authors for their excellent work.

References

1. Huang, H.-H. The Eh-pH Diagram and Its Advances. *Metals* **2016**, *6*, 23. [CrossRef]
2. Rutledge, J.; Anderson, C.G. Tannins in Mineral Processing and Extractive Metallurgy. *Metals* **2015**, *5*, 1520–1542. [CrossRef]
3. Wadnerkar, D.; Pareek, V.K.; Utikar, R.P. CFD Modelling of Flow and Solids Distribution in Carbon-in-Leach Tanks. *Metals* **2015**, *5*, 1997–2020. [CrossRef]
4. Santini, T.C.; Fey, M.V.; Gilkes, R.J. Experimental Simulation of Long Term Weathering in Alkaline Bauxite Residue Tailings. *Metals* **2015**, *5*, 1241–1261. [CrossRef]
5. Li, Y.; Qian, G.; Li, J.; Gerson, A.R. Chalcopyrite Dissolution at 650 mV and 750 mV in the Presence of Pyrite. *Metals* **2015**, *5*, 1566–1579. [CrossRef]
6. Yoon, H.-S.; Kim, C.-J.; Chung, K.W.; Jeon, S.J.; Park, I.; Yoo, K.; Jha, K. The Effect of Grinding and Roasting Conditions on the Selective Leaching of Nd and Dy from NdFeB Magnet Scraps. *Metals* **2015**, *5*, 1306–1314. [CrossRef]
7. Santos, R.M.; Van Audenaerde, A.; Chiang, Y.W.; Iacobescu, R.I.; Knops, P.; Van Gerven, T. Nickel Extraction from Olivine: Effect of Carbonation Pre-Treatment. *Metals* **2015**, *5*, 1620–1644. [CrossRef]
8. Jo, H.; Jo, H.Y.; Rha, S.; Lee, P.-K. Direct Aqueous Mineral Carbonation of Waste Slate Using Ammonium Salt Solutions. *Metals* **2015**, *5*, 2413–2427. [CrossRef]
9. Fedje, K.K.; Modin, O.; Strömvall, A.-M. Copper Recovery from Polluted Soils Using Acidic Washing and Bioelectrochemical Systems. *Metals* **2015**, *5*, 1328–1348. [CrossRef]
10. Sueoka, Y.; Sakakibara, M.; Sera, K. Heavy Metal Behaviour in Lichen-Mine Waste Interactions at an Abandoned Mine Site in Southwest Japan. *Metals* **2015**, *5*, 1591–1608. [CrossRef]
11. Castro, L.; Blázquez, M.L.; González, F.; Munoz, J.A.; Ballester, A. Exploring the Possibilities of Biological Fabrication of Gold Nanostructures Using Orange Peel Extract. *Metals* **2015**, *5*, 1609–1619. [CrossRef]
12. Paiva, A.P.; Martins, M.E.; Ortet, O. Palladium(II) Recovery from Hydrochloric Acid Solutions by N,N'-Dimethyl-N,N'-Dibutylthiodiglycolamide. *Metals* **2015**, *5*, 2303–2315. [CrossRef]
13. Lu, D.; Chang, Y.; Wang, W.; Xie. F.; Asselin, E.; Dreisinger, D. Copper and Cyanide Extraction with Emulsion Liquid Membrane with LIX 7950 as the Mobile Carier: Part 1, Emulsion Stability. *Metals* **2015**, *5*, 2034–2047. [CrossRef]
14. Saito, S.; Ohno, O.; Igarashi, S.; Kato, T.; Yamaguchi, H. Separation and Recycling for Rare Earth Elements by Homogeneous Liquid-Liquid Extraction (HoLLE) Using a pH-Responsive Fluorine-Based Surfactant. *Metals* **2015**, *5*, 1543–1552. [CrossRef]
15. Zeng, X.; Niu, L.; Song, L.; Wang, X.; Shi, X.; Yan, J. Effect of Polymer Addition on the Structure and Hydrogen Evolution Reaction Property of Nanoflower-Like Molybdenum Disulfide. *Metals* **2015**, *5*, 1829–1844. [CrossRef]
16. King, S.R.; Massicot, J.; McDonagh, A.M. A Straightforward Route to Tetrachlorauric Acid from Gold Metal and Molecular Chlorine for Nanoparticle Synthesis. *Metals* **2015**, *5*, 1454–1461. [CrossRef]
17. Inoue, K.; Gurung, M.; Xiong, Y.; Kawakita, H.; Ohto, K.; Alam, S. Hydrometallurgical Recovery of Precious Metals and Removal of Hazardous Metals Using Persimmon Tannin and Persimmon Wastes. *Metals* **2015**, *5*, 1921–1956. [CrossRef]
18. Park, K.; Jung, W.; Park, J. Decontamination of Uranium-Contaminated Sand and Soil Using Supercritical CO_2 with a TBP-HNO_3 Complex. *Metals* **2015**, *5*, 1788–1798. [CrossRef]
19. Slimi, R.; Girard, C. "High-Throughput" Evaluation of Polymer-Supported Triazolic Appendages for Metallic Cations Extraction. *Metals* **2015**. *5*, 418–427. [CrossRef]
20. Lee, S.-H.; Kwon, O.; Yoo, K.; Alorro, R.D. Removal of Zn from Contaminated Sediment by $FeCl_3$ in HCl Solution. *Metals* **2015**, *5*, 1812–1820. [CrossRef]

21. Wei, Y.-L.; Wang, Y.-S.; Liu, C.-H. Preparation of Potassium Ferrate from Spent Steel Pickling Liquid. *Metals* **2015**, *5*, 1770–1787. [CrossRef]

22. Siciliano, A. Use of Nanoscale Zero-Valent Iron (NZVI) particles for Chemical Dentrification under Different Operating Conditions. *Metals* **2015** *5*, 1507–1519. [CrossRef]

metals

MDPI

Article

The Eh-pH Diagram and Its Advances

Hsin-Hsiung Huang

Metallurgical and Materials Engineering, Montana Tech, Butte, MT 59701, USA; hhuang@mtech.edu;
Tel.: +1-406-496-4139; Fax: +1-406-496-4664

Academic Editors: Suresh Bhargava, Mark Pownceby and Rahul Ram
Received: 29 July 2015; Accepted: 28 December 2015; Published: 14 January 2016

Abstract: Since Pourbaix presented Eh *versus* pH diagrams in his "Atlas of Electrochemical Equilibria in Aqueous Solution", diagrams have become extremely popular and are now used in almost every scientific area related to aqueous chemistry. Due to advances in personal computers, such diagrams can now show effects not only of Eh and pH, but also of variables, including ligand(s), temperature and pressure. Examples from various fields are illustrated in this paper. Examples include geochemical formation, corrosion and passivation, precipitation and adsorption for water treatment and leaching and metal recovery for hydrometallurgy. Two basic methods were developed to construct an Eh-pH diagram concerning the ligand component(s). The first method calculates and draws a line between two adjacent species based on their given activities. The second method performs equilibrium calculations over an array of points (500 × 800 or higher are preferred), each representing one Eh and one pH value for the whole system, then combines areas of each dominant species for the diagram. These two methods may produce different diagrams. The fundamental theories, illustrated results, comparison and required conditions behind these two methods are presented and discussed in this paper. The Gibbs phase rule equation for an Eh-pH diagram was derived and verified from actual plots. Besides indicating the stability area of water, an Eh-pH diagram normally shows only half of an overall reaction. However, merging two or more related diagrams together reveals more clearly the possibility of the reactions involved. For instance, leaching of Au with cyanide followed by cementing Au with Zn (Merrill-Crowe process) can be illustrated by combining Au-CN and Zn-CN diagrams together. A second example of the galvanic conversion of chalcopyrite can be explained by merging S, Fe–S and Cu–Fe–S diagrams. The calculation of an Eh-pH diagram can be extended easily into another dimension, such as the concentration of a given ligand, temperature or showing the solubility of stable solids. A personal computer is capable of drawing the diagram by utilizing a 3D program, such as ParaView, or VisIt, or MATLAB. Two 3D wireframe volume plots of a Uranium-carbonate system from Garrels and Christ were used to verify the Eh-pH calculation and the presentation from ParaView. Although a two-dimensional drawing is still much clearer to read, a 3D graph can allow one to visualize an entire system by executing rotation, clipping, slicing and making a movie.

Keywords: Pourbaix diagram; Eh-pH diagram; Eh-pH applications; ligand component; equilibrium line; mass balance point; Gibbs phase rule; 3D Eh-pH diagrams; ParaView; VisIt; MATLAB

---

1. Introduction

All Eh-pH diagrams are constructed under the assumption that the system is in equilibrium with water or rather with water's three essential components, H(+1), O(−2) and e(−1); the oxidation states are presented using Arabic numbers with a + or a − sign. The diagrams are divided into areas, each of which represents a locally-predominant species. Eh represents the oxidation-reduction potential based on the standard hydrogen potential (SHE), while pH represents the activity of the hydrogen ion (H^+, also known as a proton). An Eh-pH diagram can describe not only the effects of potential and pH, but also of complexes, temperature and pressures. By convention, Eh-pH diagrams always show the thermodynamically-stable area of water by two dashed diagonal lines.

Two typical Eh-pH diagrams, both based on thermodynamic data from the NBS database [1], are presented. Figure 1 shows an Eh-pH diagram for one component (excluding three essential H(+1), O(−2) and e(−1) components) of metal, in this case manganese, Mn, while Figure 2 is that of another component of mineral acid phosphorus, P. Both diagrams show that oxidized species reside in high Eh areas, while reduced species are in low Eh areas. The metal diagram starts, at the left edge, from metal ions (Mn^{2+}) at low pH, which progressively react with OH^- as pH increases to produce metal hydroxides ($Mn(OH)_2$) or oxides. The diagram for the mineral acid starts, again from the left, with acid (H_3PO_4) and progressively deprotonates due to reactions with OH^- to finally produce phosphate ion (PO_4^{3-}) at high pH. Figure 1 also illustrates the tendencies to transition between species.

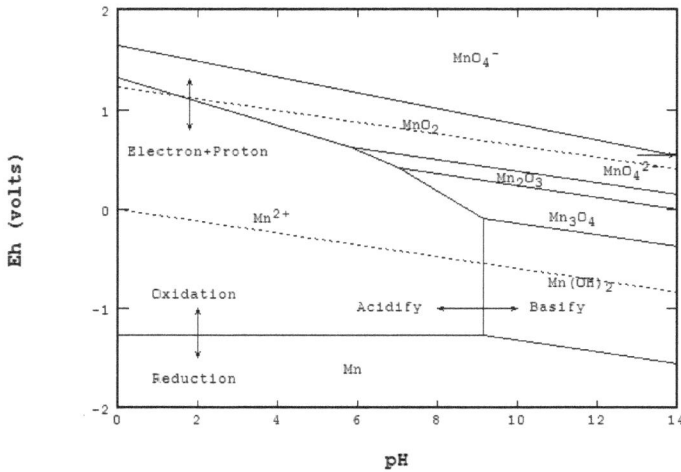

Figure 1. Eh-pH diagram of a Mn–water system. Dissolved manganese concentration, [Mn] = 0.001 M.

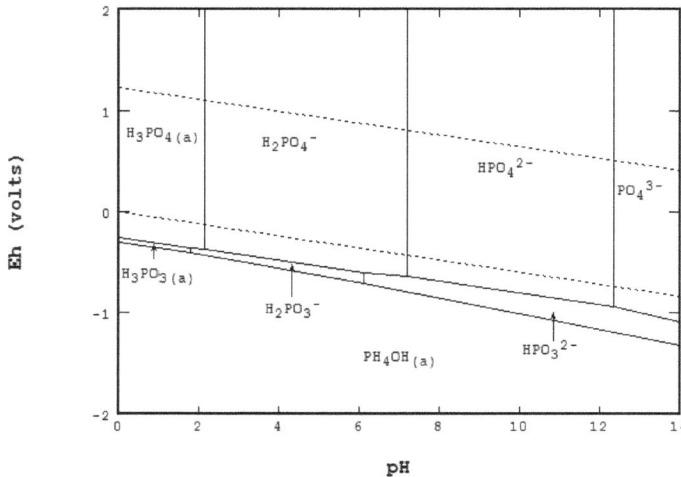

Figure 2. Eh pH diagram of a P-water system. Dissolved phosphorus species, [P] = 0.001 M.

Scope of the Paper

This paper illustrates some ways to improve a basic Eh-pH diagram for better visualization of species and stability regions. The demonstrated methods are all calculated and constructed with an ordinary PC, without a high-end graphics card, using Windows 7 or a higher version. All diagrams can be obtained in a short time. The fundamentals underlying the calculations are briefly described and/or available in the literature and listed as references. Discussions include:

1. Examples of applications: geochemical formation, corrosion and passivation, leaching and metal recovery, water treatment precipitation and adsorption.
2. Development of equilibrium line and mass balance point methods to handle ligand component(s): the theory, illustration and result comparison are presented; both methods satisfy the Gibbs phase rule derived for the Eh-pH diagram.
3. Examples by merging two or more diagrams for better illustration of the overall reactions involved in a process.
4. Demonstrations using a third party program to produce 3D diagrams with the addition of a third axis. The axis can represent the solubility of stable solids, ligand concentration or temperature. Two 3D wireframe volume plots of the Uranium-carbonate system based on a classic Garrels and Christ [2] work were used to verify the Eh-pH calculation and the presentation from ParaView.

This paper is not intended to discuss the following topics in detail:

1. Comparison among existent computer programs listed from the literature that directly or indirectly construct an Eh-pH diagram.
2. Effects from temperature, pressure, ionic strength and surface complexation for aqueous chemistry.
3. The algorithm and flow sheet to construct the diagram used by the author: they are available and referenced elsewhere; no source codes of the programs are presented.
4. Comparison or comments on third party 3D programs used by the author.

Note: The diagrams shown in this paper are solely for illustration. Unless specified, all were constructed at a temperature of 25 °C and zero ionic strength. The molarity is used for a dissolved species as [species], and Σcomponent is used to represent the sum of all mass from one component. Various thermodynamic databases were used as was convenient. Except as noted for 3D plots, all diagrams were constructed by STABCAL [3] running on the Windows operating system using win8.1 64 bit, Pentium i7, 4.3 GHz with 16 GB RAM hardware, and 1680 × 1050 resolution monitor.

2. Crucial Developments of the Eh-pH Diagram

Chapter 2 of the Pourbaix Atlas [4] presented the method of calculation and the procedure of the construction of an Eh-pH diagram. The process was relatively simple since only one component was considered.

Garrels and Christ [2] dedicated a full chapter to the Eh-pH diagrams. Several diagrams related to geochemical systems were not only presented, but also explained. They laid out a procedure to construct the diagrams when ligand(s) were involved, such as illustrated in the Fe–S and Cu–Fe–S systems. They also presented two 3D wireframe volume diagrams for the Eh-pH-CO_2 system, which will be discussed later in this paper.

A crucial development in constructing an Eh-pH diagram was in deciding how to handle a system when a ligand component was involved. Two completely different approaches were evolved.

2.1. Development of the Equilibrium Line Method

The equilibrium line method was originally used by Pourbaix for simple metal-hydroxide systems. Each line equation is derived from an electrochemical and/or acid-base reaction between species.

Garrels and Christ used Fe–S as an example to show that the same procedure presented by Pourbaix could be applied to a multicomponent system. Basically, it involved two separate steps: domain areas of ligand S were first constructed, then all Fe species (including Fe–S complexes) were distributed in each isolated area of the ligand species. Huang and Cuentas [5] presented a computer algorithm to construct this type of diagram using an early personal computer.

2.2. Development of the Mass Balance Point Method

Forssberg *et al.* [6] constructed several Eh-pH diagrams related to chalcopyrite, $CuFeS_2$, by performing equilibrium calculations for the whole system at once at each given Eh and pH. By doing so, the Cu:Fe:S ratio could be strictly maintained to 1:1:2 at all points. They used the SOLGASWATER program developed by Eriksson [7] to perform the calculation. This point-by-point mass balance method identifies the predominant species at each given point of Eh and pH. Points of the same species were combined into an area for the final diagram. The SOLGASWATER program used free energy minimization, which is commonly used for equilibrium calculation. Woods *et al.* [8] also presented diagrams for the Cu–S system using SOLGASWATER.

The mass balance method can also be computed considering the law of mass action (Huang *et al.* [9]). This approach simultaneously solves all equations, equilibria and mass balances, at each given point of Eh and pH. As with the free energy minimization method, the final diagram has to be plotted by grouping calculated results together. presented later, was reconstructed using the law of mass action for Cu–S and matched with from Woods *et al.* [8].

Besides matching the mass input, these diagrams reveal the presence of multiple solids as restricted only by the Gibbs phase rule. The key to the success of using the point-by-point method, however, is the resolution of the grids used in the calculation. Except for 3D diagrams, all mass balance diagrams in this paper were constructed using grids of at least 400×800.

3. Applications for the Diagrams

Eh-pH diagrams are widely used in many areas where an aqueous system is affected by oxidation-reduction and/or acid-base reactions, ligand complexation, temperature or pressure. The following three examples are presented to illustrate these effects.

3.1. Geochemical Formation

Copper porphyry ore deposits occur throughout the world and are very important sources of copper, silver and gold. These deposits initially consist of disseminated sulfide minerals in a rock matrix, but near-surface weathering oxidizes the sulfides and leaches dissolved metals from the residual mass. These leached metals in solution percolate downward and are often reprecipitated in an enrichment zone overlying unreacted sulfide protore. The near-surface weathered, oxidized portion of the deposit corresponds to the oxidizing region of an Eh-pH diagram, while the non-oxidizing reduced enrichment zone corresponds to the reducing diagram region. Figure 3 is a geologic sketch of an idealized porphyry deposit *versus* the depth from the surface, while Figure 4 is a copper Eh-pH diagram in which iron, sulfur and carbonate, besides copper, are considered in the calculations. The minerals predicted in the diagram, solely from thermodynamic considerations, correspond extremely well with minerals observed in these deposits and with the relationships between these minerals. In the oxidized and weathered zone, the original copper and iron sulfides are not stable, while copper carbonates (antlerite, malachite, azurite) and oxides (tenorite, cuprite) form instead. In the enrichment zone, the copper-only sulfides covellite (CuS) and chalcocite (Cu_2S) are dominant, with native copper seen to occur in both oxidized and enriched zones.

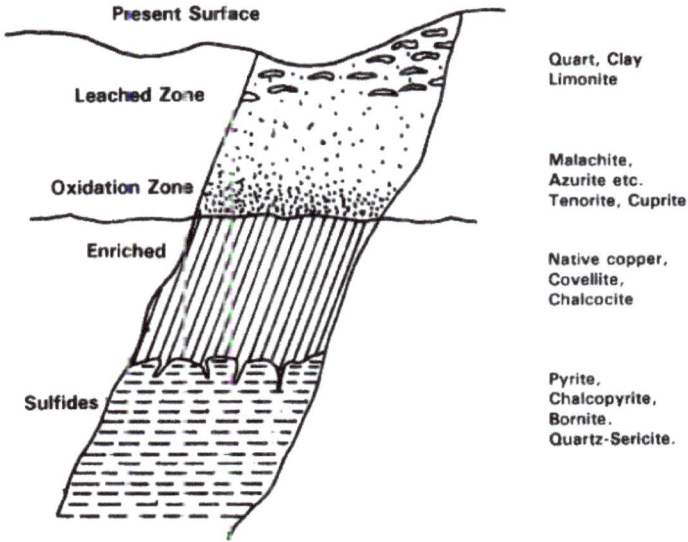

Figure 3. Illustrated copper ore deposit for comparison to the Eh-pH diagram to the right (Dudas *et al.* [10]).

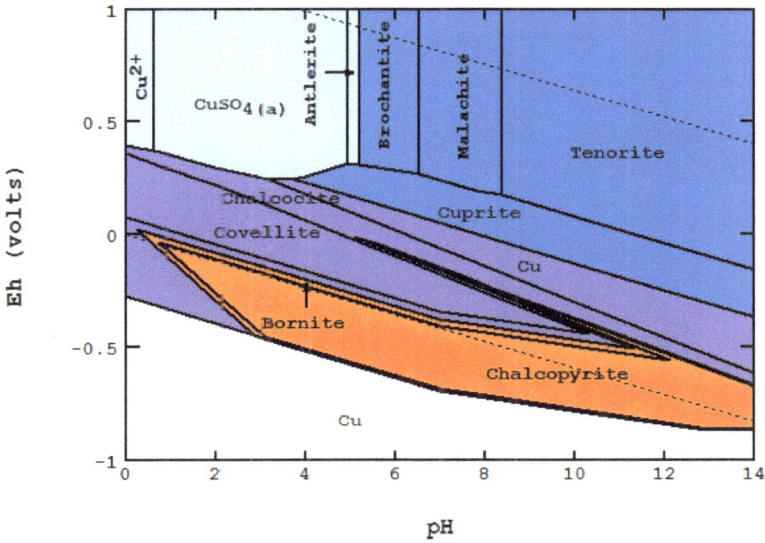

Figure 4. Eh-pH diagram Cu–CO$_2$–Fe–S in water. pCO$_2$ = 0.1 atm, [S] = 0.01 M, [Fe] = [Cu] = 0.001 M. Species were taken from the LLnL database [11].

Another geochemical example is the Eh-pH diagram modeling metamorphic conditions. In order to show the effect of high pressure, a database such as SUPCRT (Johnson *et al.* [12]) is required. See the reference from Kontny *et al.* [13] for a Fe–S diagram at 300 °C and 1500 bars pressure or Huang [14] for more calculations and examples using SUPCRT-related databases.

3.2. Corrosion and Passivation

Metallic corrosions are widespread problems of great importance in virtually all physical structures. Corrosion chiefly occurs when metal electrochemical dissolution is favored. One way to protect the metal from corrosion is to form a passivated layer, which may simply be a metal oxide. Some metal oxides, such as PbO, exhibit relatively high solubility and provide little corrosion protection.

The distribution-pH diagram (Figure 5) shows the concentrations of dissolved Pb species, as well as the solubility of PbO, *versus* pH. Formation of metal-carbonate, as shown in the Eh-pH diagram of Figure 6, offers a wider passivation region. Both diagrams were constructed using the LLnL [11] database. Pourbaix in his lectures [15] presented a similar case for using CO_2 to passivate Zn metal.

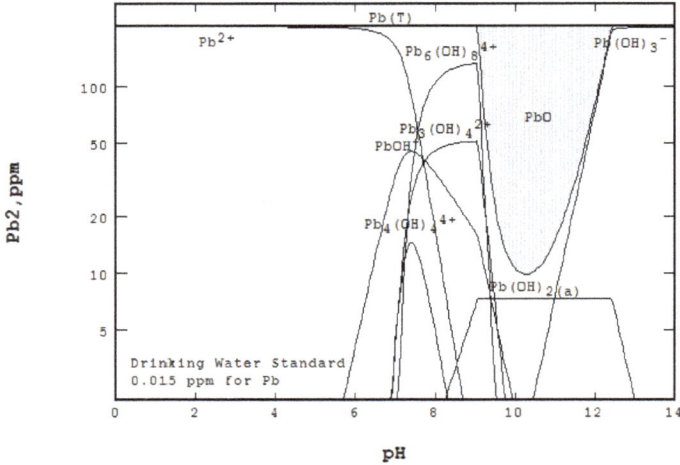

Figure 5. Solubility of PbO (shaded) *versus* pH. PbO does not provide good corrosion protection, even at elevated pHs.

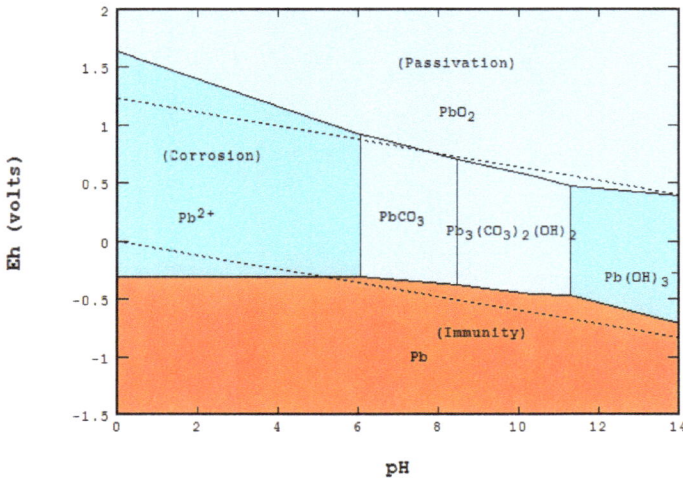

Figure 6. Eh-pH of the $PbCO_3$–water system. [Pb] = 1×10^{-6} and $[CO_3]$ = 0.001 M. Pb carbonate phases do provide corrosion resistance.

3.3. Water Treatment and Adsorption

Water discharge standards almost always include concentration limits for the acid, base and heavy metals. When feasible, precipitation of a solid, followed by a liquid-solid separation is usually the preferred means of achieving these limits, but often, stringent standards are difficult, if not impossible, to comply with by this means. Adsorption onto metal oxides/hydroxides sometimes provides an alternative means of removing these metals from the discharge solution. The adsorption of arsenic (As) by ferrihydrite is demonstrated in Figure 7 using data from Nishimura *et al.* [16]. For this particular experiment, the initial conditions were $\Sigma As = 37.5$ mg/L with a Fe/As mole ratio of 10. The source of ferric iron was dissolved $Fe_2(SO_4)_3$:$5H_2O$.

The species considered and their thermodynamic values were also taken from the LLnL database [11]. The equilibrium calculation included adsorption using a surface complexation model. Potentially adsorbed species onto ferrihydrite are three arsenates, one arsenite, two sulfates, hydrogen ion and hydroxide. Their equilibrium constants, $logK_{ads}^{int}$, were obtained from Dzombak and Morel [17]. In order to better fit the experimental data, some modifying changes were made:

1. Type 2 site density for ferrihydrite was changed from 0.2 to 0.3 mole As/mole Fe due to co-precipitation,
2. The $logK_1^{int}$ for adsorbed species $\equiv FeH_2AsO_4$ was changed from 29.31 to 31.67,
3. The adsorbed species $\equiv FeAsO_4^{2-}$ and its $logK_3^{int} = 21.404$ were added and
4. Solid scorodite ($FeAsO_4$:$2H_2O$) and its $\Delta G^0_{25C} = -297.5$ kcal/mole were included with the LLnL dbase.

Figure 7 is the resulting distribution-pH diagram, of the same type as Figure 5, for arsenate As(V). The adsorption model nicely matches the experimental data, demonstrating effectively what the arsenic removal should be. The adsorption of arsenite As(III), while not shown, also matches the experimental data. Figure 8 is presented to illustrate the Eh-pH diagram for the As–Fe–S–water system constructed using the mass-balanced (600×800 grids) method. The areas in light blue show solids and adsorbed species to a dissolved concentration less than 0.1 ppm.

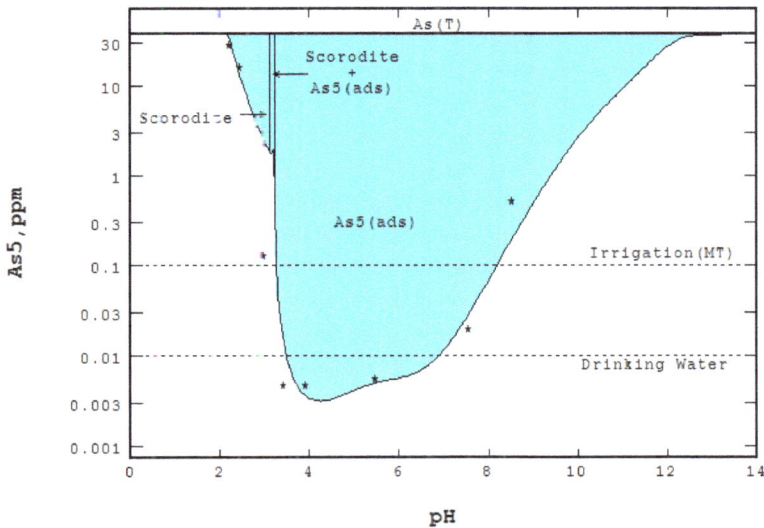

Figure 7. Distribution of As(V) *vs.* pH diagram when Fe/As = 10. Asterisks are experimentally-observed values. Drinking water standard from EPA (2001).

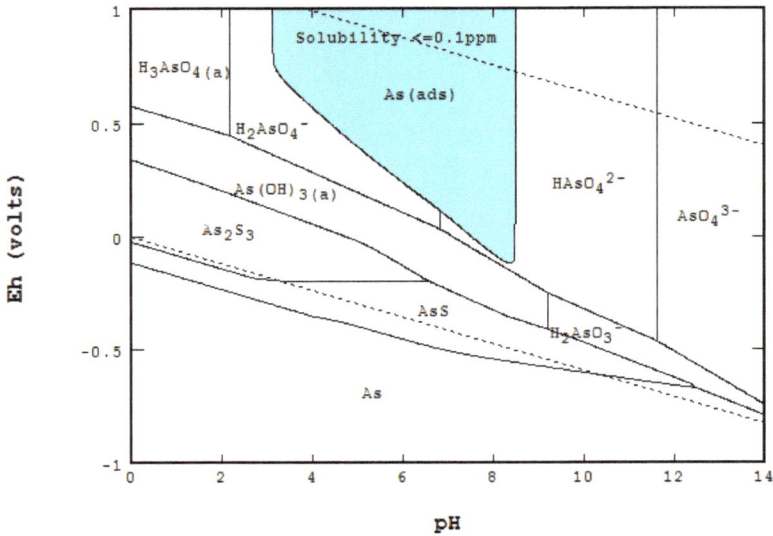

Figure 8. Eh-pH of As–Fe–S water where the mole ratio of Fe/As = 10. The colored area indicates less than 0.1 ppm concentration of As by adsorption.

3.4. Hydrometallurgical Leaching and Metal Recovery

Three applications for hydrometallurgy are presented in more detail later in the section titled "Enhancing the Eh-pH Diagrams by Merging Two or More Diagrams". These are:

1. Cyanidation of Au and cementation with Zn Metal,
2. Cementation of copper with elemental Fe, and
3. Galvanic conversion of chalcopyrite with Cu metal with two construction methods for Eh-pH diagrams to handle ligand components.

4. Descriptions and Comparison between These Two Crucial Methods

4.1. Equilibrium Equations for Eh-pH Diagrams

The chemical equation between Species A and B in the water system, with or without electron involvement, can be expressed as:

$$a\text{A} + c\text{C} \leftrightarrow b\text{B} + d\text{D} + h\text{H}^+ + w\text{H}_2\text{O}\,(+ne^-) \tag{1}$$

Species C and D are ligand and complexes produced with ligand. The stoichiometric coefficient of a species is taken as positive if it is on the right-hand side of the equation, and *vice versa*. Species H^+, H_2O and e^- may not always be on the right-had side of the equation. Because so many equations and species are involved while performing equilibrium calculations for an Eh-pH diagram, it is easier to use the free energy of formation of each involved species, ΔG_i^0, then to calculate the free energy of reaction as,

$$\Delta G_{\text{rex}} = \sum (v_i \times \Delta G_i^0) \tag{2}$$

where v_i represents the stoichiometric coefficient of species i.

Depending on whether or not the reaction involves an electron and/or hydrogen ion, the equations are:

The Nernst equation for oxidation-reduction reaction with or without acid-base:

$$\mathrm{Eh} = \mathrm{Eh}^0 + \frac{\ln(10)RT}{(n \times F)} \times \left[\log\left(\frac{\{B\}^b \{D\}^d}{\{A\}^a \{C\}^c} \right) - h\mathrm{pH} \right] \tag{3}$$

where $\mathrm{Eh}^0 = \frac{\Delta\mathrm{Grex}}{(n \times F)}$, where R is the universal gas constant, 8.314472(15) J/(K·mol); T is in kelvins; F is the Faraday constant 96,485.3399(24) J/(V·equivalent); and {A} and the others species are defined as the activities of Species A. The activities of solid and liquid are normally assumed to be one; gas is taken as the atmosphere (atm). The activity of an aqueous solution is the multiplication of the concentration in mol/L, symbolized as [A], with its activity coefficient. The coefficient can be computed from one of the appropriate models. Without having the acid-base, the "hpH" term in the equation will be dropped out.

The equilibrium equation for acid-base reaction without redox reaction:

$$\mathrm{pH} = \frac{1}{h} \times \left[\log\left(\frac{\{B\}^b \{D\}^d}{\{A\}^a \{C\}^c} \right) + \frac{\Delta\mathrm{Grex}}{\ln(10) \times RT} \right] \tag{4}$$

The equation for reaction involves neither an electron nor a hydrogen ion:

$$\log Q - \log K = \log\left(\frac{\{B\}^b \{D\}^d}{\{A\}^a \{C\}^c} \right) + \frac{\Delta\mathrm{Grex}}{\ln(10) \times RT} \tag{5}$$

Species A will be favored if $\log Q - \log K$ is positive, and *vice versa*.

As mentioned earlier, two different approaches may be used to construct an Eh-pH diagram. One is to calculate equilibrium equations between pairs of species and to construct the diagram by plotting the resulting equilibrium lines. The other is to perform equilibrium calculations from all involved species at each point in a grid, then selecting the predominant species at each point. Regardless of which method is used, these equilibrium equations have to be satisfied.

4.2. Line Method Using Equilibrium Concentration [5]

The diagram is constructed by computing the equilibrium between two adjacent species from their activities. The concentration or activity of aqueous species has to be given. Figure 9 shows the Eh-pH diagram for Cu at three different concentrations. The case where a ligand component is also involved is demonstrated in Figure 10, for the Cu–S–water system using [S] = 0.001 and [Cu] = 0.001 mol/L. The areas of predominance for the various ligand S species (labeled in light blue) were first constructed. The distribution of Cu species, including Cu–S complexes, in each S domain (such as the area of H_2S shaded with light blue) was then constructed. The final diagram of Cu species was determined by combining all of the areas from S ligands. It should be noted that the total concentration of ΣS may change depending on whether or not Cu is complexed with S. When Cu species are not complexed with S, ΣS would be 0.001 mol/L, as described. However, when Cu species are complexed with S, as in the formation of CuS, ΣS will be the molar sum of the S concentration plus the CuS concentration, which will be equal to 0.002 mol/L. In such a case, the mass of total S may not be constant, as originally assigned.

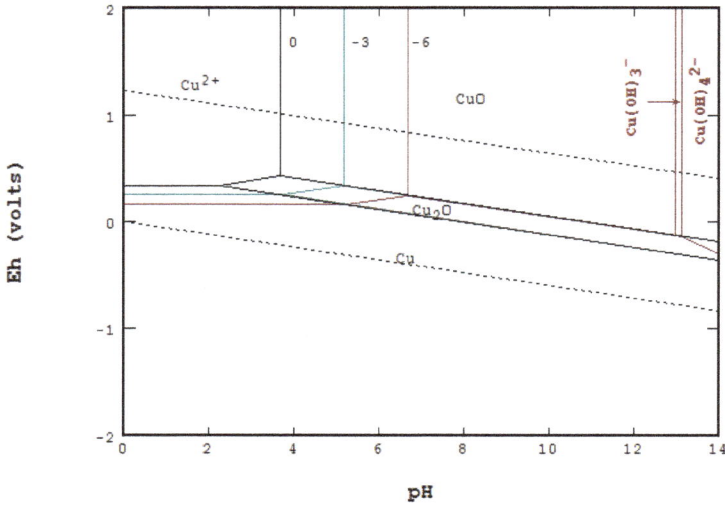

Figure 9. Eh-pH of Cu-water constructed by the line method where three concentrations in log scale are plotted. Data taken from N3S [1].

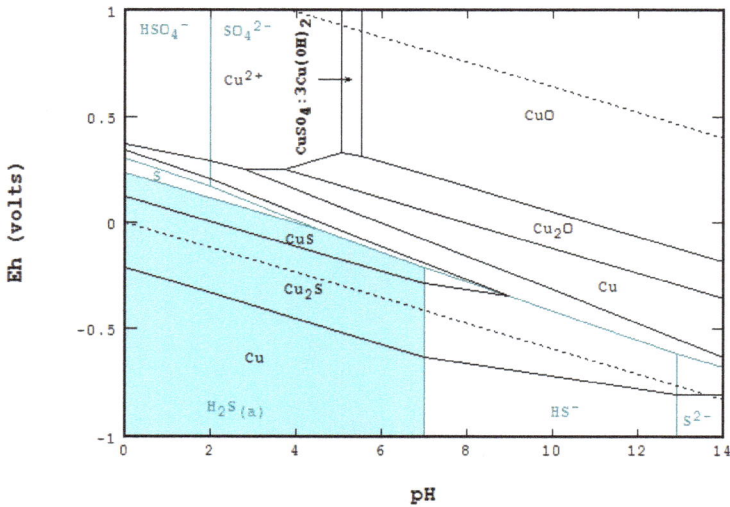

Figure 10. Eh-pH of Cu(main)–S (ligand). The line method plots the ligand first, shown in blue color. The distribution of Cu species for each domain of S is then determined.

4.3. Point-by-Point Method Using Mass Balance [9,18]

In this method, mass balances are considered and calculated with all of the equilibria from all of the components at once from every point of the grid. Unlike the line method where the concentration or activity of aqueous species is specified, this method requires knowing the total mass of each component. The calculation requires not only satisfying all equilibrium equations, but also matching all of the mass balances. The results are sorted out in order to plot the diagram for each specific component. This type of diagram is particularly important for Eh-pH diagrams, which include solids with composition ratios that are close to mineral formation ratios (such as 3:1:4 mole ratios for enargite Cu_3AsS_4)

Mass inputs, including masses of ligands, are crucial for determining critical areas of these diagrams. Figures 11 and 12 illustrate this for the Cu–S–water system (data from NBS [1]). Figure 11 shows the case where S is stoichiometrically slightly less than copper, *i.e.*, $\Sigma Cu = 0.001$ and $\Sigma S = 0.0009$ M, while Figure 12 shows the case where S is slightly in excess. The higher mass of S leads to a larger area of predominance for CuS.

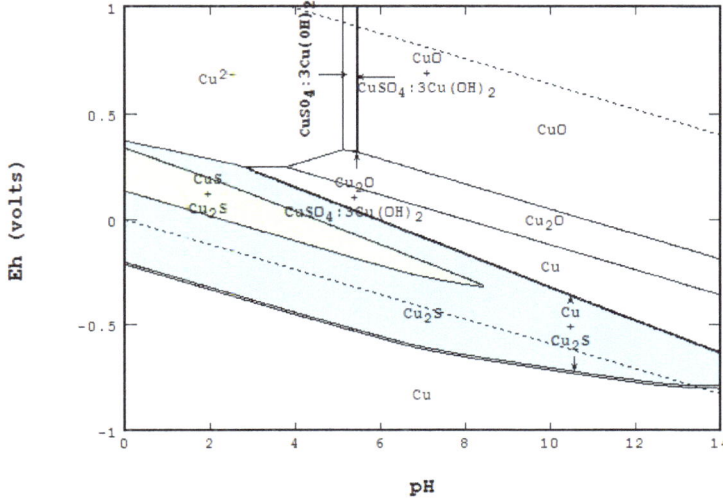

Figure 11. Mass-balanced Eh–pH diagram for the Cu–S–water system with copper slightly in excess. $\Sigma Cu = 0.001$ M and $\Sigma S = 0.0009$ M.

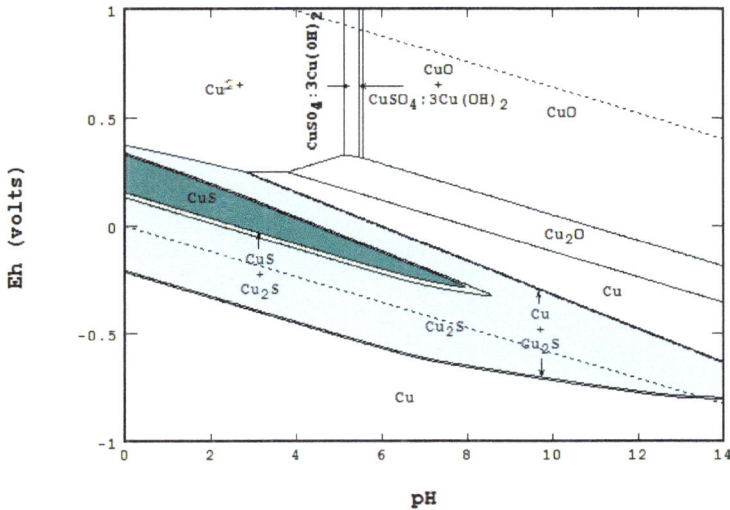

Figure 12. Mass-balanced Eh–pH diagram for the Cu–S–water system with copper slightly in deficit. $\Sigma Cu = 0.001$ M and $\Sigma S = 0.0011$ M.

4.4. Differences and Comparison between the Methods

Different results of the two methods can be seen by comparing diagrams constructed within the Cu–S system, with sulfate species not shown due to unfavorable kinetics (Woods *et al.* [8]). Figure 13 was constructed by the equilibrium line method where [Cu] = 0.118 and [S] = 0.059, and Figure 14 was constructed by the mass-balanced point method where ΣCu = 0.118 and ΣS = 0.059. Free energy data were taken from Woods *et al.* [8]. Crucial differences can be seen in the general area of Cu–S solids. As can be seen from Figure 13, even though the concentration ratio of Cu/S is specified as two to one, CuS, not Cu_2S, is the predominant species.

The mass-balanced point calculation involved points on a 400 × 800 grid to a precision of 1×10^{-10}, but took less than three minutes for a PC from creating a worksheet for input to plotting the final diagram.

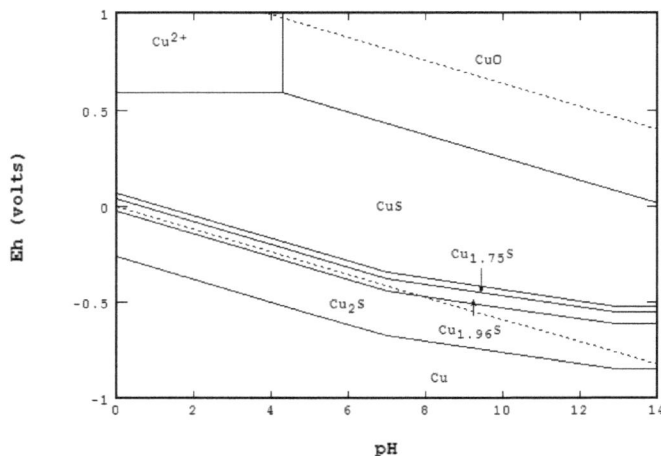

Figure 13. Eh-pH diagram for the Cu–S–water system constructed by the equilibrium line method.

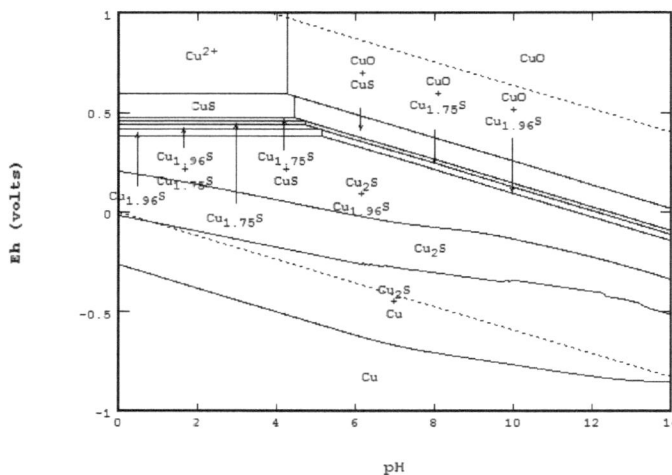

Figure 14. Eh-pH diagram for the Cu–S–water system constructed by the mass-balance point-by-point method. This diagram matches Figure 5 of Wood *et al.* [8].

The equilibrium line method was favored in the past due to its relative ease of construction. When diagrams were constructed using manual calculation (as by Pourbaix), the equilibrium line method was the only practical approach. As greater computational power became available, the mass-balanced point-by-point method came into favor. The following list includes some areas where the mass balance method should be considered over the line method.

1. When the exact composition of the system is needed: Examples include leaching and flotation studies. See Huang and Young [18] for more examples. The Eh-pH diagram of enargite (Cu_3FeS_2) (Figure 15) was constructed using data collected by Gow [19].
2. When a system is required to specify total concentration, not equilibrium concentration nor activity.
3. When the adsorption by solids, such as ferrihydrite, is considered (refer to Figure 8).
4. When multiple phases of a solid need to be shown: Figure 16 was constructed by showing the coexistence of schwertmannite with various forms of jarosites in Berkeley pit water. Water samples were taken and analyzed from 1987 to 2012 by the Montana Bureau of Mines and Geology [20], and the thermodynamic data for the solids species were regression estimated by Srivastave [21].
5. A diagram will most likely be mass balanced if a speciation program, such as PHREEQC (USGS) [22], was used to construct it. Results from the program were collected manually or electronically, then combined into an Eh-pH diagram.

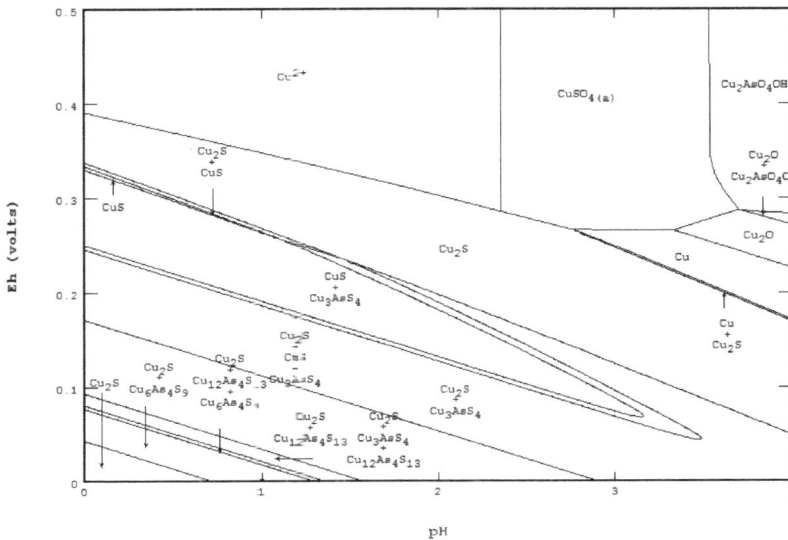

Figure 15. Mass balanced Eh-pH diagram for the enargite Cu_3AsS_4 system. The mass ratio is 0.75:0.25:1 for Cu, As and S. The diagram shows only the copper species in acid solution.

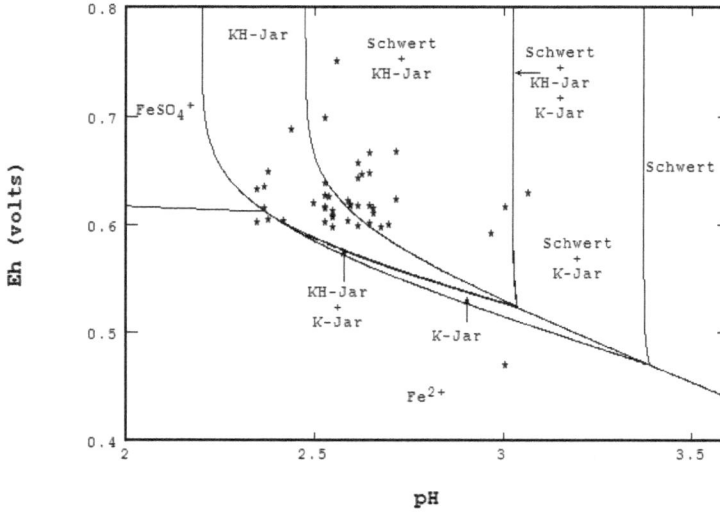

Figure 16. Mass balanced Eh-pH diagram for the Fe–K–S system at 7 °C. This diagram shows the coexistence among schwertmannite, K-jarosite and KH-jarosite. * represents the data analyzed from the sampled water.

Both methods, however, can produce identical diagrams under the following conditions:

1. A one-component system, such as Figure 1 for Mn and Figure 2 for P,
2. The concentration of ligand component(s) is much greater than the main component, such as metal corrosion by sea water, and
3. Gas is the only ligand, such as the Fe–CO_2(g) system.

4.5. Gibbs Phase Rule Applied to an Eh-pH Diagram

An Eh-pH diagram constructed using either method must follow the Gibbs phase rule. The original phase rule equation, $P + F = C + 2$, was developed for considerations of temperature and pressure. It can be refined for use in constructing an Eh-pH diagram by implementing some concepts and restrictions.

Mass-balanced method: This method calculates equilibrium from all components at once. The variables are as follows:

1. P is the total number of phases = 1 (liquid water) + 1 (gas if considered) + N (maximum number of solids/liquids),
2. F is the degree of freedom on the diagram, which is two for an open area, one on a boundary line and zero on a triple point,
3. C is the total number of components = 3 + EC (extra components). Three components are essential for Eh-pH calculation in an aqueous system. These are H(+1), O($-$2) and e($-$1). The extra components include the main component to be plotted, as well as all ligands.
4. The term of +2 is for temperature and pressure variables. Since both are considered to be constant, +2 will be dropped off. If any system involved a gaseous species, +1 should be used, but it will be canceled out with one extra gaseous phase to the equation.

The phase rule equation for an Eh-pH diagram, best expressed as the maximum number of solids plus liquids excluding the liquid phase of water, thus becomes:

$$N_{maxsolid} = C - F - 1 \tag{6}$$

Example 1, Cu and S two-component system: The incorporation of the rule is illustrated in Figure 17, in which all solids containing Cu, as well as S components are presented. Since the method computes equilibria from all components involved at once, $C = 3 + 2$ and $N_{maxsolid} = 5 - F - 1$ or $4 - F$.

1a.   In an open area of the diagram where $F = 2$, $N_{maxsolid}$ will be equal to two. The co-existence of two solids can be seen in many places on the diagram,

1b.   On a boundary line where $F = 1$, $N_{maxsolid}$ becomes three. For instance, while each of the light blue areas contains two solids, the line between them represents the presence of three: CuO, Cu_2S and $Cu_{1.96}S$,

1c.   On a triple point where three lines meet, $F = 0$, $N_{maxsolid} = 4$. At the point labeled A, for instance, even though four areas meet, only three solids are coexistent at the point: CuO, $Cu_{1.75}S$ and $Cu_{1.96}S$.

Example 2, Pb-S-KEX (potassium ethyl xanthate) three-component system: Pb-S-KEX was also used to illustrate the phase rule. Figure 18 was constructed using data taken from Pritzker and Yoon [23]. The plot illustrates a small but intricate area, Eh from -0.5 to -0.3 and pH from 10 to 13, with a resolution of 600×800. A small pink area shows three stable solids: PbS, Pb and PbX_2. This number agrees with the phase rule equation for the Eh-pH diagram of $N_{maxsolid} = 5 - F$, where F is equal to two, being inside an open area. There are four solids (PbS, Pb, PbX_2 and $Pb(OH)_2$) along the line between this pink area and the area right above it. The $N_{maxsolid}$ for all three corners of this area was no greater than five, as described by the rule.

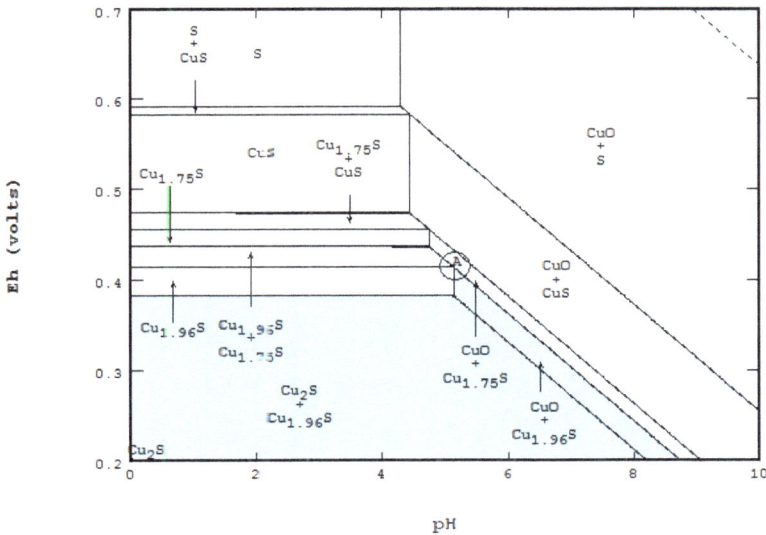

Figure 17. Mass-balanced Eh-pH diagram of the Cu–S–water system to illustrate the phase rule. This is a zoomed-in detail from Figure 14; stable solids include elemental S from the S component. $\Sigma Cu = 0.118$ and $\Sigma S = 0.059$ M.

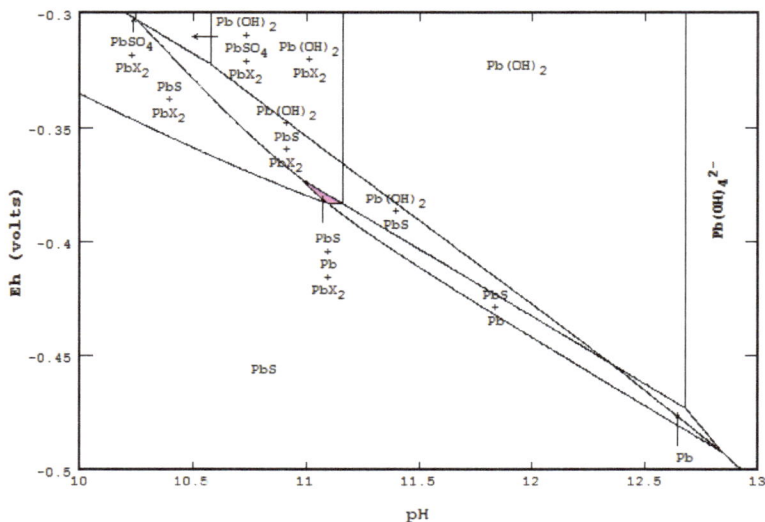

Figure 18. Mass-balanced Eh-pH diagram of the Pb–S–potassium ethyl xanthate (KEX) water system. $\Sigma Pb = \Sigma S = 0.45$ and $\Sigma X = 0.0001$ M. This zoomed-in detail has a resolution of 600×800 to verify the Gibbs phase rule.

Equilibrium line method: Since this method works on one component at a time, $N_{maxsolid} = 4 - F - 1 = 3 - F$. Referring to Figure 13, in any open area of the diagram, $N_{maxsolid} = 1$, which means only one solid is allowed. On any boundary line, $N_{maxsolid} = 2$, as shown by the line between CuO and CuS. If a triple point is formed, $N_{maxsolid}$ will be equal to three.

5. Enhancing the Eh-pH Plot by Merging Two or More Diagrams

Most Eh-pH diagrams indicate the stability of water by two dashed lines: this is a typical example of merging two diagrams together. Other examples include showing several solubilities of solid species (see Figure 9), showing dissolved species in areas dominated by solids and showing ligands in addition to the main component (see Figure 10).

An Eh-pH diagram, including the examples listed above, often shows only half of a reaction. To illustrate the whole process, merging another relevant diagram may be necessary. The combined diagram can be re-plotted or overlaid by a graphics program, such as MS PowerPoint. It is strongly suggested to use different colors for each merged diagram.

5.1. Cyanidation of Gold and Cementation with Zn Metal

The combination of a Au–CN diagram with a Zn–CN diagram is shown as Figure 19. The up-arrow indicates leaching of Au using CN as the complexing ligand and O_2 as the oxidant. The down-arrow indicates the cementation of Au replaced by Zn metal.

5.2. Cementation of Copper with Metallic Iron

Figure 20 shows the combination of a Cu–water diagram and a Fe–water diagram and illustrates cementation of copper by iron metal. Additionally, the diagram shows other reactions of interest, such as those that represent wasteful consumption of iron metal by reactions involving Fe^{3+}, $O_2(g)$ and $H^+(a)$. Although not part of the diagram calculations, the rest potential between the $Cu^{2+}/Cu-Fe^{2+}/Fe$ electrodes is also shown.

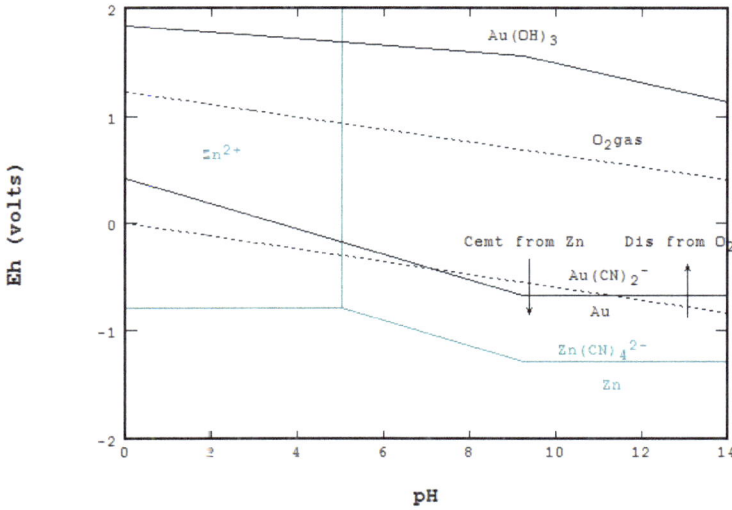

Figure 19. Combined Au–CN and Zn–CN Eh-pH diagrams would show leaching and cementation for the gold cyanidation process.

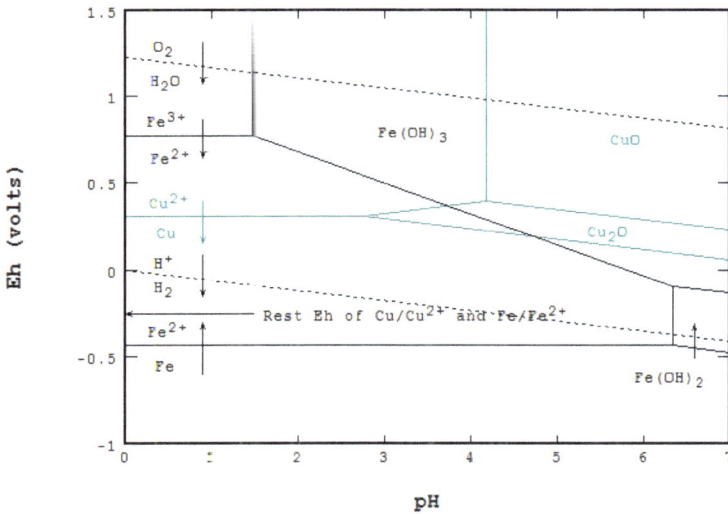

Figure 20. Combined Cu and Fe Eh-pH diagrams would show that reactions occur during copper cementation using metallic iron.

5.3. Galvanic Conversion of Chalcopyrite with Copper Metal [24]

When in contact with Cu metal, chalcopyrite reacts cathodically in an effect known as galvanic conversion:

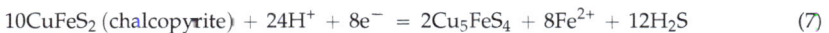

$$10CuFeS_2 \text{ (chalcopyrite)} + 24H^+ + 8e^- = 2Cu_5FeS_4 + 8Fe^{2+} + 12H_2S \tag{7}$$

$$2Cu_5FeS_4 \text{ (bornite)} + 6H^+ + 2e^- = 5Cu_2S + 2Fe^{2+} + 3H_2S \tag{8}$$

An anodic reaction takes place on the metallic copper as:

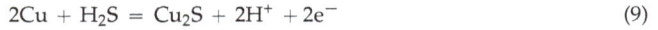

$$2Cu + H_2S = Cu_2S + 2H^+ + 2e^- \tag{9}$$

A schematic diagram for all these reactions is shown as Figure 21. In order to present all of the species involved, three Eh-pH diagrams are superimposed and shown as Figure 22.

1. The three diagrams used are: S species in cyan, Fe and Fe–S in red and Cu–Fe–S in black.
2. Areas of predominance are shown as: chalcopyrite in yellow, bornite in gray, chalcocite in light blue and metallic copper in orange.
3. The down-arrow indicates where galvanic conversion occurs down from chalcopyrite to bornite and, finally, to Cu_2S. The up-arrow indicates where the anodic reaction occurs up from metallic copper to Cu_2S.
4. The diagram indicates that both cathodic and anodic reactions lead to the formation of Cu_2S, and other final species match what Hiskey and Wadsworth [24] described.

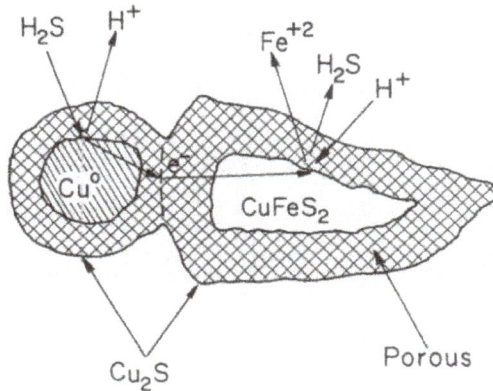

Figure 21. Schematic diagram of reactions occurring upon galvanic conversion of chalcopyrite with metallic copper [24].

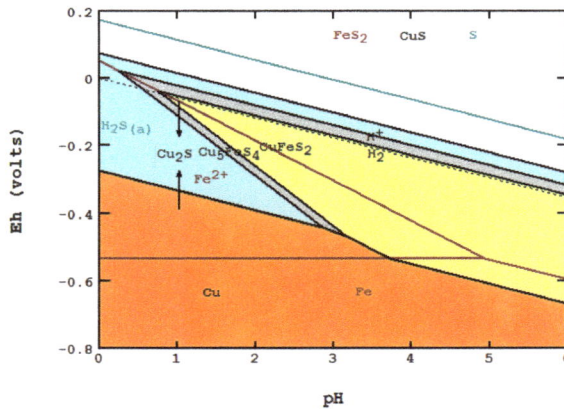

Figure 22. Combination of three Eh-pH diagrams showing the galvanic conversion reactions of chalcopyrite with metallic copper.

6. Third Dimension to an Eh-pH Diagram

Even more information may be shown by adding a third dimension to a base Eh-pH diagram. The third dimension can be the simple solubility of solid or an independent variable, such as temperature or ligand concentration. First, the data needed for a 3D Eh-pH diagram must be calculated. Thereafter, 3D programs for PC, such as ParaView [25], VisIt [26] or MATLAB [27], combine all of the data into a single diagram. These programs can also provide other functions, such as animated rotation, clipping and slicing. This section presents some 3D examples by considering extension of the Eh-pH diagram into a third dimension. Data creation and setup input files for a 3D program are briefly presented. Three areas are illustrated:

1. Eh-pH along with the solubility of stable solids; two example diagrams are illustrated: passivation of lead (Figure 6) and adsorption of As(III) and As(V) onto ferrihydrite (Figure 8).
2. Eh-pH with an extra axis for ligand CO_2: two wireframe volume diagrams of Eh-pH-CO_2 taken from Garrels and Christ [2] are used for verifying the results; these two are:

    (a) Figure 7.32b: in order to match the given ΣCO_2 for the third axis, the mass balance method has to be used; the output of 3D and discussion for this case are presented in more detail.
    (b) Figure 7.32a: since the third axis is given as the pressure of $CO_2(g)$, the equilibrium line method can be applied; the time required for the Eh-pH calculation was much less.

3. Presentation of a system in which two or more solid phases, such as CuS and Cu_2S, can coexist.

6.1. Eh-pH with Solubility

Including the solubility of solids in an Eh-pH diagram can give a much clearer view of what can happen to the solid. The following two diagrams constructed by MATLAB extend the 2D Eh-pH diagrams presented earlier. In order for MATLAB to plot 3D solubility diagrams, the Eh-pH program needs to create two files: the (name).m file contains instructions to be executed by MATLAB, and the data file contains solubility from each Eh and pH from the grid. A MATLAB plot can show the matching contour (iso-solubility) lines below the 3D feature.

Figure 23 is an extension of Figure 6, showing the solubility of Pb(II) as the third dimension. The red areas indicate conditions where corrosion can be expected to occur. The gulch area in yellow indicates conditions where Pb(II) is passivated by CO_2.

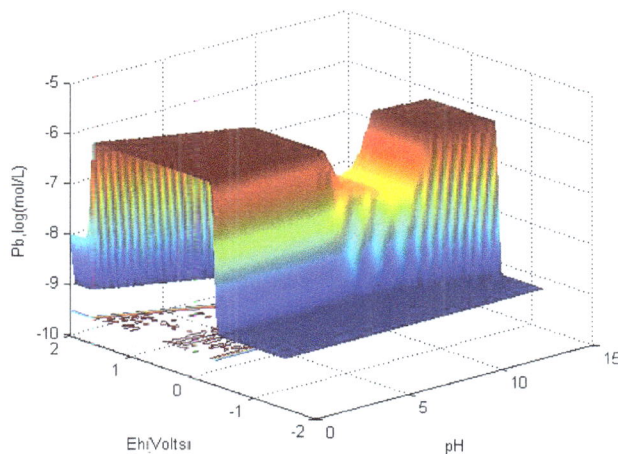

Figure 23. The Eh-pH plus solubility diagram for Pb–CO_3–water. The yellow gulch area in the middle is where Pb(II) is likely passivated by CO_3.

Figure 24 illustrates the same system as Figure 8, which showed where As(V) and (III) can be adsorbed upon the formation of ferrihydrite. The solubility diagram shows where the lowest concentrations of As(V) and (III) can be achieved. The deep blue area at the low Eh of the diagram is where arsenic metal becomes stable and immunized from corrosion.

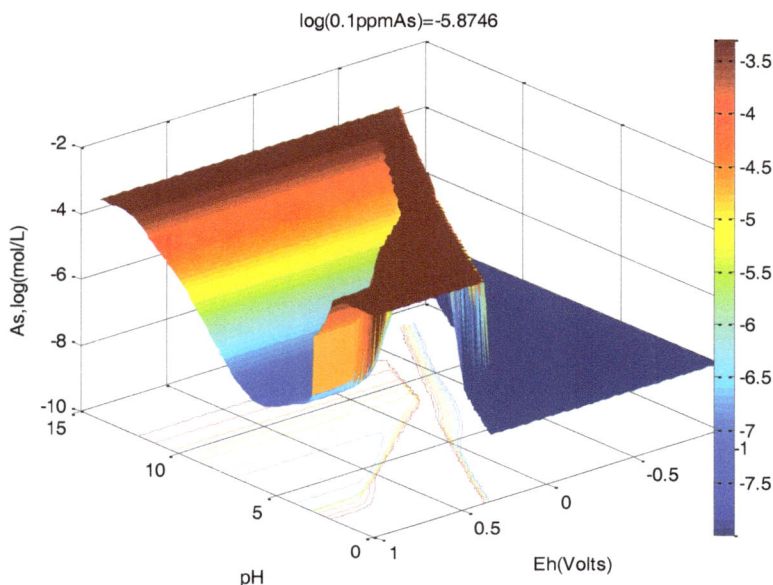

Figure 24. The Eh-pH plus solubility diagram for As–Fe–water. The valley area on the left is where the greatest adsorption of arsenic can occur.

6.2. 3D Eh-pH, Uranium with ΣCO_2: Using the Mass-Balance Point Method

Example system: Figure 7.23b from Garrels and Christ [2] is one of the earliest three-dimensional diagrams for the U–CO$_2$–water system. It is an Eh-pH diagram with the concentration of CO$_2$ used for the third dimension. See the duplicated plot from Figure 25. In it, each predominant species is enclosed by the faces of adjacent species. A semi-transparency presentation can be a more advanced graphical method, but it had not been developed at the time. A comparison to the Garrels and Christ plot was generated, considering the same species and their ΔG^0s (Free energy of formation). Other considerations required are:

1. One species in one volume: Since total carbonate is given, assuming total CO$_2$ means all carbonates, including dissolved, solids and complexes with U, the mass-balanced method for Eh-pH diagram is used. However, to match the Garrels and Christ plot, only one single solid in each volume was selected, with no regions of mixed solids allowed.
2. One missing species: One species on the Garrels and Christ diagram, indicated by a red letter A in Figure 25, seems to have been mislabeled as $UO_2(CO_3)^{4-}$. Judging from its high pH and carbonate location, and being sandwiched between U(IV) and U(VI), the species $UO_2(CO_3)_3^{5-}$ [28] seems to be a good fit. Figure 26 is the regenerated 2D Eh-pH diagram using $\log \Sigma CO_2 = -1$ M.

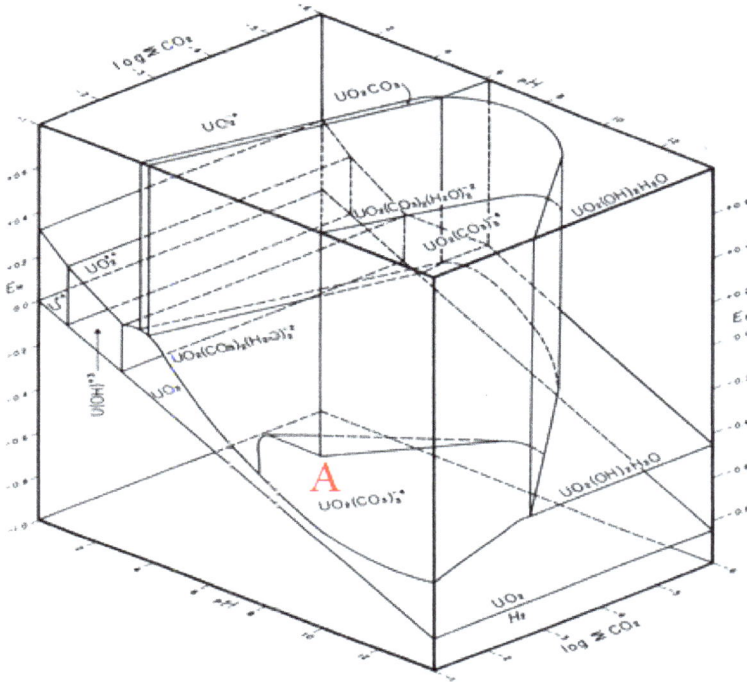

Figure 25. 3D diagram after Garrels and Christ. Label A is where the questionable species is located.

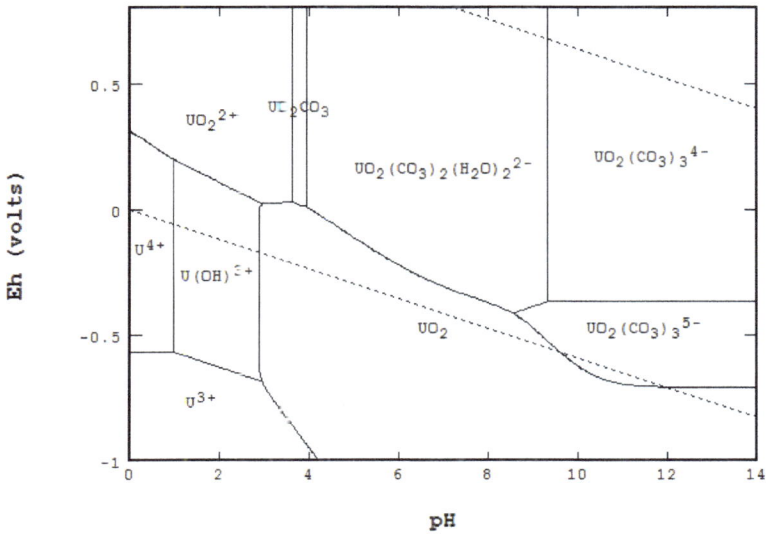

Figure 26. The Eh-pH diagram of U with $\log\Sigma CO_2 = -1$. The inclusion of $UO_2(CO_3)_3^{5-}$ clarifies a region that Garrels and Christ might have misrepresented.

Example diagrams and conditions: Figures 27–32 are three-dimensional diagrams created using ParaView (Version 4.3.1) based on the data generated by the STABCAL program. Although the complete ParaView diagram allows such functions as continuous rotation, clipping and slicing, these static diagrams demonstrate the range of features that can be achieved. A color map (bar), shown in Figure 27, indicates that the names of the species should be added at least once to one of the figures. Computational conditional limits include:

- Ranges: Eh = −1 to 0.8, pH = 0 to 14 and $\log\Sigma CO_2$ = −6 to −1,
- Grid point: 250 × 250 × 250,
- Computer: 64 bit, 3.40 GHz, 16 G of RAM,
- Program algorithm: mass balance point method using mass action law,
- Accuracy (sum of squared residual) $<1 \times 10^{-8}$ and
- Time to complete the calculation: 1:36:25 (h:mm:ss) from i7 PC or 2:00:51 from i5; contrary to using the line method, shown later, which took less than 30 s.

In order for ParaView to plot a 3D diagram, the Eh-pH program needs to create a (name).vtk file [29] that specifies the type of grid, the values of all *X*, *Y* and *Z* coordinates followed by all of the point data that indicate which species are to be plotted for each point on all three axes.

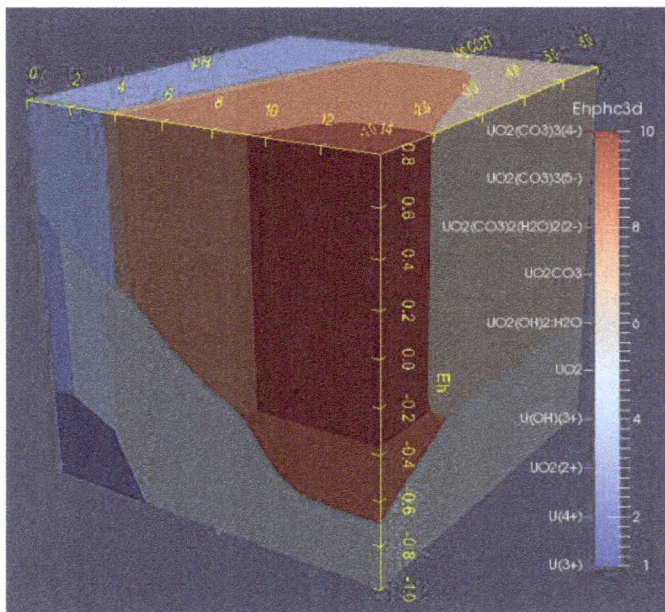

Figure 27. Eh-pH-$\log\Sigma CO_2$ of uranium where the species are shown by colors. A color bar is inserted to show the corresponding species.

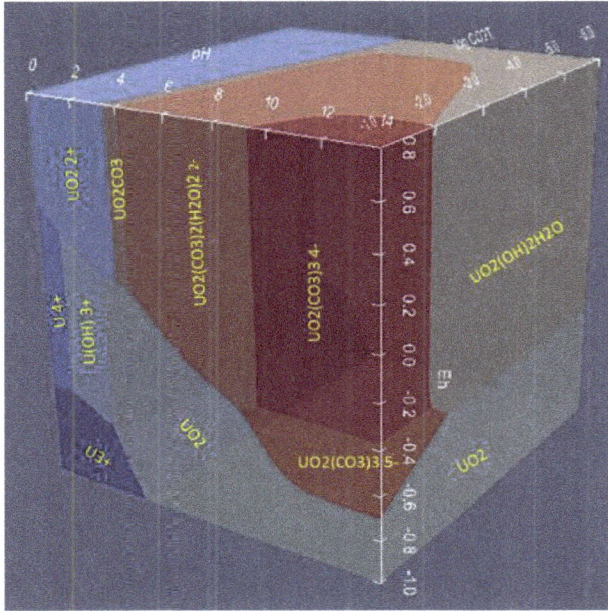

Figure 28. The transparent view of the 3D plot. Each colored area is labeled with the name of the species.

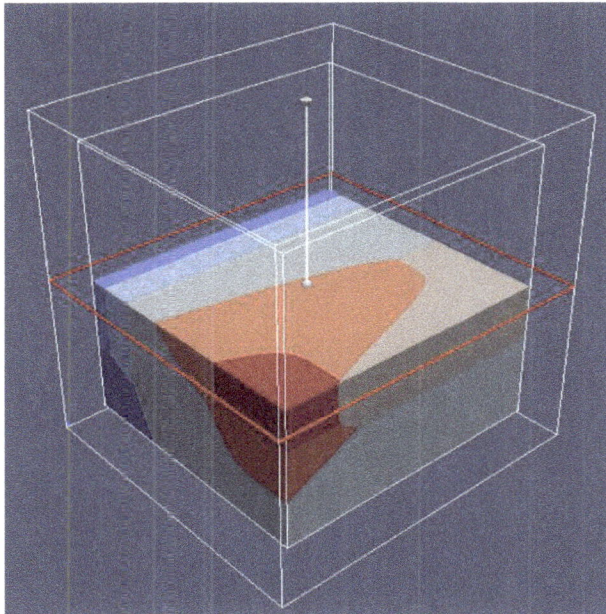

Figure 29. The cl p plot shows only regions below the constraint of Eh −0.1 V.

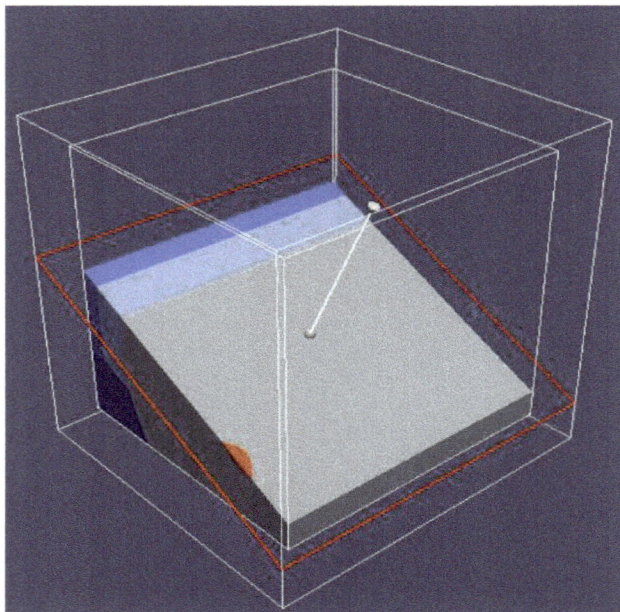

Figure 30. The clip plot shows species below the plane of theoretic water stability between H_2O and $H_2(g)$. Compare to the lower dashed water line in Figure 26 or the lower plane where no species were shown in Figure 25.

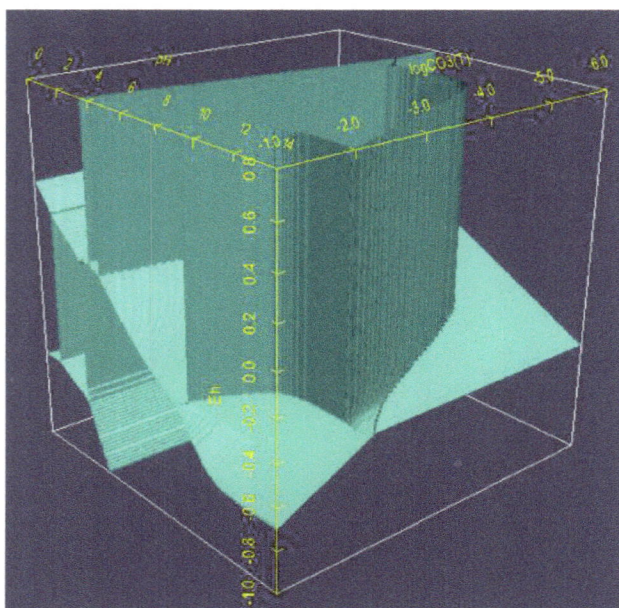

Figure 31. 3D plot that shows the boundaries (contour) between species.

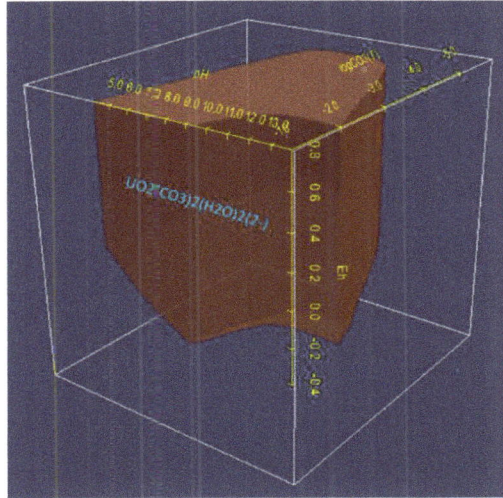

Figure 32. Semi-transparent plot that shows the stability region of $UO_2(CO_3)_2(H_2O)_2{}^{2-}$ species.

A three-dimensional plot has the great advantage of being able to show the effects of multiple variables at once. Tools, such as rotate, slice and clip, can easily identify areas of particular interest. For a complicated system, however, a three-dimensional image may not be as clear as a two-dimensional drawing due to the memory and screen resolution imposed by an ordinary PC. It is therefore wise to choose or combine the use of 2D and 3D to have the best presentation.

A simple way to turn a set of line-drawn diagrams into a three-dimensional object is to arrange six two-dimensional surface plots (two from each of Eh-pH, Eh-ligand and pH-ligand) into a cube. It may be necessary to reverse the *x*- or *y*-axis direction for this purpose. Simple plastic cubes, intended for use with photographs, are readily available. See Figure 33, where the section on the left (three plots combined) is the top-front view and the section on the right is the bottom-rear view. The two sections are jointed along the Eh axis, at pH = 14 and log ΣCO_3 = −1 M. Rotating the right sections to the left will make a complete cube. To show the continuity between plots, species UO_2 was purposely colored in light blue.

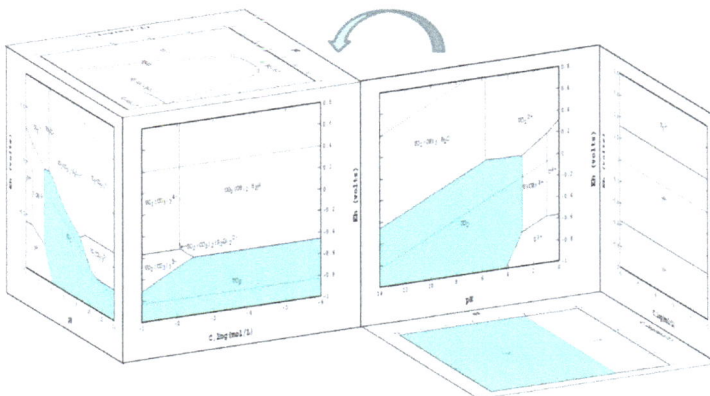

Figure 33. Combination of six surface plots of the U Eh-pH-logΣCO_2 system to form a cube.

6.3. 3D Eh-pH, Uranium with Pressure CO$_2$(g): Using the Equilibrium Line Method

This example used Figure 7.32a from Garrels and Christ for U with the CO$_2$ system. Since the extra axis is the pressure of CO$_2$(g), the equilibrium line method was used. The Eh-pH program took less than 30 s to create necessary data for making the (name).vtk file for ParaView. Without having to repeat the same features presented earlier, only the semi-transparent surfaces and the boundaries between species are shown on Figures 34 and 35 respectively. These two plots agree with the original diagram presented by Garrels and Christ.

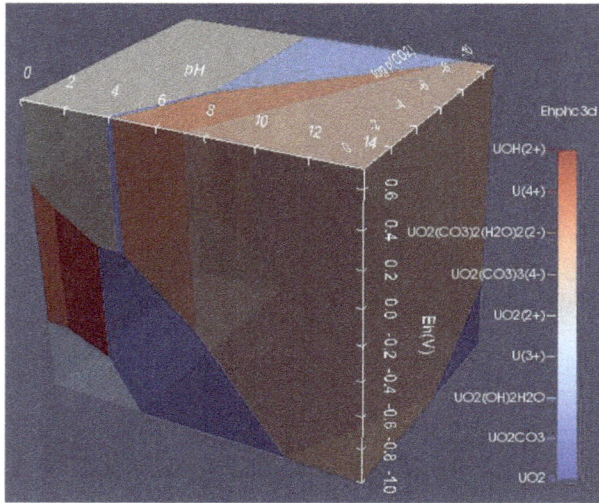

Figure 34. Eh-pH-logpCO$_2$(g) of uranium where the species are shown by colors. This diagram matches Figure 7.32a of Garrels and Christ.

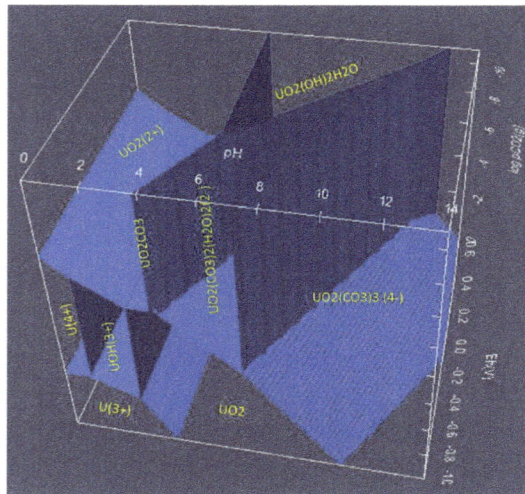

Figure 35. 3D plot that shows the boundaries (contour) between species. This diagram used data from the line method where logpCO$_2$(g) was given.

6.4. How to Show Two or More Solids in One Area

ParaView is capable of showing areas occupied by more than one solid phase. As with single solid regions, multi-phase areas are also colored and shown on the color map. Using the same conditions as for Figures 11 and 12, Figures 36 and 37 are plotted by ParaView indicating four two-solid phases. The areas occupied by CuS plus Cu_2S are indicated in red and pointed to by an arrow.

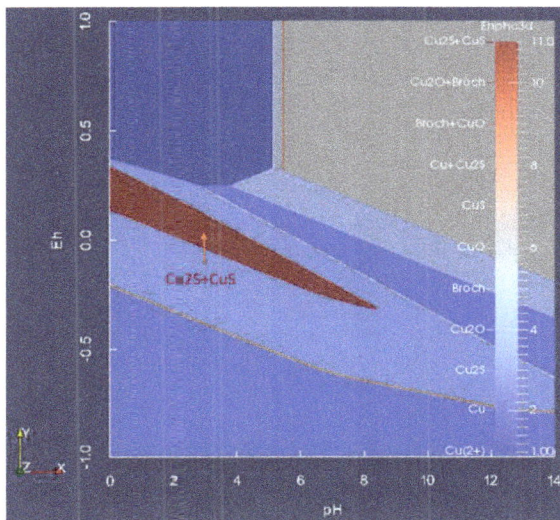

Figure 36. ParaView of Cu–S–water, where $\Sigma Cu = 0.001$ and $\Sigma S = 0.0009$ M. See the 2D plot of Figure 11 for a comparison.

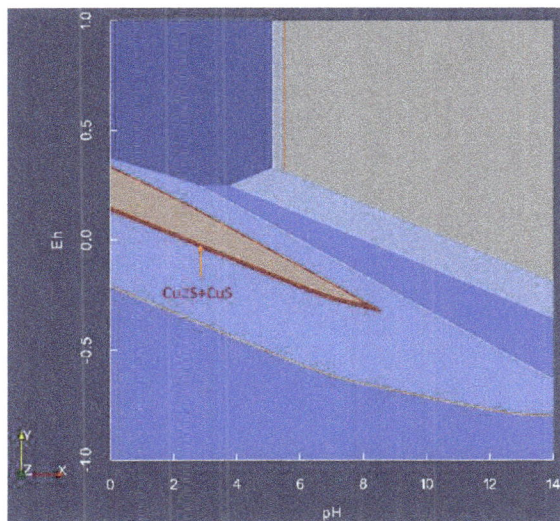

Figure 37. ParaView of Cu–S–water, where $\Sigma Cu = 0.001$ and $\Sigma S = 0.0011$ M. See the 2D plot of Figure 12 for a comparison.

7. Summary

An Eh-pH diagram, commonly known as a Pourbaix diagram, is an effective way of presenting the effects from oxidation-reduction potential, acid and base, complexing ligands, temperature and pressure for an aqueous system. It can be used in many scientific fields, including hydro- and electro-metallurgy, geo and solution chemistry and corrosion science. Diagrams describing natural copper deposits, lead corrosion prevention and arsenic adsorption by ferrihydrite and leaching and metal recovery were illustrated. The fundamental principles behind the diagram were briefly described.

To handle ligand components for these diagrams, two separate methods developed over time, the line of equilibrium concentration and the point of complete mass balances, were described and illustrated. Both satisfy the Gibbs phase rule in their own way of calculation. The comparison and applications from these two methods are mentioned and illustrated.

Many advances of the Eh-pH diagrams are presented. These are:

Merging diagrams: Most of the Eh-pH diagrams describe only half of the reactions. Merging several same-sized Eh-pH diagrams together can better illustrate the overall system. Examples include cyanidation of Au plus cementation with Zn metal, cementation of Cu with metallic Fe and galvanic conversion of chalcopyrite with metallic Cu.

Creating 3D diagrams: Expanding the Eh-pH program to create data from an extra axis is relatively simple. However, a more professional third party 3D program is the best choice for drawing the final diagram. The author has tested ParaView, VisIt and MATLAB for 3D Eh-pH diagrams. Most of these programs can perform animations, such as rotation, clipping and slicing. The diagram can be semi-transparent or show the boundary for better illustration. The following examples are demonstrated:

1. An Eh-pH diagram that shows solubility. Examples include: lead corrosion prevention with CO_2 and concentrations of arsenic adsorption by ferrihydrite; MATLAB was used for more colorful pictures.
2. An Eh-pH with an independent variable; diagrams created by ParaView were illustrated for its functionality, and example are:

    The mass balance point method for the Uranium system where the extra axis is $\log(\Sigma CO_3)$;
    The equilibrium line method for the Uranium system where the third axis is $\log p(CO_2(g))$;
    The Cu–S system for showing the coexistence of two or more solids in one volume.

When a 3D picture becomes too complex to label all of the species involved, two-variable diagrams (Eh-pH, Eh-ligand, Eh-temperature or pH-ligand) can be presented side-by-side for clarity.

Acknowledgments: The author would like to thank Dave Tahija, Hecla Greens Creek Mining, for editing this manuscript and Chen-Luh Lin, Metallurgical Engineering University of Utah, for introducing the use of ParaView. Thanks to colleagues in the Department of Metallurgical and Materials Engineering at Montana Tech for their support, in particular Courtney A. Young, Chairman of the Department, for introducing the mass balance method and encouragement for continuing development of the STABCAL program, and to Steve McGrath for his valuable discussion on the 3D Uranium-carbonate diagrams from Garrels and Christ.

Conflicts of Interest: The author declares no conflict of interest.

References

1. Wagman, D.D.; Evans, W.H.; Parker, V.B.; Schumm, R.H.; Halow, I.; Bailey, S.M.; Churney, K.L.; Nuttall, R.L. The NBS tables of chemical thermodynamic properties. *J. Phys. Chem. Ref. Data* **1982**. [CrossRef]
2. Garrels, R.M.; Christ, C.L. Eh-pH. In *Solutions, Minerals, and Equilibria*; Freeman, Cooper & Company: New York, NY, USA, 1975; Chapter 7; pp. 172–277.
3. *STABCAL*, version 2015; Stability Calculation for Aqueous and Nonaqueous System; Montana Tech: Butte, MT, USA, 2015.
4. Pourbaix, M. *Atlas of Electrochemical Equilibria in Aqueous Solution*, 1st ed.; Pergamon Press: Bristol, UK, 1966.

5.   Huang, H.H.; Cuentas, L. Construction of Eh-pH and Other Stability Diagrams of Uranium in a Multicomponent system with a Microcomputer—I. Domains of Predominance Diagram. *Can. Metall. Q.* **1989**, *28*, 225–234. [CrossRef]

6.   Forssberg, E.; Antti, B.-M.; Palsson, B. Computer-assisted Calculations of thermodynamic equilibria in the Chalcopyrite-ethyl xanthate system. In *Reagents in the Minerals Industry*; Jones, M.J., Oblatt, R., Eds.; The Institution of Mining and Metallurgy: London, UK, 1984; pp. 251–264.

7.   Eriksson, G.A. An algorithm for the computation of aqueous multicomponent, multiphase equilibria. *Anal. Chim. Acta* **1979**, *112*, 375–383. [CrossRef]

8.   Woods, R.; Yoon, R.H.; Young, C.A. Eh-pH diagrams for stable and metastable phases in copper-sulfur-water system. *Int. J. Miner. Process.* **1987**, *20*, 109–120. [CrossRef]

9.   Huang, H.H.; Twidwell, L.G.; Young, C.A. Speciation for aqueous system—An equilibrium calculation approach. In *Computational Analysis in Hydrometallurgy—35th Annual Hydrometallurgy Meeting*; Dixon, D.G., Dry, M.J., Eds.; CIM: Calgary, AB, Canada, 2005; pp. 295–310.

10. Dudas, L.; Maass, H.; Bhappu, R. Role of mineralogy in heap and *in situ* leaching of copper ores. In *Solution Mining Symposium*; AIME: New York, NY, USA, 1974; pp. 193–201.

11. *LLnL database*, version V8.R6.230. Lawrence Livermore National Laboratory: Livermore, CA, USA, 2010.

12. Johnson, J.; Oelkers, E.H.; Helgeson, H.C. SUPCRT92: A software package for calculating the Standard molal thermodynamic properties of minera., gases, aqueous species, and reactions from 1 to 5000 bar and 0 to 1000 °C. *Comput. Geosci.* **1992**, *18*, 899–947. The database for the program has be updated as slop98.dat from Geopig group. Available online: https://www.asu.edu/sites/default/files/slop98.dat (accessed on 4 July 2006). [CrossRef]

13. Kontny, A.; Friedrich, A.; Behr, H.J.; de Wall, H.; Horn, E.E. Formation of ore minerals in metamorphic rocks of the German continental deep drilling site (KTB). *J. Geophys. Res.* **1997**, *102*, 18323–18336. [CrossRef]

14. Huang, H.H. The Application of Revised HKF Model for Thermodynamically Describing Elevated Pressure and Temperature processes such as Treatment of Gold Bearing Materials in Autoclaves. In *Hydrometallurgy 2008—Proceedings of the Sixth International Sixth International Symposium*; Young, C.A., Taylor, P.R., Anderson, C.G., Choi, Y., Eds.; Society for Mining, Metallurgy, and Exploration (SME): Littleton, CO, USA, 2008; pp. 1066–1077.

15. Pourbaix, M. *Lectures on Electrochemical Corrosion*; Plenum Press: New York, NY, USA, 1973; pp. 143–144.

16. Wang, Q.; Nishimura, T.; Umetsu, Y. Oxidative Precipitation for Arsenic Removal in Effluent Treatment. In *SME Minor Elements 2000 Processing and Environmental Aspects of As, Sb, Se, Te and Bi*; Young, C.A., Ed.; SME: Littleton, CO, USA, 2000; pp. 39–52.

17. Dzombak, D.A.; Morel, F.M.M. *Surface Complexation Modeling: Hydrous Ferric Oxide*; John Wiley & Sons: New York, NY, USA, 1990.

18. Huang, H.H.; Young, C.A. Mass-Balanced Calculations of Eh-pH diagrams using STABCAL. In *Mineral and Metal Processing IV*; Woods, R., Ed.; Electrochemical Society: Pennington, NJ, USA, 1996; pp. 227–233.

19. Gow, R.N.V. Spectroelectrochemistry and Modelling of Enargite (Cu_3AsS_4) Reactivity under Atmospheric Conditions. Ph.D. Thesis, The University of Montana, Butte, MT, USA, June 2015.

20. Duaime, T.E.; Tucci, N.J. Butte Mine Flooding Operable Unit: Water-Level Monitoring and Water-Quality Sampling 2011. Available online: http://www.pitwatch.org/download/mbmgannual/BMF-2011.pdf (accessed on 25 March 2014).

21. Srivastave, R. Estimation and Thermodynamic Modeling of Solid Iron Species in the Berkeley Pit water. M.Sc. Thesis, The University of Montana, Butte, MT, USA, June 2015.

22. Parkhust, D.L.; Appelo, C.A.A. PHREEQC Computer Program for Speciation, Batch-Reaction, One-Dimensional Transport, and Inverse Geochemical Calculations version 3.3.3. Available online: http://wwwbrr.cr.usgs.gov/projects/GWC_coupled/phreeqc/ (accessed on 13 March 2014).

23. Pritzker, M.D.; Yoon, R.H. Thermodynamic Calculations on Sulfide Flotation System: I. Galena-Ethyl Xanthate System in the Absence of Metastable Species. *Int. J. Miner. Process.* **1984**, *12*, 95–125. [CrossRef]

24. Hiskey, J.B.; Wadsworth, M.E. Galvanic Conversion of Chalcopyrite. *Metall. Trans. B* **1975**, *6*, 183–190. [CrossRef]

25. ParaView version 4.3.1 64-bit. Available online: http://www.paraview.org/download/ (accessed on 11 February 2015).

26. VisIt 2015 Version 2.4.2. Available online: https://wci.llnl.gov/simulation/computer-codes/visit/ (accessed on 23 April 2012).
27. *MATLAB*, Version R2013a; Mathworks Computer program: Natick, MA, USA, 2013.
28. Grossmann, K.; Arnold, T.; Lkeda-Ohno, A.; Steudtner, R.; Geipel, G.; Bernhard, G. Fluorescence properties of a uranyl(V)-carbonate species [U(V)O$_2$(CO$_3$)$_3$]$^{5-}$ at low temperature. *Spectrochim. Acta A* **2009**, *72*, 449–453. [CrossRef] [PubMed]
29. VTK Format. The VTK User's Guide, Version 4.2, Kitware. 2010. Available online: http://www.vtk.org/wp-content/uploads/2015/04/file-formats.pdf (accessed on 19 February 2015).

metals

MDPI

Review

Tannins in Mineral Processing and Extractive Metallurgy

Jordan Rutledge and Corby G. Anderson *

Kroll Institute for Extractive Metallurgy, Colorado School of Mines, Golden, CO 80401, USA; jrutledg@mines.edu
* Author to whom correspondence should be addressed; cganders@mines.edu; Tel.: +1-303-273-3580;
  Fax: +1-303-273-3795.

Academic Editor: Suresh Bhargava
Received: 28 July 2015; Accepted: 21 August 2015; Published: 27 August 2015

Abstract: This study provides an up to date review of tannins, specifically quebracho, in mineral processing and metallurgical processes. Quebracho is a highly useful reagent in many flotation applications, acting as both a depressant and a dispersant. Three different types of quebracho are mentioned in this study; quebracho "S" or Tupasol ATO, quebracho "O" or Tupafin ATO, and quebracho "A" or Silvafloc. It should be noted that literature often refers simply to "quebracho" without distinguishing a specific type. Quebracho is most commonly used in industry as a method to separate fluorite from calcite, which is traditionally quite challenging as both minerals share a common ion—calcium. Other applications for quebracho in flotation with calcite minerals as the main gangue source include barite and scheelite. In sulfide systems, quebracho is a key reagent in differential flotation of copper, lead, zinc circuits. The use of quebracho in the precipitation of germanium from zinc ores and for the recovery of ultrafine gold is also detailed in this work. This analysis explores the wide range of uses and methodology of quebracho in the extractive metallurgy field and expands on previous research by Iskra and Kitchener at Imperial College entitled, "Quebracho in Mineral Processing".

Keywords: quebracho; tannin; flotation; fluorite; germanium; precipitation; Tupasol

1. Introduction

Tannins are organic, wood derived compounds that have many industrial applications including leather production, chemical and petroleum processes, and froth flotation. The most commonly used form of tannins comes from two types of trees that grow in southeastern South America. Figure 1 displays one of the species of quebracho trees native to Argentina.

This valuable material, quebracho, is extracted from the inner core of the tree, or the heartwood. The heartwood is chipped, heated to around 230 °F under pressure and evaporated under a vacuum. This produces Tupafin ATO, the most basic form of quebracho, which is soluble in warm water. If treated with the addition of sodium bisulfate the compound becomes Tupasol ATO, quebracho that is soluble at all ranges of pH and temperatures. Silvafloc is a quebracho with added amine groups [2]. There are two chemically distinct tannin groups: hydrolysable and condensed. Quebracho is a member of the condensed tannin group and will not break down to form other compounds in when acids, alkali or enzymes are introduced [3]. Chemically, quebracho is made up of carbon, oxygen, and hydrogen atoms. The quebracho compounds are made up of phenol and carboxylic groups. When ionized, these groups provide the adsorption onto cationic surfaces and with the addition of hydrogen bonding, the mineral surfaces become hydrophilic [2]. Quebracho adsorption occurs in a variety of ways including hydrogen bonding, complex formation between OH groups and cations, charge neutralization with OH- meets a positively charged surface, and from electrostatic attraction between negatively charged quebracho micelles and positively charged mineral surfaces [3].

Figure 1. A cut from the Quebracho Colorado tree displays the deep red-brown heartwood [1].

2. Flotation

2.1. Jamesonite

The use of tannins for the polymetallic ore containing tin, antimony lead, and zinc was studied at the Changpo flotation plant in China. The optimum flotation conditions for jamesonite ($Pb_4FeSb_6S_{14}$) were at a pH below 8.5 with addition of lime. In these conditions it is problematic to depress sulfides including marmatitie, arsenopyrite, pyrite, and pyrrhotite. Five different types of tannins were used to determine the depressing affect—larch, bayberry, valonia, acacia mangium, and emblic.

Lab scale bulk flotation tests were carried out to recover the antimony, lead, and zinc into a single concentrate. The concentrate was reground and floated with lime, sodium cyanide, zinc sulfate, and tannins. All tannin varieties produced positive results on jamesonite flotation except bayberry. Larch and emblic are condensed tannins with flavonoids and were the most effective as opposed to the hydrolysable tannins that contained carboxyl and hydroxyl groups like the bayberry tannin. In addition to this factor, tannins with more total color correlated to more quinines and thus better selectivity. The presence of larch tannins in the lab flotation circuit improved the grade significantly from 18.5% to 23.5%, while the recovery remained constant at 86%.

X-ray diffraction was used to compare the patterns between concentrates with and without the larch tannin; these are shown in Figure 2. The intensity of the peaks was compared to determine the depressing effect of tannin on each mineral. Calcite and quartz were the most strongly depressed, followed by pyrite, marcasite, sphalerite, pyrrhotite and arsenopyrite. However, the jamesonite intensity was greatly increased with the addition of tannins—thus, tannins have no depressing affect.

In addition to lab scale work, an industrial test was also completed at the Changpo plant. The circuit had a capacity of 2000 t/day and contained a rougher cell, eight cleaner cells, and four scavenger cells. The larch tannins were added to the second, fourth, and sixth cleaner cells. The results were positive for the two-month trial with an increase in antimony grade from 13.1% to 17.6% and increase in lead grade from 14.8% to 19.9% while the recoveries of both remained constant.

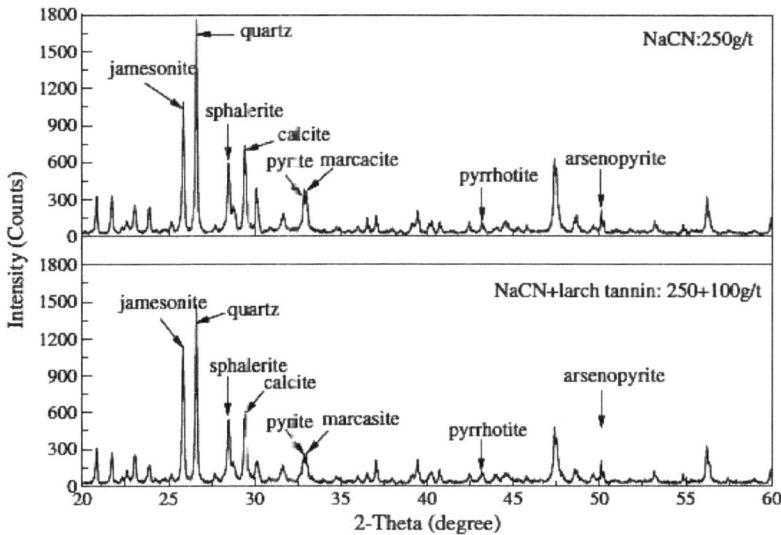

Figure 2. X-Ray diffraction peaks show the quebracho effect on the concentrates [4].

The results of this study indicate that the use of tannins, particularly larch tannin, which is condensed and has a higher content of tannin species, helps to depress minerals to aid the flotation of Jamesonite [4].

2.2. Strontium

Historically, the main application for strontium was in the production of cathode ray tubes for color televisions; however, these televisions are becoming obsolete and other applications such as ferric magnets, zinc reduction, specialty alloys and fireworks now take up the majority of the market. The most common sources of strontium are celestite ($SrSO_4$) and strontianite ($SrCO_3$).

There are several common methods for the recovery of strontium. The first involves the black ash method whereby finely ground celestite is combined with coal and heated to 1100 °C to reduce the strontium sulfate into a soluble strontium sulfide. Strontium carbonate is precipitated out as a final product by passing soda ash through the sulfide solution. The soda ash method involves treating ground celestite with soda ash and steam for several hours. Strontium carbonate is formed and sodium sulfate goes into solution [5].

The main setback in the production of a strontium concentrate is the separation of the calcareous gangue, primarily calcite, from the celestite mineral as both minerals are similar in physical and chemical properties. Flotation proves to be a viable method with the addition of quebracho as a means to depress calcite.

In the flotation studies done by F. Hernainz Bermudez de Castro and M. Calero de Hoces, quebracho was used as a key reagent in addition to sodium silicate and sodium oleate. Sodium silicate, an inorganic modifier, was compared with the organic quebracho reagent to depress the calcite. Both of these depressants act by decreasing the adsorption of the collector, sodium oleate, on the calcite surface. In addition to these reagents, pine oil was used as a frother, with hydrochloric acid and sodium hydroxide used as pH modifiers. The study indicated that quebracho did have a depressing effect on the calcite; however, it also depressed the celestite and resulting in poor selectivity. For a given concentration of collector this affect was largely independent of pH. However, concentration of collector changed the outcome drastically. With an increase in concentration of sodium oleate, the depression action of quebracho was significantly lowered. Thus, the interaction of both the collector and the depressor is an important factor to watch. A competition between the two reagents occurs for the adsorption on the calcite surface [6]. While the quebracho was not a selective reagent under these conditions, further work should be done possibly with combining depressors, to fully understand the limitations of quebracho.

2.3. Rare Earth Oxides

A study on the Mt. Weld deposit in Western Australia looked into the recovery of rare earth oxides from limonitic siltstone ore rich in both iron and calcium. Two different upgrading methods were tested, one, which focused on multiple flotation, stages following fine grinding with a P_{80} of 110 μm. Sodium silicate was used for desliming, with sodium sulfide, starch and fatty acid used in the rougher and scavenger cells. Quebracho was utilized in the cleaners to depress any iron and calcium minerals remaining in the rougher concentrate While the results displayed the upgrading was substantial with a grade of 49.5% in 16 percent of the total weight, a large portion of hematite and calcium oxide remained in the final concentrate. The study concluded that optimizing grinding conditions and modifying the depressant usage in the cleaners would be needed [7].

2.4. Scheelite

The separation of scheelite ($CaWO_4$) from calcite ($CaCO_3$) proves to be quite difficult given the physical and chemical similarities that both minerals share—including hardness, solubility, and specific gravity [8]. Thus, flotation was researched as a means for the selective recovery of scheelite. An investigation was conducted by the U.S. Bureau of Mines to determine a procedure to efficiently float calcareous scheelite ores. The ores had varying assays of scheelite ranging from 0.16 to 0.54 weight percent and from 24.3 to 32.0 weight percent calcite. All of the experiments consisted of three separate flotation phases. A traditional reagent scheme of fatty acid collectors, sodium silicate as a depressant and sodium carbonate for pH control was used in the addition to quebracho. First, the sulfide minerals in the ore were floated off prior to scheelite flotation—modified dixanthogen and cresylic acid had the highest recoveries of sulfides while pulling almost no scheelite. The pulp was then conditioned with sodium carbonate, quebracho, and trisodium phosphate to depress the gangue minerals and finally a rougher and one cleaner produced the final scheelite concentrate. Alternatives for quebracho such as lignin sulfonate salts, sodium cyanide, and other tannin extracts were tested yet quebracho remained the most effective for calcite depression. The Table 1 shows the optimal reagent additions from the batch tests.

Table 1. Depressant Dosing Amounts [9].

Reagent	Amount (Pounds per Ton)
Linoleic-oleic acid	1.2
Sodium silicate	3
Sodium carbonate	4
Quebracho	1–2 *

* The amount of quebracho needed was directly related to the amount of calcite in the ore.

The preliminary work studied many factors of flotation including water purity, pulp density, temperature, and reagents. A more detailed study was completed on the factors that were considered "critical in achieving optimal results." These factors were found to be the role of quebracho as a depressant and the control of iron. The role of quebracho in the flotation of scheelite was critical. While the recovery is only slightly affected by the amount of quebracho in the system, the grade is significantly improved with the addition of quebracho. Quebracho has a marked effect on the selectivity of the collector and allows the scheelite to be preferentially chosen over the calcite during flotation. It was shown that a deficiency in quebracho was met with a much lower grade concentrate, while excess quebracho had little effect.

It was shown that the order in which the reagents were added for conditioning and the conditioning time were important variables. While the reagents can be added simultaneously, the quebracho should be added before the collector to allow the scheelite surfaces to become preferential to the calcite surfaces. Quebracho primarily influences the grade of the concentrate. This was also the case with the conditioning time, as seen in Table 2. A marked difference in grade and an insignificant difference in recovery were seen by changing time. The optimum time for conditioning was 3 to 5 min for the quebracho followed by a 5 to 10 min conditioning with the collector. The reason for declining grade with increasing time was explained that as time went on the quebracho film was removed from the calcite surfaces [9].

Table 2. Conditioning Time Data [9].

Quebracho Conditioning Time (minutes)	Collector Conditioning Time (minutes)	% WO$_3$ in Concentrate
3	3	6.6
3	5	8.9
3	10	7.7
3	20	2.2
5	3	6.1
10	3	2.1
20	3	1.2

The work done by the Imperial College suggests some further interpretations of the 1964 Bureau of Mines study. The primary advantage of using the sodium carbonate in addition to controlling the pH is to limit the amount of soluble calcium and magnesium ions within the system. When excess ions exist within the system, the quebracho is more likely to bind to the scheelite surfaces and displace the fatty acid collector [3].

In another study done at the Istanbul Technical University, multiple modifiers were used in addition to the oleoyl sarcosine collector including quebracho, oxine, and alkyl oxine. Zeta potential measurements were taken over a range of modifier concentrations to display electrokinetic data. For quebracho alone, in concentrations from 10 ppm to 150 ppm, the zeta potential for scheelite was more negative than that of calcite, which indicates the scheelite would be depressed. In higher concentrations above 150 ppm, the zeta potentials were comparable, consequently no selectivity would occur. However, when conditioning with alkyl oxine was done prior to adding quebracho, the zeta potential of calcite surfaces was greater than that of the scheelite surfaces—providing selective depression. Competitive adsorption occurs on the scheelite surfaces between alkyl oxine and quebracho; alkyl oxine is chemically adsorbed onto the surface and decreased the negative charge on the surface. The zeta potentials of scheelite and calcite were significantly different enough to feasibly separate via flotation. Both the recovery and grade with the oxines were higher than without [8].

2.5. Fluorite

Although fluorite is often regarded as a gangue mineral for metals production, high-grade fluorite has value in a variety of markets. The most notable use for fluorite ore is in the production of hydrofluoric acid used for the majority of fluorine compounds. It is also used in uranium production, aluminum production and as a flux. Figure 3 displays the relative amounts of fluorspar used in different applications over time.

End Uses of Fluorspar

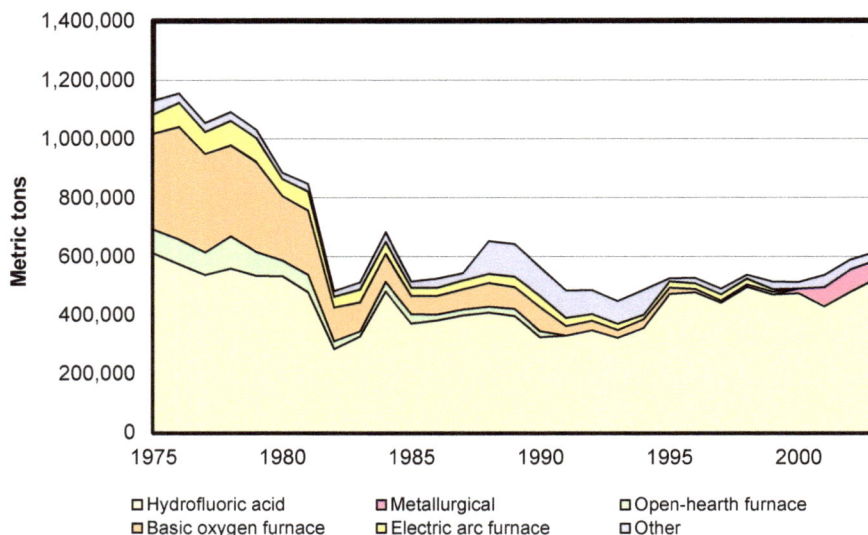

Figure 3. United States Geological Survey data showing the uses of fluorspar over the last few decades is illustrated [10].

The traditional means to recover fluorspar involved utilizing ores that did not contain significant amounts calcium carbonate. Means to separate fluorite from silica were well established; however up until the 1940s, calcareous deposits were untreatable. Cullen and Lavers revealed a new method for fluorite flotation in 1936. Anderson, Stengl, and Trewartha set out to produce a high quality fluorite concentrate (no less than 97% CaF_2) from calcareous type deposits. The process involved grinding the material to around 150 microns, floating at a pH of 8 with fatty acid collectors and quebracho to depress the calcite and other gangue [11]. Currently, quebracho and tannin extracts are the depressant of choice for generating high purity acid-grade fluorite concentrates.

The positive advantage of adding quebracho into the flotation circuit was seen in the study done by Clemmer and Anderson. The head assay of zinc and lead flotation tailings was 0.13% Pb, 0.45% Zn, 26.92% CaF_2, and 42.98% $CaCO_3$. The material was ground to below 100 mesh and treated with oleic acid and quebracho in a cell. The final concentrate assayed at 99% CaF_2, 0.03% Pb, and 0.05% Zn [12]. With an enrichment ratio of 3.68, it is a very formidable process and proves to be viable for the production of fluorite in multiple applications.

The basic principle behind the success of quebracho in the fluorspar flotation circuit lies in the fact that it depresses the calcite, silicates, and metal sulfides much stronger than the fluorite. With respect to calcite, it is particularly challenging to separate from fluorite because both minerals have Ca^{2+} cations in lattice. When quebracho is adsorbed, the surface becomes hydrophilic. Many attempts to describe and characterize the selective nature of quebracho have been undertaken. One of the primary insights involved the depression variance between tannin derivatives. It is clear through many studies that the condensed tannins, like that of quebracho, are far more selective than the hydrolysable tannins like that of tannic acid. The hydrolysable tannins have a greater depression affect to a disadvantage compared to the condensed tannins; the depression on the hydrolysable tannins is too strong and thus provides a less selective depression [3].

The method of bonding also affects the behavior of tannins within a system. Plaskin and Shrader had several insights on how tannins interacted with the surfaces of minerals being floated with oleic acid. They determined the tannins to be semi-colloidal with hydrogen bonding being present. Tannins carry a negative surface charge due to the phenolic-OH groups present. Tannins thus react with the surfaces of minerals via chemisorption. This reaction isn't exclusive to the mineral surface itself but also interacts with the collector being used. The electrokinetic potential of the mineral is slightly negative with the addition of small amounts of tannin. This negative surface charge repels anionic collectors causing depression, but attracts cationic collectors [3].

The use of quebracho in a circuit relies heavily on the other reagents in the flotation scheme and their concentrations. In general, fatty acid collectors are used exclusively in fluorspar flotation with quebracho. Common collectors include oleic acid, sulphated soap, spermol, and oleate. The general range for collector addition is around 200–500 grams per ton while the tannins are similar at around 100–500 grams per ton. The Imperial College study determined that both the quebracho and the collector compete for adsorption on the surfaces of calcite and fluorite. Adsorption was observed for all scenarios of calcite and fluorite in the presence of quebracho and oleate. When only oleate was present in the system, fluorite adsorbed almost double the amount than the calcite. Adding only quebracho, the adsorption levels onto the calcite and the fluorite were very similar. Thus, it was shown that both the oleate and the quebracho played a significant role in the depression of calcite. With calcite holding less oleate than fluorite and quebracho adsorbing similarly onto both minerals, the fluorite floats and the calcite is depressed. In this manner, the order of conditioning is important. It was found that when the quebracho was added before the oleate, a greater depression of calcite took place and when the oleate was added prior to the quebracho, an increase in calcite flotation was observed [3].

The effect of pH on the flotation of fluorite with quebracho has been studied quite thoroughly. In practical applications, the pH ranges from 8 to 9.5. This accounts for multiple factors of adsorption of the quebracho and the oleate, the acid consumption, and practical recovery and grade balances. The Imperial College study indicated that quebracho remains a depressant up until a pH of around 10. The optimal adsorption pH for quebracho on fluorite surface is around 7, while the dissolution of calcite requires the pH be at or above 8 [3]. The work done by Clarence Thom suggests that pre-conditioning the ore at a pH less alkaline than that of optimal flotation levels improves the separation. Thom had optimum recovery when the quebracho conditioning was carried out at a lower pH (ideally below 8) before soda ash and oleic acid were added [13]. While the pH of the cell has been studied extensively, temperature could be a new parameter of interest.

Another aspect that has been studied with regards to fluorite flotation is the effect of soluble ions within the system—specifically calcium ions. As both fluorite and calcite are marginally soluble, it can be concluded that a small amount of calcium ions will be present in the aqueous solution. The Imperial College work acknowledged that adding different concentrations of calcium ions into the solution, by means of calcium chloride, affected the flotation. At concentrations greater than 10^{-4} M calcium ions, the fluorite was severely depressed, keeping all other factors constant. However, when the concentration of calcium is slightly lower, on the order of 10^{-5} M, the calcite is heavily depressed, while the fluorite floats. In industry, it is common practice to use softened water to avoid the negative affect of dissolved calcium ions on the flotation of fluorite [3].

In the patent from Allen and Allen, the generation of acid grade fluorspar, 97% CaF_2 or greater, was investigated. The main research done involved the addition of ferrous sulfate salts and the effect of quebracho in the circuit. It was shown that a crude fluorspar ore assaying at 87.23% CaF_2, 4.68% SiO_2, and 4.66% $CaCO_3$ was significantly upgraded to 97.48% CaF_2 with a recovery of 91 percent. In this example, the material was crushed and fed to a wet ball mill unit running at 70 percent solids where 0.275 pounds of ferrous sulfate was added. The material was reduced to 89% passing 200 mesh, was flocculated, thickened and transferred to a conditioning tank. Reagents on the amounts of 2.2 pounds per ton of stearic fatty acid, 3.3 pounds per ton Acintol FA2 tall oil fatty acid, and 0.97 pounds of sodium carbonate were added to the conditioning cell at 207 °F for 20 min. Quebracho was added at the conditioner tank discharge at 0.13 pounds per ton of crude ore in addition to 3.67 pounds per ton of sodium silicate. This was fed to the first flotation cells, which produced a heavily mineralized fluorspar froth while maintaining the pH between 8 and 9. Quebracho was further used in the subsequent two cleaning stages at 0.37 pounds per ton. From this study, it was concluded that the addition of soluble ferrous sulfate salts (on the order of 0.2 to 0.8 pounds per ton) in combination with quebracho aided in depressing gangue minerals present including silica and calcium carbonate, whilst improving the flotation of fluorite.

Allen and Allen also delved into the effect of different types of fatty acids. Fatty acids are the most common type of collectors for fluorite flotation. This study suggested that the use of saturated fatty acids was critical in the reagent scheme. They argued that using high concentrations of the traditional unsaturated fatty acid collectors, like oleic acid and tall oil acids, was not the best practice. Instead, it was recommended having a collector composition between 30 and 60 percent saturated fatty acids. The saturated fatty acids were preferably to have between 12 and 18 carbon atoms like those of stearic acid, lauric acid, and palmitic acid. Increased recoveries of fluorite were seen with the use of saturated fatty acids.

Allen and Allen also discovered that the use of guar within the system in addition to the reagents discussed above helped to further improve the flotation of fluorite. It was shown that when guar was added to the pulp the gangue minerals including barium sulfate, pyrite, and clay slimes were significantly depressed. It was suggested that the guar be introduced into the circuit in the mill or the conditioner when the ferrous sulfate and depressants are added [14].

While fluorite flotation is conventionally carried out in a standard bank of cells with multiple stages comprised of roughers, scavengers and cleaners, column flotation also proves to be a successful method. The Fish Creek fluorite processing plant in Eureka County Nevada proved column flotation to be superior due to increased recovery [15]. The work done by the Gujarat Mineral Development Corporation in Kadipani proved column flotation to be more advantageous than that of the traditional route. The original processing circuit consisted of a rougher, scavenger and six stages of cleaning while the modified column flotation eliminated four of these stages. The differences between column and conventional flotation can be seen in Table 3. The findings of this study showed that the use of tannins in a column flotation scheme as opposed to a traditional flotation circuit improved the overall separation of fluorite from the gangue consisting of $CaCO_3$, SiO_2, and P_2O_5 [16].

Specific column flotation parameters were researched in fluorite flotation on an ore from the Nossa Senhora do Carmo Mining Company. The work focused on using a short column, described as the collection zone being 8 m tall, with a negative bias regime. Negative bias is obtained by having a smaller tailing output flow than the feed input flow. The cell was kept at 30% solids with a pH of 10 and sodium silicate, cornstarch, quebracho, and tall oil were added in sequence during the conditioning stage. The study concluded that increased recoveries were observed in this environment due to the negative bias allowing loaded air bubbles to more readily rise to create froth. While this scheme created a good environment for the recovery of fluorite, it should be used as a rougher as the grade of the concentrate is fairly low [17].

A very common issue regarding flotation as a means of separation involves slime coatings. The heterocoagulation of a fluorite and gangue minerals was observed for a feed fluorite ore of the Minera de las Cuevas concentration plant in Mexico. The work focused on the flotation circuit with oleic acid as a collector and quebracho as a depressant with the addition of CMC and water glass as dispersants. SEM (scanning electron microscope) was used to observe the slime coatings of the gangue minerals, calcium carbonate and quartz, on the fluorite surface. It was found that strong heterocoagulation occurred around pH 9, which is close to general operating parameters for pH in a fluorite flotation circuit. The study suggested that this phenomenon may be due to strong electrical double layer attraction yet weak double layer repulsion between the gangue particles and the fluorite particles at these conditions. The addition of the dispersants CMC and water glass improved the flotation circuit by reducing the slime coatings; an increased recovery of fluorite from 72% to 78.5% was seen with the grade remaining continuous at 98% [18].

Table 3. Kadipani Flotation Data [16].

Feed Assay %						
	Sample	CaF$_2$	CaCO$_3$	SiO$_2$	P$_2$O$_5$	
	1	65.84	5.62	16.68	3.14	
	2	74.51	5.10	10.82	2.16	
	3	78.8	4.36	10.60	1.88	
	4	84 69	4.67	4.57	1.78	
	5	89 42	4.23	2.56	0.89	
	6	90.58	2.43	3.00	1.20	

	Concentrate Assay, % Conventional Flotation				**Concentrate Assay, % Column Flotation**			
Sample	CaF$_2$	CaCO$_3$	SiO$_2$	P$_2$O$_5$	CaF$_2$	CaCO$_3$	SiO$_2$	P$_2$O$_5$
1	76.80	5.62	6.73	2.40	82.80	5.12	3.70	3.20
2	80.86	6.60	6.16	2.13	88.82	4.86	1.47	1.41
3	80.01	2.40	9.80	2.01	92.56	1.65	1.53	1.35
4	88.77	4.64	4.57	1.78	91.80	3.40	1.06	1.07
5	91.75	3.31	1.64	0.72	95.21	1.59	0.60	0.18
6	92.10	3.61	1.95	0.68	94.35	2.12	1.07	0.58

The stability of the feed into a flotation circuit is a critical parameter to control, particularly as it affects the grades and recoveries of the valuable mineral in question. The work done by Schubert, Baldauf, Kramer, and Schoenherr on the development of fluorite flotation examined the effects of changing feed in high calcareous ores with the addition of quebracho, sodium hexametaphosphate, and oleoyl sarcosine. Oleoyl sarcosine with around 20% oleic acid was favored over the traditional collector containing 60%–70% oleic acid The calcium carbonate content in the ores tested ranged from 7 percent to 45 percent while the fluorite ranged from 29 to 52 percent. As noted previously, the most important parameter to control was the concentrations of the reagents; specifically the amount of quebracho added. This study illustrated that quebracho had two major effects: when at low concentrations, depression is the primary rule and at higher concentrations the flotation of fluorite becomes activated. It was also explained that quebracho acts as a dispersant in the system particularly at high concentrations. The most favorable recoveries occurred in the pH range from 8–9 and in highly flocculated systems with a particle size from 0.04 to 0.16 mm [19].

2.6. Sulfides: Copper-Lead-Zinc-Nickel Ores

To effectively separate polymetallic sulfide ores, depressants including lime, sulfites and cyanide in conjunction with xanthate collectors are traditionally used. Quebracho proves to be a promising alternative to these depressants and is favored for operating conditions, as the tannins are completely non-toxic and work at moderate pH levels [3]. The order of depression by quebracho is shown below for sulfides:

$$\text{pyrite} > \text{pyrrhotite} > \text{sphalerite} > \text{copper sulfides} \quad [2] \tag{1}$$

Imperial College provided a comprehensive study on the effects of quebracho as a depressant for sulfides. The depressant action of quebracho was found to be a function of the –OH group content [3]. Before this study, the quebracho form most commonly used for sulfides was Tupasol ATO; however, Tupafin ATO proved to be the most effective quebracho depressant form.

The primary aim in nearly all sulfide separation includes removing the pyrite content from the rest of the valuable sulfides. An important distinction of the form of pyrite was taken into account. As pyrite experiences high surface oxidation, both non-oxidized and oxidized pyrites were tested. In the presence of xanthate, the non-oxidized pyrite was sufficiently depressed by all three forms of quebracho. In the case of the oxidized pyrite, it is clear that quebracho is less effective for depression. Figures 4 and 5 display the differences in depression that quebracho has on non-oxidized and oxidized pyrite. Most notably the range of pH that the pyrite is depressed is much smaller in the case of the oxidized pyrite where large fluctuations occur between pH around 6 to 9. In the case of the non-oxidized pyrite, a pH from around 4 to 9.5 was sufficient for the depression of pyrite. These figures also illustrate how the three forms of quebracho behave differently. Silvafloc (quebracho A) has the best depressing effect when the pyrite is non-oxidized, but Tupafin ATO (quebracho O) provides the best results when dealing with the oxidized pyrite. Since it is feasible that there will be a mix of both oxidized and non-oxidized pyrite in a sulfide flotation cell, the preferred quebracho type is Tupafin ATO. The reasoning behind the differences in depression effect between oxidized and non-oxidized pyrite was explained through testing with ferric salts. It was shown that when increasing amounts of Fe^{3+} ions were added to the solution quebracho no longer depressed pyrite.

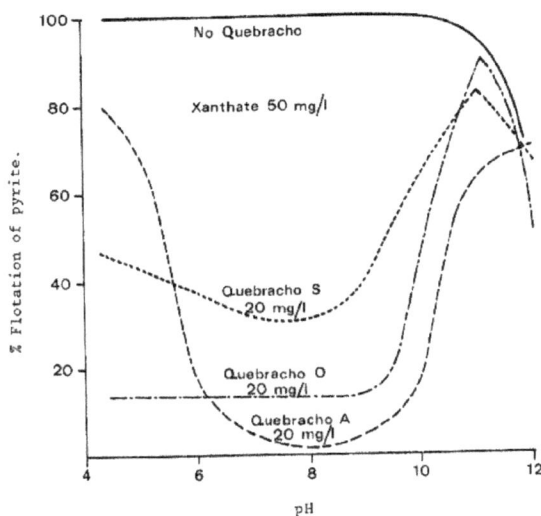

Figure 4. Depression of non-oxidized pyrite with addition of xanthate and quebracho [3].

Quebracho is known to complex with trivalent iron especially above pH 6; thus, it was determined that as the levels of ferric iron in the pulp rose, as was the case with oxidized pyrite, the depressing effect of quebracho decreased. It should be noted that ferrous iron does not form a complex with quebracho however it readily oxidizes to ferric iron [3]. A pH 6 was shown to provide the best conditions for depressing pyrite.

It is interesting to note that quebracho does not always behave the same way in terms of adsorption. With separation of calcite from fluorite, quebracho and the collector oleate are competitive adsorbates. Both fluorite and calcite adsorb the quebracho and oleate by the same mechanisms via pulp and lattice calcium ions. However, in the case of sulfide systems, it was shown that this was not the case. Quebracho and xanthate do not act as competitive adsorbates within the system; xanthate is adsorbed onto the mineral surfaces at very similar values with and without the addition of quebracho. This indicates that the depressing effect of quebracho is completed by rendering the surface hydrophilic [3].

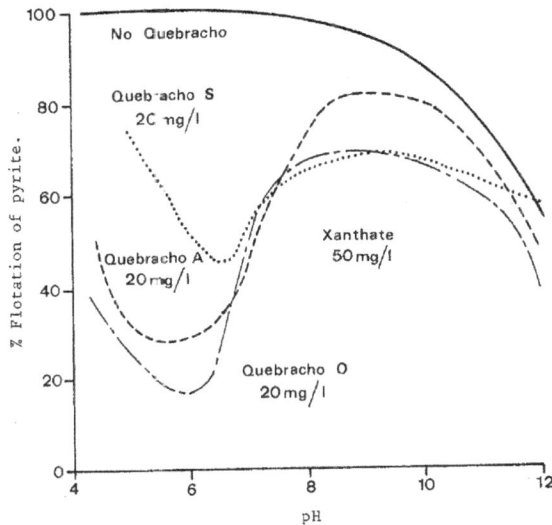

Figure 5. Depression of oxidized pyrite with addition of xanthate and quebracho [3].

Laboratory and pilot plant testing was done on the Cayeli-Riz massive sulfide deposit in Turkey to compare two different depressant systems. The deposit has two ores; the yellow ore is mostly copper with fine-grained pyrite while the black ore contains both copper and zinc that is finely disseminated. Both ores are very high in pyrite and thus required a specific depressant system that would depress the gangue sulfides without depressing the valuable sulfides. The copper rougher flotation was taken out at a pH of 10.5–11.5, the copper cleaning at a lower pH 4.5–5.5, and the zinc flotation was at a high pH above 11.5. The first depressant system used a 1:1 mixture of cyanide and thiourea at 120 g/t in the copper rougher to depress zinc and lime was used to depress the pyrite at 8000 g/t with controlled addition of SO_2. The second system used the same additions of lime and SO_2 in addition to 2H-BC, which is a mixture of quebracho and dextrin-maleic acid. The new addition of 2H-BC was successful at depressing the ultrafine pyrite and increased both the grade and recovery of the copper and zinc concentrates by several percentage points [20].

A recent study taken out by the National University of San Juan in Argentina set out to prove that tannins and quebracho are a cleaner alternative to the conventional flotation reagents used in copper sulfide flotation. The work specifically set out to show that the Cu/Fe ratio could be improved by depressing pyrite with the organic depressant. Hallimond tube flotation tests were taken out on both pure pyrite and pure chalcopyrite minerals. The pyrite was floated at a pH 8 and showed good depression (around 40% pyrite depressed) with increasing tannin concentration up to around 0.25 g/L where it leveled off. The effect of pH on the recovery of pyrite was shown to favor in the alkaline and acidic range where only 5%–10% pyrite was recovered with a mixture of tannins and lime. It was also illustrated that the presence of Ca^{2+} ions (in the form of lime) improved the depression effect of the tannins on the pyrite system. Similar tests were carried out in the Hallimond tube for pure chalcopyrite. The recovery of pyrite and chalcopyrite with and without the addition of tannin is shown in Figure 6 [21].

Figure 6 clearly illustrates that the use of tannins have a marked depression effect on pyrite while having little influence on chalcopyrite recovery. Flotation work was carried out in a Denver cell where the pH was set constant at 10 with the addition of lime and a varying amount of tannins between 0 and 600 g/t. With increasing tannin addition, the pyrite was drastically reduced in the concentrate with the best results at 600 g/t. These tests concluded that the addition of tannins in the system significantly increased the grade of the copper while only sacrificing 1 percent of recovery of copper due to the depression of pyrite [21].

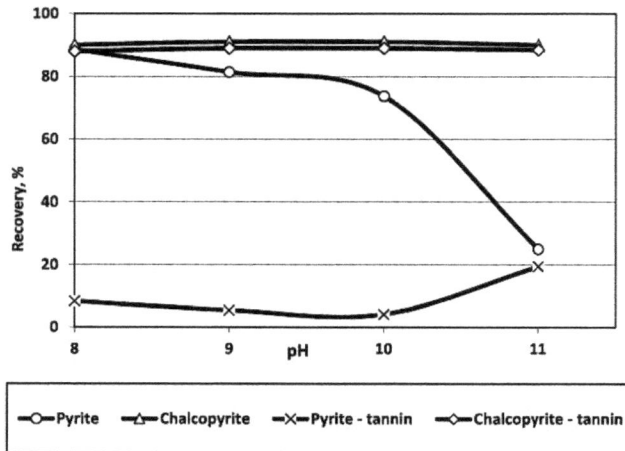

Figure 6. The effect of tannins on pure chalcopyrite and pyrite Hallimond tube flotation [21].

A United States patent titled "Separation of Polymetallic Sulphides by Froth Flotation" by Srdjan Bulatovic and Robert S. Salter investigates the use of quebracho as a depressant in complex sulfide ores. Bulatovic and Salter's work focuses on comparing a newly invented depressant mixture to the traditional depressant system on several polymetallic sulfides. The new depressant is made by dissolving quebracho into solution with either dextrin or guar gum. This mixture undergoes a second reaction with a lignin sulphonate to make a water-soluble polymer. The polymer is then partially monomerized by at least one of the following reagents: alkali metal cyanide, alkaline earth metal cyanide, metal sulfate, and/or a sulfite. For the experimental trials, the lignin sulphonate polymers LS7 and LS8 are used; LS7 is the compound that is reacted with sodium cyanide while LS8 is the compound that has been reacted with zinc sulfate/sodium cyanide complex.

The first set of experiments focused on the clean separation of copper and zinc from a massive sulfide deposit in Canada. The Cu–Zn ore contains valuable amounts of silver and the primary gangue minerals are quartz, pyrite and pyrrhotite. Four sets of experiments were conducted on two ore types, high and low copper, to compare the effects of the LS8 depressant *versus* sodium cyanide on the recovery and grade of the concentrates. The first set of experiments was laboratory scale while the second set was tested on a full-scale commercial plant. The LS8 depressant was added during the grinding process and in the copper cleaner flotation circuit, with a total of 170 g/t added. In the lab scale trials, the LS8 depressant made large improvements over the sodium cyanide on the rejection of zinc and iron sulfides from the copper concentrate by several percentage points. The sequential zinc concentrate grade and recovery was also improved with the addition of LS8. In the plant trails, similar results were obtained whereby the zinc recovery in the copper circuit was reduced, copper concentrate grade was increased, and zinc recovery in the zinc circuit was improved.

The second segment of laboratory experiments was conducted on two different massive sulfide ores containing mostly copper and nickel. The first ore, from British Columbia, contained copper and nickel as well as significant amounts of platinum and palladium. The comparison tests between the LS8 depressant and sodium cyanide proved that the LS8 reagent had a positive effect on the circuits. With the use of LS8 the copper was recovered to a higher degree in the copper flotation circuit, while the nickel was effectively depressed. The nickel recovery in the nickel circuit was increased and overall, the platinum and palladium recovery was improved. The second ore, from Northern

Ontario, also showed improved results with the use of LS8. Most significantly, the nickel recovery in the nickel cleaner concentrate was increased by more than 50 percent. These examples prove that the LS8 depressant is highly effective for depressing nickel in the flotation of copper-nickel ores.

The third installment of experiments involved a massive sulfide with mostly lead and zinc with large amounts of pyrite. Lime was compared to the LS7 depressant in laboratory testing. LS7 was added during the grinding stage at 250 g/t while the lime was added in the lead circuit at 750 g/t and the zinc circuit at 3500 g/t. The recovery of lead in the lead concentrate remained the same, while the grade improved over 30 percent with the addition of LS7. Zinc grade and recovery in the zinc concentrate was also improved with the use of LS7. In plant operations, testing was done to compare the addition of 300 g/t of LS7 to no addition of LS7. By using the LS7 reagent, grade of the lead concentrate was increased with no recovery loss.

The work done by Bulatovic and Salter proved that the invention of a new series of quebracho containing depressants greatly improved the differential flotation of polymetallic sulfides in contrast to traditional reagent schemes [22].

3. Precipitation

Germanium

Germanium is primarily consumed as a vital element for the production of semiconductors for electronics use. It is most closely associated with the elements C, Zn, Si, Cu, Fe, Ag, and Sn. Argyrodite, germanite, canfieldite, and renierite germanium containing minerals; however, these minerals do not form specific ore deposits. Thus, the majority of germanium is sourced from trace and minor amounts found in coal, iron ore, and Cu–Pb–Zn sulfide deposits [23]. Sphalerite ores with low amounts of iron prove to be the most important source of germanium containing a few hundred ppm that is contained within the lattice of the zinc sulfide [24]. As a minor element in zinc ores, the challenge comes in finding a way to selectively remove the germanium from the rest of the residual metals and compounds.

There are a variety of different methods that germanium is recovered from zinc circuits. While germanium can be collected pyrometallurgically, the concerns of volatile germanium oxides and sulfides prove hydrometallurgical methods to be the most viable route. There are two main methods to recover germanium hydrometallurgically; the first is by precipitation with tannins and the second is solvent extraction (generally used for higher concentrations of germanium in solution) [25]. Tech Cominco's process uses the latter. A substantial circuit is used whereby the zinc concentrate is calcined by two Lurgi Fluo-solid roasters and a suspension roaster. Zinc oxide is the product of this process, with the sulfides being taken off as gas oxides. ZnO from the roasting process is sent through a ball mill, cyclones, and electrostatic precipitators prior to being sent to the leaching plant. Leaching tanks contain the calcined zinc material at a pH from 1.7 to 3.5. The residue from the leaching is leached one more time with sulfuric acid to take the germanium and indium into solution. The germanium is recovered and concentrated by selective solvent extraction [26]. Another zinc operation in Russia, the Joint Stock Company Chelyabinsk Electrolytic Zinc Plant, also possesses germanium as a byproduct. This plant utilizes fluidized bed reactors to roast the sulfides and convert them to oxides that are leached. The underflow from the leaching process is dried and sent to another processing facility to recover the germanium from the residue [27].

The recovery of germanium from Cu–Pb–Zn ore was studied at the Cinkur Zinc Plant in Kayseri, Turkey. This plant produces zinc electrolytically and requires that the leach solution be refined to remove impurity metals like Ge, Cd, Ni, Co, and Cu before electrolysis occurs. Figure 7 shows the operating flow sheet.

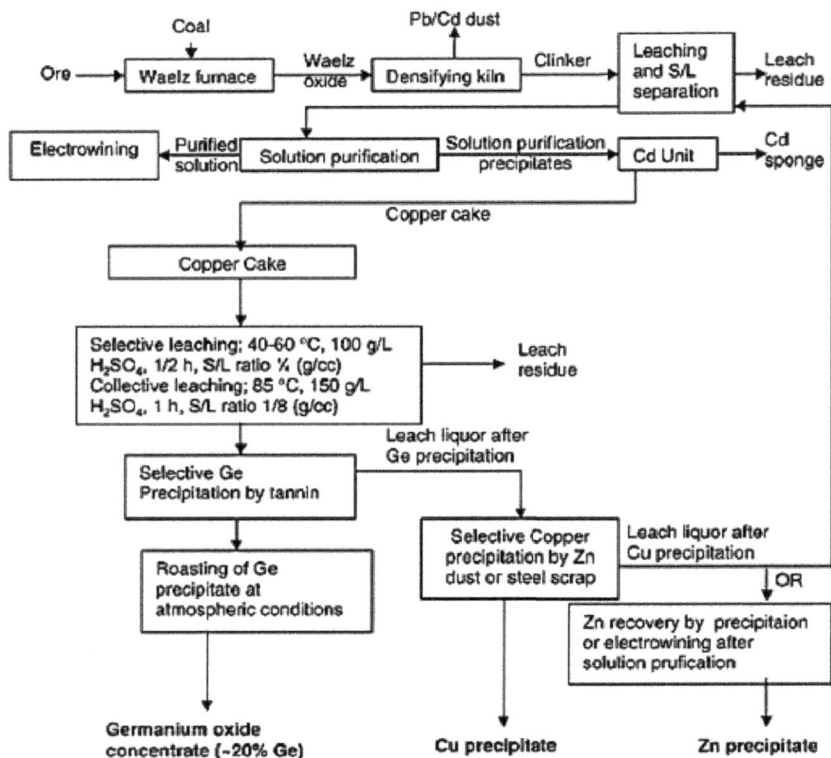

Figure 7. Flow sheet of the Cinkur plant in Turkey [25].

Precipitation was used to hydrometallurgically recover these metals in solution. First, the cobalt and nickel were precipitated using $CuSO_4 \cdot 5H_2O$ and As_2O_3 and zinc powder. The second precipitation utilized $CuSO_4 \cdot 5H_2O$, Sb_2O_3 and zinc to remove the cadmium from solution. These precipitated cake solutions provide the largest source of germanium at the plant with a concentration of over 700 ppm. The cake was leached with sulfuric acid and the leach solution treated with tannins to precipitate out germanium. Two different leaching schemes were created; one for selective leaching and one for collective leaching. The selective leaching found optimum operating conditions with a temperature from 40–60 °C, 100 grams per liter sulfuric acid, 30 min leach time, a solid to liquid ratio of 1/4, and no air pumping. It was found that 78% of the germanium was recovered under these conditions. With regards to the collective leaching approach a temperature of 85 °C, 150 grams per liter sulfuric acid, 1 h of leach time, and a solid to liquids ratio of 1/8 with air pumping was advantageous. The collective leaching method resulted in 92.7% germanium recovery although the solution had significant amounts of Cu, Zn, Ni, Co, and Cd. The tannin precipitation study indicated that 94% of the germanium could be selectively recovered from leach solution [25]. The primary reaction for the precipitation of germanium with tannins is shown below:

$$Ge^{4+} + O_2 + H_6T \rightarrow GeO_2 \cdot H_6T \tag{2}$$

where T is tannic acid ($C_{14}H_{10}O_9$) [25]. The study suggested that a more effective germanium precipitation from the zinc electrolytic solution could be done with tannins rather than the copper and zinc additives mentioned above.

Precipitation methods as sulfides, hydroxides, or with tannic acid are a viable option for obtaining germanium hydrometallurgically. The role of tannins in the precipitation of germanium from zinc production has been given more weight in recent years. Tannins work my forming chelates that complex with metal ions in solution. Optimization of tannins use is of great interest as the reagents are expensive. A study to understand the behavior of tannins chelation was done on the Chinese Yunnan Chihong Zinc and Germanium plant, the largest germanium producer in the world. The experiment was two-stage with the first part looking into the binding of tannins to metal ions individually and second, the effect of binding in a polymetallic system. It was shown through the experiments that the metals with higher valences (like that of germanium and ferric iron) had the best binding ability with tannins. This is explained by the electrostatic attraction that favors ions with larger valence. Figures 8 and 9 show the binding ability, noted as the precipitation yield for metals in both a single metal solution and a polymetallic solution. The conditions for complexation were a stirring rate of 10 rps, temperature of 373 K, and a duration of 500 s [28].

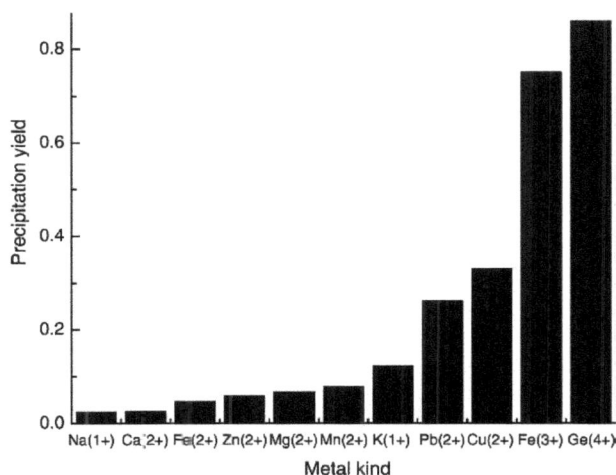

Figure 8. Binding ability of various metals in a single metal solution with tannins [28].

The complexation reaction of metals with tannins is described as follows:

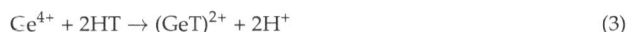

$$Ge^{4+} + 2HT \rightarrow (GeT)^{2+} + 2H^+ \tag{3}$$

where T is tannins ($H_2(GeO_2C_{76}H_{52}O_{46} \cdot nH_2O)$ [29].

Additional research was also done to determine if adding tannins in stages produced different results than adding tannins in one step. In the study the same amount of tannins was introduced, 50 g, divided equally among the number of stages to the 1 L leach solution. It was shown that yield increased dramatically with additional stages. The single step tannin addition resulted in a percent yield of 85.2 and steadily increased each time as stages were added to the five stage addition that had a yield of 99.8 percent. A possible explanation for this behavior is diffusion; with the single step addition, diffusion had a shorter time to react with more difficulty. In addition, the tannins concentration may affect the binding ability of metal ions [28].

45

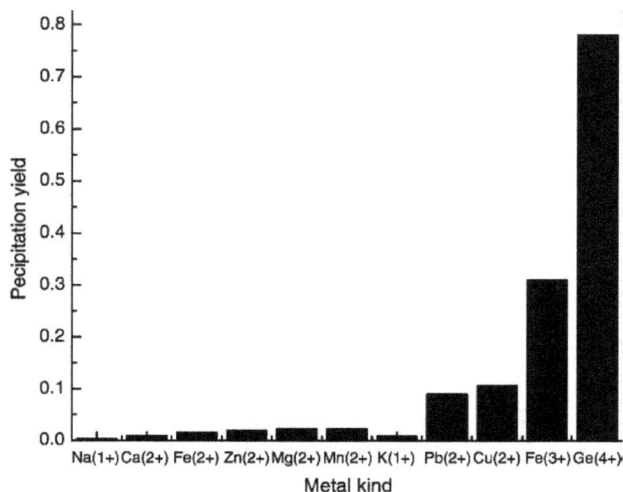

Figure 9. Binding ability of various metals in polymetallic system with tannins [28].

4. Precious Metals Recovery

4.1. Gold Bacterial Oxidation

The Fosterville Gold Mine in central Victoria, Australia, treats primary sulfide refractory gold by means of bacterial oxidation, cyanidation, and froth flotation. Nearly 80% of the gold is contained within the pyrite and the remainder in the arsenopyrite and stibnite, while the gangue consists of carbonates, quartz and other silicates. The gold occurs as a solid solution within the sulfide mineral; thus, it cannot be separated by direct cyanide leaching. The BIOX® biooxidation system was implemented to oxidize and break down the minerals from sulfides to sulfate and make iron and arsenic soluble. The following counter-current decantation washes the sulfate, iron, and arsenic in solution and the solids remaining are leached with cyanide.

While the majority of gold remains in the solids, part of it is lost to the waste liquor stream. The gold grade has been up to 10 ppm in this stream and represents around two to four percent gold loss in total. It was found that the majority of the gold lost was under 0.2 microns, in the ultrafine range, and that it occurred as colloidal gold. Several methods to recover the colloidal gold were researched and tested including coagulation and gold capture by means of creating a solid surface for the gold to adsorb onto. Both limestone and Portland cement precipitation generated undesirable affects within the circuit.

Quebracho was investigated as a means to precipitate a surface for the colloidal gold to adsorb to. The tannins formed fine precipitates from the soluble iron in the liquor and thus created a large surface area for the collection of the ultrafine gold. Multiple series of tests were completed, in all cases the addition of quebracho decreased the gold loss. Full-scale plant trials proved successful, reducing gold loss from the waste liquor by 40%. Given the results, the plant installed a tannins mixing and addition station to implement it long-term. With the implementation of tannins into the circuit, the overall gold recovery in the Fosterville plant has increased by 2.6 percent [29].

Golden Star Resources Limited in Bogoso, Ghana follows a similar processing scheme to that of Fosterville. The primary source of gold is submicron particles entrapped within pyrite and arsenopyrite. The BIOX® bacterial oxidation process is used to oxidize the sulphides and free up the gold. Counter-current decantation (CCD) is done to collect the solids, which contains the majority of gold and is further processed with cyanide. The gold that remains in the CCD solution is not soluble, but colloidal gold. The Golden Star plant saw 0.57 g/L gold go to tailings with the CCD solution. Quebracho tannin supplied by Nowata Mining Chemical Manufacture and Supply was tested.

Varied concentrations, between 0.05 g/L and 1 g/L, of quebracho tannin were prepared at 10 percent strength and were tested at a laboratory scale. Total suspended solids (TSS) and reduction in gold loss were the two main end goals of the study. It was found that the quebracho tannin was optimum at a level of 0.05 g/L where there was a 21.25% reduction in the TSS and a 45.79% gold loss reduction. These results warranted the industrial scale test, which resulted in 53.51% reduction in TSS and an additional recovery of 0.13 grams of gold per tonne of ore [30].

4.2. Tannin Gel Adsorption

The use of tannins to create a gel adsorbent to recover precious metals from aqueous solutions has been studied by several groups. One of the first to study adsorption behavior with tannin gels was Takeshi Ogata and Yoshio Nakano from the Tokyo Institute of Technology. Their work focused on the synthesization of a tannin gel derived from wattle tannin for the application of recovering gold from electronic waste as an alternative to other hydrometallurgical operations like solvent extraction and ion exchange. Ogata and Nakano developed a method to immobilize the tannin particles, which are naturally soluble in water. The tannin powder was dissolved into a sodium hydroxide solution before adding formaldehyde to act as a cross-linking agent. The solution was allowed to gelatinize for 12 h at an elevated temperature of 353 K. The tannin gel was then crushed to 125–250 micron particles. The particles were added to a gold solution containing 100 ppm gold prepared from hydrogen tetrachloroaurate tetrahydrate. With varying pH, different chlorogold complexes were created. In the pH range from 2–3.8 the predominant complex was $AuCl_4^-$. Between pH 2–3.8, there was no dependence on initial pH for the recovery of gold. This was interpreted by Ogata and Nakano to be explained by the conversion of chlorogold complexes. It was shown that the adsorption rate of gold increased with increasing temperature. Additional tests were completed with 5 mg of tannin gel in 100 ppm gold solution to obtain the maximum adsorption capacity of the gold onto the gel. An extremely high value of 8000 mg of gold per 1 g of tannin gel was found for the adsorption capacity. Through XRD and FT-IR spectra research, it was shown that the hydroxyl groups on the tannin gel particles were oxidized to carbonyl groups during the adsorption testing. Trivalent gold in solution is reduced to elemental gold on the surface of the tannin gel particles [31].

Work done by a group at Saga University in Japan and Memorial University of Newfoundland in Canada expanded on the initial study done by Ogata and Nakano. This collaborative study focused on the use of tannin gel to selectively recover precious metals from a polymetallic acidic leach solution of circuit boards from spent mobile phones. The tannin gel was prepared to the procedure specified by Ogata and Nakano. Persimmon tannin was used to create the original tannin gel. This was further processed to create a chloromethylated persimmon tannin, which was treated to obtain the final tannin gel with bisthiourea functional groups. This tannin gel was labeled BTU-PT. It was shown that the adsorption behavior varied for different metal ions. Precious metals in the hydrochloric acid solution, particularly gold and palladium, had a drastically higher percent adsorption onto the tannin gel than the base metals like copper, iron, zinc and nickel. Maximum adsorption capacities of the tannin gel were found to be 5.18 mol kg^{-1} for Au(III) ions, 1.80 mol kg^{-1} for Pd(II) ions, 0.67 mol kg^{-1} for Pt(IV) ions. The mechanism behind the adsorption was studied with FT-IR technology. It was found that ion exchange, electrostatic interactions, and thiocarbonyl coordination were all mechanisms of adsorption for precious metals on the BTU-PT gel. It was shown the that tannin gel could be successfully regenerated for up to 5 cycles of use. Finally, an industrial study was done using actual circuit board leach liquor and as indicated in the initial laboratory work, the BTU-PT tannin gel acted as a successful adsorbent for precious metals while not taking up base metals in solution [32].

Work done by Yurtsever and Sengil at Sakarya University in Turkey explored the use of a valonia tannin gel for the removal of silver ions from solution. The valonia tannin was sources from a Turkish leather company and the gel was prepared similarly to Ogata and Nakano's procedure. It was shown in the study that the major factors influencing the adsorption of silver ions onto the tannin gel were solution pH, temperature of solution, and concentration of silver ions, and contact time. The peak of adsorption was seen at a pH 5. As temperature increased, from 20 to 90 degrees Celsius, the amount of silver ions adsorbed onto the tannin gel decreased. In addition, the more silver ions initially in the system lead to a higher amount of adsorption [33].

5. Pressure Leaching

Zinc and Nickel

In the 1960s, Gordon Sherritt Mines Limited filed a number of patents for the extraction of zinc from iron containing sulfide ores. In this process, elevated temperature and pressure with the addition of oxygen and sulfuric acid, leached zinc. Up until the 1970s, the pressure leaching of zinc complexes was done below the melting point of sulfur at around 120 °C to avoid the formation of elemental sulfur at higher temperatures. However, it was found that if the temperature was increased, a more advantageous rate of reaction occurred. With the formation of elemental sulfur at temperatures above 120 °C, the zinc sulfide particles were wetted by the molten sulfur and thus were not available for leaching. To provide the best kinetics and complete extraction of zinc, an additive was needed to prevent the unreacted zinc sulfide particles from being occluded by the molten sulfur. The additives found to be beneficial were those tannin compounds like quebracho, lignins, and lignosulphonates. The amount of additives needed was found to be around 0.1–0.3 g per liter. In a series of tests conducted at 150 °C, it was proven that the additives significantly improved the extraction of zinc. By adding 0.1 g per liter calcine lignosulphonate and 0.2 g per liter quebracho, the zinc extraction was increased from 54.4% to 90.0% in one test and 63.3% to 97.8% in another. The use of additives such as quebracho greatly increased the metal recovery in the elevated temperature leaching and became part of the commercialized Sherritt Zinc Pressure Leach Process [34].

Similar work investigating the effects of these additives was done on other sulfide systems. Libin Tong and David Dreisinger conducted research at the University of British Columbia on pentlandite, nickeliferous pyrrhotite, pyrrhotite, and chalcopyrite. Studies were conducted on the work of adhesion which is the work required to separate liquid sulfur from the mineral surface. This is expressed by the Dupre equation:

$$W_a = \gamma_{MA} + \gamma_{SA} - \gamma_{MS} = -\Delta G_W / \alpha \tag{4}$$

where W_a is the work of adhesion, γ are interfacial tension, ΔG_W is the surface free energy and α is the surface area of the mineral.

It was shown that in three different systems, pentlandite, nickeliferous pyrrhotite, and pyrrhotite that the work of adhesion was significantly lowered with the addition of quebracho, lignosulphonate, and humic acid. This implies that the additives should be good dispersing agents of the molten sulfur [35].

6. Conclusions

This study has explored the role of tannins in extractive metallurgy, specifically flotation and precipitation. Quebracho has proved to be an efficient depressant in many flotation systems, both sulfidic and calcareous, improving the overall selectivity of the collectors. Quebracho has also shown to aid in the precipitation of metals from solution including that of germanium and gold. While it is known that quebracho can be used for many different flotation and precipitation applications, the fundamentals of how quebracho functions requires additional investigation.

Acknowledgments: This work was funded by the Silvateam through collaboration with the Colorado School of Mines Kroll Institute for Extractive Metallurgy in the George S. Ansell Department of Metallurgical and Materials Engineering.

Author Contributions: Jordan Rutledge and Corby Anderson researched, wrote and revised this review paper as requested by and on behalf of Silvateam, the project sponsor.

Conflicts of Interest: The authors declare no conflict of interest.

References

1. Quebracho Colorado tree. Photograph. Britannica Online for Kids. Available online: http://kids.britannica. com/comptons/art-173288 (accessed on 1 July 2015).
2. Leja, J. *Surface Chemistry of Froth Flotation*; Plenum Press: New York, NY, USA, 1982.
3. Iskra, J.; Fleming, M.G.; Kitchener, J.A. *Quebracho in Mineral Processing*; Imperial College London: London, UK, 1980.
4. Chen, J.; Li, Y.; Long, Q.; Wei, Z.; Chen, Y. Improving the selective flotation of jamesonite using tannin extract. *Int. J. Miner. Process.* **2011**, *100*, 54–56. [CrossRef]
5. Ober, J. *Strontium*; US Geological Survey-Minerals Information: Reston, VI, USA, 1996.
6. De Castro, F.H.B.; de Hoces, M.C. Influence of quebracho and sodium silicate on flotation of celestite and calcite with sodium oleate. *Int. J. Miner. Process.* **1993**, *37*, 283–298. [CrossRef]
7. Guy, P.; Bruckard, W.; Vaisey, M. Beneficiation of Mt Weld Rare Earth Oxides by Gravity Concentration, Flotation and Magnetic Separation. In Proceedings of the Seventh Mill Operators' Conference, Kalgoorlie, WA, Australia, 12–14 October 2000
8. Ozcan, O.; Bulutcu, A. Electrokinetic, infrared and flotation studies of scheelite and calcite with oxine, alkyl oxine, oleoyl sarcosine and quebracho. *Int. J. Miner. Process.* **1993**, *39*, 275–290. [CrossRef]
9. Dean, K.C.; Schack, C.H. *Flotation of Calcareous Scheelite Ores*; US Bureau of Mines: Washington, DC, USA, 1963.
10. Kelly, T.D.; Matos, G.R. 2005 Fluorspar statistics, Historical statistics for mineral and material commodities in the United States: US Geological Survey Data Series 140. US Geological Survey. Available online: http://pubs.usgs.gov/ds/2005/140/ (accessed on 1 July 2015).
11. Mahoning Mining Company. Method of Concentrating Fluorspar Ores. US 2263552, 1941.
12. Mahoning Mining Company. Method of Concentrating Fluorspar Ores. US 2168762, 1938.
13. Thom, C. Method of Concentrating Fluorspar Ores. US 3207304, 1962.
14. Allied Chemical Corporation. Beneficiation of Fluorspar Ores. US 3536193, 1970.
15. McKay, J.; Foot, D., Jr.; Shirts, M. Parameters Affecting Column Flotation of Fluorite. In Proceedings of SME Annual Meeting, Denver, CO, USA, 24–27 February 1987.
16. Raju, G.; Prabhakar, S. Beneficiation of fluorspar by column flotation. *Miner. Metall. Process.* **2000**, *17*, 167–172.
17. Aliaga, W.; Sampaio, C.; Brum, I.; Ferreira, K.; Batistella, M. Flotation of high-grade fluorite in a short column under negative bias regime. *Miner. Eng.* **2006**, *19*, 1393–1396. [CrossRef]
18. Song, S.; Lopez-Valdivieso, A.; Martinez-Martinez, C.; Torres-Armenta, R. Improving fluorite flotation from ores by dispersion processing. *Miner. Eng.* **2006**, *19*, 912–917. [CrossRef]
19. Schubert, H.; Baldauf, H.; Kramer, W.; Schoenherr, J. Further development of fluorite flotation from ores containing higher calcite contents with oleoylsarcosine as collector. *Int. J. Miner. Process.* **1990**, *30*, 185–193. [CrossRef]
20. Bulatovic, S.; Wyslouzil, D. Selection and evaluation of different depressants systems for flotation of complex sulphide ores. *Miner. Eng.* **1995**, *8*, 63–75. [CrossRef]
21. Sarquís, P.; Menéndez-Aguado, J.; Mahamud, M.; Dzioba, R. Tannins: The organic depressants alternative in selective flotation of sulfides. *J. Clean. Prod.* **2014**, *84*, 723–726. [CrossRef]
22. Falconbridge Limited. Separation of Polymetallic Sulphides by Froth Flotation. US 2952329, 1990.
23. Höll, R.; Kling, M.; Schroll, E. Metallogenesis of germanium—A review. *Ore Geol. Rev.* **2007**, *30*, 145–180. [CrossRef]
24. Krebs, R.E. *The History and Use of Our Earth's Chemical Elements*; Greenwood: Westport, CT, USA, 1998.
25. Kul, M.; Topkaya, Y. Recovery of germanium and other valuable metals from zinc plant residues. *Hydrometallurgy* **2008**, *92*, 87–94. [CrossRef]
26. MetSoc of CIM. Welcome! MetSoc of CIM | Metallurgy & Materials Home. Available online: http://www. metsoc.org/virtualtour/processes/zinc-lead/oxide.asp (accessed on 18 May 2015).
27. Moskalyk, R.R. Review of germanium processing worldwide. *Miner. Eng.* **2004**, *17*, 393–402. [CrossRef]

28. Liang, D.; Wang, J.; Wang, Y.; Wang, F.; Jiang, J. Behavior of tannins in germanium recovery by tannin process. *Hydrometallurgy* **2008**, *93*, 140–142. [CrossRef]

29. Symes, R. Recovery of Colloidal Gold from Oxidised Concentrate Wash Liquor. In Proceedings of the Tenth Mill Operators' Conference, Adelaide, SA, Australia, 12–14 October 2009.

30. Buah, W.K.; Asamoah, R.K.; Boadi, I. Effects of Quebracho Tannin on Recovery of Colloidal Gold from Bioleached Wash Liquor. *Ghana Min. J.* **2015**, *15*, 44–49.

31. Ogata, T.; Nakano, Y. Mechanisms of gold recovery from aqueous solutions using a novel tannin gel adsorbent synthesized from natural condensed tannin. *Water Res.* **2005**, *39*, 4281–4286. [CrossRef] [PubMed]

32. Gurung, M.; Adhikari, B.B.; Kawakita, H.; Ohto, K.; Inoue, K.; Alam, S. Selective Recovery of Precious Metals from Acidic Leach Liquor of Circuit Boards of Spent Mobile Phones Using Chemically Modified Persimmon Tannin Gel. *Ind. Eng. Chem. Res.* **2012**, *51*, 11901–11913. [CrossRef]

33. Yurtsever, M.; Sengil, A. Adsorption and desorption behavior of silver ions onto valonia tannin resin. *Trans. Nonferrous Met. Soc. China* **2012**, *22*, 2846–2854. [CrossRef]

34. Gordon Sherritt Mines Limited. Recovery of Zinc from Zinc Sulphides by Direct Pressure Leaching. US 3867268, 1975.

35. Tong, L.; Dreisinger, D. Interfacial properties of liquid sulfur in the pressure leaching of nickel concentrate. *Miner. Eng.* **2009**, *22*, 456–461. [CrossRef]

![metals logo]

Article

CFD Modelling of Flow and Solids Distribution in Carbon-in-Leach Tanks

Divyamaan Wadnerkar, Vishnu K. Pareek and Ranjeet P. Utikar *

Department of Chemical Engineering, Curtin University, Perth, WA 6102, Australia;
d.wadnerkar@curtin.edu.au (D.W.), v.pareek@curtin.edu.au (V.K.P.)
* Author to whom correspondence should be addressed; r.utikar@curtin.edu.au; Tel.: +61-8-9266-9837;
   Fax: +61-8-9266-2681.

Academic Editors: Suresh Bhargava, Mark Pownceby and Rahul Ram
Received: 29 July 2015; Accepted: 23 October 2015; Published: 28 October 2015

Abstract: The Carbon-in-Leach (CIL) circuit plays an important role in the economics of a gold refinery. The circuit uses multiphase stirred tanks in series, in which problems such as dead zones, short-circuiting, and presence of unsuspended solids are detrimental to its efficiency. Therefore, the hydrodynamics of such a system is critical for improving the performance. The hydrodynamics of stirred tanks can be resolved using computational fluid dynamics (CFD). While the flow generated by the impellers in the CIL tanks is complex and modelling it in the presence of high solid concentration is challenging, advances in CFD models, such as turbulence and particle-fluid interactions, have made modelling of such flows feasible. In the present study, the hydrodynamics of CIL tanks was investigated by modelling it using CFD. The models used in the simulations were validated using experimental data at high solid loading of 40 wt. % in a lab scale tank. The models were further used for examining the flow generated by pitched blade turbine and HA-715 Mixtec impellers in lab scale CIL tanks with 50 wt. % solids. The effect of design and operating parameters such as off-bottom clearance, impeller separation, impeller speed, scale-up, and multiple-impeller configuration on flow field and solid concentrations profiles was examined. For a given impeller speed, better solids suspension is observed with dual impeller and triple impeller configurations. The results presented in the paper are useful for understanding the hydrodynamics and influence of design and operating parameters on industrial CIL tanks.

Keywords: CIL tanks; CFD; hydrodynamics; HA-715 impeller; impeller configuration; design

1. Introduction

Carbon-in-Leach (CIL) circuit is a process that concentrates gold from 2.5 to 3.5 g/t in ore to 10,000 to 15,000 g/t on carbon. It is a process of continuous leaching of gold from ore to liquid and counter-current adsorption of gold from liquid to carbon particles in a series of tanks. The tanks used in the CIL circuit are continuously stirred tanks and contain high concentration of ores. Efficient operation of CIL tanks requires suspension of the ore particles in the leaching solution and hence, to provide maximum contact between ore and leaching solution. However, problems such as settled solids and presence of dead zones are detrimental to the efficiency, and could be identified and solved by understanding the flow field and solid distribution in the system. Furthermore, reducing the energy consumption to achieve a higher contact is always desired and can be achieved by proper design and optimization while investigating the hydrodynamics.

While, some information on the hydrodynamics such as residence time distribution (RTD) can be obtained from experiments using tracer studies; detailed quantitative measurement of the flow field inside the CIL tank is challenging. In such a scenario, computational fluid dynamics (CFD) can prove to be an inexpensive and viable solution. With the availability of improved models for turbulence, interphase drag force, particle-particle interaction models, *etc.* and advances in computational speeds, resolving the complex multiphase flows phenomenon is possible using CFD.

In the present work, models for simulating high solid loading stirred tanks are validated with experimental data. The flow field generated by a pitched blade turbine and a HA-715 impeller is compared. The effect of design parameters such as off-bottom clearance, impeller separation, impeller speed, scale-up, and multiple-impeller configuration is examined by modelling the flow and estimating the suspension quality for each case.

2. Literature Review

Carbon-in-leach tanks are high solid loading stirred tanks of large diameter (~10–15 m) with solid concentration of up to 50% by weight (~28% by volume). Such a high concentration renders opacity to the system which makes its hydrodynamic investigation challenging even at a small scale. With the advent of radioactive experimental techniques such as Computer Aided Radioactive Particle Tracking (CARPT), Positron Emission Particle Tracking (PEPT), *etc.*, reliable quantitative data can be obtained [1–3]. The data can now be used to validate the computational models for the high solid loading regimes in turbulent flows and further advance the design and optimization of CIL tanks.

While simulating high solid concentration (20% by volume) stirred tanks, Altway *et al.* [4] found major discrepancy while validating the solid concentration profile using data from Yamazaki *et al.* [2]. Micale *et al.* [5] used Multiple Reference Frame (MRF) and Sliding Grid (SG) approach to study the clear liquid layer and the suspension height for dense solid-liquid systems. In their simulations, the power numbers were 2.98, 2.74, and 2.68 for N = 5, 6.33 and 8 RPS respectively (particle loading of 9.6% v/v), which were significantly smaller than the experimental values of 4.59, 4.37, and 4.23. They attributed the imperfection in the solid suspension prediction to second order effects (particle drag modifications due to liquid turbulence, presence of other particles, particle-particle direct interactions, *etc.*) that were neglected in the study. Ochieng and Lewis [6], Fradette, *et al.* [7], Ochieng and Onyango [8], Kasat, *et al.* [9], Fletcher and Brown [10], Tamburini, *et al.* [11–13], conducted simulations for volume fractions below 20%, and validated using non-local properties like cloud height, suspension quality, *etc.* In another similar study, Gohel, *et al.* [14] used qualitative and quantitative data for cloud height to validate the simulation of high concentration solids suspension in stirred tanks. In these studies, while the parameters, for example cloud height, were accurately predicted, the errors in the predictions of local hydrodynamics were not verified in the absence of data. In the direction of resolving the local hydrodynamics of the stirred tanks, Liu and Barigou [15] used local velocity field and solids concentration data for model validation and found that even though a good agreement in the axial concentration profile was observed, the local concentration predictions could still be very poor and could vary from experiments by several folds. The inaccuracy was attributed to inadequate models for particle sedimentation, lift-off, and particle-particle interaction in their CFD model. They suggested incorporating particle-particle interaction in the solids pressure term given by Gidaspow [16], but did not use it in their models due to convergence problems. The highest solid loading for which the simulation results in stirred tanks are reported is 20% (by volume), and the simulations are not able to predict the local concentration due to limitations of models [4,5].

In the CIL tanks, Dagadu, *et al.* [17] used a radioactive tracer to investigate the RTD in the CIL tanks and developed a model to predict it. While the results provided an overview of complexity of the system and suggested the presence of stagnant and active volumes, the details such as solid accumulation, dead-zones, uniformity, *etc.* are still unknown. To address these issues, Dagadu *et al.* [18,19] investigated the mixing in the CIL tanks by conducting CFD simulations and drawing inferences based on the flow field and eddy viscosity. In both of these instances, neither the relationship of the flow field or turbulent viscosity with the solid distribution is quantified, nor the solid distribution in the tanks are presented. While validation with experimental data is missing, the scope of both studies is limited to evaluating the ability of the models for predicting the qualitative flow behavior in such systems. Drawing conclusions for critical design and operating parameters requires comprehensive validation of models and controlled investigation of influence of these parameters on the flow field and solid distribution.

The current study focuses on the extensive validation of local solid concentration distribution in high solid loading stirred tanks. The shortcomings of previous investigations are addressed by using the appropriate constitutive models for turbulence, drag, particle-particle interactions, *etc.* The validated model is then used for investigation of critical parameters in the CIL tanks.

3. Model Description

3.1. Governing Equations

The hydrodynamic simulations are conducted using Eulerian-Eulerian multiphase model. In this model, each phase is treated as an interpenetrating continuum represented by a volume fraction at each point of the system. Reynolds averaged mass and momentum balance equations are solved for each phase. The governing equations are:

Continuity equation:

$$\frac{\partial}{\partial t}\left(\alpha_q \rho_q\right) + \nabla.\left(\alpha_q \rho_q \bar{u}_q\right) = 0 \tag{1}$$

Momentum equation:

$$\frac{\partial}{\partial t}\left(\alpha_q \rho_q \vec{u}_q\right) + \nabla\left(\alpha_q \rho_q \vec{u}_q \vec{u}_q\right) = -\alpha_q \nabla p + \nabla.\bar{\bar{\tau}}_q + \alpha_q \rho_q \vec{g} + \vec{F}_{td,q,vm,lift} + \vec{F}_{12} \tag{2}$$

The rotation of impeller can be simulated using sliding grid approach (SG) or Multiple Reference Frame (MRF). While MRF provides reasonably accurate steady state solution, SG is found to be more accurate compared to MRF approach [20,21]. Therefore, SG will be used in the paper to simulate the impeller rotation.

Turbulence is not resolved in the RANS simulations and therefore, it needs to be modelled. Standard k-ε is the most commonly used model in the RANS simulations of stirred tanks [4,9,21–25]. However, it finds limitation in modelling anisotropic turbulence in the impeller discharge region and under-predicts turbulent kinetic energy in flow impingement region. RSM model predicts the Reynolds stresses by explicitly solving their governing equations. Hence, it resolves the anisotropic turbulence and results in improved predictions of turbulence in such regions [26]. Large eddy simulation (LES) also resolves the larger anisotropic turbulence scales while requiring simulation of complete domain with fine mesh, which is computationally expensive [27]. For the same scale, the computational requirement using LES can be 100 times higher than that for RSM simulation. Therefore, in this paper, RSM model is used for modelling turbulence.

The equations of Reynolds Stress Model (RSM) model for turbulence are given below.

$$\frac{\partial}{\partial t}\left\langle u_i' u_j' \right\rangle + \nabla.\left(\vec{u}_c u_i' u_j'\right) = \nabla.\left(C_s \frac{k\left\langle u_i' u_j' \right\rangle}{\varepsilon} \nabla \left\langle u_i' u_j' \right\rangle\right) - C_1 \frac{\varepsilon}{k}\left(\left\langle u_i' u_j' \right\rangle - (2/3)\delta_{ij}k\right)$$
$$- C_2\left(P_{ij} - (2/3)\delta_{ij}P - S_{ij}\right) - (2/3)\delta_{ij}\varepsilon \tag{3}$$

$$\frac{\partial}{\partial t}\left(\rho_q \varepsilon\right) + \nabla\left(\rho_q \vec{u}_q \varepsilon\right) = \nabla.\left(C_\varepsilon \frac{\rho k}{\varepsilon}\left\langle u_i' u_i' \right\rangle \nabla \varepsilon\right) + \frac{\varepsilon}{k}\left(C_{1\varepsilon} G_{k,q} - C_{2\varepsilon} \rho_m \varepsilon\right) \tag{4}$$

where C_s, C_1, C_2, $C_{1\varepsilon}$, $C_{1\varepsilon}$, and $C_{2\varepsilon}$ are constants. And, $P_{ij} = -\left\langle u_i' u_k' \right\rangle \nabla.u_j - \left\langle u_j' u_k' \right\rangle \nabla.u_i$; $P = \frac{1}{2}P_{ii}$;

$\rho_q = \sum\limits_{i=1}^{N} \alpha_i \rho_i$; $\vec{u}_q = \dfrac{\sum\limits_{i=1}^{N} \alpha_i \rho_i \vec{u}}{\sum\limits_{i=1}^{N} \alpha_i \rho_i}$.

Eddy viscosity is computed from

$$\mu_{t,q} = \rho_q C_\mu \frac{k^2}{\varepsilon} \tag{5}$$

Evaluation of generation of turbulent kinetic energy is consistent with Boussinesq hypothesis and is computed as

$$G_{k,q} = \mu_{t,q} \left(\nabla \vec{u}_q + \nabla \vec{u}_q^T \right) : \nabla \vec{u}_q \tag{6}$$

The equation of conservation of granular temperature is given as:

$$\frac{3}{2} \left[\frac{\partial}{\partial t} (\alpha_s \rho_s \Theta_s) + \nabla \cdot \left(\alpha_s \rho_s \vec{u}_s \Theta_s \right) \right] = - \left(p_s \bar{\bar{I}} + \nabla \cdot \bar{\bar{\tau}}_s \right) : \nabla \vec{u}_s + \nabla \cdot (k_{\Theta_s} \nabla \Theta_s) - \gamma_{\Theta_s} + \varphi_{ls} \tag{7}$$

where $- \left(p_s \bar{\bar{I}} + \nabla \cdot \bar{\bar{\tau}}_s \right) : \nabla \vec{u}_q$ is the generation of energy by the solid stress tensor, $\nabla \cdot (k_{\Theta_s} \nabla \Theta_s)$ is the diffusion of energy, γ_{Θ_s} is the collisional dissipation of energy and φ_{ls} is the energy exchange between fluid and solid phase.

Table 1. Constitutive equations.

Description	Equations
Liquid-phase stress tensor	$\bar{\bar{\tau}}_q = \alpha_q \mu_q \left(\nabla \vec{u}_q + \nabla \vec{u}_q^T \right) + \alpha_q \left(\lambda_q - \frac{2}{3} \mu_q \right) \nabla \vec{u}_q \bar{\bar{I}}$
Solid-phase stress tensor	$\bar{\bar{\tau}}_s = \alpha_s \mu_s \left(\nabla \vec{u}_s + \nabla \vec{u}_s^T \right) + \alpha_s \left(\lambda_s - \frac{2}{3} \mu_s \right) \nabla \vec{u}_s \bar{\bar{I}}$
Solid shear viscosity	$\mu_s = \mu_{s,col} + \mu_{s,kin} + \mu_{s,fr}$
Collisional viscosity	$\mu_{s,col} = \frac{4}{5} \alpha_s \rho_s d_s g_{0,ss} (1 + e_{ss}) \left(\frac{\Theta_s}{\pi} \right)^{0.5} \alpha_s$
Kinetic viscosity	$\mu_{s,kin} = \frac{\alpha_s \rho_s d_s}{6(3 - e_{ss})} \left[1 + \frac{2}{5} (1 + e_{ss})(3 e_{ss} - 1) \alpha_s g_{0,ss} \right]$
Frictional viscosity	$\mu_{s,fr} = \frac{p_s \sin \varphi}{2 \sqrt{I_{2D}}}$
Solids pressure	$p_s = \alpha_s \rho_s \Theta_s + 2 \rho_s (1 + e_{ss}) \alpha_s^2 g_{0,ss} \Theta_s$
Radial distribution function	$g_{0,ss} = \left[1 - \left(\frac{\alpha_s}{\alpha_{s,max}} \right)^{1/3} \right]^{-1}$
Diffusion coefficient of granular temperature	$k_{\Theta s} = \frac{15 \alpha_s \rho_s d_s \sqrt{\Theta_s \pi}}{4(41 - 33\eta)} \left[\begin{array}{l} 1 + \frac{12}{5} \eta^2 (4\eta - 3) \alpha_s g_{0,ss} \\ + \frac{16}{15\pi} (41 - 33\eta) \eta \alpha_s g_{0,ss} \end{array} \right]$ where $\eta = 0.5(1 + ess)$
Collision dissipation energy	$\gamma_{\Theta s} = \frac{12(1 - e_{ss}^2) g_{0,ss}}{d_s \sqrt{\pi}} \rho_s \alpha_s^2 \Theta_s^{3/2}$

The stress-strain tensor in the momentum transfer equation is due to viscosity and Reynolds stresses that include the effect of turbulent fluctuations. Boussinesq's eddy viscosity hypothesis is used for the closure of momentum transfer equation. The particle-particle interaction is modelled using kinetic theory of granular flow by assuming that its behavior similar to dense gas. Similar to the thermodynamic temperature of gases, the granular temperature is used to model the fluctuating velocity of particles [16]. The constitutive equations for momentum equation are given in Table 1.

3.2. Turbulent Dispersion Force

Turbulent fluctuations result in dispersion of phases from high volume fraction regions to low volume fraction regions. The turbulent dispersion force is significant when the size of turbulent eddies is larger than the particle size [9]. The effect of turbulence dispersion force on the hydrodynamics in CIL tanks, is incorporated using the Burns, *et al.* [28] model. The model equations for turbulence dispersion force are given below:

$$\vec{F}_{td,q} = C_{TD} k_{pq} \frac{D_{t,q}}{\sigma_{pq}} \left(\frac{\nabla \alpha_p}{\alpha_p} - \frac{\nabla \alpha_q}{\alpha_q} \right) \tag{8}$$

3.3. Interphase Drag Force

Interphase drag is the resultant force experienced by the particle in the direction of relative motion due to a moving fluid. Since, the solids and liquid phases are treated as inter-penetrating, an inter-phase momentum exchange term is required. The interphase exchange force is calculated using the following expression given by Syamlal *et al.* [29]:

$$F_d = \frac{3\alpha_l \alpha_s \rho_l C_D \mathrm{Re}}{4 d_p V_r^3} \left(\vec{u}_l - \vec{u}_s \right) \qquad (9)$$

where, $V_r = 0.5 \left(A - 0.06\mathrm{Re} + \sqrt{(0.06\mathrm{Re})^2 + 0.12\mathrm{Re}(2A - B) + (A)^2} \right)$; $A = \alpha_g^{5.14}$; $B = 0.8\alpha_g^{2.28}$, for $\alpha_g \leq 0.85$; $B = \alpha_g^{3.65}$, for $\alpha_g > 0.85$.

The drag coefficient is calculated using an expression that has a form derived by Dalla Valle and Kenning [30]:

$$C_D = \left(0.63 + \frac{4.8}{\sqrt{\mathrm{Re}/V_r}} \right)^2 \qquad (10)$$

4. Methodology and Boundary Conditions

4.1. Vessel Geometry

In this paper, a flat bottomed cylindrical tank was simulated (see Figure 1). The shaft of the impeller was concentric with the axis of the tank. A Mixtec HA-715 was used as impeller for CIL tanks, but for validation 45° six-bladed pitched blade turbine pumping downwards (PBTD) impeller was employed. For validation, vessel geometry dimensions, material properties, *etc.* were taken from paper of Guida *et al.* [3]. The dimensions of tank and impeller are given in Table 2. The fluid and particle properties used in the simulation are also tabulated in the same table. Conditions such as solid concentrations, impeller speed, and the Reynolds number in the tank are tabulated in Table 3. Simulations were carried out at just suspension speed for each case, which was determined by Guida *et al.* [3] following Zwietering [31] criterion. According to this criterion, no particle should remain stationary on the base of the vessel for longer than 1–2 s.

Table 2. Dimensions of domain and properties of materials used in this study.

	Tank (m)		PBTD (m)		HA-715
T	0.288, 10	D	$T/2$	D	$T/2, T/3$
H	T	B_l	0.055	D_{shaft}	0.01152
W	$T/10$	B_w	0.041	-	-
C_i	$T/2, T/3, T/4,$ $T/6, T/8$	D_{shaft}	0.01	-	-
		D_{hub}	0.034	-	-

Table 3. Conditions in stirred tanks used for simulations.

Name of Case	X (wt. %)	$N = N_{js}$ (RPM)	ρ_l (kg/m^3)	μ_l (Pa·s)	ρ_p (kg/m^3)	d_p (mm)
PBTD-Validation	40	589.8	1150	0.001	2585	3
CIL tanks (Lab Scale)	50	200–700	1000	0.001	2550	0.075
CIL tanks (Full Scale)	50	22.15	1000	0.001	2550	0.075

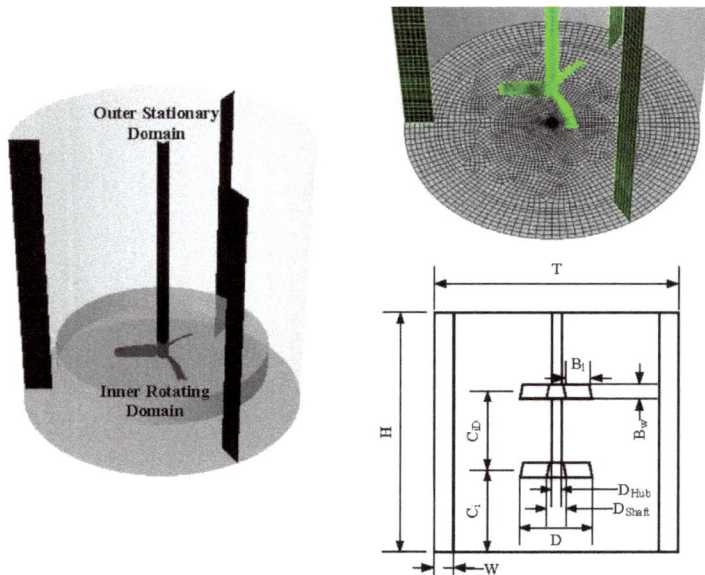

Figure 1. Computational domain, grid distribution, and schematic of the stirred tank.

4.2. Numerical Simulations

The stirred tank for validation consists of six PBT blades and four baffles and that of CIL tanks consists of three HA-715 blades and three baffles. In all the cases studied in the paper, simulation of the full tank is conducted. For the case of PBTD, the moving zone with dimensions $r = 0.06$ m and $0.036 < z < 0.137$ is created (where z is the axial distance from the bottom). For the case of HA-715, the moving zone of height $T/10$ and $r = 3T/4$ is created, with top and bottom surfaces equidistant from the center of impeller. The impeller rod outside this zone is considered as a moving wall. Impellers used in all the cases simulated in the study are operated in the down-pumping mode. The top of the tank is open, so it is defined as a wall of zero shear. The solids in the tank are assumed to be settled in the beginning of simulations. Therefore, a solid volume fraction of 0.6 is patched in the stirred tank up to a height such that the volume averaged solid concentration is the same as the uniformly distributed solids specified in Table 3. For modelling the turbulence, a RSM turbulence model is used. The standard model parameters are C_μ: 0.09, C_s: 0.22, C_1: 1.8, C_2: 0.6, $C_{1\varepsilon}$: 1.44, $C_{2\varepsilon}$: 1.92, σ_k: 1.0 and σ_ε: 1.3. In the present work, SIMPLE scheme is used for Pressure-Velocity coupling along with the standard pressure interpolation scheme. To avoid any numerical diffusion and unphysical oscillations, a third order Quadratic Upstream Interpolation for Convective Kinetics (QUICK) discretization scheme is used for momentum, volume fraction, turbulent kinetic energy, and turbulent dissipation rates. The convergence of the simulation is verified by monitoring residual values as well as additional parameters namely turbulence dissipation over the volume, turbulence dissipation at the surface right below impeller and torque on the shaft. Once the residuals and additional parameters are constant, a simulation is deemed to be converged. For all the cases presented in the paper, a time-step of 0.001 s is less than the time which the impeller needs to sweep by a single computational cell (one cell time step ~0.003 s). One cell time step is considered appropriate for the CFD simulations in stirred tanks [12]. But using a time-step of 0.001 s resulted in divergence in the initial time-steps. Therefore, the time-step size initially used in the simulations is 0.0001s, which is gradually increased to 0.001 s. For the full scale geometry, the one cell time step increased due to the reduction in the impeller speed. Therefore, the maximum time-step used for full-scale simulation is 0.01 s (one cell time step ~0.025 s). The results of lab scale simulations approached transient steady values after 12 s. Therefore, the data used for results and discussion is time averaged for the last 3 s. For full scale simulations, the time taken for reaching the transient steady value is 900 s, and the data is time-averaged for 300 s. The numerical solution of the system is obtained by using the commercial CFD solver ANSYS FLUENT 15.0 using 48 cores of Magnus with Cray XC40, Intel Xeon E5-2690V3 "Haswell" processors running at 2.6 GHz system at Pawsey supercomputing center. Each time step takes an average of 20 s of the wall clock time which reduces after approaching convergence when residuals reach at 10^{-5}.

The finite volume method solves the partial differential equations in a spatially discretized domain. The discretization should be fine enough to resolve the physics while not incurring excess computational power. For the purpose, a grid independency test is conducted initially on the single phase flow and is presented. The models used in the paper are validated with the experimental results at 40 wt. % solid loading stirred tank published by Guida *et al.* [3]. The validated model is used for analyzing the flow field generated by HA-715 impeller and assessing its efficacy compared to the PBTD. The influence of different impeller speeds, impeller off-bottom clearance and impeller diameter on the solid suspension, homogeneity and power consumption are examined. The appropriate stirred tank design parameters are used for scale-up and simulations and are conducted to gain an insight on the flow developed. Flow generated by single and multiple impeller systems in the full-scale CIL tanks is also investigated.

5. Results and Discussion

5.1. Grid Independency and Validation

For testing the grid-independency, the experimental values for axial, tangential, and radial velocities at impeller discharge plane for single phase flow were compared with the simulation results using computational grids with 32,000 (coarse), 140,000 (medium) and 900,000 (fine) cells (see Figure 2). The results using mesh with 32,000 cells are accurate at $z = 0.2$ H plane. The predictions improve slightly while using 140,000 cells, and beyond that the improvement is marginal. Furthermore, the power for the stirred tanks is calculated by integrating the turbulence dissipation rate over the volume. For the coarse, medium, and fine meshes, the values of power number are 1.34, 1.53, and 1.57, respectively, which re-emphasizes that a mesh of 140,000 cells is suitable for simulation and further refinement will only lead to marginal improvement at significant computational cost. Therefore, the mesh with 140,000 cells is used for the rest of the simulations in the study. For the full scale geometry, the computational grid of 140,000, 900,000, and 3,000,000 cells are investigated based on the same parameters. The values of power number obtained for the three cases are 1.27, 1.49, and 1.55, respectively. Mixing time is another parameter that is sensitive to the number of computational cells in the domain. For the full scale tank, the mixing time for the three cases are 75 s, 113 s, and 111 s respectively to achieve 99% homogeneity. It is evident from the analysis that the results of 900,000 cells and 3,000,000 cells are similar. Therefore, the grid of 900,000 cells was selected for the simulation of full scale CIL tanks.

The models used in the study are validated by comparing the flow field generated by PBTD and the concentration profiles obtained at 40 wt. % solid loading. The axial, tangential, and radial velocity plots describe the flow generated by the PBTD. The PBTD pumps the fluid downwards leading to highly negative axial velocities at the impeller plane. The jet leaving the impeller flows down to the bottom of the stirred tank, is then redirected towards the periphery and circulates back to the top along the walls. Due to such motion, a flow loop is formed near the impeller. It results in decreasing axial velocity when moving radially outwards in impeller plane that eventually increases due to the upwards flow near the walls. A 45° inclination of the impeller blade imparts momentum in the tangential direction resulting in moderate values of tangential velocity. Due to the downward flow developed by a PBTD, the magnitude of the radial component of velocity is the lowest. The maximum values of radial, tangential, and axial velocity in the impeller plane are 0.1 U_{tip}, 0.28 U_{tip}, and 0.45 U_{tip} respectively. Both radial and tangential components gain value with the increase in radius as the velocity increases radially along the impeller blades. The maximum value is attained by both of these components close to the impeller tip after which a sudden decline is observed due to no momentum source in the absence of impeller. These characteristics are well represented in the velocity profiles shown in Figure 2.

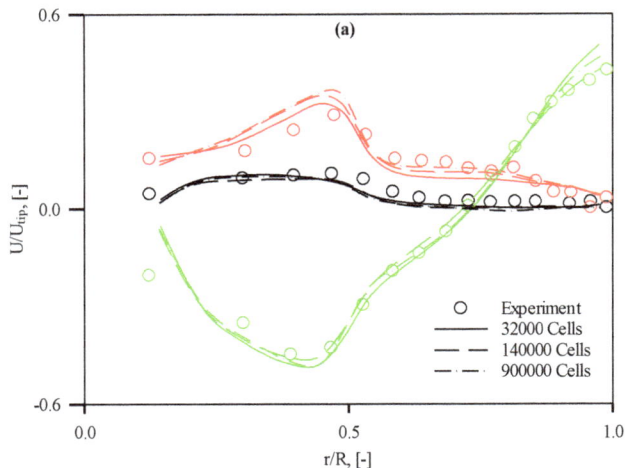

Figure 2. Comparison of dimensionless axial (green), radial (black) and tangential (red) velocity at z = 0.2H plane for computational grid with different resolution.

The average axial concentration profile is compared with the experimental data of Guida *et al.* [3] in Figure 3a. In some cases, the average axial concentration does not provide the information for local variation of concentration. Therefore, radial concentration profiles at different heights are also compared with the experiments. As is evident from Figure 3, the results obtained are in agreement with the experimental data. As the impeller speed used is the speed of just suspension, the solid concentration near the bottom of the tank is only double the average concentration. The jet from the PBTD moves downwards and strikes the bottom around the $r/R = 0.5$. Therefore, the concentration at this location is low. However, the solid is accumulated at the bottom center and the bottom corner of the tanks due to low velocity fields in both of these regions. Near 0.29 Z, the low concentration in the loop of the eye formed by the jet circular loop is also well predicted by the simulations. With increasing height, the axial velocity near the wall and impeller diminish and therefore, the concentration profile also becomes inverted with low concentration near the periphery and impeller rod. This shape of concentration profile is maintained till the top of the stirred tank with the magnitude of concentration decreasing with height.

The validity of the CFD simulations is further tested by comparing the results of cloud height in stirred tanks. For quantifying the cloud height, Hicks *et al.* [32] conducted experiments and reported the values at various suspension speeds, D/T, power, torque, and suspension ratios. Bittorf and Kresta [33] developed a model for predicting the cloud height based on the analysis of liquid jet velocity and just suspension speed. Due to inherent complexity due to several variables such as solid concentration, suspension speed, particle diameter, D/T ratio, C/T ratio, experimental errors, *etc.*, error in the predictions lie in the range of ±16%. Therefore, the simulations are conducted using 10.6% solid concentration (by weight) at N_{js} with D/T of 0.5 and C/T of 0.25. These results were compared with the experimental results of Hicks *et al.* [32]. The cloud height obtained from CFD simulations is 0.93 compared to the experimental value of 0.94. For the solid concentration of 20% and 40%, the cloud height calculated from CFD simulation is 0.935 and 0.942 respectively. This aligns with the observation by Hicks *et al* [32] that the cloud height remains nearly constant between solid concentration of 10% and 40%. It is evident that the models used in the simulations are able to predict the local hydrodynamics for high solid loading stirred tank systems accurately and can be used for investigating hydrodynamics in CIL tanks.

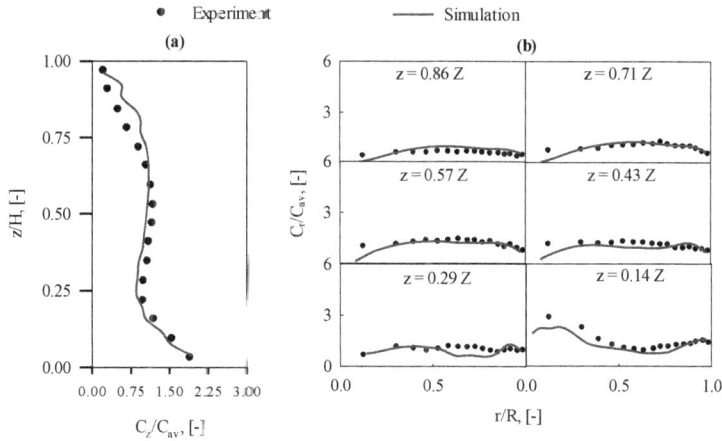

Figure 3. (**a**) Averaged axial concentration profiles and (**b**) radial concentration profiles plotted at different heights in stirred tanks with 40 wt. % solid loading using PBTD.

5.2. Flow Field

The simulations of lab scale stirred tank with PBTD and HA-715 were conducted to compare the hydrodynamics in both the cases. Only for these set of simulations, four baffles were employed for the HA-715 case, which is equal to that of PBTD cases. The flow field generated by both axial impellers viz. PBTD and HA-715 are shown in Figure 4. The asymmetry seen in the case of HA-715 impellers is due to the use of three impeller blades which cannot simultaneously appear in the mid-baffle plane. Both of these impellers generate strong axial flow with similar flow field profile. Similar to the flow generated by PBDT, the jet originates from the impeller blades for HA-715 impeller and strikes the bottom of the tank. Such a motion is effective for the suspension of ore particles at the bottom of the tank. After hitting the bottom, the fluid changes direction and moves upwards along the wall of the stirred tank. This results in higher axial velocities near the bottom-wall region of the stirred tanks, where both radial and tangential velocities diminish. The value of axial velocity near the wall gradually diminishes with the ascent due to the counter effect of gravity. The height to which axial velocity remains significant is proportional to the magnitude of velocity at the bottom, which exerts force against gravity. Such a behavior is also evident from Figure 4. It can be observed from contours that for the same impeller rotation speed, the flow generated by PBTD is dominant compared to that generated by HA-715. However, the flow generated at a particular impeller rotation speed is not the sole criterion for determining the efficacy of impellers. Rather than the impeller speed, the power consumed is a more reasonable criterion, where the desired flow needs to be generated by supplying a specific amount of power. While the power consumed for the PBTD case is 20.79 W, its value for the HA-715 case at 300 RPM is only 3.23 W. Therefore, the power used in HA-715 was increased to 12.89 W by doubling the impeller speed. This change resulted in a flow field that is far more effective than that generated by the PBTD, while using 60% of the power. The efficiency of mixing for both the cases is quantified by determining the mixing time using an 8% solute patched at the top of stirred tank. The value of mixing time for PBTD is 4.2 s, while the same for HA-715 at 660 RPM is 2.9 s. Therefore, for the given power consumption, HA-715 shows better mixing characteristics than PBTD.

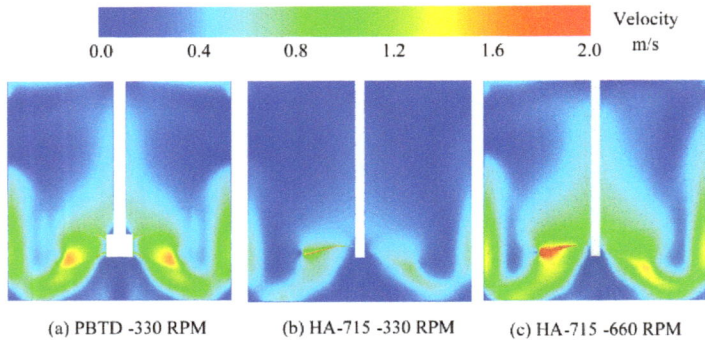

(a) PBTD -330 RPM   (b) HA-715 -330 RPM   (c) HA-715 -660 RPM

Figure 4. Flow field contours in a stirred tank generated by PBTD and HA-715.

5.3. Concentration Profiles

The CIL stirred tanks at 50 wt. % solid loading are simulated at lab scale. The influence of impeller speed, diameter, and off-bottom clearance is analyzed by plotting the concentration profiles for each of the cases in Figure 5. The variation of off-bottom clearance and impeller speed presented in this paper is between $0.125\ T$ and $0.5\ T$, and 200 and 700, respectively. These variations are studied for impeller diameters of $0.5\ T$ and $0.33\ T$.

At low off-bottom clearance for the impeller diameter of $0.5\ T$, the concentration profile remains uniform in the bottom half of the tank. It suggests the lower off-bottom clearance contributes to suspension of the settled solids. Therefore, the rise in the solid concentration near the bottom is not visible for $C = T/8$ case. As the clearance is gradually increased, the velocity of jet hitting the bottom also decreases, and hence the possibility of the presence of settled solids increases. The presence of settled solid at the bottom is dependent on the velocity of jet when it hits the bottom. Also this velocity is governed by the distance between the impeller and the bottom, and the impeller rotation speed. For high off-bottom clearance and low impeller rotation speeds, the jet velocity hitting the bottom is low, and hence a higher solid concentration can be observed near the bottom of the tank. With the increase in the off-bottom clearance, solid concentration assumes a vertically flipped *S*-shaped profile. This profile is an indication of significant variation in the solid concentration with height. Such a profile means, the concentration of solid is high at the bottom, it lowers around the region of impeller, then gradually increases above the impeller, and then finally diminishes near the top of the stirred tanks. Comparatively lower concentration in the region near the impeller is because the jet carrying solids circulates after hitting the bottom, and forms an eye with very low velocity at the center, resulting in lower solid accumulation in this region. In all the cases in which the lower concentration is observed, a circular loop formed by the jet is present. Therefore the presence of circular loop and the magnitude of the jet velocity determines the formation of a low concentration zone near the impeller. This circular loop is not formed in two cases: first, at low impeller speeds and high off-bottom clearance as the velocity diminishes before the jet reaches the bottom of the tank; and second, at low off-bottom clearance, where the jet hits the bottom but cannot get enough distance to form a loop. In both of these cases, a flat profile near the impeller is observed.

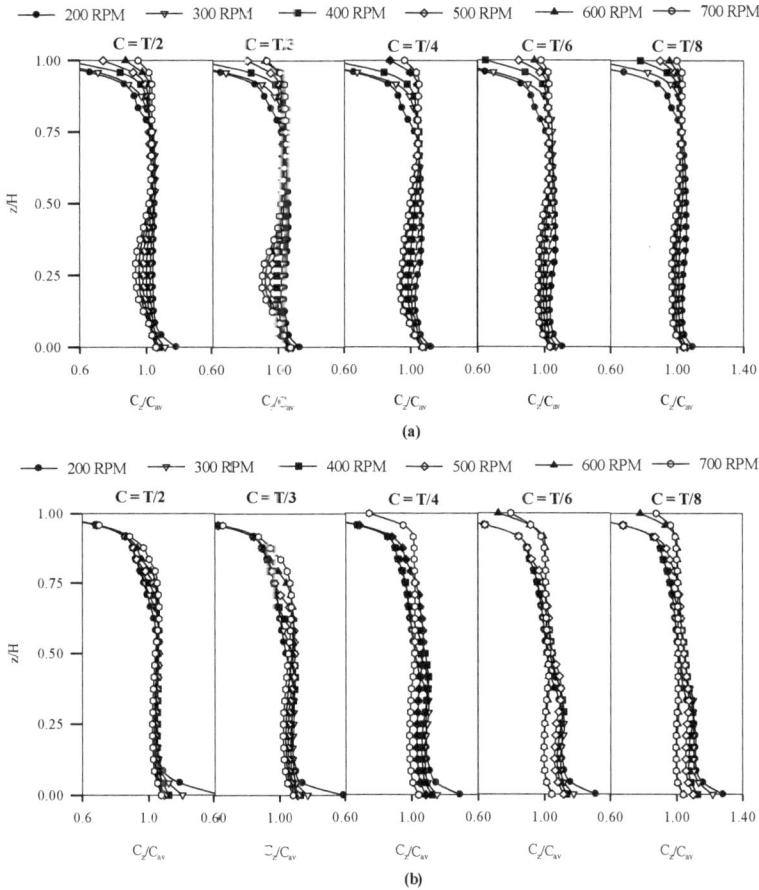

Figure 5. Axial concentration profiles at different stirrer speeds and off-bottom clearance for impeller to tank diameter ratio of (**a**) 0.5 and (**b**) 0.33.

When the impeller diameter is decreased to 0.33 *T*, the flipped *S*-shaped concentration profile is not dominant. On the contrary, the concentration is high near the bottom, which decreases to average concentration while reaching the height of the impeller, and then gradually decreases when reaching near the top. Clearly, the conditions for strong circular loop formed in the 0.5 *T* case is not met in the 0.33 *T* case. With the decrease in the impeller diameter and similar rotation velocity, the impeller tip speed is reduced and therefore, the velocity magnitude of the jet is not sufficient to form a circular loop of solids. Therefore, the profiles for all the cases are almost identical, with strong influence of impeller speed. At low impeller speeds, a high accumulation of solids at the bottom and low concentration at the top is observed. At high impeller speeds, the concentration values showed less variation with height. However, the influence of off-bottom clearance on the accumulation of solid near the bottom remained the same as in the case of 0.5 *T*. A low off-bottom clearance is found beneficial to suspend the settled solids.

5.4. Suspension Quality and Power Consumption

The suspension quality and power consumption are the two important parameters for determining the efficiency of a stirred tank. A high homogeneity (high suspension quality) achieved at a low power consumption is the objective of design and optimization. Suspension quality in a stirred tank can be defined in two different ways: first, by defining the variation between the minimum and maximum concentration [34] and second, by determining the homogeneity by variation in local concentration compared to the average concentration [35]. Mathematical representation of both these methods is given in equations below:

$$\Delta C_{\max} = \frac{C_{\max} - C_{\min}}{C_{\text{average}}} \times 100 \tag{11}$$

$$H_{\text{susp}} = 1 - \sqrt{\frac{1}{n}\sum_{i=1}^{n}\left(\frac{C_i}{C_{\text{average}}} - 1\right)^2} \tag{12}$$

The results obtained from both of these equations are plotted in Figure 6. While the results from the H_{susp} equation suggest that the homogeneity increases with the increase in impeller speed and reaches a plateau value, the results of the $\Delta C_{\max}$ equation appear quite erratic. Such an observation is a result of error in the $C_{\min}$ or $C_{\max}$ values due to local errors during computation. Therefore, a conclusion cannot be drawn from these plots with confidence. H_{susp} provides insight with respect to the homogeneity prevailing in the stirred tanks.

For the case of $D_i/T = 0.5$, the homogeneity is low at low impeller speeds, which improves with increasing impeller speed and reaches a plateau. However, for an impeller off-bottom clearance of $T/3$, the homogeneity decreases at very high velocity because of the generation of a dominant flipped S-shaped concentration profile (see Figure 5a). For a clearance higher than $T/3$ or lower than $T/3$, the flipped S-shaped profile is not dominant and therefore, the decrease in homogeneity at very high impeller speed is also not observed. The highest deviation in the homogeneity occurs due to the suspended solids at the bottom. Decreasing the off-bottom clearance allows the high velocity jet to force the solids to suspend. Therefore, at off-bottom clearance of $T/8$, the homogeneity improves at lower impeller speed compared to other cases.

The results of cases $D_i/T = 0.33$ are far different from those of $D_i/T = 0.5$. In these cases, the improvement in the homogeneity is not significant with the increase in the impeller rotation speed. Exceptions are the case of off-bottom clearance of $T/4$, $T/6$, and $T/8$, where the homogeneity crosses 0.9 beyond the impeller rotation speed of 600, 500, and 500 RPM respectively. This suggests that the higher off-bottom clearance and low impeller speeds result in dead zones in the stirred tank, hence reducing the homogeneity. When the velocities in the lower part of the tank are sufficiently high, the solids suspend and a higher value of homogeneity is achieved.

For a particular impeller speed, the power consumption in a stirred tank system is proportional to D_i^5. Therefore, the power consumption for an impeller diameter of $D_i/T = 0.5$ is higher than that of $D_i/T = 0.33$ at a given RPM (see Figure 6). For an impeller rotation speed of 700 RPM, the power consumption for cases of $D_i/T = 0.33$ is less than 9 W. The impellers with $D_i/T = 0.5$ can operate only up to 400 RPM using this power, and beyond 400 RPM, these impellers will require additional power. However, at the same time, homogeneity higher than a value of 0.9 can only be achieved for an impeller rotation speed above 400 RPM using this impeller diameter, while the same can be achieved using an impeller diameter of $D_i/T = 0.33$ and clearance of $T/6$ or $T/8$ using 70% of the power requirement at 600 RPM. Therefore, $D_i/T = 0.33$ is more efficient than $D_i/T = 0.5$ for obtaining high suspension quality for a given power consumption. The power consumption also varies with the off-bottom clearance [36]. The closer the impeller is to the bottom of the tank, the higher the turbulence and power dissipation. Although only minor, the power consumption of $T/8$ off-bottom clearance is more than that of $T/6$. Therefore out of all the cases discussed, the impeller diameter of $D_i/T = 0.33$

with clearance of $T/6$ and impeller rotation speed of 600 RPM is appropriate for efficiently suspending high concentration of solids in a stirred tank using an HA-715 impeller.

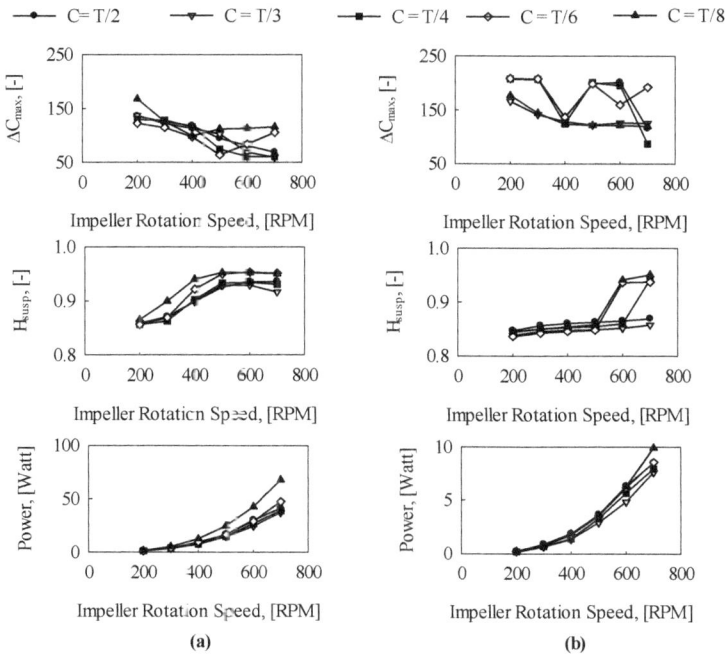

Figure 6. Suspension quality and power consumption plotted at different stirrer speeds and off-bottom clearance for impeller to tank diameter ratio of (**a**) 0.5 and (**b**) 0.33.

5.5. Scale-Up

Different scale-up criteria are available based on attaining homogeneity or complete suspension condition in the stirred tanks [37–40]. Buurman *et al.* [37] suggested a scaling up rule of $N_{js} \propto D^{-0.666}$ for the complete suspension in stirred tanks assuming the turbulence to be isentropic. Barresi and Baldi [40] suggested that $N_{js} \propto D^{-0.666}$ is strictly valid in a homogeneous isotropic turbulent field and devised a new rule of $N_{js} \propto D^{-0.666}$ by applying the turbulence theory to the solids distribution. Magelli *et al.* [38] investigated the solids distribution in high aspect ratio stirred tanks under batch and semibatch conditions and found that a constant $ND^{0.93}$ was more appropriate. Montante *et al.* [39] analyzed the scale-up criteria for stirred tanks based on the dependency of minimum impeller speed as a function of impeller diameter and the dependency of settling velocity on the λ/dP, and backed the scale-up criterion of a constant $ND^{0.93}$. Montante *et al.* [41] further investigated the role of turbulence on the particle settling and suspension in scaled-up vessels, and arrived at the conclusion that intermediate turbulent fluctuation maintains the vertical solid concentration. Therefore, a value intermediate of constant tip speed $N_{js} \propto D^{-1}$ and constant power per unit volume ($N_{js} \propto D^{-0.666}$), *i.e.*, $N \propto D^{-0.93}$ is appropriate for scaling-up. Based on this criterion, an impeller velocity of 22.15 RPM used for a stirred tank with $T = 10$ m.

The axial concentration profiles in laboratory scale and full scale stirred tanks are shown in Figure 7. The homogeneity in solid concentration observed in the laboratory scale stirred tank at 600 RPM is not extrapolated to the full scale stirred tank at 22.15 RPM. The presence of low concentration on the top can be explained by the very low velocity field present in the top part of the full scale tank (see Figure 8). The velocity imparted by the impeller is observed to be localized in the lower part of the stirred tank and has apparently no influence on the top. Therefore, the velocity vectors between the middle and top of the impeller are nearly invisible showing low velocity zones. The average value of turbulent kinetic energy in the lower half of the tank is 0.033 m^2/s^2, which reduces to 0.0008 m^2/s^2 in the lower half of the tank. Without sufficient kinetic energy in the flow field in this zone, the settling of solids is evident. As a result, the solid concentration drops sharply to zero after 8.5 m in the full scale tank. Due to the same low velocity field in the middle to top zone of the stirred tank, the remaining solid stays suspended in the bottom half of the stirred tank.

From the above discussion it is clear that the power provided in the lower half of the stirred tank is sufficient to suspend the solids. However, due to the absence of a power source in the upper half, maintaining the homogeneity in the stirred tank is not possible. Therefore, a multiple impeller system is desired to achieve homogeneity.

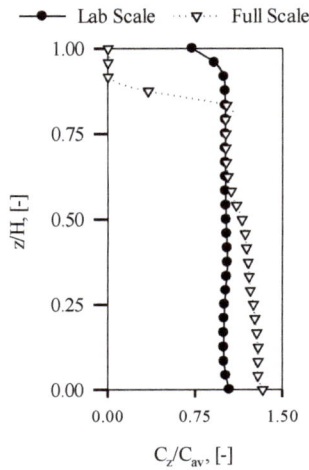

Figure 7. Axial concentration profiles for lab scale ($T = 0.288$ m) and full scale ($T = 10$ m) stirred tanks with $C = T/6$ and $D_i = T/3$.

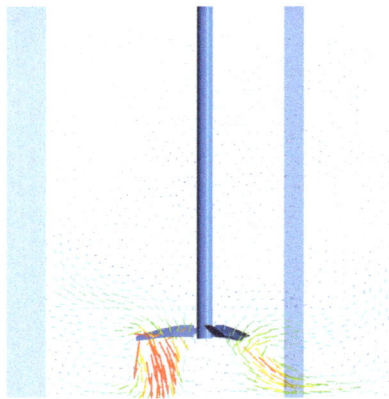

Figure 8. Velocity vectors on baffle-impeller blade plane in a full scale stirred tank.

5.6. Multiple-Impeller Systems

As the homogeneity was not achieved with the single impeller and additional power for solid suspension near the top is required, more impellers are required to be installed in the full scale CIL tanks. Therefore, three configurations in multi-impeller CIL systems were investigated, the details of which are given in Table 4.

Table 4. Details of impeller configuration in multi-impeller CIL tanks.

Variable	Twin-CT6	Twin-CT3	Triple-CT4
Off-bottom clearance, C_i	$T/6$	$T/3$	$T/4$
Distance between impellers, C_{iD}	$2T/3$	$T/3$	$T/4$

The solid concentration contours on the baffle-impeller plane in the full-scale CIL tanks for single impeller and multi-impeller cases are given in Figure 9. The inhomogeneity in the single and Twin-CT6 cases are evident from the contour plot. A continuous gradient is observed from bottom-to-top in these two cases. This gradient can be categorized into three zones of very low, medium, and high solid concentration. The use of a second impeller in the Twin-CTby6 case forms a jet that draws the low concentration liquid from the top and circulates the high concentration fluid along the periphery of the tank. Such a flow is favorable for increasing the suspension quality near the top of the CIL tanks, as a result of which the extent of the low concentration zone is reduced. The zone of high solid concentration is the largest for the single impeller system (extends to $z = 0.6\ T$), which is reduced by using a second impeller which disperses the high solid concentration present near the middle of the CIL tanks (extends to $z = 0.5\ T$). However, the limited range of the jet formed by this impeller does not significantly affect the high concentration zone which still extends to half of the tank. Nonetheless, the use of a second impeller is partially effective in improving the homogeneity.

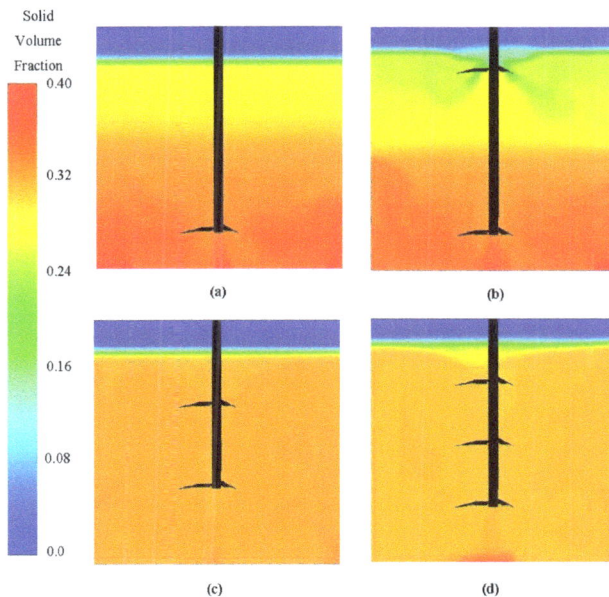

Figure 9. Concentration profiles in baffle-impeller plane for (**a**) Single impeller; (**b**) Twin-CT6; (**c**) Twin-CT3 and (**d**) Triple-CT4.

While, the presence of impeller near the gradient at the top was found beneficial, a similar approach near the interface between medium and low concentration zones can be applied. Therefore, Twin-CT3 and Triple-CT4 cases were investigated specifically targeting at the interface of the gradients between the zones. The strategy appears to be beneficial as the interface between the medium and high concentration zone disappears and an intermediate concentration is observed to be prevalent in the majority of tank. The average concentration values of intermediate concentration zones for Twin-CT3 and Triple-CT4 cases are 0.32 and 0.31, respectively. The increased distance from the top in Twin-CT3 case results in the weakening of drawing forces at the interface of low and medium concentration zones. Therefore, compared to Twin-CT6 case, the low concentration zone in Twin-CT3 case has extended to single impeller value of 0.9 T. This value again increases to 0.95 T for Triple-CT4 case, where the distance of impeller from the interface of low and medium concentration zones is reduced. The concentration profiles indicate accumulation of solid particles at the bottom center of the CIL for off-bottom clearance of $T/6$ and $T/4$, which is not present in the off-bottom clearance of $T/3$. A closer scrutiny of the flow near the bottom of the CIL tanks indicate that the secondary loops play a vital role in the suspension of solid particles at the bottom center of the tank. Ibrahim and Nienow [42] made a similar observation while investigating the solid suspension using different impellers and Newtonian fluids with different viscosities. For $T/3$ clearance, the secondary loop formed near the bottom is strong which results in the values of axial velocity required for solid suspension at the bottom. For all other cases, the secondary loop is not strong and the values of axial velocity also approach zero. This results in accumulation of solids near the bottom center.

Twin-CT3 case and Triple-CT6 case present feasible configurations for achieving homogeneity in CIL tanks. Comparison of the suspension quality and power requirement may provide further information on the efficiency of these two cases. In the case of Twin-CT3, the combination of lower intermediate concentration and extended low concentration zone results in a low suspension quality of 0.72, nearly the same of 0.78 for the Triple-CT6 case. The power requirements for single impeller, Twin-CT3 case and Triple-CT4 cases are 6.5 kW, 12 kW, and 18 kW, respectively. Therefore, the increased homogeneity is achieved at the expense of 6 kW of power per additional impeller.

6. Conclusions

CFD was used to simulate the high solid loading carbon-in-leach tanks. The averaged and local concentration profiles in high solid loading systems were validated with the available experimental data. The validated models were used to simulate the lab scale and full scale CIL tanks. The influence of design and operating parameters such as off-bottom clearance, impeller diameter, and impeller rotation speed were investigated. Several impeller configurations in the full scale CIL tanks for attaining homogeneity in the CIL tanks were also assessed. The conclusions deduced from the study are as follows:

1. The Euler-Euler simulation approach with KTGF, Syamlal drag model, RSM turbulence model and turbulent dispersion force model appropriately predict the local hydrodynamics in high solid loading stirred tank systems.
2. For a given power consumption, the flow generated by the HA-715 impeller is more dominant than the PBTD.
3. The low off-bottom clearance is favorable in achieving homogeneity at low impeller speed for lab scale CIL tanks.
4. For scale-up, multiple impeller systems are necessary for providing kinetic energy in the upper half of the CIL tanks.
5. While a low off-bottom clearance is suitable for solid suspension, solids can however accumulate at the bottom center in full scale CIL tanks due to weak secondary loops.

6. The dual impeller configuration with $T/3$ clearance and triple impeller configuration with $T/4$ clearance minimize the problems encountered in the CIL tanks. Additional impellers require approximately 6 kW of extra power in CIL tanks.

The insight of the prevailing hydrodynamics in the CIL tanks in this study was useful for the conclusions drawn for its design. The models can be further applied to other industrial high solid loading stirred tank systems to investigate the optimal design and operating parameters.

Acknowledgments: This work was supported by resources provided by the Pawsey Supercomputing Centre with funding from the Australian Government and the Government of Western Australia.

Author Contributions: D.W. and R.P.U. conceived and designed the CFD simulations. D.W. conducted simulation and acquired data. D.W., R.P.U. and V.K.P. analyzed and interpreted the data. D.W. drafted the manuscript and R.P.U. critically revised it.

Conflicts of Interest: The authors declare no conflict of interest.

Nomenclature

B_l	blade length, m
B_w	blade width, m
C_i	impeller clearance, m
C_{iD}	impeller-impeller distance, m
C	concentration in volume percent, (-)
C_{av}	average concentration in volume percent, (-)
C_D	drag coefficient, (-)
C_{Do}	particle drag coefficient in still fluid
C_H	cloud height, m
D or D_i	impeller diameter, m
D_{shaft}	shaft diameter, m
D_{hub}	hub diameter, m
d_P	particle diameter, m
$\vec{F}_{td}$	force due to turbulent dissipation, kg·m/s^2
$\vec{F}_q$	external force, kg·m/s^2
$\vec{F}_{Lft}$	lift force, kg·m/s^2
$\vec{F}_{vm}$	virtual mass force, kg·m/s^2
$\vec{F}_{12}$	interphase interaction force, kg·m/s^2
g	acceleration due to gravity, m/s^2
G_k	production of turbulence kinetic energy, kg·m^2/s^3
H	tank height, m
$\overline{\overline{I}}$	unit stress tensor, Pa
k	turbulence kinetic energy per unit mass, m^2/s^2
M	torque, N·m
N	impeller speed, 1/min
N_{js}	speed of just suspension, 1/min
N_{Re}	Reynolds number, (-)
N_P	power number, (-)
N_Q	pumping number, (-)
p	pressure and is shared by both the phases, Pa
P	power delivered to the fluid, W
T	tank diameter, m
$\vec{u}$	velocity vector, m/s
$\vec{u}_{dr}$	drift velocity, m/s

U_{tip}	Impeller tip velocity, m/s
	Greek Letters
α	volume fraction
γ	shear rate, 1/s
ε	turbulence dissipation rate, m^2/s^3
ε_b	bulk turbulence dissipation rate, m^2/s^3
λ	Kolmogorov length scale, m
μ	shear viscosity, Pa·s
μ_t	turbulent viscosity, m^2/s
ρ	density kg/m^3
σ	Prandtl numbers
σ_{sl}	dispersion Prandtl number
τ	shear stress, Pa
$\bar{\bar{\tau}}$	stress tensor, Pa
θ_m	mixing time, s
υ	bulk viscosity
Subscripts	
1 or l	continuous or primary phase
2 or s	dispersed or secondary phase
m	mixture properties
z	axial point

References

1. Barigou, M. Particle tracking in opaque mixing systems: An overview of the capabilities of PET and PEPT. *Chem. Eng. Res. Des.* **2004**, *82*, 1258–1267. [CrossRef]
2. Stevenson, R.; Harrison, S.T.L.; Mantle, M.D.; Sederman, A.J.; Moraczewski, T.L.; Johns, M.L. Analysis of partial suspension in stirred mixing cells using both MRI and ERT. *Chem. Eng. Sci.* **2010**, *65*, 1385–1393. [CrossRef]
3. Guida, A.; Nienow, A.W.; Barigou, M. PEPT measurements of solid-liquid flow field and spatial phase distribution in concentrated monodisperse stirred suspensions. *Chem. Eng. Sci.* **2010**, *65*, 1905–1914. [CrossRef]
4. Altway, A.; Setyawan, H.; Winardi, S. Effect of particle size on simulation of three-dimensional solid dispersion in stirred tank. *Chem. Eng. Res. Des.* **2001**, *79*, 1011–1016. [CrossRef]
5. Micale, G.; Grisafi, F.; Rizzuti, L.; Brucato, A. CFD simulation of particle suspension height in stirred vessels. *Chem. Eng. Res. Des.* **2004**, *82*, 1204–1213. [CrossRef]
6. Ochieng, A.; Lewis, A.E. Nickel solids concentration distribution in a stirred tank. *Miner. Eng.* **2006**, *19*, 180–189. [CrossRef]
7. Fradette, L.; Tanguy, P.A.; Bertrand, F.; Thibault, F.; Ritz, J.-B.; Giraud, E. CFD phenomenological model of solid-liquid mixing in stirred vessels. *Comput. Chem. Eng.* **2007**, *31*, 334–345. [CrossRef]
8. Ochieng, A.; Onyango, M.S. Drag models, solids concentration and velocity distribution in a stirred tank. *Powder Technol.* **2008**, *181*, 1–8. [CrossRef]
9. Kasat, G.R.; Khopkar, A.R.; Ranade, V.V.; Pandit, A.B. CFD simulation of liquid-phase mixing in solid-liquid stirred reactor. *Chem. Eng. Sci.* **2008**, *63*, 3877–3885. [CrossRef]
10. Fletcher, D.F.; Brown, G.J. Numerical simulation of solid suspension via mechanical agitation: Effect of the modelling approach, turbulence model and hindered settling drag law. *Int. J. Comput. Fluid Dyn.* **2009**, *23*, 173–187. [CrossRef]
11. Tamburini, A.; Cipollina, A.; Micale, G.; Ciofalo, M.; Brucato, A. Dense solid-liquid off-bottom suspension dynamics: Simulation and experiment. *Chem. Eng. Res. Des.* **2009**, *87*, 587–597. [CrossRef]
12. Tamburini, A.; Cipollina, A.; Micale, G.; Brucato, A.; Ciofalo, M. CFD simulations of dense solid-liquid suspensions in baffled stirred tanks: Prediction of suspension curves. *Chem. Eng. J.* **2011**, *178*, 324–341. [CrossRef]

13. Tamburini, A.; Cipollina, A.; Micale, G.; Brucato, A.; Ciofalo, M. CFD simulations of dense solid-liquid suspensions in baffled stirred tanks: Prediction of the minimum impeller speed for complete suspension. *Chem. Eng. J.* **2012**, *193–194*, 234–255. [CrossRef]
14. Gohel, S.; Joshi, S.; Azhar, M.; Horner, M.; Padron, G. CFD modeling of solid suspension in a stirred tank: Effect of drag models and turbulent dispersion on cloud height. *Int. J. Chem. Eng.* **2012**. [CrossRef]
15. Liu, L.; Barigou, M. Numerical modeling of velocity field and phase distribution in dense monodisperse solid-liquid suspensions under different regimes of agitation: CFD and PEPT experiments. *Chem. Eng. Sci.* **2013**, *101*, 837–850. [CrossRef]
16. Gidaspow, D. *Multiphase Flow and Fluidization: Continuum and Kinetic Theory Descriptions*; Academic Press: San Diego, CA, USA, 1994.
17. Dagadu, C.P.K.; Akaho, E.H.K.; Danso, K.A.; Stegowski, Z.; Furman, L. Radiotracer investigation in gold leaching tanks. *Appl. Radiat. Isot.* **2012**, *70*, 156–161. [CrossRef] [PubMed]
18. Dagadu, C.P.K.; Stegowski, Z.; Furman, L.; Akaho, E.H.K.; Danso, K.A. Determination of flow structure in a gold leaching tank by CFD simulation. *J. Appl. Math. Phys.* **2014**, *2*, 510–519. [CrossRef]
19. Dagadu, C.P.K.; Stegowski, Z.; Sogbey, B.J.A.Y.; Adzaklo, S.Y. Mixing analysis in a stirred tank using computational fluid dynamics. *J. Appl. Math. Phys.* **2015**, *3*, 637–642. [CrossRef]
20. Aubin, J.; Fletcher, D.F.; Xuereb, C. Modeling turbulent flow in stirred tanks with CFD: The influence of the modeling approach, turbulence model and numerical scheme. *Exp. Therm. Fluid Sci.* **2004**, *28*, 431–445. [CrossRef]
21. Montante, G.; Magelli, F. Modelling of solids distribution in stirred tanks: Analysis of simulation strategies and comparison with experimental data. *Int. J. Comput. Fluid Dyn.* **2005**, *19*, 253–262. [CrossRef]
22. Fan, L.; Mao, Z.; Wang, Y. Numerical simulation of turbulent solid-liquid two-phase flow and orientation of slender particles in a stirred tank. *Chem. Eng. Sci.* **2005**, *60*, 7045–7056. [CrossRef]
23. Khopkar, A.R.; Kasat, G.R.; Pandit, A.B.; Ranade, V.V. Computational fluid dynamics simulation of the solid suspension in a stirred slurry reactor. *Ind. Eng. Chem. Res.* **2006**, *45*, 4416–4428. [CrossRef]
24. Ljungqvist, M.; Rasmuson, A. Numerical simulation of the two-phase flow in an axially stirred vessel. *Chem. Eng. Res. Des.* **2001**, *79*, 533–546. [CrossRef]
25. Micale, G.; Montante, G.; Grisafi, F.; Brucato, A.; Godfrey, J. CFD simulation of particle distribution in stirred vessels. *Chem. Eng. Res. Des.* **2000**, *78*, 435–444. [CrossRef]
26. Murthy, B.N.; Joshi, J.B. Assessment of standard k-ε, RSM and LES turbulence models in a baffled stirred vessel agitated by various impeller designs. *Chem. Eng. Sci.* **2008**, *63*, 5468–5495. [CrossRef]
27. Derksen, J.; Akker, V.D.; Harry, E.A. Large eddy simulations on the flow driven by a rushton turbine. *AIChE J.* **1999**, *45*, 209–221. [CrossRef]
28. Burns, A.D.; Frank, T.; Hamill, I.; Shi, J.-M. The favre averaged drag model for turbulent dispersion in eulerian multi-phase flows. In Proceedings of the 5th International Conference on Multiphase Flow, ICMF, Yokohama, Japan, 30 May–4 June 2004. Paper No. 392.
29. Syamlal, M.; Rogers, W.; O'Brien, T.J. Mfix Documentation: Theory Guide. Available online: https://mfix.netl.doe.gov/documentation/Theory.pdf (accessed on 26 October 2015).
30. Del Valle, V.H.; Kenning, D.B.R. Subcooled flow boiling at high heat flux. *Int. J. Heat Mass Transfer* **1985**, *28*, 1907–1920. [CrossRef]
31. Zwietering, T.N. Suspending of solid particles in liquid by agitators. *Chem. Eng. Sci.* **1958**, *8*, 244–253. [CrossRef]
32. Hicks, M.T.; Myers, K.J.; Bakker, A. Cloud height in solids suspension agitation. *Chem. Eng. Commun.* **1997**, *160*, 137–155. [CrossRef]
33. Bittorf, K.J.; Kresta, S.M. Prediction of cloud height for solid suspensions in stirred tanks. *Chem. Eng. Res. Des.* **2003**, *81*, 568–577. [CrossRef]
34. Zhao, H.-L.; Lv, C.; Liu, Y.; Zhang, T.-A. Process optimization of seed precipitation tank with multiple impellers using computational fluid dynamics. *JOM* **2015**, *67*, 1451–1458. [CrossRef]
35. Tamburini, A.; Cipollina, A.; Micale, G.; Brucato, A.; Ciofalo, M. CFD simulations of dense solid-liquid suspensions in baffled stirred tanks: Prediction of solid particle distribution. *Chem. Eng. J.* **2013**, *223*, 875–890. [CrossRef]
36. Armenante, P.M.; Mazzarotta, B.; Chang, G.-M. Power consumption in stirred tanks provided with multiple pitched-blade turbines. *Ind. Eng. Chem. Res.* **1999**, *38*, 2809–2816. [CrossRef]

37. Buurman, C.; Resoort, G.; Plaschkes, A. Scaling-up rules for solids suspension in stirred vessels. *Chem. Eng. Sci.* **1986**, *41*, 2865–2871. [CrossRef]
38. Magelli, F.; Fajner, D.; Nocentini, M.; Pasquali, G. Solid distribution in vessels stirred with multiple impellers. *Chem. Eng. Sci.* **1990**, *45*, 615–625. [CrossRef]
39. Montante, G.; Pinelli, D.; Magelli, F. Scale-up criteria for the solids distribution in slurry reactors stirred with multiple impellers. *Chem. Eng. Sci.* **2003**, *58*, 5363–5372. [CrossRef]
40. Barresi, A.; Baldi, G. Solid dispersion in an agitated vessel. *Chem. Eng. Sci.* **1987**, *42*, 2949–2956. [CrossRef]
41. Montante, G.; Bourne, J.R.; Magelli, F. Scale-up of solids distribution in slurry, stirred vessels based on turbulence intermittency. *Ind. Eng. Chem. Res.* **2008**, *47*, 3438–3443. [CrossRef]
42. Ibrahim, S.; Nienow, A.W. Comparing impeller performance for solid-suspension in the transitional flow regime with newtonian fluids. *Chem. Eng. Res. Des.* **1999**, *77*, 721–727. [CrossRef]

metals

MDPI

Article

Experimental Simulation of Long Term Weathering in Alkaline Bauxite Residue Tailings

Talitha C. Santini [1,2,3,]*, Martin V. Fey [4] and Robert J. Gilkes [3]

1 School of Geography, Planning, and Environmental Management, the University of Queensland, Brisbane, QLD 4072, Australia

2 Centre for Mined Land Rehabilitation, the University of Queensland, Brisbane, QLD 4072, Australia

3 School of Earth and Environment, University of Western Australia, 35 Stirling Highway, Crawley, WA 6009, Australia; bob.gilkes@uwa.edu.au

4 Department of Plant Production and Soil Science, University of Pretoria, Private Bag X20, Hatfield, Pretoria 0002, South Africa; martirvennfey@gmail.com

* Author to whom correspondence should be addressed; t.santini@uq.edu.au; Tel.: +61-7-3346-1647; Fax: +61-7-3365-6899.

Academic Editor: Suresh Bhargava

Received: 18 June 2015; Accepted: 7 July 2015; Published: 14 July 2015

Abstract: Bauxite residue is an alkaline, saline tailings material generated as a byproduct of the Bayer process used for alumina refining. Developing effective plans for the long term management of potential environmental impacts associated with storage of these tailings is dependent on understanding how the chemical and mineralogical properties of the tailings will change during weathering and transformation into a soil-like material. Hydrothermal treatment of bauxite residue was used to compress geological weathering timescales and examine potential mineral transformations during weathering. Gibbsite was rapidly converted to boehmite; this transformation was examined with *in situ* synchrotron XRD. Goethite, hematite, and calcite all precipitated over longer weathering timeframes, while tricalcium aluminate dissolved. pH, total alkalinity, and salinity (electrical conductivity) all decreased during weathering despite these experiments being performed under "closed" conditions (*i.e.*, no leaching). This indicates the potential for auto-attenuation of the high alkalinity and salinity that presents challenges for long term environmental management, and suggests that management requirements will decrease during weathering as a result of these mineral transformations.

Keywords: weathering; pressure vessels; *in situ* XRD; bauxite residue; reaction kinetics

1. Introduction

1.1. Tailings as a Soil Parent Material

Tailings are the residual mixture of solids and liquids remaining after extraction of a mineral or energy resource from its ore. Around 7 Gt of tailings are produced worldwide each year, with almost half of this total derived from copper extraction alone [1]. Tailings have typically been treated solely as waste materials; however, tailings are increasingly being recognized as substrates for soil development [2–4]. The annual mass of tailings produced globally is approximately one-third of the mass of soil lost through erosion globally [1,5]. Understanding how tailings materials weather into soils over extended timescales is therefore important not only for managing long term environmental impacts associated with tailings (e.g., acid mine drainage), but also for predicting future ecosystem properties and land use capabilities given the substantial and growing contribution of tailings to parent material inputs for global soil development.

   Bauxite residue is an alkaline, saline-sodic byproduct of the Bayer process for alumina refining. Most economic reserves of bauxite worldwide are hosted in deeply weathered lateritic deposits [6], and the bauxite residues produced during the Bayer process contain a characteristic mineral assemblage dominated by weathering-resistant minerals, such as quartz (SiO_2), goethite (FeOOH), hematite (Fe_2O_3), anatase (TiO_2) and rutile (TiO_2). Precipitates formed during the Bayer process, such as the desilication products sodalite ($Na_8(AlSiO_4)_6Cl_2$) and cancrinite ($Na_6Ca_2(AlSiO_4)_6(CO_3)_2$), calcite ($CaCO_3$), tricalcium aluminate ($Ca_3Al_2(OH)_{12}$), and perovskite ($CaTiO_3$), also report to the residue fraction during processing, as well as minor amounts of undigested aluminium oxides (gibbsite ($Al(OH)_3$) and boehmite (AlOOH)). Each year, 120 Mt of bauxite residue are produced globally, adding to the 3 Gt already stored in tailings facilities [7]. After deposition in tailings storage facilities, management tasks include collection and treatment of alkaline, saline leachates, and rehabilitation of the bauxite residue surface. Rehabilitation aims to stabilize the surface and improve visual amenity by establishing a vegetation cover. Insofar as the objective of rehabilitation is to establish a vegetation cover, rehabilitation strategies aim to accelerate pedogenesis (soil formation) in bauxite residue and enhance natural mechanisms of alkalinity and salinity attenuation such as precipitation, dissolution, and leaching. The weathering trajectory of bauxite residue, therefore, needs to be determined in order to identify any attenuation mechanisms that can be targeted by applied treatments. Furthermore, an understanding of the potential weathering trajectories of bauxite residue may aid in minimizing amounts and types of applied treatments. Refinery scale Bayer processing commenced in 1894, which gives 120 years exposure to weathering in the very oldest residue deposits [7]. The longest weathering timeframe for bauxite residue reported in studies to date was 40 years under a tropical climate in Guyana, in which dissolution of sodalite and calcite were observed [3]. This reflects the fact that weathering of primary minerals, an essential part of soil formation, can be a slow process in the field which is partially constrained by reaction kinetics. Limited evidence of mineral weathering has also been reported in other studies of soil development in bauxite residue [8,9]. Laboratory simulation of weathering is therefore essential in order to understand how this material behaves over longer weathering timescales and ensure that management strategies are effectively designed to minimize adverse environmental impacts over the long term.

1.2. Laboratory Simulation of Weathering

   Compressing the geological timescales over which mineral weathering occurs requires manipulation of environmental conditions to accelerate reaction rates. Increasing temperature and/or pressure can accelerate attainment of chemical equilibrium according to the modified Arrhenius Equation (Equation (1)):

$$K = A \left(\exp \left(-(E_a + P\Delta V)/RT \right) \right) \tag{1}$$

where K is reaction rate (s^{-1}; for a first-order reaction), A is the pre-exponential factor (s^{-1}; for a first-order reaction), E_a is activation energy ($J \cdot mol^{-1}$), P is partial pressure of reactant ($mol \cdot cm^{-3}$), ΔV is activation volume ($cm^3 \cdot mol^{-1}$), R is the universal gas constant ($8.314 \ J \cdot mol^{-1} \cdot K^{-1}$) and T is temperature ($^\circ K$).

   Soxhlet extractors have been used to simulate weathering for a variety of geological materials [10–14]; however, the very fine texture of bauxite residue (median particle diameter 130 μm for bauxite residue used in this study; [15]) inhibited leaching. Furthermore, the Soxhlet extraction procedure is slow and labour intensive when conducting experiments at multiple temperatures to investigate reaction kinetics. Pressure vessels (also known as pressure bombs, Teflon bombs, Parr bombs) have been used to accelerate reactions and rapidly bring materials to chemical equilibrium, because they apply both elevated temperatures and pressures to the material under investigation. Their previous use in weathering studies has mostly concerned the alteration of single minerals during pedogenesis [16–18]. Reaction kinetics can be derived from these experiments and extrapolated to predict reaction rates under field conditions [17,19]. The drawback of pressure vessels compared to Soxhlet extractors for weathering simulations is that they do not allow for the effect of leaching, being a "closed" system. Allowing reaction products to accumulate may create a local equilibrium which slows further reactions. Different equilibrium states could therefore be identified for the same material under various leaching conditions. Bauxite residue has a very low hydraulic conductivity [20], which limits leaching; and under low leaching rates, equilibration with pore water may proceed to chemical equilibrium. The use of pressure vessels to predict the long term equilibrium composition of bauxite residue may therefore be a reasonable representation of reactions that occur under field conditions.

The aims of this study were to: (a) analyse the chemical and mineralogical composition of bauxite residue over time under simulated weathering conditions in pressure vessels; (b) extrapolate results from the pressure vessel studies to temperatures and pressures typical of field conditions in bauxite residue deposit areas (tailings storage facilities); and (c) determine whether pressure vessels provide a suitable method for rapidly determining potential weathering trajectories of bauxite residue under field conditions. Under hydrothermal conditions, we expected to observe some conversion of mineral phases that are commonly associated with Bayer process conditions, such as conversion of gibbsite to boehmite [21], goethite to hematite [22], tricalcium aluminate to calcite [23], and quartz to desilication products [24]. Dissolution of sodalite and calcite has been observed during weathering under leaching conditions [3,25,26]. However, dissolution of these minerals was not expected within this experiment because no leaching was possible

2. Experimental Section

2.1. Study Design

This study used pressure vessels to elevate temperature and pressure in order to rapidly determine the likely weathering trajectory of bauxite residue. Two types of pressure vessels were used: PTFE-lined stainless steel pressure vessels; and fused silica capillaries, sealed and pressurized with N_2 to avoid boiling. These vessels were used in two experiments to provide complementary information. The stainless steel pressure vessels allowed sampling and analysis of both solid and liquid phases, but required 90 min to reach the required temperature in an oven. This limited temporal resolution, and complete transformation of some minerals (e.g., gibbsite to boehmite) was observed between the initial ($t = 0$) and second ($t = 90$ min) sampling times. Use of the fused silica capillary reaction vessels at the Australian Synchrotron Powder Diffraction beamline 10BM1 resolved this problem, as they heated quickly (less than 40 s in our experiment) and full X-ray diffraction (XRD) patterns can be collected *in situ* on a rapid basis (20 s in this experiment) with the Mythen detector array. Results are unreplicated; that is, only one sample was analysed for each time and temperature combination. This is common in mineral dissolution and precipitation studies where reaction rates are studied in response to variables, such as temperature or pH [17,22,27–34] to derive reaction kinetics. A single time and temperature combination in the pressure vessel experiments (100 °C, 909 h) was conducted in triplicate to estimate variability; this approach has also been used elsewhere [35].

2.2. Bauxite Residue Sample

Bauxite residue was sourced from Alcoa of Australia Limited's Kwinana refinery. Iron oxides (goethite and hematite) were the major mineral phases present, with smaller amounts of quartz, and calcite and sodalite formed during the Bayer process (Table 1). In 1:1 bauxite residue to water suspensions, initial pH was 13.07, electrical conductivity (EC) was 15.52 dS·m^{-1}, and total alkalinity (TA) was 7.35 g/L CaCO$_3$.

2.3. Stainless Steel Pressure Vessel Experiment

Five hydrothermal treatment temperatures (100, 140, 165, 200 and 235 °C) were employed in order to extract kinetic data for observed reactions. Subsamples (4 g) of bauxite residue were added to 5 mL of ultrapure deionised water to create bauxite residue slurries inside 25 cm^3 internal volume PTFE-lined stainless steel pressure vessels (Parr Instrument Company, Moline, IL, USA) and heated to the designated temperature in a rotating oven. One vessel was removed after 1.5, 3.75, 21, 28, and 48 h and allowed to cool to room temperature. Additional samples of the 100 °C treatment were also collected at 170, 336 and 909 h. Slurry samples were taken from each vessel after cooling (<2 h).

Table 1. Mineral concentrations in bauxite residue prior to hydrothermal treatment, as determined by quantitative mineral analysis by Rietveld refinement of an X-ray diffraction pattern.

Mineral	Formula	Concentration (% Weight) [a]
Goethite	FeOOH	41
Hematite	Fe_2O_3	11
Calcite	$CaCO_3$	11
Quartz	SiO_2	9.4
Boehmite	AlOOH	8.4
Muscovite	$KAl_2Si_3AlO_{10}(OH)_2$	8.4
Gibbsite	$Al(OH)_3$	3.4
Sodalite	$Na_8(AlSiO_4)_6Cl_2$	2.8
Anatase	TiO_2	1.7
Ilmenite	$FeTiO_3$	1.7
Tricalcium aluminate	$Ca_3Al_2(OH)_{12}$	0.4
Amorphous/unidentified	-	0.8

[a] X-ray diffraction analysis was performed as described for hydrothermally-treated samples in Section 2.4. using Rietveld refinement implemented in TOPAS-Academic (v. 4.1; [36] Coelho Software, Brisbane, Australia) without running in sequential refinement mode.

Cooled slurries were centrifuged at 3000 rpm for 5 min to separate liquid and solid phases. Supernatants were immediately filtered through a 0.2 μm cellulose acetate filter and analysed for pH, EC, and total alkalinity by titration with 0.1 M HCl to endpoints pH 8.5 and 4.5 [37]. Subsamples of supernatants were diluted 1:10, acidified with HCl to pH < 2 and stored at 4 °C prior to analysis by Inductively Coupled Plasma Optical Emission Spectrometry (ICP-OES; Optima 5300DV, PerkinElmer, Waltham, MA, USA) to determine Al, Ca, Fe, K, Na, Si, and Ti concentrations.

Treated solids were shaken with acetone in an end-over-end shaker for 90 min and then centrifuged at 3000 rpm for 5 min. The supernatant was decanted and the solids allowed to dry at ambient laboratory temperature for 48 h. Dried solids were hand ground with an agate mortar and pestle and then packed into 0.3 mm internal diameter glass capillaries for XRD analysis (wavelength 0.8271 Å, collection time 5 min) at the Powder Diffraction beamline (10BM1) at the Australian Synchrotron, Victoria, Australia. Peak locations and areas for selected minerals were then calculated using Traces [38]. Anatase, which is naturally present in the bauxite residue, was used as an internal standard for the purpose of comparing peak areas of selected minerals between XRD patterns. Anatase has been found to be only sparingly soluble in the presence of Bayer liquor at 90 °C under ambient pressure [39].

2.4. Fused Silica Capillary Pressure Vessel Experiment

In situ XRD experiments were conducted at the Australian Synchrotron's beamline 10BM1, with slurries heated in Norby cells [40], consisting of fused silica capillaries (internal diameter: 0.8 mm; wall thickness: 0.1 mm, typical working volume: 0.02 mL) heated with a hot air blower, continuously rotated through 270° along the longitudinal axis of the capillary, and held under N_2 pressure during heating to various temperatures (140–170 °C) to avoid boiling. The N_2 regulator could not provide sufficient pressure to avoid boiling above 170 °C. Slurries were prepared <2 h before analysis at the same bauxite residue: Deionised water ratio as in the stainless steel pressure vessel experiments (44% wt solids). Diamond powder was added to the dry bauxite residue at 5% wt (2.2% wt in slurries) as an internal standard for the purposes of quantitative phase analysis by Rietveld refinement. Diamond was selected for its low reactivity under the experimental conditions, and for its few diffraction peaks within the 2θ angular range of interest, which minimizes overlaps with other phases. Diffraction patterns were collected continuously every twenty seconds using the Mython microstrip detector, capable of collecting 80° 2θ simultaneously, to allow calculation of mineral transformation kinetics. XRD patterns were collected first at ambient temperature with the required N_2 pressure applied, during heating, and for the duration of the experiment. Operating temperature was reached in 20–40 s. Run times varied between 1.5 h and 4 h, depending on factors such as loss of synchrotron beam, capillary failure, and observed reaction rates.

To quantify changes in mineral concentrations and amorphous content during this experiment, quantitative mineral analysis of X-ray diffractograms was conducted using TOPAS-Academic (v. 4.1; [36] Coelho Software, Brisbane, Australia) for Rietveld refinement. Compared with the peak area based data analysis approach, Rietveld refinement removes the potential for errors from peak overlaps, reports mineral concentrations on an absolute rather than relative scale, and allows quantification of amorphous content. Data were analysed using a sequential refinement approach, which used the refined (output) mineral concentrations for dataset x as the input mineral concentrations for the Rietveld refinement of dataset x − 1, where x is any dataset in the collected sequence of XRD patterns for bauxite residue at a particular temperature. Reported mineral concentrations (on a percent weight basis) were corrected for amorphous content (including X-ray amorphous solids, fused silica capillary, and water in the slurry) using the diamond internal standard and Equations (2) and (3):

$$Abs_i = (w_i STD_{true})/(STD_{meas}) \qquad (2)$$

$$w_{amorph} = 1 - (\sum_{i=1}^{n} Abs_i) \qquad (3)$$

where Abs_i is the absolute weight fraction of phase i, w_i is the weight fraction of phase i determined by Rietveld refinement, STD_{true} is the known weight fraction of the diamond standard, STD_{meas} is the calculated weight fraction of the diamond standard by Rietveld refinement and w_{amorph} is the total absolute weight fraction of amorphous and unidentified phases.

2.5. Reaction Kinetics Calculations

An extent of reaction parameter, α, was calculated from either peak areas (stainless steel pressure vessel experiment) or mineral concentrations (fused silica capillary experiment), where I_t was the observed peak area or mineral concentration at a sampling time, t, and I_{max} was the maximum peak area or mineral concentration attained (Equation (4)):

$$\alpha = (I_t)/(I_{max}) \qquad (4)$$

Kinetic data was then extracted using an approach identical to that detailed by Murray *et al.* [22]. The Sharp-Hancock equation (Equation (5); [41]), a linearised version of the Avrami-Erofe'ev Equation (Equation (6); [42–45]), was used to calculate kinetic parameters n (reaction order) and k (reaction rate) by plotting $\ln(-\ln(1 - \alpha))$ against $\ln(t)$ and fitting a least-squares linear equation.

$$\ln(-\ln(1 - \alpha)) = n \ln(t - t_0) + n \ln(k) \qquad (5)$$

$$\alpha = 1 - \exp(-k \, (t - t_0))^n \qquad (6)$$

Nucleation time (t_0) was assumed to be zero as no new minerals were formed and growth sites therefore existed for all phases. Values for $\ln(k)$ were then plotted against $1/T$ in an Arrhenius plot to determine activation energy, E_a, and the pre-exponential factor, A.

3. Results and Discussion

3.1. Supernatant Chemistry

Decreases in pH, EC, and total alkalinity (TA) were observed over time in bauxite residue in the 200 °C and 235 °C treatments (Figure 1), indicating a potential for "auto-attenuation" of alkalinity and salinity within the bauxite residue. This is in sharp contrast with the weathering trajectories of tailings derived from sulfidic ores, in which initially low pH and high salinity are exacerbated during weathering due to oxidation of residual iron sulfides [2,4]. Some increases in EC and TA were observed in 100 °C and 135 °C treatments whereas EC and TA initially increased and then sharply decreased in the 165 °C, 200 °C, and 235 °C treatments. This suggests that the lower temperature hydrothermal treatments progress along the same reaction trajectory as the higher temperature treatments, with slower reaction rates in the lower temperature treatments.

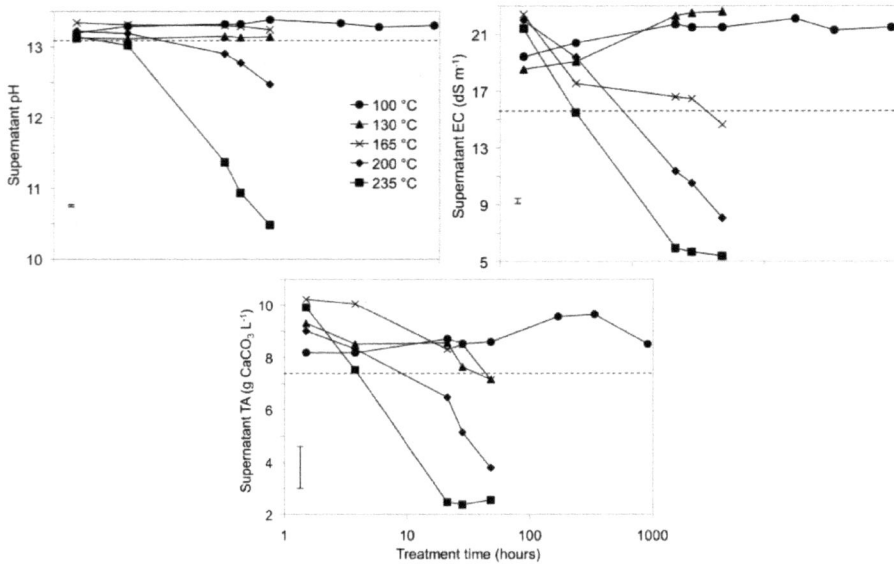

Figure 1. pH, EC, and total alkalinity (TA) of supernatants from bauxite residue during pressure vessel treatments at various temperatures. Dashed lines indicate $t = 0$ values. Note that treatment time (X axis) is graphed on a logarithmic scale. Error bars indicate 95% confidence interval for the mean of the 100 °C treatment at 909 h.

Supernatant concentrations of Al, K, Na, and S all decreased over time to below initial concentrations in the 200 and 235 °C treatments, suggesting removal from solution by mineral precipitation. Concentrations of Al, Na, and S increased and then decreased to concentrations slightly above or below initial concentrations in the 165 °C treatment. This suggests that similar reactions are responsible for observed changes in concentrations of Al, Na, and S at temperatures from 165 to 235 °C, and that reaction rates decrease with decreasing temperature. At 100 and 130 °C, Al, K, Na, and S concentrations increased and remained above initial concentrations during hydrothermal treatment; this suggests that precipitation reactions occur very slowly or do not occur at all at lower temperatures. Silicon concentrations in the supernatant increased over time in all but the 235 °C treatment. This could be a result of dissolution of reactive Si (from muscovite or finely divided quartz), with precipitation of sodalite or cancrinite only occurring to an appreciable extent at 235 °C. Iron concentrations in solution remained near or below the detection limit for all treatment temperatures except 235 °C (Figure 2).

The assumption of anatase insolubility was supported by ICP-OES data, which indicated that little Ti was released to solution, except at 235 °C. Titanium and iron concentrations increased simultaneously and were present in solution at an approximate Fe:Ti molar ratio of 8:1, which is inconsistent with ilmenite dissolution and rather may indicate dissolution of a minor titanium-substituted iron oxide phase, such as titanohematite. Ca concentrations were below detection limits (0.35 mg/L) in all samples and are therefore not shown in Figure 2.

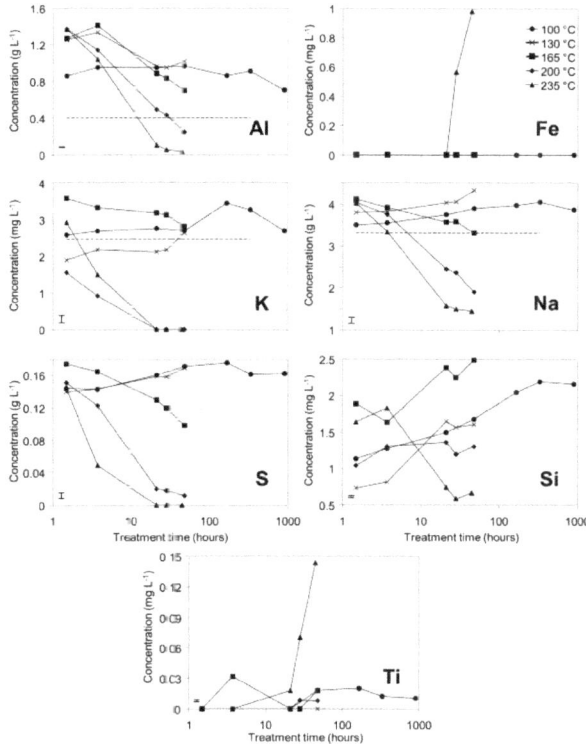

Figure 2. Concentrations of selected elements (mg or $g \cdot L^{-1}$) in supernatant from bauxite residue after treatment in pressure vessels. Dashed lines indicate concentrations at $t = 0$. Where no dashed line is visible, concentrations at $t = 0$ were below detection limits. Note that treatment time (X axis) is graphed on a logarithmic scale. Error bars indicate 95% confidence interval for the mean of the 100 °C treatment at 909 h.

3.2. Solids Mineralogy

The rapid collection of XRD patterns from the bauxite residue slurry in fused silica capillary reaction vessels allowed construction of time sequence graphs which illustrate the temperature dependent transformation of gibbsite to boehmite over short timescales (Figures 3–6). Gibbsite and boehmite dynamics reflect a dissolution-precipitation mechanism based on observed rates of transformation in Figures 3–6, and previous studies of hydrothermal conversion [46,47]. This would aid in lowering pH, EC, and TA by diluting the supernatant (Equation (7)). No Al would be released to the supernatant, although some Al initially present in solution may also be precipitated as boehmite.

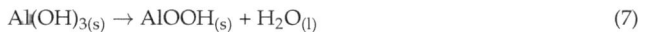

$$Al(OH)_{3(s)} \rightarrow AlOOH_{(s)} + H_2O_{(l)} \tag{7}$$

The gibbsite to boehmite reaction could only be observed by *in situ* synchrotron XRD because the transformation time (<50 min at 165–170 °C) was faster than the required time for the stainless steel pressure vessels to heat to temperature. The gibbsite to boehmite transformation is known to occur above 100 °C [22]. Transformation of gibbsite to boehmite was very slow at temperatures ≤150 °C in this study and increased markedly above this temperature. Elevated pressure slows the transformation, because the sum of the molar volumes of the products is greater than that of the reactants (Equation (7)). At lower temperatures, the inhibitory effect of elevated pressure is likely to have slowed the transformation; however, this effect appeared to have less influence on reaction rates at higher temperatures. A similar conclusion was drawn by Mehta and Kalsotra [48], who observed conversion of gibbsite to boehmite under hydrothermal conditions only at temperatures ≥190 °C. Higher alumina to caustic ratios (A/C; Al_2O_3:Na_2O ratio (g/L) in the supernatant) also raise the minimum transformation temperature [49,50]; the alumina:caustic ratio was consistently below 0.5 in this experiment and was therefore unlikely to have inhibited transformation of gibbsite to boehmite.

Goethite precipitated to a minor extent at 170 °C (Figure 6). Anatase concentrations appeared stable according to quantitative mineral analysis graphs (Figures 3–6), which supports interpretations based on Ti concentrations in the supernatants (Figure 2). Amorphous content was stable during the transformation of gibbsite to boehmite; given that calculated amorphous content includes contributions from water in the slurry, amorphous content was expected to increase as a result of the water released during this transformation. The water released during transformation of gibbsite (present in the bauxite residue slurry at approximately 2% wt) to boehmite may have been insufficient to increase the amorphous content detectable within the error of the Rietveld-based quantification approach. The quality of the diffractograms was compromised by the short collection time; this hindered accurate quantification of gibbsite concentrations below 0.5% wt, and quantification of tricalcium aluminate and ilmenite concentrations. Diffractogram quality may also have been too poor to detect small changes in amorphous content of the solid phase. Longer collection times (1–2 min) are required for complex mineral assemblages if quantification of minor phases, or small changes in concentration of major phases are of interest.

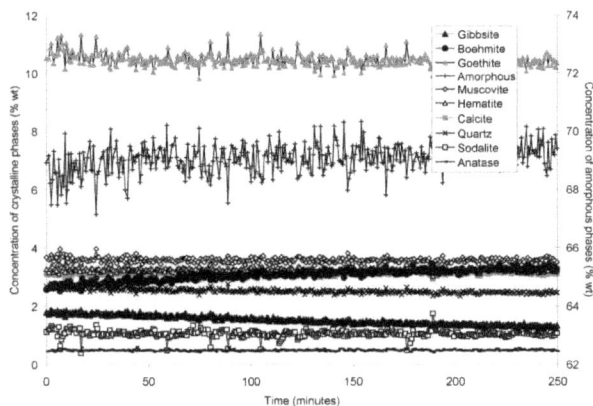

Figure 3. Concentration of crystalline and amorphous phases in bauxite residue slurry in capillary reaction vessels as a function of treatment time at 140 °C, as determined by Rietveld-based quantitative mineral analysis of X-ray diffractograms. Note that amorphous content is plotted on the secondary Y axis and represents contributions from X-ray amorphous minerals in bauxite residue, water in the slurry, and the fused silica capillary.

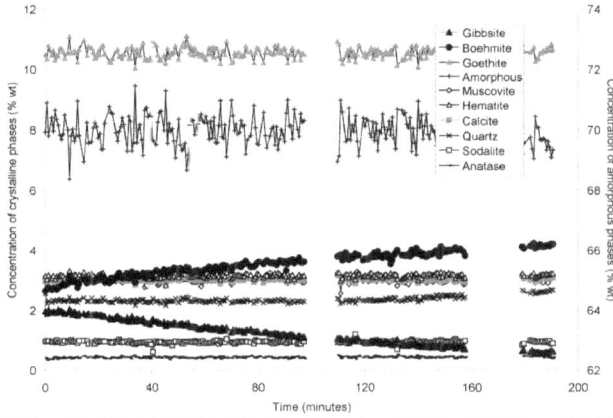

Figure 4. Concentration of crystalline and amorphous phases in bauxite residue slurry in capillary reaction vessels as a function of treatment time at 150 °C, as determined by Rietveld-based quantitative mineral analysis of X-ray diffractograms. Note that amorphous content is plotted on the secondary Y axis and represents contributions from X-ray amorphous minerals in bauxite residue, water in the slurry, and the fused silica capillary. Gaps in data from 100–110, and 160–180 min were caused by beam loss at the synchrotron. Heating of the slurry continued during beam loss.

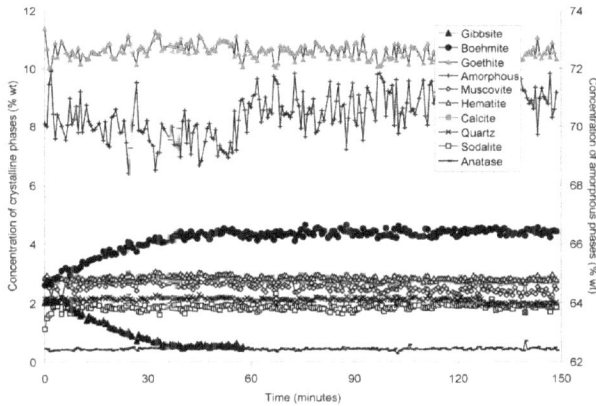

Figure 5. Concentration of crystalline and amorphous phases in bauxite residue slurry in capillary reaction vessels as a function of treatment time at 165 °C, as determined by Rietveld-based quantitative mineral analysis of X-ray diffractograms. Note that amorphous content is plotted on the secondary Y axis and represents contributions from X-ray amorphous minerals in bauxite residue, water in the slurry, and the fused silica capillary.

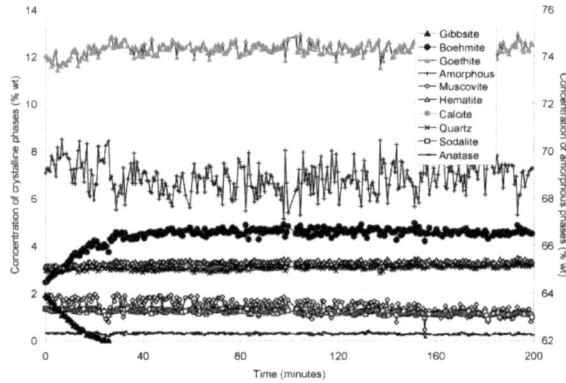

Figure 6. Concentration of crystalline and amorphous phases in bauxite residue slurry in capillary reaction vessels as a function of treatment time at 170 °C, as determined by Rietveld-based quantitative mineral analysis of X-ray diffractograms. Note that amorphous content is plotted on the secondary *Y* axis and represents contributions from X-ray amorphous minerals in bauxite residue, water in the slurry, and the fused silica capillary.

During extended hydrothermal treatment in stainless steel pressure vessels, boehmite, goethite, hematite, and calcite appeared to precipitate (Figure 7; Table 2), along with minor precipitation of sodalite; whereas gibbsite and tricalcium aluminate dissolved (Figure 7). As discussed above, gibbsite is likely converted to boehmite during hydrothermal treatment. Although conversion of goethite to hematite has been observed under Bayer process conditions [22], goethite dissolution was not observed during this experiment, perhaps due to the inhibitory effect of anatase [22]. The precipitation of both goethite and hematite was likely limited by low dissolved Fe concentrations. $Fe(OH)_4^-{}_{(aq)}$ is thermodynamically predicted to be the dominant form of dissolved iron at pH values observed in these experiments, and goethite and hematite precipitation would therefore have a minimal effect upon pH, EC, and TA (Equations (8) and (9)) even if precipitation occurred to a greater extent:

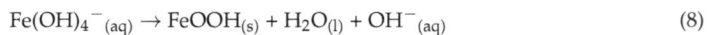

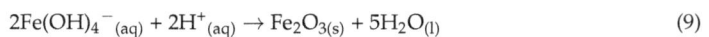

$$Fe(OH)_4^-{}_{(aq)} \rightarrow FeOOH_{(s)} + H_2O_{(l)} + OH^-{}_{(aq)} \tag{8}$$

$$2Fe(OH)_4^-{}_{(aq)} + 2H^+{}_{(aq)} \rightarrow Fe_2O_{3(s)} + 5H_2O_{(l)} \tag{9}$$

Tricalcium aluminate dissolved more slowly than gibbsite and the majority of Al and Ca supplied to solution from dissolution of both minerals is likely to have been precipitated as boehmite and calcite given that Al and Ca concentrations did not increase substantially during dissolution (Equation (10); Figure 2). This conclusion is consistent with the conversion mechanisms described elsewhere [23,46,47].

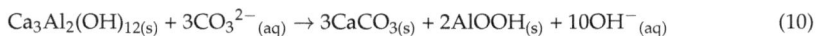

$$Ca_3Al_2(OH)_{12(s)} + 3CO_3^{2-}{}_{(aq)} \rightarrow 3CaCO_{3(s)} + 2AlOOH_{(s)} + 10OH^-{}_{(aq)} \tag{10}$$

Calcite and sodalite precipitation were not observed in the fused silica capillary experiments; changes in calcite:anatase ratio in Table 2 indicate that this is a slower reaction and smaller in magnitude than the boehmite precipitation reaction. Precipitation of calcite may have only occurred to a minor extent during the fused silica capillary experiments, and therefore may not have been detected. A similar explanation may account for the observations of goethite and hematite precipitation in the longer term stainless steel pressure vessel experiment. Both minerals precipitated more slowly and to a lesser extent than boehmite (Table 2), and precipitation may therefore not have been detectable over timescales of 150–250 min at temperatures of 140–165 °C.

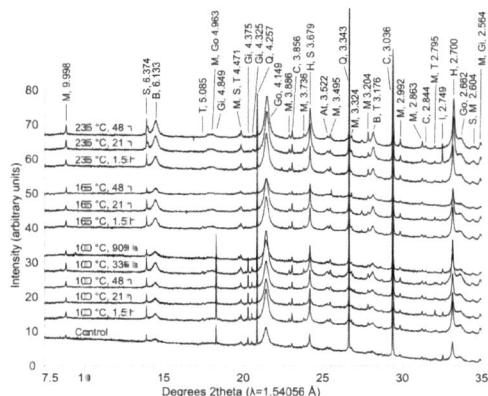

Figure 7. X-ray diffraction patterns from bauxite residue before and after hydrothermal treatment. Treatment time and temperature are indicated for each pattern. Peaks between 7.5°–35° 2θ are labeled with d-spacings (Å) and abbreviations for minerals as follows: M—muscovite; S—sodalite; B—boehmite; T—tricalcium aluminate; Go—goethite; Gi—gibbsite; Q—quartz; C—calcite; H—hematite; At—natase; I—ilmenite.

Muscovite appeared to dissolve at 165–170 °C in the fused silica capillary experiments but was stable in the stainless steel pressure vessel experiments. As the only K-bearing mineral in bauxite residue, muscovite behaviour could be expected to mirror supernatant K concentrations (Figure 2), dissolving at temperatures ≤165 °C and precipitating at temperatures ≥200 °C. This is not fully supported by the XRD data from the fused silica capillary or the stainless steel pressure vessel experiments. All XRD patterns were collected from powders or slurries that were rotated during collection of XRD patterns; preferred orientation is therefore unlikely to cause changes in peak areas between patterns. Muscovite has been observed to dissolve under alkaline conditions [28,31], although dissolution rates decrease as aqueous Al and Si concentrations increase [32]. The minor dissolution observed in fused silica capillary experiments at 165–170 °C may be due to decreases in aqueous Al in supernatant solution at temperatures ≥165 °C (Figure 2). It is unclear why muscovite dissolution was not observed by *ex situ* XRD of hydrothermally treated solids.

Table 2. Percentage change in mineral:anatase ratio during hydrothermal treatment of bauxite residue, relative to untreated bauxite residue. Ratio was calculated from primary peak areas of each mineral from synchrotron XRD scans. Mineral names are abbreviated as follows: Gi—gibbsite; T—tricalcium aluminate; B—boehmite; C—calcite; Go—goethite; H—hematite; S—sodalite. Mineral:anatase ratios of muscovite, anatase, and quartz did not indicate dissolution or precipitation in response to temperature or treatment time.

Temperature (°C)	Treatment Time (h)	Mineral						
		Gi	T	B	C	Go	H	S
100	1.5	54	41	82	30	47	15	47
100	21	47	24	102	43	48	29	62
100	48	46	−2	153	44	52	30	64
100	336	43	−2	313	88	107	89	106
165	1.5	−100	−14	300	53	71	55	68
165	21	−100	−100	468	95	119	109	118
165	48	−100	−100	584	106	138	117	190
235	1.5	−100	−35	348	85	99	90	123
235	3.75	−100	−100	526	88	114	119	150
235	21	−100	−100	587	94	141	177	294
235	48	−100	−100	643	111	156	198	389

3.3. Reaction Kinetics

Peak areas of minerals in XRD patterns of residues from the 100, 165, and 235 °C treatments in the stainless steel pressure vessel experiment (Table 2) were used to calculate extents of reaction (α) and kinetic parameters (Equations (4)–(6)). The rapid dissolution of gibbsite and tricalcium aluminate precluded calculation of reaction kinetics for these phases from the stainless steel pressure vessel experiment data. Fused silica capillary reaction vessel data were used for the calculation of gibbsite reaction kinetics, at hydrothermal treatment temperatures of 140, 150, 165, and 170 °C, using absolute mineral concentration as calculated by Rietveld refinement of XRD patterns rather than peak area of minerals relative to the internal standard. The quality of the XRD patterns from the fused silica capillary experiment was insufficient to allow accurate quantification of tricalcium aluminate and reaction kinetics were therefore not calculated for this mineral.

Sharp-Hancock analysis of mineral transformations within these experiments indicated that the plots of $\ln(-\ln(1 - \alpha))$ against $\ln(t)$ were well described by linear functions because r^2 values were generally >0.80 (Table 3), except for goethite at 100 °C. This indicates that a single reaction mechanism dominated over the course of the dissolution or precipitation reactions for most minerals. The n values of 0.8–1.2 observed for gibbsite dissolution (Table 4) are consistent with first-order reaction kinetics ($n = 1$). Contracting area kinetics, with an n value of 1.04, and contracting volume kinetics, with an n value of 1.08, would also be feasible explanations for the observed n values and are consistent with the transformation of gibbsite to boehmite as a dissolution-precipitation mechanism [46,47].

Table 3. Coefficients of determination (r^2) for Sharp-Hancock lines of best fit ($\ln(-\ln(1 - \alpha))$ *vs.* $\ln(t)$) to mineral transformations in bauxite residue. Mineral names are abbreviated as follows: Gi—gibbsite; B—boehmite; C—calcite; Go—goethite; H—hematite; S—sodalite.

Temperature (°C)	Gi	Temperature (°C)	B	C	Go	H	S
140	0.90	100	0.86	0.84	0.64	0.91	0.91
150	0.98	165	0.94	0.99	0.99	0.99	0.92
165	0.94	235	0.86	1.00	1.00	1.00	0.97
170	0.99						

Table 4. Reaction orders (n) derived from Sharp-Hancock lines of best fit ($\ln(-\ln(1 - \alpha))$ *vs.* $\ln(t)$) to mineral transformations in bauxite residue. Mineral names are abbreviated as follows: Gi—gibbsite; B—boehmite; C—calcite; Go—goethite; H—hematite; S—sodalite.

Temperature (°C)	Gi	Temperature (°C)	B	C	Go	H	S
140	0.804	100	0.289	0.281	0.200	0.363	0.162
150	1.072	165	0.357	0.441	0.361	0.301	0.331
165	0.991	235	0.403	0.096	0.314	0.499	0.508
170	1.185						

The E_a value calculated from this study (Table 5) match well with values from Mehta and Kalsotra'sstudy [48] of boehmite precipitation under hydrothermal, alkaline conditions for boehmite (E_a 69.63–73.46 kJ·mol^{-1}). Under alkaline, heated (60–85 °C) conditions, at ambient pressure, the E_a for gibbsite dissolution is 107 kJ·mol^{-1} [51]. This is lower than the value observed in this study (Table 5), which suggests that pressure inhibited the gibbsite-boehmite transformation [21] The E_a value determined from this study for sodalite precipitation (Table 5) was much higher than that determined in other studies for first-order reaction kinetics involving precipitation from solution (38.2–48.4 kJ·mol^{-1}; [52,53]). This suggests that sodalite is not directly precipitated from solution, but that precipitation is reliant upon the dissolution or transformation of another mineral. A similar elevation in activation energy for cancrinite precipitation under Bayer conditions was observed when cancrinite formed from sodalite rather than directly precipitating from solution [54]. Muscovite or quartz could be sources of reactive silica for sodalite precipitation; however, neither were observed to dissolve to an appreciable extent within this study. No values for activation energies of precipitation of calcite, goethite, or hematite under similar reaction conditions were available in the literature.

The low values of n returned from Sharp-Hancock analysis ($0.10 < n < 0.85$) for minerals other than gibbsite suggest that precipitation of minerals proceeded via three-dimensional (3D) diffusion controlled reaction mechanisms (Table 3), because these reactions occur with an ideal n value of 0.57 [55]. However, the activation energies calculated from Arrhenius plots of $\ln(k)$ against $1/T$ ranged from 76–131 kJ·mol^{-1} (Table 5). These activation energies are far higher than expected for diffusion controlled reactions (generally <21 kJ·mol^{-1}; [56]), and therefore suggest that precipitation reactions were in fact surface controlled.

Table 5. Activation energies (E_a; kJ·mol^{-1}), pre-exponential ("frequency") factors (A; s^{-1}), and reaction rates at 165 °C (k; s^{-1}) calculated from Arrhenius plots for mineral transformations in bauxite residue. Mineral names are abbreviated as follows: Gi—gibbsite; B—boehmite; C—calcite; Go—goethite; H—hematite; S—sodalite.

Parameter	Gi	B	C	Go	H	S
E_a	190.79	88.03	99.58	75.54	75.71	131.15
$\ln A$	49.00	12.80	18.45	9.79	8.38	21.07
$\log(k)$, 165 °C	-1.47	-4.94	-3.86	-4.76	-5.39	-6.50

In an attempt to distinguish which kinetic models best describe observed behaviour, lines were fitted to the reaction data for $[\ln(1 - \alpha)]$ against t, which should result in a straight line with gradient k if the reaction is controlled by first-order kinetics, and $[1 - (1 - \alpha)^{1/3}]^2$ against t, which should result in a straight line with gradient k if the reaction is controlled by 3D diffusion [55]. Gibbsite data was also fitted to $[1 - (1 - \alpha)^{1/2}]$ against t, which should result in a straight line with gradient k if the reaction proceeds at a decelerating rate proportional to remaining mineral surface area (contracting area), and $[1 - (1 - \alpha)^{1/3}]$ against t, which should result in a straight line with gradient k if the reaction proceeds at a decelerating rate proportional to remaining mineral volume (contracting volume). Coefficient of determination values (r^2) were used to compare the linear fits achieved by each model. Fits achieved were similar, although r^2 values for first-order reaction models were generally better than r^2 values for 3D diffusion models for precipitating minerals (Table 6). The gibbsite dissolution rate appears to be constant at 140 and 150 °C (Figures 3 and 4), but appears to decelerate at 165 and 170 °C (Figures 5 and 6). However, the first-order, contracting area, and contracting volume models fit gibbsite dissolution at 150 and 170 °C equally well (Table 6). Overall, the fits to first-order, contracting area, and contracting volume models for gibbsite dissolution were almost identical, which does not allow the most appropriate model amongst these three to be identified.

Reaction half lives (time required for half of the reactant(s) in a chemical reaction to be consumed), as predicted by the Avrami-Erofe'ev Equation and using the values of n expected for first-order surface controlled and 3D diffusion controlled reactions, were compared with observed reaction half lives, calculated by interpolation of Sharp-Hancock lines of best fit. Precipitation reaction half lives were predicted far more accurately by the first-order surface control model (Table 7). This information, combined with the high activation energies which are inconsistent with the 3D diffusion mechanism, suggests that the rates of these precipitation reactions were controlled by first-order surface kinetics. It also seems unlikely that 3D diffusion would control reaction rates in a continuously stirred system. However, there was more than an order of magnitude difference between observed and predicted reaction half lives for boehmite, goethite, muscovite, and sodalite using first-order kinetics, which suggests that first-order kinetics do not accurately model reaction kinetics in all cases.

Table 6. Coefficient of determination (r^2) values for least square lines fitted to mineral precipitation or dissolution data under a first-order surface control kinetic model (FO), 3D diffusion control kinetic model (3D), contracting area model (CA), and contracting volume model (CV). Mineral names are abbreviated as follows: Gi—gibbsite; B—boehmite; C—calcite; Go—goethite; H—hematite; S—sodalite.

Mineral	Kinetic Model	Temperature (°C)		
		100	165	235
B	FO	0.985	0.998	0.785
	3D	0.999	0.989	0.813
C	FO	0.995	0.989	0.938
	3D	0.996	0.994	0.944
Go	FO	0.994	0.985	0.988
	3D	0.991	0.996	0.992
H	FO	0.992	0.819	0.994
	3D	0.995	0.854	0.999
S	FO	0.963	0.993	0.999
	3D	0.989	0.962	0.998

Mineral	Kinetic Model	Temperature (°C)			
		140	150	165	170
Gi	FO	0.907	0.987	0.948	0.976
	3D	0.890	0.954	0.937	0.916
	CA	0.907	0.986	0.939	0.986
	CV	0.907	0.987	0.943	0.984

Table 7. Reaction half lives ($t_{1/2}$) as calculated by interpolation of Sharp-Hancock lines of best fit to observed data (OB) and the Avrami-Erofe'ev equation predicted for first-order surface control (FO) and 3D diffusion control (3D) for mineral transformations in bauxite residue. Mineral names are abbreviated as follows: Gi—gibbsite; B—boehmite; C—calcite; Go—goethite; H—hematite; S—sodalite.

Temperature	Kinetic Model	Gi	B	C	Go	H	S
	OB	0.0061	2	2	2	17	83
165°C (h)	FO	0.0057	16.78	1.40	11	47	595
	3D	0.0570	67870	869	32412	416986	35509858
25 °C (years)	FO	32098	164	61	21	94	1513193

Using the kinetic data presented in Tables 4 and 5, and assuming that reactions are controlled by a first-order surface mechanism (*i.e.*, $n = 1$), reaction rates were extrapolated to determine reaction half lives at field conditions (1 atm pressure, 25 °C). Calcite, boehmite, goethite and hematite reach their reaction half lives within two centuries; however, gibbsite and sodalite take substantially longer to reach their half lives. Boehmite precipitation was predicted to occur more rapidly than gibbsite dissolution at field conditions, which indicates that gibbsite dissolution would control the gibbsite to boehmite transformation. Although auto-attenuation of residue alkalinity and salinity under closed system conditions is theoretically possible, it would take millions of years under field conditions if controlled by these mineral reactions. No mineral transformations or reactions were observed during simulated weathering that would be potentially deleterious to rehabilitation and long term environmental management, although the slow dissolution of calcite and sodalite (which precipitated to a minor extent under closed system conditions) during rainfall leaching would contribute to the maintenance of high alkalinity and salinity in bauxite residue pore water and leachates. Given the mineral transformations observed here, the soil forming from bauxite residue is likely to have similar mineralogical properties to deeply weathered soils characteristic of tropical environments, such as Nitisols and Ferralsols [57], owing to the dominance of iron oxides and absence of high-activity clay minerals. These soils are not alkaline or saline so substantial leaching of residue would be required to remove soluble salts, especially sodium which remains in pore water at substantial concentrations (≥ 1.5 g·L^{-1}) even at the highest treatment temperature.

The reactions observed in this experiment may not occur under field conditions. Non-linearity in the response of activation energies and reaction rates to temperature and pressure can occur due to different reaction mechanisms being favoured [56], which hinders the use of observed activation energies and rates to predict behaviour outside the observed range. Diffusion does not appear to be a significant rate limiting process in the observed reactions, but is dependent on pressure, and the direction of this dependence (and thus the effect of pressure on reaction rates according to Equation (1)) cannot be predicted without determining activation volumes of the reactions. The effect of pressure on reaction rates is considered to be small relative to temperature for pressures found in the Earth's crust (1–10,000 kbar) [56], and is, therefore, unlikely to cause substantial variation between observed and predicted reaction half lives within this study. However, the effect of temperature on activation energies, rates, and reaction mechanisms could result in a different suite of reactions occurring in bauxite residue within storage areas under ambient temperature and pressure compared to those observed at elevated temperatures and pressures. Results from this study illustrate a potential weathering trajectory for bauxite residue; but this would need to be validated against observations of mineral transformations in the long term for field weathered bauxite residues to confirm that the predicted reactions occur in the field. Further, trace element speciation and mobility, which were beyond the scope of this study, also merit evaluation under both simulated laboratory weathering and field weathering scenarios to ensure that the weathering reactions observed here do not result in pulses of elements, such as As, Se, Mo, and V, being released.

4. Conclusions

Hydrothermal treatment of bauxite residue slurry resulted in a decrease in pH, EC, and total alkalinity, as well as decreases in concentrations of Al, K, Na, and S, in the liquid component of the slurry; these decreases were inversely related to treatment temperature and time. Conversion of gibbsite to boehmite, and the dilution from water released to solution during this reaction, appeared to be the main mechanism responsible for decreases in pH, EC, TA, and element concentrations. Dissolution and precipitation kinetics were adequately described by first-order kinetics. Extrapolation of reaction rates and activation energies from hydrothermal conditions imposed in this study to standard temperature and pressure indicated that calcite, goethite, hematite, and boehmite precipitation could be expected to occur within a few centuries; but that dissolution of gibbsite and precipitation of sodalite would take substantially longer. Auto-attenuation of alkalinity and salinity in bauxite residue under closed conditions is theoretically possible according to the reaction kinetics determined in this study, although it would occur over lengthy timescales. Based on the observed mineral transformations, soil developing from bauxite residue is likely to have characteristics associated with Nitisols or Ferralsols. Extrapolation of reaction kinetics observed under hydrothermal conditions to standard temperatures and pressures may be invalid because of different reaction mechanisms being favoured at different temperatures and pressures. Temperature is likely to be more important than pressure in this regard. Validation of observed mineral behaviour in this study against field weathered residues is required to determine if mineral transformations observed under hydrothermal conditions provide an accurate representation of weathering reactions under field conditions.

Acknowledgments: The authors thank Stephen Leavy, Tina Matheson, and Alan Jones (Alcoa of Australia Limited) for assistance with the stainless steel reaction vessel experiment; Michael Smirk (University of Western Australia) and Bree Morgan (Monash University) for assistance with the fused silica capillary reaction vessel experiment; and Ian Madsen (CSIRO) and Matthew Rowles (Curtin University) for assistance with Rietveld refinement of X-ray diffractograms. Part of this research was undertaken on the Powder Diffraction beamline (10BM1) at the Australian Synchrotron, Victoria, Australia, and the authors would like to thank Kia Wallwork for assistance with the fused silica capillary reaction vessel experiment. This research was supported by an Australian Postgraduate Award, and funding from the Minerals and Energy Research Institute of Western Australia, Alcoa of Australia Ltd., and BHP Billiton Worsley Alumina Pty Ltd.

Author Contributions: T.S. and M.F. conceived the experiments; T.S. designed experiments with contributions from M.F. and R.G.; T.S. performed experiments, data analysis and interpretation, and wrote the paper.

Conflicts of Interest: The authors declare no conflict of interest. This study was supported by funding from Alcoa of Australia Ltd. and BHP Billiton Worsley Alumina Ltd.; however, funders had no role in the design of the study or interpretation of results.

References

1. Mudd, G.M.; Boger, D.V. The ever growing case for paste and thickened tailings—Towards more sustainable mine waste management. *J. Aust. Inst. Min. Metall.* **2013**, *2*, 56–59.
2. Uzarowicz, L.; Skiba, A. Technogenic soils developed on mine spoils containing iron sulphides: Mineral transformations as an indicator of pedogenesis. *Geoderma* **2011**, *163*, 95–108. [CrossRef]
3. Santini, T.C.; Fey, M.V. Spontaneous vegetation encroachment upon bauxite residue (red mud) as an indication and facilitator of *in situ* remediation processes. *Environ. Sci. Technol.* **2013**, *47*, 12089–12096. [CrossRef] [PubMed]
4. Hayes, S.M.; Root, R.A.; Perdrial, N.; Maier, R.M.; Chorover, J. Surficial weathering of iron sulfide mine tailings under semi-arid climate. *Geochim. Cosmochim. Acta* **2014**, *141*, 240–257. [CrossRef] [PubMed]
5. Wilkinson, B.H.; McElroy, B.J. The impact of humans on continental erosion and sedimentation. *Geol. Soc. Am. Bull.* **2007**, *119*, 140–156. [CrossRef]
6. Bardossy, G.; Aleva, G.J.J. *Lateritic Bauxites*; Elsevier: Amsterdam, The Netherlands, 1990.
7. Power, G.; Grafe, M.; Klauber, C. Bauxite residue issues: I. Current management, disposal and storage practices. *Hydrometallurgy* **2011**, *108*, 33–45. [CrossRef]
8. Santini, T.C.; Fey, M.V. Fly ash as a permeable cap for tailings management: Pedogenesis in bauxite residue tailings. *J. Soils Sediments* **2015**, *15*, 552–564. [CrossRef]
9. Santini, T.C.; Fey, M.V. Assessment of Technosol formation and *in situ* remediation in capped alkaline tailings. *Catena* **2015**. in review.
10. Henin, S.; Pedro, G. The laboratory weathering of rocks. In *Experimental Pedology*; Hallsworth, E.G., Crawford, D.V., Eds.; Butterworths: London, UK, 1965; pp. 15–22.
11. Williams, C.; Yaalon, D.H. An experimental investigation of reddening in dune sand. *Geoderma* **1977**, *17*, 181–191. [CrossRef]
12. Singleton, G.A.; Lavkulich, L.M. Adaptation of the Soxhlet extractor for pedologic studies. *Soil Sci. Soc. Am. J.* **1978**, *42*, 984–986. [CrossRef]
13. Sullivan, P.J.; Sobek, A.A. Laboratory weathering studies of coal refuse. *Environ. Geochem. Health* **1982**, *4*, 9–16. [CrossRef]
14. Hodson, M.E. Experimental evidence for mobility of Zr and other trace elements in soils. *Geochim. Cosmochim. Acta* **2002**, *66*, 819–828. [CrossRef]
15. Taylor, S.; Pearson, N. *Properties of Bayer Process Solids from Alcoa WA Refineries and Their Component Minerals*; Alcoa of Australia Ltd.: Perth, Australia, 2001; p. 86.
16. Hurst, V.J.; Kunkle, A.C. Dehydroxlyation, rehydroxylation, and stability of kaolinite. *Clays Clay Miner.* **1985**, *33*, 1–14. [CrossRef]
17. Imasuen, O.I.; Tazaki, K.; Fyfe, W.S.; Kohyama, N. Experimental transformation of kaolinite to smectite. *Appl. Clay Sci.* **1989**, *4*, 27–41. [CrossRef]
18. Mosser-Ruck, R.; Cathelineau, M. Experimental transformation of Na, Ca smectite under basic conditions at 150 °C. *Appl. Clay Sci.* **2004**, *26*, 259–273. [CrossRef]
19. Su, C.; Harsh, J.B. Dissolution of allophane as a thermodynamically unstable solid in the presence of boehmite at elevated temperatures and equilibrium vapor pressures. *Soil Sci.* **1998**, *163*, 299–312. [CrossRef]
20. Nikraz, H.R.; Bodley, A.J.; Cooling, D.J.; Kong, P.Y.L.; Soomro, M. Comparison of physical properties between treated and untreated bauxite residue mud. *J. Mater. Civ. Eng.* **2007**, *19*, 2–9. [CrossRef]
21. Wefers, K.; Misra, C. *Oxides and Hydroxides of Aluminium*; Alcoa Technical Paper No. 19 (revised); Aluminium Company of America: Pittsburgh, PA, USA, 1987; p. 92.
22. Murray, J.; Kirwan, L.; Loan, M.; Hodnett, B.K. *In situ* synchrotron diffraction study of the hydrothermal transformation of goethite to hematite in sodium aluminate solutions. *Hydrometallurgy* **2009**, *95*, 239–246. [CrossRef]
23. Whittington, B.I. The chemistry of CaO and $Ca(OH)_2$ relating to the Bayer process. *Hydrometallurgy* **1996**, *43*, 13–35. [CrossRef]

24. Smith, P. The processing of high silica bauxites—Review of existing and potential processes. *Hydrometallurgy* **2009**, *98*, 162–176. [CrossRef]
25. Thompson, T.L.; Hossner, L.R.; Wilding, L.P. Micromorphology of calcium carbonate in bauxite processing waste. *Geoderma* **1991**, *48*, 31–42. [CrossRef]
26. Menzies, N.W.; Fulton, I.M.; Kopitke, R.A.; Kopittke, P.M. Fresh water leaching of alkaline bauxite residue after sea water neutralization. *J. Environ. Qual.* **2009**, *38*, 2050–2057. [CrossRef] [PubMed]
27. Hawkins, D.B.; Roy, R. Experimental hydrothermal studies on rock alteration and clay mineral formation. *Geochim. Cosmochim. Acta* **1963**, *27*, 1047–1054. [CrossRef]
28. Knauss, K.G.; Wolery, T.J. Muscovite dissolution kinetics as a function of pH and time at 70 °C. *Geochim. Cosmochim. Acta* **1989**, *53*, 1493–1501. [CrossRef]
29. Huang, W.L. Stability and kinetics of kaolinite to boehmite conversion under hydrothermal conditions. *Chem. Geol.* **1993**, *105*, 197–214. [CrossRef]
30. Frogner, P.; Schweda, P. Hornblende dissolution kinetics at 25 °C. *Chem. Geol.* **1998**, *151*, 169–179. [CrossRef]
31. Kuwahara, Y. *In situ* observations of muscovite dissolution under alkaline conditions at 25–50 °C by AFM with an air/fluid heater system. *Am. Mineral.* **2008**, *93*, 1028–1033. [CrossRef]
32. Oelkers, E.H.; Schott, J.; Gauthier, J.M.; Herrero-Roncal, T. An experimental study of the dissolution mechanism and rates of muscovite. *Geochim. Cosmochim. Acta* **2008**, *72*, 4948–4961. [CrossRef]
33. Gudbrandsson, S.; Wolff-Boenisch, D.; Gislason, S.R.; Oelkers, E.H. An experimental study of crystalline basalt dissolution from $2 \leq pH \leq 11$ and temperatures from 5 to 75 °C. *Geochim. Cosmochim. Acta* **2011**, *75*, 5496–5509. [CrossRef]
34. Saldi, G.D.; Schott, J.; Pokrovsky, O.S.; Gautier, Q.; Oelkers, E.H. An experimental study of magnesite precipitation rates at neutral to alkaline conditions and 100–200 °C as a function of pH, aqueous solution composition and chemical affinity. *Geochim. Cosmochim. Acta* **2012**, *83*, 93–109. [CrossRef]
35. Heimann, R.B.; Vandergraaf, T.T. Cubic zirconia as a candidate waste form for actinides: Dissolution studies. *J. Mater. Sci. Lett.* **1988**, *7*, 583–586. [CrossRef]
36. Coelho, A. *TOPAS-Academic v 4.1*, Coelho Software: Brisbane, Australia, 2007.
37. Vogel, A.I. *A Textbook of Quantitative Inorganic Analysis*; Longman Group: New York, NY, USA, 1978, p. 925.
38. GBC Scientific Equipment Pty. Ltd. *Traces: X-ray Diffraction Screen Processing Software and Accessories, v 6.7.20*, GBC Scientific Equipment: Hampshire, NH, USA, 2006.
39. Chester, R.; Jones, F.; Loan, M.; Oliviera, A.; Richmond, W.R. The dissolution behaviour of titanium oxide phases in synthetic Bayer liquors at 90 °C. *Hydrometallurgy* **2009**, *96*, 215–222. [CrossRef]
40. Norby, P. Hydrothermal conversion of zeolites: An *in situ* synchrotron X-ray powder diffraction study. *J. Am. Chem. Soc.* **1997**, *119*, 5215–5221. [CrossRef]
41. Sharp, J.H.; Hancock, J.D. Method of comparing solid state kinetic data and its application to the decomposition of kaolinite, brucite, and $BaCO_3$. *J. Am. Ceram. Soc.* **1972**, *55*, 74–77.
42. Avrami, M. Kinetics of phase change. I. General theory. *J. Chem. Phys.* **1939**, *7*, 1103–1113. [CrossRef]
43. Avrami, M. Kinetics of phase change. II. Transformation-time relations for random distribution of nuclei. *J. Chem. Phys.* **1940**, *8*, 212–225. [CrossRef]
44. Avrami, M. Granulation, phase change, and microstructure kinetics of phase change. III. *J. Chem. Phys.* **1941**, *9*, 177–185. [CrossRef]
45. Erofe'ev, B.V. Generalised equation of chemical kinetics and its application in reactions involving solids. *C. R. (Dokl.) Acad. Sci. URSS* **1946**, *52*, 511–514.
46. Tsuchida, T. Hydrothermal synthesis of submicrometer crystals of boehmite. *J. Eur. Ceram. Soc.* **2000**, *20*, 1759–1764. [CrossRef]
47. Gong, X.; Nie, Z.; Qian, M.; Liu, J.; Pederson, L.A.; Hobbs, D.T.; McDuffle, N.G. Gibbsite to boehmite transformation in strongly caustic and nitrate environments. *Ind. Eng. Chem. Res.* **2003**, *42*, 2163–2170. [CrossRef]
48. Mehta, S.K.; Kalsotra, A. Kinetics and hydrothermal transformation of gibbsite. *J. Ther. Anal. Calorim.* **1991**, *367*, 267–275. [CrossRef]
49. Dash, B.; Tripathy, B.C.; Bhattacharya, I.N.; Das, S.C.; Mishra, C.R.; Pani, B.S. Effect of temperature and alumina/caustic ratio on precipitation of boehmite in synthetic sodium aluminate liquor. *Hydrometallurgy* **2007**, *88*, 121–126. [CrossRef]

50. Dash, B.; Tripathy, B.C.; Bhattacharya, I.N.; Das, S.C.; Mishra, C.R.; Mishra, B.K. Precipitation of boehmite in sodium aluminate liquor. *Hydrometallurgy* **2009**, *95*, 297–301. [CrossRef]
51. Grenman, H.; Salmi, T.; Murzin, D.Y.; Addai-Mensah, J. The dissolution kinetics of gibbsite in sodium hydroxide at ambient pressure. *Ind. Eng. Chem. Res.* **2010**, *49*, 2600–2607. [CrossRef]
52. Adamson, A.N.; Bloore, E.J.; Carr, A.R. *Extractive Metallurgy of Aluminium—Volume 1*; Interscience: New York, NY, USA, 1963; pp. 23–58.
53. Cresswell, P.J. Factors affecting desilication of Bayer process liquors. In Proceedings of the Chemeca 84: The 12th Australian Chemical Engineering Conference, Melbourne, Australia, 26–29 August 1984.
54. Barnes, M.C.; Addai-Mensah, J.; Gerson, A. A methodology for quantifying sodalite and cancrinite phase mixtures and the kinetics of the sodalite to cancrinite phase transformation. *Microporous Mesoporous Mater.* **1999**, *31*, 303–319. [CrossRef]
55. Francis, R.J.; O'Brien, S.; Fogg, A.M.; Halasyamani, P.S.; O'Hare, D.; Loiseau, T.; Ferey, G. Time-resolved *in situ* energy and angular dispersive X-ray diffraction studies of the formation of the microporous gallophosphate ULM-5 under hydrothermal conditions. *J. Am. Chem. Soc.* **1999**, *121*, 1002–1015. [CrossRef]
56. Lasaga, A.C. *Kinetic Theory in the Earth Sciences*; Princeton University Press: Princeton, NJ, USA, 1998.
57. IUSS Working Group WRB. *World Reference Base for Soil Resources 2014*; World Soil Resources Reports No. 106; Food and Agriculture Organisation of the United Nations: Rome, Italy, 2014.

metals

MDPI

Article

"High-Throughput" Evaluation of Polymer-Supported Triazolic Appendages for Metallic Cations Extraction

Riadh Slimi and Christian Girard *

Unité de Technologies Chimique et Biologiques pour la Santé (UTCBS), UMR 8258 CNRS, U 1022 Inserm, Ecole Nationale Supérieure de Chimie de Paris (Chimie ParisTech)/PSL Research University, 11 rue Pierre et Marie Curie, 75005 Paris, France; riadhslimi82@yahoo.fr

* Author to whom correspondence should be addressed; christian.girard@chimie-paristech.fr;
  Tel.: +33-144-276-748; Fax: +33-144-276-496.

Academic Editor: Hugo F. Lopez

Received: 25 February 2015; Accepted: 7 March 2015; Published: 12 March 2015

Abstract: The aim of this work was to find and use a low-cost high-throughput method for a quick primary evaluation of several metal extraction by substituted piperazines appendages as chelatants grafted onto Merrifield polymer using click-chemistry by the copper (I)-catalyzed Huisgen's reaction (CuAAC) The polymers were tested for their efficiency to remove various metal ions from neutral aqueous solutions (13 cations studied: Li^+, Na^+, K^+, Mn^{2+}, Fe^{3+}, Co^{2+}, Ni^{2+}, Cu^{2+}, Cd^{2+}, Ba^{2+}, Ce^{2+}, Hg^+ and Pb^{2+}) using the simple conductimetric measurement method. The polymers were found to extract all metals with low efficiencies ($\leq$40%), except for Fe^{3+} and Hg^+, and sometimes Pb^{2+}. Some polymers exhibited a selectivity for K^+, Cd^{2+} and Ba^{2+}, with good efficiencies. The values obtained here using less polymer, and a faster method, are in fair correspondence (average difference $\pm$16%) with another published evaluation by atomic absorption spectroscopy (AAS).

Keywords: polymer functionalization; click chemistry; CuAAC; complexants; metallic cations; complexation; depollution; catalysis

1. Introduction

Water pollution by metallic ions and other pollutants is becoming an increasing concern nowadays. This modification of the water, in all his reservoirs, is mainly due to Human activities with uncontrolled rejects of such pollutants. The pollution has a strong impact onto the global ecosystem as well as drinkable water sources. There is thus a strong need for methods to analyze traces and to remove the pollutants from water. Usual methods for removing metallic salts from water range from distillation to the use of engineered materials such as zeolites, polymers, membranes, *etc.* The long time known ion-exchange resins can be used for this purpose [1–3]. Usually, the polymers are engineered is such a way that their nature can be hydrophilic [4–7] or phobic [8,9], to meet the requirements for their use [10–14]. Many polymers have been designed in order to include chelatants to fix metal ions to be used in applications such as purification, depollution or catalysis [15–20].

Due to our interest into metal chelation, supported catalysis and Huisgen's reaction, we became interested into the preparation of polymers based on this approach [21–23]. We thus started to use the "click-chemistry" concept for polymer functionalization and especially copper (I)-catalyzed Huisgen's cycloaddition ("copper (I)-catalyzed azide/alkyne cycloaddition" or CuAAC) [24–29]. The use of CuAAC has the advantage to give a quick access of controlled substitutions onto the polymer by the use of its azided version and alkynes with various substituents [30–35]. This CuAAC is linking the azided polymer and the substituents bearing alkyne by forming the 1,4-triazole linkage. All the introduced substituents and the triazole can be implicated into chelation through a "triazole design", or a "pendant design", or both parts implicated into an "integrated design". The chelation can be a mono- or multi-dentate mode due to the vicinity of other chelating entities and the flexible structure of the polymer chains. (Figure 1) [34,35].

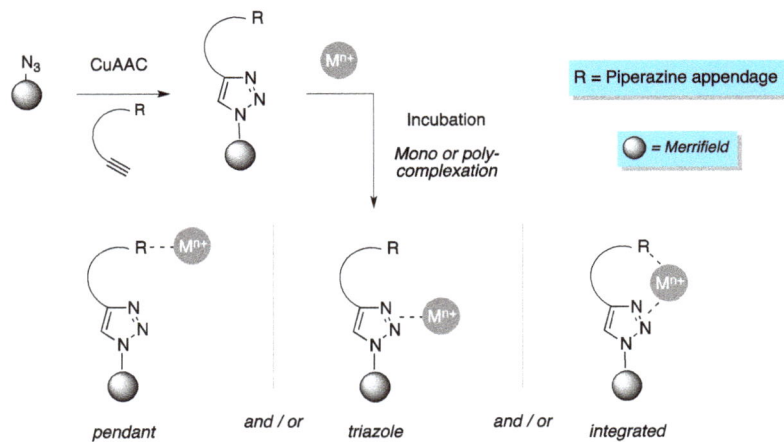

Figure 1. Functionalization of azido-Merrifield polymer by complexing appendages using copper (I)-catalyzed azide/alkyne cycloaddition (CuAAC) for a 1,2,3-triazole linkages and possible chelation modes.

The goal of this work was to try to find a faster method, using less polymer, for the evaluation of several metal cations complexation evaluation. We present in this communication the use of the less sensitive conductimetric method for the study of piperazine-triazole-substituted poly(styrenes). The polymers were tested for their ability to extract metal cations salts (Li^+, Na^+, K^+, Mn^{2+}, Fe^{3+}, Co^{2+}, Ni^{2+}, Cu^{2+}, Cd^{2+}, Ba^{2+}, Ce^{3+}, Hg^+ and Pb^{2+}) from neutral aqueous solutions. The results were found to be fair enough to be used for a primary evaluation at a "high-throughput" level when compared to our previous atomic absorption spectroscopy measurements (within ±16% average difference) [36].

2. Experimental Section

2.1. N-Substituted Piperazine Propargylcarbamates and Polymers

Poly(azidomethylstyrene) was prepared from Merrifield polymer as already reported. N'-propargylcarbamates of N-substituted piperazine, and the corresponding polymers containing triazole-linked piperazines preparations were described in a another publication as depicted in Figure 2 [36,37]. Typical procedures are indicated below.

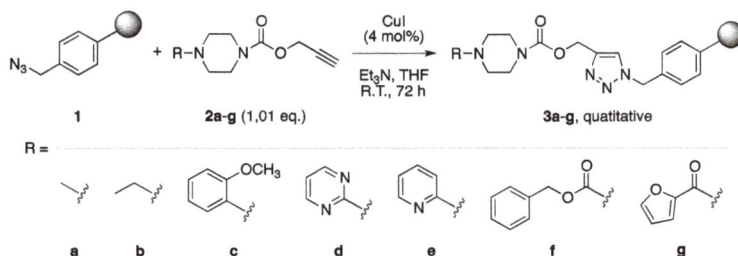

Figure 2. Synthesis of the polymers by CuAAC procedure.

2.1.1. General Procedure for the Synthesis of Propargylcarbamates Derivatives of Piperazines **2a–2g**

To a solution of the required N-substituted piperazine **1a** to **1g** (12.0 mmol) in acetonitrile (45 mL) was added Na_2CO_3 (1.27 g, 12.0 mmol, 1 eq.). Propargyl chloroformate (1.42 g, 1.17 mL, 12.0 mmol, 1 eq.) was then added dropwise. The reaction mixture was stirred for 48 h at room temperature and then filtered and evaporated under vacuum. The resulting carbamates **2a–2g** were sufficiently pure to be used without further purification.

2.1.2. General procedure for the Synthesis of Polymers **3a–3g**

Coupling reactions onto the polymer using CuAAC were conducted accordingly to the general procedure indicated below in round bottom flasks equipped with a reflux condenser. To a suspension of 3.00 g of azidomethyl polystyrene **A** (1.82 mmol N_3 g^{-1}, 5.46 mmol N_3) in THF (60 mL) was added 6.30 mmol (1.15 eq.) of the alkyne (**2a–2g**), 9.00 mL of triethylamine (6.75 g, 66,7 mmol 12.2 eq.) and 2.40 mg of copper (I) iodide (12.6 μmol, 4 mol%). The suspension was slowly stirred at room temperature 72 h. After this time, the complete disappearance of the IR band of the azide of the polymer **A** (2103 cm^{-1}) was observed. The resulting polymer was filtered on sintered glass and washed sequentially with CH_2Cl_2, pyridine, and MeOH (60 mL each), the sequential washings being repeated two other times. The resulting polymers **3a–3g** were finally dried overnight in an oven at 50 °C.

2.2. Conductimetric Quick Primary Evaluation of Metals Complexation

2.2.1. Extraction

Neutral aqueous solutions of metal salts were prepared with 50 mg of LiCl (1.179 mmol), NaCl (0.856 mmol), KCl (0.671 mmol), $MnCl_2 \cdot 4H_2O$ (0.253 mmol), $Fe(NO_3)_3 \cdot 9H_2O$ (0.124 mmol), $CoCl_2 \cdot 6H_2O$ (0.210 mmol), $NiCl_2 \cdot 6H_2O$ (0.210 mmol), $CuCl_2 \cdot 2H_2O$ (0.293 mmol), $CdCl_2 \cdot H_2O$ (0.248 mmol), $BaCl_2 \cdot 2H_2O$ (0.205 mmol), $CeCl_3 \cdot 7H_2O$ (0.134 mmol), $HgNO_3$ (0.190 mmol) and $Pb(NO_3)_2$ (0.151 mmol) in 1 L of distilled water.

Aliquots of the each polymers (100 mg; **3a**: 0.137 mmol, **3b**: 0.134 mmol, **3c**: 0.121 mmol, **3d**: 0.126 mmol, **3e**: 0.126 mmol, **3f**: 0.117 mmol and **3g**: 0.123 mmol piperazine) were incubated in triplicate with 20 mL (50 mg L^{-1}, equals to 1 mg of salt) of each metal ion solution (23.6 μmol Li^+, 17.1 μmol Na^+, 13.4 μmol K^+, 5.05 μmol Mn^{2+}, 2.47 μmol Fe^{3+}, 4.20 μmol Co^{2+}, 4.21 μmol Ni^{2+}, 5.87 μmol Cu^{2+}, 4.97 μmol Cd^{2+}, 4.09 μmol Ba^{2+}, 2.68 μmol Ce^{3+}, 3.81 μmol Hg^+ and 3.02 μmol Pb^{2+}) at 25 °C for 24 h. The suspension was then filtrated on filter paper, which was previously washed with distilled water until no difference in conductimetry was observed between the washes and water.

2.2.2. Conductimetric Measurements

Evaluation of the chelated metal was done by conductimetric measurements on a bench Conductivity/TDS/°C Meter CO 3000 L, pHenomenal® by VWR (Paris, France) on the filtrate by comparison with the conduction of the initial solution of the metal salt. The results, average of three experiments, were expressed as percentages of extraction of the metal (Figure 3 (below) and Figure 4, Section 3.2).

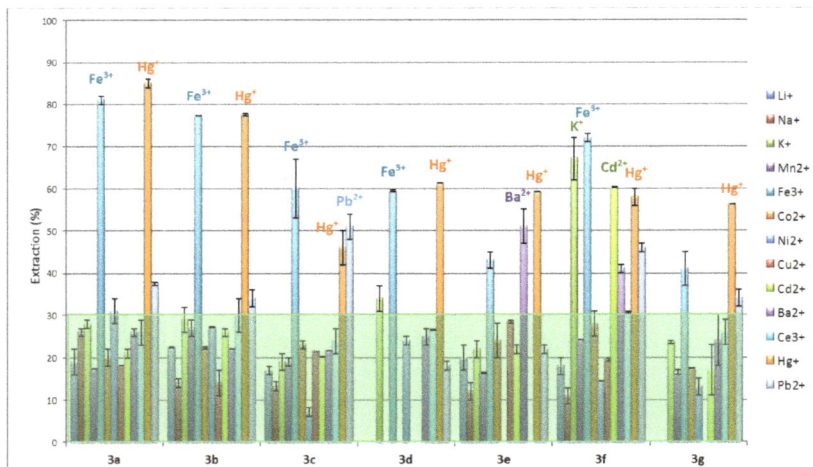

Figure 3. Extraction efficiencies of metal cations by the triazolic piperazine polymers **3a–3g** (cations extracted at levels $\geq$50% are indicated).

3. Results and Discussion

3.1. Polymers Preparation

The polymers needed for this study have been prepared elsewhere [36]. They were easily accessible, in different substitution motifs, using poly(azidomethylstyrene) (**1**) and various *N*-substituted-piperazine-*N*′-propargylcarbamates (**2a–2g**). A CuAAC procedure afforded a quantitative reaction to form the triazole linkage onto the polymers **3a–3g** (Figure 2, Section 2.1) [36,37].

3.2. Extraction Results

After 24 h incubation on a 100 mg scale of the polymers in 20 mL of 50 mg L^{-1} solutions of the salts (1 mg of salt, see Section 2.2.1 for details), the percentages of extraction for each metal were calculated by conductimetric differences between the initial and final solutions, each experiment having been carried out in triplicate. The results for each polymer as a function of the metallic cations are presented in Figure 3 (Section 2.2.2).

When looking at the whole results, we can observe than most of the cations were poorly extracted, at 30% or below, but exceptions. Since the polymeric structure differs only by the R substituent on the nitrogen (Figure 2, Section 2.1), an analysis has been done to understand the influence of the substituent's nature on the extraction.

In the alkyl series, polymer **3a**, R = Me, was a very good extractant for Fe^{3+} (81% $\pm$ 1%) and Hg^+ (85% $\pm$ 1%), less for Pb^{2+} (37.5% $\pm$ 0.3%), and not very good for other metals ($\leq$30%). When changing R for Et, in polymer **3b**, the same extractive properties were found, only at a little lower level, for Fe^{3+} (77.35% $\pm$ 0.05%) and Hg^+ (77.5% $\pm$ 0.3%), less for Pb^{2+} (34% $\pm$ 2%), the other metals being around or below 30%, as for the previous one.

When entering the aryl series, an obvious decrease in the extraction potentials was observed, alongside some other interesting behaviors. For the uncited cations in the following paragraph, they were extracted $\leq$30%. For R = 4-methoxyphenyl in **3c**, Fe^{3+} (60% $\pm$ 7%) and Hg^+ (46% $\pm$ 4%) were less extracted. However, Pb^{2+} (51% $\pm$ 3%) was better extracted. By replacing the phenyl group by a nitrogen containing aromatic one, like in the 2-pyrimidyl (**3d**), Fe^{3+} (59.4% $\pm$ 0.2%) was extracted at the same level, lower than the alkyl family (**3a**, **3b**). Hg^+ (61.25% $\pm$ 0.05%) was a little less extracted than by *N*-alkyl substituted polymers **3a** a–d **3b** but better than the 4-methoxyphenyl polymer **3c**. In the case of polymer **3d**, Pb^{2+} (18% ± 1%) extraction dropped, but K^+ (34% ± 3%) uptake increased. It is interesting to point out that some metals were not at all extracted by **3d**: Li^+, Na^+, Mn^{2+}, Co^{2+}, Cu^{2+} and Cd^{2+}. When the substituent borne by the nitrogen was 2-pyridyl (**3e**), the extraction level dropped for Fe^{3+} (43% ± 2%), but stayed similar for Hg^+ (59.19% ± 0.04%) and Pb^{2+} (22% ± 1%). Interestingly, another metal was extracted, Ba^{2+} (51% ± 4%), while Ni^{2+} and Ce^{3+} were not extracted.

By changing the nature of the substituent by introducing a carbamate function in **3f** (R = C(O)OCH$_2$Ph), the polymer became less selective. This polymer (**3f**) still extracted Fe^{3+} (72% ± 1%), but at the same level as the *N*-alkyl substituted ones (**3a**, **3b**), it extracted Pb^{2+} (46% ± 1%) at the level of the 4-methoxyphenyl polymer **3c**, and stayed in the same range as the previous one for Hg^+ (58% ± 2%) and Ba^{2+} (41% ± 1%). Furthermore, the polymer **3f** was a good extractant for K^+ (67% = 5%) and Cd^{2+} (60.3% ± 0.1%).

Finally, when the piperazine nitrogen onto the polymer was an amide of 2-furoic acid (**3g**), The extraction behavior came back to the triade Fe^{3+} (41% ± 4%), Hg^+ (56.23% ± 0.04%), and Pb^{2+} (34% ± 2%). The levels were similar as the ones of **3e**. Once again some cations were not extracted like Li^+, Na^+, and Cu^{2+}.

From this first analysis, based only on the difference of the R group, it seems that the presence of an alkyl substituent (**3a**, R = Me, **3b**, R = Et) is giving the best extraction levels for Fe^{3+} and Hg^+, probably due to the increased electronegativity of the amine. Inductive and steric hindrance effects can explain the differences between **3a** and **3b**. The introduction of a 2-methoxyphenyl group on the nitrogen (**3c**) changes it to an aniline, less basic, which is extracting less Fe^{3+} and Hg^+. However, the presence of the methoxy group in *ortho* position seems to help in Pb^{2+} extraction. It is possible that this oxygenated group is implicated into the chelation of this metal. Replacement of this aromatic by a 2-pyrimidyl (**3d**) and 2-pyridyl (**3e**), less and more basic respectively when compared one to the other, still gives polymers capable of extracting Fe^{3+} and Hg^+, with similar levels as **3c**. Special features of these polymers are higher extraction of K^+ for **3d** and Ba^{2+} for **3e**, as well as exclusion of some cations: Li^+, Na^+, Mn^{2+}, Co^{2+}, Cu^{2+} and Cd^{2+} for **3d**, and Ni^{2+} and Ce^{3+} for **3e**.

The electronegativities and chelating capabilities of these amino-R groups are however difficult to put in relation with their extracting properties. The most puzzling effect is the presence of a benzylcarbamate onto the nitrogen of the polymer (**3f**). This group is totally modifying the properties of the polymer. In this case, more metallic ions are extracted, with the classical Fe^{3+} and Hg^+. This includes K^+, Cd^{2+}, Ba^{2+}, and Pb^{2+}. This may suggest another chelation mode introduced by the presence of the carbamate. Finally, the presence of a derivative of furoic acid as an amide on the nitrogen (**3g**) do not seems to helps since extraction levels are going down with extraction of the usual Fe^{3+}, Hg^+, and Pb^{2+}, and exclusion of Li^+, Na^+, Cu^{2+}.

The electronegativities of the substituted nitrogen of the piperazine can in part explain some of the relative extraction efficiencies. However, it is difficult to draw a clear conclusion. We have also tried to rationalize the interactions between the polymers and the metal ions based on their electropositivities, ionic radii and water solvatation. Once again, no clear link can be drawn about the extraction efficiencies based on the metal cation properties. The only difference that can explain, once again in part, the preference of the polymers for Fe^{3+} and Hg^+, and in some cases Pb^{2+}, is the counter-ion of the salt used for the study. All metallic salts were chlorides except for Fe^{3+}, Hg^+, and Pb^{2+}, which were used as their nitrates.

The final analysis we have tried to make is to try to find out the chelation mode from the piperazine-triazole in relation with the metal cations. For the best extraction results, over 40%, it was not possible to make a clear discrimination between the three modes (pendant, triazole and integrated). The ratios piperazine-triazole:metal cation varied from 13:1 (**3f** and K^+, 67% ± 5%) to 118:1 (**3e** and Fe^{3+}, 43% ± 2%), since the chelating moieties were in large excess. This cannot gives a clear hit on the chelation mode, which can be of polydentate type or simply a statistical repartition on the chelation sites, without knowledge of the chelation type.

All the results and analyses cannot clearly identify the discrete complexation behavior of the metal by the polymers at the solid/liquid interface. However, we were able to obtain better results than with our first series of triazolic polymers based on propargyl amides and propiolic anilides onto poly(styrene) [38].

Figure 4 presents our results by cation absorption to help to find the best extractant. By drawing a cut-off for selection at 40% extraction level, it is clear that none of the polymers is very efficient for the removal of Li^+, Na^+, K^+, Co^{2+}, Ni^{2+}, Cu^{2+} and Ce^{3+}. For K^+, polymer **3f** is the best with 3.51 mg K^+ g^{-1}.

- In the case of Fe^{3+}, as said before, all polymers are complexing this ion. The best results are 1.12 mg Fe^{3+} g^{-1} **3a**, 1.07 mg Fe^{3+} g^{-1} **3b** and 1.00 mg Fe^{3+} g^{-1} **3f**; followed by 0.83 mg Fe^{3+} g^{-1} **3c** and 0.82 mg Fe^{3+} g^{-1} **3d**, while **3e** and **3g** were borderline.
- Cd^{2+} and Ba^{2+} are more efficiently removed by **3f** (3.37 mg Cd^{2+} g^{-1}, 2.31 mg Ba^{2+} g^{-1}) and **3e** (2.87 mg Ba^{2+} g^{-1}).
- For Hg^+, as for Fe^{3+}, all polymers can be used. In order, by sorption capacities, are **3a** (6.49 mg Hg^+ g^{-1}), **3b** (5.92 mg Hg^+ g^{-1}), **3d** (4.68 mg Hg^+ g^{-1}), **3e** (4.52 mg Hg^+ g^{-1}), **3f** (4.43 mg Hg^+ g^{-1}), **3g** (4.30 mg Hg^+ g^{-1}) and **3c** (3.51 mg Hg^+ g^{-1}).
- Finally, in the case of Pb^{2+}, polymers **3c** (3.19 mg Pb^{2+} g^{-1}) and **3f** (2.88 mg Pb^{2+} g^{-1}) are the more efficient.

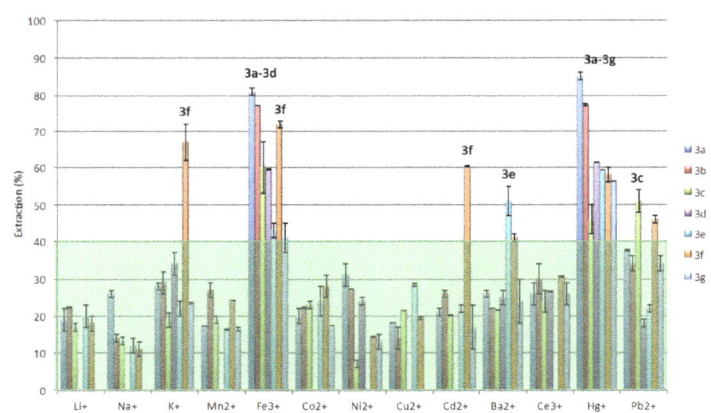

Figure 4. Extraction efficiencies of metal cations by the triazolic piperazine polymers **3a–3g** (polymers extracting at levels ≥50% are indicated).

4. Conclusions

In this work, we studied chemical grafting of piperazine chelating units onto commercial poly[styrene] (Merrifield resin) using CuAAC procedure between the azided polymer and selected piperazine-*N*-propargylcarbamates. The synthesized polymers were characterized by FTIR. They were then tested for their efficiency to extract metallic ions from aqueous solution (Li^+, Na^+, K^+, Mn^{2+}, Fe^{3+}, Co^{2+}, Ni^{2+}, Cu^{2+}, Cd^{2+}, Ba^{2+}, Ce^{3+}, Hg^+ and Pb^{2+}). All polymers were found to extract most of the ions at low level (≤40%), with the exception of Fe^{3+}, Hg^+ and Pb^{2+}. Some polymers showed selectivity for K^+, Cd^{2+} and Ba^{2+}. Extraction efficiencies reached up 85%, with the highest sorption capacity at 6.49 mg Hg^+ g^{-1} of polymer (**3a**).

The conductimetric method used, even having a less precise reputation, was good enough to have a quick evaluation of several cations removal. This method has been selected both for its lower cost in appartatus when compared to AAS and ICP methods. It is also faster and easier to do the measurements in order to speed up the process to find the best and highly selective extractant for a range of engineered polymers.

Even if no clear interpretation can be done with the results for the interfacial chelation process, the good extraction properties encourage us to continue polymers modifications using CuAAC in order to find new polymeric complexants for depollution and catalytic applications. Further studies will be reported in due course.

Acknowledgments: This project was financed by Tunisian 05/UR/12-05 and French UMR 8258 CNRS/U1022 INSERM grants. R. S. is grateful to the Tunisian Ministry of Research and Education for fellowship.

Author Contributions: This is a part of the Ph.D. thesis of R. Slimi under the supervision of C. Girard. Experiments were conducted by R. Slimi. R. Slimi and C. Girard analyzed the data.

Conflicts of Interest: The authors declare no conflict of interest.

References

1. Rifi, E.H.; Leroy, M.J.F.; Brunette, J.P. Schloesser-Becker, C. Extraction of copper, cadmium and related metals with poly(sodium acrylate–acrylic acid) hydrogels. *Solvent Extr. Ion Exch.* **1994**, *12*, 1003–1119. [CrossRef]
2. Hodgkin, J.H.; Eibl, R. Gold extraction with poly(diallylamine) resins. *React. Polym. Ion Exch. Sorb.* **1988**, *9*, 285–291. [CrossRef]
3. Rivas, B.L.; Klahenhoff, D.; Perich, I.M. Kinetics of uranium sorption from acidic sulphate solutions onto crosslinked polyethyleneimine based resins. 10. *Polym. Bull.* **1990**, *23*, 219–223. [CrossRef]
4. Anspach, W.M.; Marinsky, J.A. Complexing of nickel(II) and cobalt(II) by a polymethacrylic acid gel and its linear polyelectrolyte analog. *J. Phys. Chem.* **1975**, *79*, 433–439. [CrossRef]
5. Nichide, H.; Oki, N.; Suchida, E.T. Complexation of poly(acrylic acid)s with uranyl ion. *Eur. Polym. J.* **1982**, *18*, 799–802. [CrossRef]
6. Pollack, G.H. Water, energy and life: Fresh views from the water's edge. *Int. J. Des. Nat. Ecodyn.* **2010**, *5*, 27–29. [CrossRef]
7. Rifi, E.H.; Rastegar, F.; Brunette, J.P. Uptake of cesium, strontium and europium by a poly(sodium acrylate-acrylic acid) hydrogel. *Talanta* **1995**, *42*, 811–816. [CrossRef] [PubMed]
8. Wang, X.; Weiss, R.A. A facile method for preparing sticky, hydrophobic polymer surfaces. *Langmuir* **2012**, *28*, 3298–3305. [CrossRef] [PubMed]
9. Loret, J.F.; Brunette, J.P.; Leroy, I.F.M.; Candau, S.G.; Prevost, M. Liquid-lipophilic gel extraction of precious metals. *Solv. Extr. Ion Exch.* **1988**, *6*, 585–603. [CrossRef]
10. Nghiem, L.D.; Mornane, P.; Potter, I.D.; Pereira, J.M.; Cattrall, R.W.; Kolev, S.D. Extraction and transport of metal ions and small organic compounds using polymer inclusion membranes (PIMs). *J. Membr. Sci.* **2006**, *281*, 7–41. [CrossRef]
11. Peterson, J.; Nghiem, L.D. Selective extraction of cadmium by polymer inclusion membranes containing PVC and Aliquat 336: Role base polymer and extractant. *Int. J. Environ. Technol. Manag.* **2010**, *12*, 359–368. [CrossRef]
12. Kebiche Senhadji, O.; Sahi, S. Kahloul, N.; Tingry, S.; BenAmor, M.; Seta, P. Extraction du Cr(VI) par membrane polymère à inclusion. *Sci. Technol. A* **2008**, *27*, 43–50.
13. Guibaud, G.; Baudu, M.; Dollet, P. Condat, M.L.; Dagot, C. Role of extracellular polymers in cadmium adsorption by activated sludges. *Environ. Technol.* **1999**, *20*, 1045–1054. [CrossRef]
14. Upitis, A.; Peterson, J.; Lukey, C.; Nghiem, L.D. Metallic ion extraction using polymer inclusion membranes (PIMs): optimising physical strength and extraction rate. *Desalin. Water Treat.* **2009**, *6*, 41–47. [CrossRef]
15. Zotti, G.; Zecchin, S.; Schiavon, G.; Berlin, A.; Penso, M. Ionochromic and potentiometric properties of the novel polyconjugated polymer from anodic coupling of 5,5'-Bis(3,4-(ethylenedioxy)thien-2-Yl)-2,2'-Bipyridine. *Chem. Mater.* **1999**, *11*, 3342–3351. [CrossRef]
16. Brembilla, A.; Cuny, J.; Roizard, D.; Lochon, P. Un nouveau polymère catalyseur bifonctionnel: Le polyvinyl-5 (6)benzimidazoleméthanethiol Synthèse et catalyse de l'hydrolyse de l'acétate de *p*-nitrophényle. *Eur. Polym. J.* **1983**, *19*, 729–735. [CrossRef]

17. Rhazi, M.; Desbrières, J.; Tolaimate, A.; Rinaudo, M.; Vottero, P.; Alagui, A.; El Meray, M. Influence of the nature of the metal ions on the complexation with chitosan: Application to the treatment of liquid waste. *Eur. Polym. J.* **2002**, *38*, 1523–1530. [CrossRef]

18. Kozlowski, C.; Walkowiak, W. Applicability of liquid membranes in chromium(VI) transport with amines as ion carriers. *J. Membr. Sci.* **2005**, *266*, 143–150. [CrossRef]

19. Kozlowski, C.; Apostoluk, W.; Walkowiak, W.; Kita, A. Removal of Cr(VI), Zn(II) and Cd(II) ions by transport across polymer inclusion membranes with basic ion carriers. *Physicochem. Probl. Min. Process.* **2002**, *36*, 115–122.

20. Toy, P.H.; Janda, K.D. Soluble polymer-supported organic synthesis. *Acc. Chem. Res.* **2000**, *33*, 546–554. [CrossRef] [PubMed]

21. Girard, C.; Önen, E.; Aufort, M.; Beauvière, S.; Samson, E.; Herscovici, J. Reusable polymer-supported catalyst for the [3+2] Huisgen cycloaddition in automation protocols. *Org. Lett.* **2006**, *8*, 1689–1692. [CrossRef] [PubMed]

22. Jlalia, I.; Meganem, F.; Herscovici, J.; Girard, C. "Flash" solvent-free synthesis of triazoles using a supported catalyst. *Molecules* **2009**, *14*, 528–539. [CrossRef] [PubMed]

23. Jlalia, I.; Beauvineau, C.; Beauvière, S.; Önen, E.; Aufort, M.; Beauvineau, A.; Khaba, E.; Herscovici, J.; Meganem, F.; Girard, C. Automated synthesis of a 96 product-sized library of triazole derivatives using a solid phase supported copper catalyst. *Molecules* **2010**, *15*, 3087–3120. [CrossRef] [PubMed]

24. Li, C.; Finn, M.G. Click chemistry in materials synthesis. II. Acid-swellable crosslinked polymers made by copper-catalyzed azide–alkyne cycloaddition. *J. Polym. Sci. Part A: Polym. Chem.* **2006**, *44*, 5513–5518. [CrossRef]

25. Tasdelen, M.A.; Yilmaz, G.; Iskin, B.; Yagci, Y. Photoinduced free radical promoted copper(I)-catalyzed click chemistry for macromolecular syntheses. *Macromolecules* **2011**, *45*, 56–61. [CrossRef]

26. Liu, Y.; Díaz, D.D.; Accurso, A.A.; Sharpless, K.B.; Fokin, V.V.; Finn, M.G. Click chemistry in materials synthesis. III. Metal-adhesive polymers from Cu(I)-catalyzed azide–alkyne cycloaddition. *J. Polym. Sci. Part A* **2007**, *45*, 5182–5189. [CrossRef]

27. Dag, A.; Durmaz, H.; Demir, E.; Hizal, G.; Tunca, U. Heterograft copolymers via double click reactions using one-pot technique. *J. Polym. Sci. Part A: Polym. Chem.* **2008**, *46*, 6969–6977. [CrossRef]

28. Opsteen, J.A.; Brinkhuis, R.P.; Teeuwen, R.L.M.; Lowik, D.W.P.M.; van Hest, J.C.M. "Clickable" polymersomes. *Chem. Commun.* **2007**, 3136–3138. [CrossRef]

29. Riva, R.; Schmeits, S.; Jérôme, C.; Jérôme, R.; Lecomte, P. Combination of ring-opening polymerization and "click chemistry": Toward functionalization and grafting of poly(ε-caprolactone). *Macromolecules* **2007**, *40*, 796–803. [CrossRef]

30. Löber, S.; Rodriguez-Loaiza, P.; Gmeiner, P. Click linker: Efficient and high-yielding synthesis of a new family of SPOS resins by 1,3-dipolar cycloaddition. *Org. Lett.* **2003**, *5*, 1753–1755. [CrossRef] [PubMed]

31. Iskin, B.; Yilmaz, G.; Yagci, Y. ABC type miktoarm star copolymers through combination of controlled polymerization techniques with thiol-ene and azide-alkyne click reactions. *J. Polym. Sci. Part A: Polym. Chem.* **2011**, *49*, 2417–2422. [CrossRef]

32. Xue, X.; Zhu, J.; Zhang, Z.; Cheng, Z.; Tu, Y.; Zhu, X. Synthesis and characterization of azobenzene-functionalized poly(styrene)-*b*-poly(vinyl acetate) via the combination of RAFT and "click" chemistry. *Polymer* **2010**, *51*, 3083–3090. [CrossRef]

33. Shin, J.-A.; Lim, Y.-G.; Lee, K.-H. Synthesis of polymers including both triazole and tetrazole by click reaction. *Bull. Korean Chem. Soc.* **2011**, *32*, 547–552. [CrossRef]

34. Struthers, H.; Mindt, T.L.; Schibli, R. Metal chelating systems synthesized using the copper(I) catalyzed azide-alkyne cycloaddition. *Dalton Trans.* **2010**, *39*, 675–696. [CrossRef] [PubMed]

35. Urankar, D.; Pinter, B.; Pevec, A.; De Proft, F.; Turel, I.; Košmrlj, J. Click-triazole N_2 coordination to transition-metal ions is assisted by a pendant pyridine substituent. *Inorg. Chem.* **2010**, *49*, 4820–4829. [CrossRef]

36. Slimi, R.; Ben Othman, R.; Sleimi, N.; Ouerghui, A.; Girard, C. Polystyrene-supported triazolic substituted piperazines for metal ions extraction. *React. Funct. Polym.* **2015**. submitted.

37. Arseniyadis, S.; Wagner, A.; Mioskowski, C. A straightforward preparation of amino-polystyrene resin from Merrifield resin. *Tetrahedron Lett* **2002**, *43*, 9717–9719. [CrossRef]
38. Ouerghui, A.; Elamari, H.; Ghammouri, S.; Slimi, R.; Meganem, F.; Girard, C. Polystyrene-supported triazoles for metal ions extraction: Synthesis and evaluation. *React. Funct. Polym.* **2014**, *74*, 37–45. [CrossRef]

Article

Effect of Polymer Addition on the Structure and Hydrogen Evolution Reaction Property of Nanoflower-Like Molybdenum Disulfide

Xianwen Zeng, Lijing Niu, Laizhou Song *, Xiuli Wang, Xuanming Shi and Jiayun Yan

College of Environmental and Chemical Engineering, Yanshan University, Qinhuangdao 066004, China;
ydzxw7903@163.com (X.Z.); nljing0321@163.com (L.N.); xlwang7904@ysu.edu.cn (X.W.);
sxming3231@163.com (X.S.); yanjiayun@163.com (J.Y.)
* Author to whom correspondence should be addressed; songlz@ysu.edu.cn; Tel./Fax: +86-335-8061569.

Academic Editors: Suresh Bhargava, Mark Pownceby and Rahul Ram
Received: 8 September 2015; Accepted: 25 September 2015; Published: 9 October 2015

Abstract: Nano-structured molybdenum disulfide (MoS_2) catalysts have been extensively developed for the hydrogen evolution reaction (HER). Herein, a novel hydrothermal intercalation approach is employed to fabricate nanoflower-like 2H–MoS_2 with the incorporation of three polymers, polyvinylpyrrolidone (PVP), polyvinyl alcohol (PVA), and polyethylenimine (PEI). The as-prepared MoS_2 specimens were characterized by techniques of scanning electron microscope (SEM), transmission electron microscopy (TEM), X-ray diffraction (XRD), together with Raman and Fourier transform infrared spectroscopy (FTIR). The HER properties of these lamellar nanoflower-like composites were evaluated using electrochemical tests of linear sweep voltammetry (LSV) and electrochemical impedance spectroscopy (EIS). The existent polymer enlarges the interlayer spacing of the lamellar MoS_2, and reduces its stacked thickness. The lamellar MoS_2 samples exhibit a promoting activity in HER at low additions of these three polymers (0.04 g/g MoS_2 for PVA and PEI, and 0.08 g/g MoS_2 for PVP). This can be attributed to the fact that the expanded interlayer of MoS_2 can offer abundant exposed active sites for HER. Conversely, high additions of the polymers exert an obvious interference in the HER activity of the lamellar MoS_2. Compared with the samples of MoS_2/PVP–0.08 and MoS_2/PEI–0.04, the MoS_2/PVA–0.04 composite exhibits excellent activity in HER, in terms of higher current density and lower onset potential.

Keywords: nanoflower-like molybdenum disulfide; hydrogen evolution reaction; polymer intercalation; electrochemical test

1. Introduction

Hydrogen as an ideal, clean and efficient secondary energy resource serves as one of the most competent candidates for replacing petroleum fuels for the future. The electrolysis of water is the most mature and promising method for the production of hydrogen [1,2]. Thus, a variety of nonmetal and metal materials such as carbides [3,4], and Pt-group alloys [5,6] employed as catalysts for the hydrogen evolution reduction (HER) have been extensively explored. Up to date, it has been difficult to push the industrial applications of them because of the poor activities in HER or high costs. Compared with the above-mentioned materials, chalcogenides [7–9], particularly, molybdenum disulfide (MoS_2), possess the merits of acceptable cost, acidic stability, easy fabrication, and a higher electrocatalytic property [10,11]. The suitability of MoS_2 as an excellent catalyst for HER, is mainly due to the catalytic activity at the edge of this lamellar crystal [7,12,13]. It has been proved by theoretical and experimental studies that the active sites of 2H-MoS_2 for HER locate the (010) and (100) planes with the existence of unsaturated molybdenum and sulfur atoms, while the (002) basal plane is inactive [7,12,14].

In general, there are three pathways to obtain MoS_2 catalyst with an effective activity in HER. They are briefly described as follows: (1) increase in the intrinsic activity of active sites, for instance, Ni, B, and Fe metals have been incorporated to improve the catalytic activity of MoS_2 [15,16]; (2) improvement in the electrical contact between the catalytic sites, such that graphite [17], mesoporous carbon [18], and reduced graphene oxide [19,20] have been employed to improve electrical conductivity; (3) enhancement in the number of active sites, thus the decrease in crystal size and stacked thickness have been taken into account. It is well known that the HER activity of MoS_2 decreases with the increase of stacked thickness, because of the Van der Waals force interactions within its interlayer [21–23]. Therefore, most studies in the electrocatalytic enhancements of MoS_2 have focused on improving the density of the active sites and reducing the stacked thickness. Based on this consideration, the chemical exfoliation process [24] and the low pressure chemical vapor deposition method [25] were employed for the fabrications of exfoliated $1T–MoS_2$ and monolayer MoS_2. In view of the enhancement in the HER of nano-structured MoS_2, various intercalation techniques have been employed to extend the interlayer space and to increase the active sites of MoS_2 specimens. Compared with intercalations of graphene oxide and some other materials, the process of polymer intercalation can be easily performed. To the best of our knowledge, however, as for the improvement in HER of MoS_2-based catalyst, insufficient efforts for the intercalation of polymers have been devoted to the increase in the exposed unsaturated edge sites and the reduction in the stacked thickness. Thus, the enhancement in HER of MoS_2 catalyst merits extensive exploration.

The aim of this study is to reveal the effects of intercalated polymers on the structure, density of active edge sites, and HER activity of nanoflower-like MoS_2. Thus, three polymers, *i.e.*, polyvinylpyrrolidone (PVP), polyvinyl alcohol (PVA), and polyethylenimine (PEI) were added to the aqueous solutions employed for the fabrication of nanoflower-like MoS_2 via a solvothermal synthesis approach. Herein, the prepared MoS_2 specimens were characterized using scanning electron microscope (SEM), transmission electron microscopy (TEM), X-ray diffraction (XRD), as well as Raman and Fourier transform infrared spectroscopy (FTIR). The electrocatalytic performances in HER of theses lamellar catalysts were evaluated by the techniques of linear sweep voltammetry (LSV) and electrochemical impedance spectroscopy (EIS). This work sheds insight into the enhancement in HER of the MoS_2-based material.

2. Material and Methods

2.1. Material

Analytical grade reagents of ammonium molybdate tetrahydrate $((NH_4)_6Mo_7O_{24}\cdot4H_2O$, AHM, 99 wt. %), thiourea $(CH_4N_2S$, 99 wt. %), polyvinyl alcohol (PVA, average molecular weight of 77087.5, 93 wt. %), polyethylenimine (PEI, average molecular weight of 600, 99 wt. %), polyvinylpyrrolidone (PVP, average molecular weight of 10,000, 95 wt. %), and sulfuric acid $(H_2SO_4$, 98 wt. %) were purchased from Jingchun Scientific Co. Ltd. (Shanghai, China). The 5 wt. % of Nafion solution was provided by Alfa Aesar Chemicals Co. Ltd. (Shanghai, China). The above mentioned reagents were used as received and without further purification.

2.2. Synthesis of Nanoflower-Like MoS_2

The preparation process of nanoflower-like MoS_2 was as follows: 1.41 g of AHM and 0.26 g of thiourea were dissolved in 20 mL of distilled water, and the mixed solution was stirred to form a homogeneous solution. Then, a certain amount of polymer (PVA in the range of 0.04–0.4 g/g MoS_2, PVP in the range of 0.08–2.0 g/g MoS_2, PEI in the range of 0.04–0.4 g/g MoS_2) was added to the mixed solution. This solution was stirred for another 30 min at room temperature to guarantee the homogeneity of the solution. Afterward, the mixed solution was transferred to a 25 mL Teflon-lined stainless steel autoclave placed in an electric cooker (WRN-010, Eurasian, Tianjin, China). Then the temperature of this mixed solution was increased to 200 °C and maintained for 24 h. When the temperature of the solution naturally cooled from 200 °C to room temperature, the solution was centrifuged and then washed with distilled water and ethanol to remove residual reactants. Finally, the obtained powders were dried at 60 °C under a vacuum atmosphere. With the addition of PVA, PVP, and PEI, the prepared samples were described as MoS_2/(PVA)–X, MoS_2/(PVP)–X and MoS_2/(PEI)–X (Table 1), respectively; where X denotes the concentration of polymer (g/g MoS_2) in the solution. Without the presence of polymer, the pure MoS_2 was fabricated by a similar process to that mentioned above.

Table 1. Additions of reagents for MoS_2/polymers catalysts with the hydrothermal synthesis process. Polyvinyl alcohol (PVA), polyvinylpyrrolidone (PVP), polyethylenimine (PEI), Thiourea, ammonium molybdate tetrahydrate (AHM) and H_2O are listed.

Sample	PVA (g/g MoS_2)	PVP (g/g MoS_2)	PEI (g/g MoS_2)	Thiourea (g)	AHM (g)	H_2O (mL)
Pure MoS_2	–	–	–	0.26	1.41	20
MoS_2/PVA–0.04	0.04	–	–	0.26	1.41	20
MoS_2/PVA–0.2	0.2	–	–	0.26	1.41	20
MoS_2/PVA–0.4	0.4	–	–	0.26	1.41	20
MoS_2/PVP–0.08	–	0.08	–	0.26	1.41	20
MoS_2/PVP–0.4	–	0.4	–	0.26	1.41	20
MoS_2/PVP–2.0	–	2.0	–	0.26	1.41	20
MoS_2/PEI–0.04	–	–	0.04	0.26	1.41	20
MoS_2/PEI–0.2	–	–	0.2	0.26	1.41	20
MoS_2/PEI–0.4	–	–	0.4	0.26	1.41	20

2.3. Characterization

An X-ray diffractometer (XRD, Smartlab, Tokyo, Japan) with a Cu Kα radiation (λ = 1.5418 Å) was employed to investigate the structures of the prepared catalysts. Raman spectroscopy was conducted using a Raman Microscope (E55+FRA106, Bruker, Karlsruhe, Germany) with an excitation wavelength of 514.5 nm. The morphology analysis was performed using a scanning electron microscope (SEM, S-4800, Hitachi, Tokyo, Japan) with an accelerating voltage of 5 kV. Transmission electron microscopy (TEM) and high resolution TEM (HRTEM) images were performed with a microscope (JEM-2010, JEOL, Tokyo, Japan) at a voltage of 200 kV. A Nicolet iS5 Fourier transform infrared spectrometer (Thermo Fisher Scientific, Madison, WI, USA) was adopted to determine the FTIR spectra of the specimens with 16 scans at a resolution of 4 cm^{-1} interval.

2.4. Fabrication of the Electrodes

The fabrication process of the tested electrode was described as follows. First, 5 mg of pure MoS_2 or MoS_2/polymer powder and 30 μL of Nafion solution (5 wt. %) were dispersed in 1 mL of ethanol-water solution at room temperature (the volume ratio between absolute ethanol and distilled water is 1:3). Then the mixed solution was dispersed for 1 h using an ultrasonic cleaner (YJ5120-10, Kun Shan Ultrasonic Instruments Co. Ltd., Kunshan, China). Second, 5 μL of the dispersed solution was dropped onto the surface of a glassy carbon electrode (GCE) with a diameter of 3 mm (Aida Hengsheng Technology Co. Ltd., Tianjin, China). Finally, the modified GCE with a powder loading of 0.34 mg/cm^2 was dried at room temperature for 24 h. The obtained electrodes were denoted as MoS_2/PVA–X, MoS_2/PVP–X and MoS_2/PEI–X (Table 1), where X denotes the different addition of PVA, PVP, and PEI. The pure MoS_2 electrode was fabricated exactly as the above process, except for pure MoS_2 powder applied to replace the MoS_2/polymer powder.

2.5. Electrochemical Measurements

All the electrochemical measurements were performed in a 0.5 $mol·L^{-1}$ H_2SO_4 solution at 298 K under atmospheric pressure, using an electrochemical workstation (CHI 650C, Chenhua Co. Ltd.,

Shanghai, China). A three stand electrode cell was employed, where the fabricated electrode mentioned above, Ag/AgCl, and Pt wire were employed as working electrode, reference electrode, and auxiliary electrode. Before the tests, all the electrolytes were deaerated by purging with pure N_2 gas for 40 min before the tests.

Linear sweep voltammetry (LSV) was conducted between -0.25 and 0.1 V at 2 mV·s^{-1}. Electrochemical impedance spectroscopy (EIS) was measured at a cathode overpotential of 200 mV; the employed amplitude of the sinusoidal signal was 5 mV. All EIS tests were always carried out from high frequency to low frequency, and the frequency ranged from 10^5 to 10^{-2} Hz. During the test, all potentials were collected *versus* the Ag/AgCl electrode. Then the potentials (*versus* Ag/AgCl) were calibrated to a reversible hydrogen electrode (RHE) as follows: $E_{RHE} = E_{Ag/AgCl} + 0.059pH + 0.209$ V. All the measurements were performed in triplicate to guarantee the reproducibility of the experimental data.

3. Results and Discussions

3.1. Characterization of Nanoflower-Like MoS$_2$

3.1.1. X-Ray Diffraction (XRD) Spectra

The XRD patterns of the prepared MoS$_2$-based nanoflower-like samples are presented in Figure 1. For the purpose of comparison, the standard pattern of the pristine 2H-MoS$_2$ (JCPDS No. 37-1492) is also displayed within Figure 1. Three peaks appearing at $2\theta = 15.0°$, $33.8°$, and $57.1°$ correspond to the (002), (100), and (110) planes of the pristine 2H–MoS$_2$, respectively. It is noteworthy that, compared to that of pure MoS$_2$, the locations of the composites corresponding to the (002) peaks shift to lower angles. The diffraction peaks of composites centered at around $2\theta = 12.5°$ correspond to the interspacing (*d*) of 7.08 Å obtained via the Bragg equation. Herein, the interspacing thicknesses of the three MoS$_2$ composite specimens are identical, which may be due to the low additions of PVA, PVP, and PEI. The interspacing of 7.08 Å for MoS$_2$/polymer composites is much larger than that of the pure MoS$_2$ (5.9 Å), indicating the expansion of lattice and the intercalation of polymers [26,27]. In addition, the weak characteristic of these three peaks for the prepared samples is validated, suggesting their poor crystallizations. Meanwhile, with a decrease in concentrations of polymers, the shapes of these reflection peaks become sharper, suggesting enhancement of crystallinities. It is noteworthy that the decline of the (002) diffraction peak indicates a low stacking height [28], which is effective for offering more active sites and lower intrinsic resistance for HER [21].

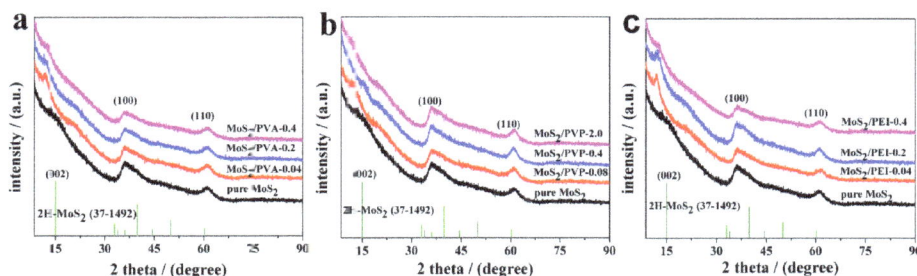

Figure 1. X-ray diffraction (XRD spectra of the as-prepared MoS$_2$-based nanoflower-like catalysts: (**a**) MoS$_2$/polyvinyl alcohol (PVA) composite; (**b**) MoS$_2$/polyvinylpyrrolidone (PVP) composite; (**c**) MoS$_2$/polyethylenimine (PEI) composite.

3.1.2. Raman Spectra

The Raman spectra of MoS_2/PVA composites are reported in Figure 2a. The two characteristic peaks of pure MoS_2 at 372 and 402 cm^{-1} match well with the E_{2g}^1 and A_{1g} vibrational models of the hexagonal MoS_2 [29]. With the increase in concentration of PVA, intensities of E_{2g}^1 and A_{1g} peaks become low and weak, demonstrating the decrease in crystallinities of the composites. Herein, Raman spectra of MoS_2/PVP and MoS_2/PEI composites with different additions of PVP and PEI are not given, because of their similarities to that of MoS_2/PVA. In general, the difference in frequency between E_{2g}^1 and A_{1g} is helpful in determining the number of stacked layers [30,31]. As demonstrated in Figure 2a,b, for MoS_2/PVA, MoS_2/PVP, MoS_2/PEI composites, a red shift of the E_{2g}^1 band and a blue shift of the A_{1g} band are observed; thus the smaller differences between E_{2g}^1 and A_{1g} can be validated for these three composites. In contrast with that of pure MoS_2, we confirmed that the intercalations of polymers reduce the stacking thicknesses of the nanoflower-like MoS_2 composites. It should be pointed out that Raman spectra of the composite specimens strongly depend on the stoichiometry of their components, thus the different HER activity between the pure MoS_2 and the composites might be confirmed.

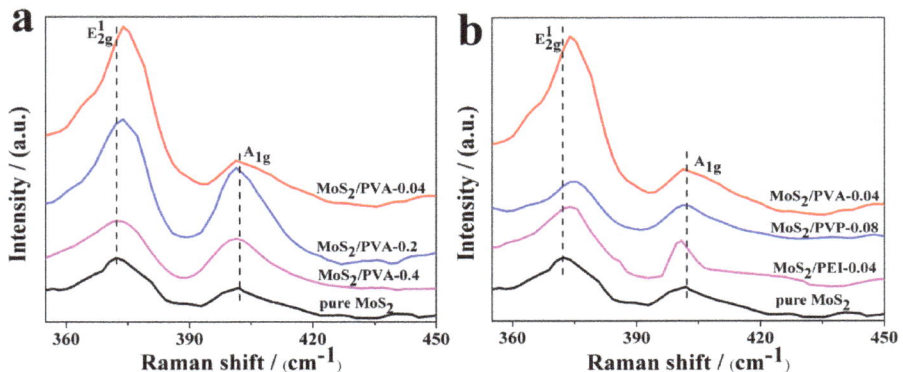

Figure 2. Raman spectra of various MoS_2-based nanoflower-like catalysts: (**a**) MoS_2/PVA composite; (**b**) MoS_2/PVA–0.04, MoS_2/PVP–0.08, and MoS_2/PEI–0.04 composites.

3.1.3. FTIR Spectra

FTIR spectra of MoS_2/PVA composites (Figure 3a), MoS_2/PVP composites (Figure 3b), and MoS_2/PEI (Figure 3c) composites were measured. As shown in Figure 3, the strong absorbance peaks appearing at 3000 to 3600 cm^{-1} indicate the presences of intermolecular and intramolecular hydrogen bonds; nevertheless, to some extent, these peaks are overlapped by the absorbance of water molecules. The characteristic peaks of O–H, CH_2, and C–H groups for PVA (Figure 3a) with the wave number in the range of 1330–1648 cm^{-1} can be observed [32], while the peaks of the C–O stretch and O–H bending can be found at 1096 cm^{-1}. The characteristic peaks of the MoS_2/PVA composites are consistent with that of pure PVA. As for the pure PVP and MoS_2/PVP composites (Figure 3b), the peaks corresponding to C=O stretching vibration, and C–N stretching vibration are observed at 1659 cm^{-1}, and 1284 cm^{-1} [33]. The peak located at 3380 cm^{-1} (Figure 3c) corresponds to the –NH_2 group in PEI. In addition, peaks appearing at 2948, 2840, and 1459 cm^{-1} can be assigned to asymmetric stretching vibrations of C–H bond, while the peak at 1106 cm^{-1} is attributed to the stretching vibration of the C–N group [34]. Thus, the MoS_2/PVP and MoS_2/PEI composites present the same characteristic peaks as the pure PVP and PEI polymers. Based on the above analyses, the existence of PVA, PVP, and PEI can be validated, thereby indicating theses three polymer molecules can be intercalated in the MoS_2 gallery space during the hydrothermal process.

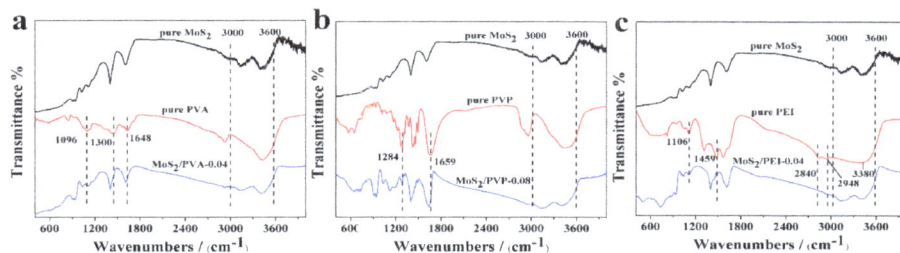

Figure 3. Raman and Fourier transform infrared spectroscopy (FTIR) spectra of (**a**) MoS$_2$/PVA–0.04 composite; (**b**) MoS$_2$/PVP–0.08 composite; (**c**) MoS$_2$/PEI–0.04 composite.

3.1.4. SEM Morphologies

The morphologies of the pure MoS$_2$ and the synthesized MoS$_2$/PVA composites are illustrated in Figure 4. All samples are assembled by the lamellar nanoflower-like MoS$_2$ with diverse lateral sizes ranging from 100 to 400 nm. With the increase in PVA addition (Figure 4b–d), large nano-size flowers with vague edges can be found, suggesting insufficient crystallization of these composites, consistent with the results of XRD. It should be noted that a similar tendency in the presence of PVP and PEI can also be confirmed (Figure 5). Among the MoS$_2$/polymer composite samples, the MoS$_2$/PVA–0.4, MoS$_2$/PVP–2.0 and MoS$_2$/PEI–0.4 samples exhibit remarkable aggregations (Figures 4d and 5c,f), which may be attributed to the excessive adsorption of polymers at the surface and edge sites of the MoS$_2$ composites. In addition, with the decrease in polymer addition, the edges of MoS$_2$ composites (Figures 4b and 5a,d) can be easily observed, thereby indicating the high crystallinity of the samples. Therefore, a nanoflower-like MoS$_2$ sample with a high crystallinity and small particle size can be obtained with low additions of polymers (0.04 g/g MoS$_2$ for PVA and PEI, and 0.08 g/g MoS$_2$ for PVP).

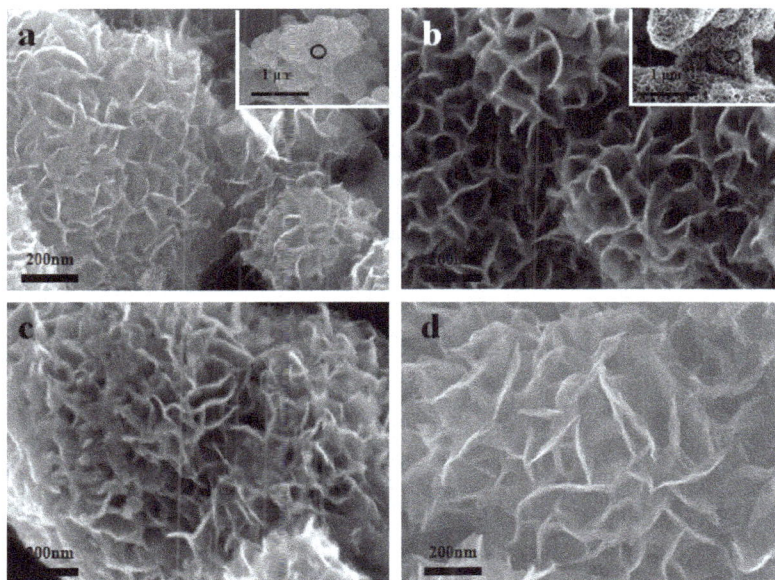

Figure 4. Scanning electron microscope (SEM) images of (**a**) pure MoS$_2$; (**b**) MoS$_2$/PVA–0.04 composite; (**c**) MoS$_2$/PVA–0.08 composite; and (**d**) MoS$_2$/PVA–0.4 composite.

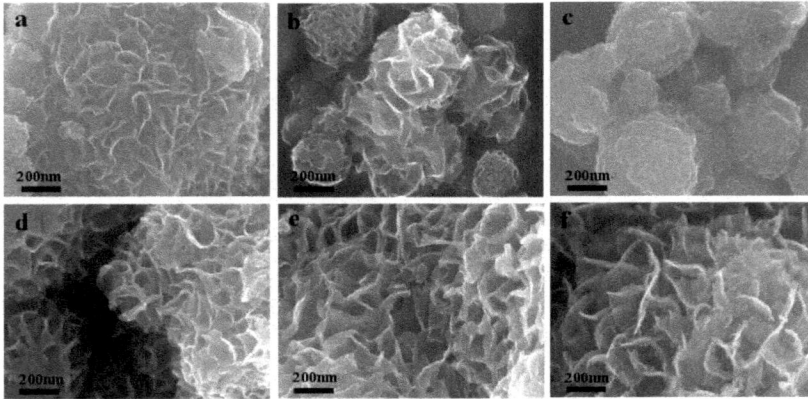

Figure 5. SEM images (**a–f**) of the MoS$_2$/polymer composites synthesized at various concentrations, where **a–f** represent MoS$_2$/PVP–0.08, MoS$_2$/PVP–0.4, MoS$_2$/PVP–2.0, MoS$_2$/PEI–0.04, MoS$_2$/PEI–0.2, and MoS$_2$/PEI–0.4 composites.

3.1.5. TEM

To further elucidate the morphologies of the obtained composites, transmission electron microscopy (TEM) and high-resolution TEM (HRTEM) measurements were performed. From the fabricated composites, MoS$_2$/PVA–0.04, MoS$_2$/PVP–0.08, and MoS$_2$/PEI–0.04 three samples were selected for the tests. Also, the pure MoS$_2$ sample was analyzed for comparison. As displayed in Figure 6, these samples with a diameter of about 100 nm are assembled by lamellar MoS$_2$ powders. Nano-size flowers with a thickness of 2–5 nm can be observed, which are notably thinner than that of the pure MoS$_2$ (~10 nm). The lattice inter spaces of 7.08 Å for MoS$_2$/PVA–0.04, MoS$_2$/PVP–0.08 and MoS$_2$/PEI–0.04 composites are much larger than that of pure MoS$_2$ (5.9 Å), which is also consistent with the results of XRD. This may be attributed to the intercalations of polymers, and the interactions between the oxygen-containing functional groups of the polymers and the MoS$_2$ precursors [19]. Thus, the polymers intercalated into the MoS$_2$ interlayer can afford a number of exposed edges for HER.

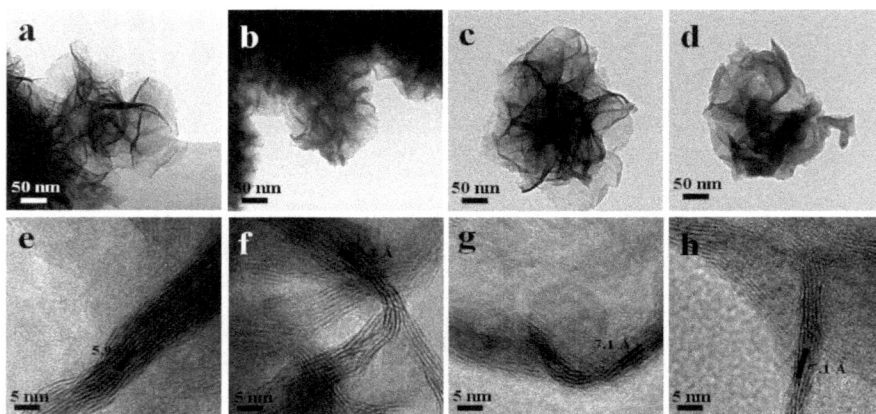

Figure 6. Transmission electron microscopy (TEM) (**a–d**) and high resolution transmission electron microscopy (HR-TEM) (**e–h**) of the various samples, represented for pure MoS$_2$, MoS$_2$/PVA–0.04, MoS$_2$/PVP–0.08, and MoS$_2$/PEI–0.04 composites.

3.2. HER Activity of the Nanoflower-Like MoS$_2$/Polymer

3.2.1. Linear Sweep Voltammetry (LSV)

To evaluate the catalytic activity in HER of the MoS$_2$/polymer composite, LSV measurements were performed in 0.5 mol·L^{-1} H$_2$SO$_4$ solution with a scanning rate of 2 mV·s^{-1} at room temperature. Cathodic polarization curves (Figure 7a) and corresponding Tafel plots (Figure 7d) of MoS$_2$/PVA–0.04, MoS$_2$/PVA–0.08, MoS$_2$/PVA–0.4 were obtained. In comparison with the pure MoS$_2$ sample, it can be observed that all the MoS$_2$/PVA composites present much lower onset overpotentials (η); among them, the MoS$_2$/PVA–0.04 sample exhibits the lowest value of η (~40 mV), suggesting the excellent HER activity. The onset potentials for other MoS$_2$/PVA composites are in the range of 180–250 mV, these values are still lower than that of the pure MoS$_2$ (300 mV). As shown in Figure 7a, compared with other specimens, the MoS$_2$/PVA–0.04 sample exhibits the largest cathodic current density of 20 mA·cm^{-2} at (η = 250 mV), which is 80 times that of the pure MoS$_2$ (0.25 mA·cm^{-2}). Similarly, the enhancements in HER for MoS$_2$/PVP and MoS$_2$/PEI composites can also be validated (Figure 7b,c). The current densities of MoS$_2$/PVP–0.08 and MoS$_2$/PEI–0.04 (η = 250 mV) are 9.04 mA·cm^{-2} and 9.60 mA·cm^{-2}, respectively; which are 36-fold and 38-fold levels higher than that of the pure MoS$_2$.

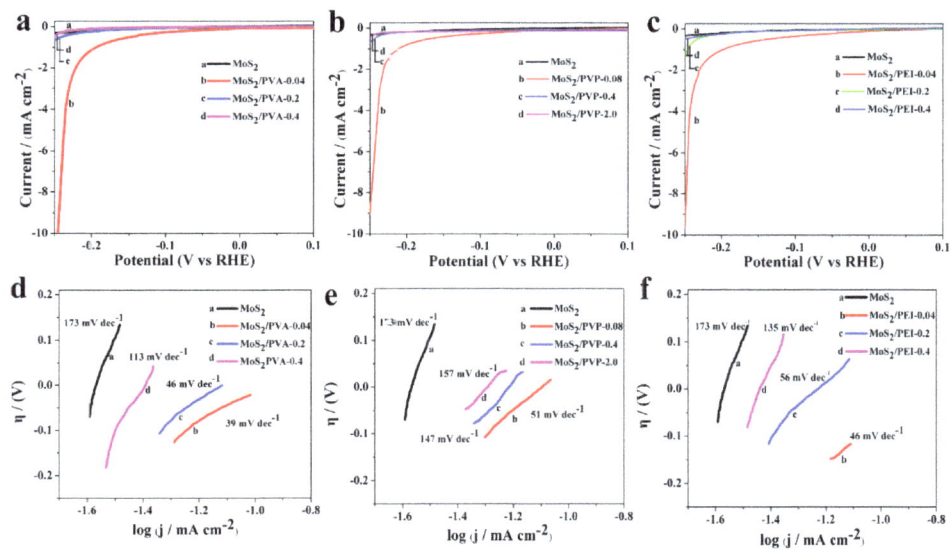

Figure 7. Polarization curves (**a**–**c**) and corresponding Tafel plots (**d**–**f**) of the MoS$_2$/PVA, MoS$_2$/PVP, MoS$_2$/PEI composites.

It is well known that the smaller the Tafel slope of the sample, the higher the HER activity [35,36] will be. The Tafel plots (Figure 7d–f) fit the following Tafel equation:

$$\eta = b \times \log j + a \tag{1}$$

where, η (mV) is the overpotential, and j (mA·cm^{-2}) is the current density. The values of Tafel slope are ~173, ~113, ~46, and ~39 mV·dec^{-1} for MoS$_2$, MoS$_2$/PVA–0.4, MoS$_2$/PVA–0.2, and MoS$_2$/PVA–0.04, which are very consistent with published reports [37,38]. Among the samples, MoS$_2$/PVA–0.04 with the smallest Tafel slope exhibits the highest HER activity. The small Tafel plot compares to that of the other reported polymer [39] Although Tafel slopes of MoS$_2$/PVA–0.2, MoS$_2$/PVA–0.4 samples are larger than that of MoS$_2$/PVA–0.04, their HER activities are still higher than that of the pure MoS_2 because of their smaller Tafel slopes. Similarly, the same can be concluded for MoS_2/PVP and MoS_2/PEI composites. Among these two kinds of composite, the MoS_2/PVP–0.08 and MoS_2/PEI–0.04 samples possess notable HER activities due to smaller Tafel slopes than those of other corresponding samples (Figure 7e,f). However, these two composites for HER activity are inferior to the MoS_2/PVA–0.04 sample.

Based on the above mentioned results, we realize that all the prepared composites present quite different HER activities due to the effects of different concentrations of polymers. Considering the feasibilities of intercalation polymers in the interlayer of MoS_2 catalysts and excessive adsorption of them on the surface and edge sites of MoS_2, the HER activities of these three kinds of composites are reduced with increase in the additions of the polymers. Thus, the enhancement in HER for nanoflower-like MoS_2 can be achieved by adjusting the polymer addition.

3.2.2. Electrochemical Impedance Spectroscopy (EIS)

The kinetics of MoS_2/polymer composites in HER can be further illustrated by EIS, and the results are reported in Figure 8. The EIS data were fitted using a modified Randles equivalent circuit shown in Scheme 1. The values of the circuit elements obtained by fitting the experimental results are shown in Table 2. Herein, a constant phase element (CPE) was adopted to model the electrical double layer of the electrode/electrolyte interface. The necessity of CPE in place of pure capacitance is due to the dispersed distribution of Nyquist diagrams in the high frequency domain [40]. The double layer capacitance (C_{dl}) and exchange current density (i_0) of the electrode were estimated using Equations (2) and (3) [41]:

$$C_{dl} = Y_0 \times \left(R_s^{-1} + R_{ct}^{-1} \right)^{n-1} \tag{2}$$

$$i_0 = \frac{RT}{nFR_{ct}} \tag{3}$$

where, Y_0 ($mS \cdot s \cdot cm^{-2}$) regarded as a capacity parameter, is the CPE coefficient; the dimensionless CPE exponent (n) is related to the constant phase angle. R_s ($\Omega \cdot cm^2$), and R_{ct} ($\Omega \cdot cm^2$) are the solution resistance, and the charge transfer resistance, respectively.

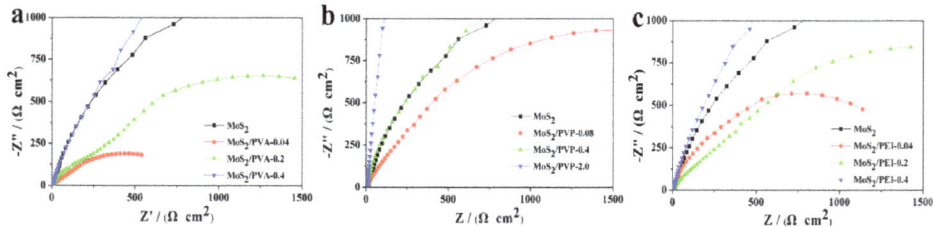

Figure 8. Nyquist plots of MoS_2-based nanoflower-like composite catalysts: (**a**) MoS_2/PVA composite; (**b**) MoS_2/PVP composite; (**c**) MoS_2/PEI composite.

As indicated by Figure 8 and Table 2, compared with that of the MoS_2 pure sample, the Nyquist plots reveal a decrease in charge transfer resistance (R_{ct}) (from 7.55 $k\Omega \cdot cm^2$ to 0.84 $k\Omega \cdot cm^2$), demonstrating the excellent conductivity of the composites (Table 2). R_{ct} of MoS_2/polymer composites follow the sequence of MoS_2/PVA < MoS_2/PEI < MoS_2/PVP. R_{ct} of the MoS_2/PVA–0.04 is smaller than that of other specimens, suggesting a faster charge transfer of this composite. The double layer capacitance (C_{dl}) is calculated to evaluate the exposed active surface area of the electrode. As shown in Table 2, MoS_2/PVA–0.04 exhibits the largest C_{dl} compared to the other ones. Furthermore, MoS_2/PVA–0.04 exhibits a significant exchange current density (i_0) of 5.89×10^{-5} $mA \cdot cm^{-2}$, which is 2–11 times larger than that of the other composites. The large i_0 further confirms the excellent activity for HER catalysis. Thus, among the composite samples, the $MoS_2/PVA-0.04$ composite with the lowest R_{ct}, highest C_{dl} and i_0 provides sufficient active sites, and thereby guarantees excellent HER activity.

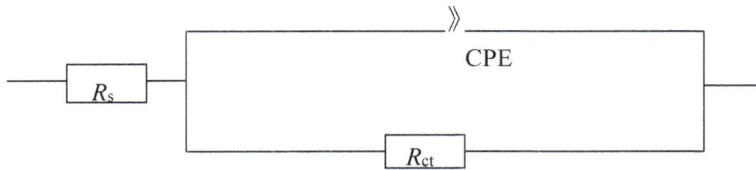

Scheme 1. Randles electrical equivalent circuit compatible with the Nyquist diagrams shown in Figure 8. R_s, R_{ct} and CPE are the solution resistance, the charge transfer resistance, and a constant phase element, respectively.

Table 2. Analyzed parameters of electrochemical impedance spectroscopy (EIS) spectra for the composites.

Sample	$R_s/(\Omega\cdot cm^2)$	$R_{ct}/(k\Omega\cdot cm^2)$	$Y_0/(mS\cdot s\cdot cm^{-2})$	n	$C_{dl}/(mF\cdot cm^{-2})$	$i_0/(\times10^{-5}mA\cdot cm^{-2})$
MoS_2	6.13	7.55	0.16	0.67	0.29	0.50
$MoS_2/PVA-0.04$	4.87	0.84	1.00	0.52	2.13	5.89
$MoS_2/PVA-0.2$	5.24	1.96	0.25	0.72	0.64	1.82
$MoS_2/PVA-0.4$	5.86	6.05	0.19	0.79	0.39	0.54
$MoS_2/PVP-0.08$	5.72	2.62	0.27	0.71	0.85	1.38
$MoS_2/PVP-0.4$	6.54	3.04	0.20	0.87	0.40	0.97
$MoS_2/PVP-2.0$	7.58	6.23	0.02	0.81	0.34	0.51
$MoS_2/PEI-0.04$	6.22	1.45	0.30	0.83	0.87	2.14
$MoS_2/PEI-0.2$	6.45	2.47	0.21	0.66	0.62	1.58
$MoS_2/PEI-0.4$	5.71	6.18	0.20	0.80	0.37	0.52

Y_0 regarded as a capacity parameter, is the CPE coefficient. the dimensionless CPE exponent (n) is related to the constant phase angle. R_s and R_{ct} are the solution resistance and the charge transfer resistance. The double layer capacitance (C_{dl}) and exchange current density (i_0) are listed in the table.

3.3. Stability

In addition to the HER activity, stability plays a crucial role in the potential application for a MoS_2-based catalyst. In order to reveal the stability of this catalyst in acidic solution, a long-term cycling test for the $MoS_2/PVA-0.04$ composite was conducted. This test was performed with the potential ranging from -0.25 to 0.1 V (*vs* RHE) at a scan rate of 20 mV·s^{-1}. It can be observed that the current density of the composite (Figure 9) shows a slight decrease (less than 4%) after 1000 CV cycles. Thus, the $MoS_2/PVA-0.04$ composite exhibits excellent durability in the HER process.

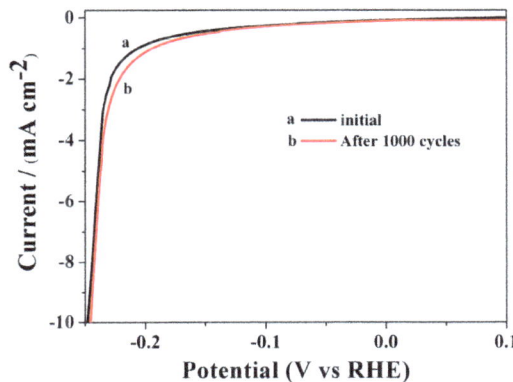

Figure 9. Stability test for the $MoS_2/PVA-0.04$ composite. RHE = reversible hydrogen electrode.

4. Conclusions

In this study, three kinds of MoS_2/polymer composites were fabricated by a facile hydrothermal method. The effects of three polymers PVA, PVP, and PEI on the stacking height and the quantity of exposed active sites of the nanoflower-like MoS_2 were evaluated. The intercalated or adsorbed polymers provide carbonyl and hydroxyl functional groups for attachment of the MoS_2 precursor, and enlarge the interlayer spacing of MoS_2 particles. The expanded interlayers of nanoflower-like MoS_2 catalysts lead to the enhancement in active sites, resulting in a high electrocatalytic activity for HER. Of all the composites, the MoS_2/PVA–0.04 composite exhibits the best HER activity, due to a small overpotential (~40 mV), a large cathodic current (20 mA·cm^{-2}), and a small Tafel slope (~39 mV·dec^{-1}).

Acknowledgments: This work was supported by the Scientific Research Foundation of the Hebei Higher Education Institutions of China (Grant No. ZD2015120). Herein, we are also very grateful to the editors and the reviewers for giving the valuable and instructive comments.

Author Contributions: Laizhou Song designed the study. Xianwen Zeng, Lijing Niu, Xiuli Wang, Xuanming Shi and Jiayun Yan performed the experiments. Xianwen Zeng, Lijing Niu and Laizhou Song wrote the paper. Xianwen Zeng, Lijing Niu, and Laizhou Song reviewed and edited the manuscript. All authors read and approved the manuscript.

Conflicts of Interest: The author declares no conflict of interest.

References

1. Mallouk, T.E. Water electrolysis: Divide and conquer. *Nat. Chem.* **2013**, *5*, 362–363. [CrossRef] [PubMed]
2. Raimondi, F.; Scherer, G.G.; Kötz, R.; Wokaun, A. Nanoparticles in energy technology: Examples from electrochemistry and catalysis. *Angew. Chem. Int. Ed.* **2005**, *44*, 2190–2209. [CrossRef] [PubMed]
3. Liu, Y.; Kelly, T.G.; Chen, J.G.G.; Mustain, W.E. Metal carbides as alternative electrocatalyst supports. *ACS Catal.* **2013**, *3*, 1184–1194. [CrossRef] [PubMed]
4. Hinnemann, B.; Moses, P.G.; Bonde, J.; Jørgensen, K.P.; Nielsen, J.H.; Honch, S.; Chorkendorff, I.; Nørskov, J.K. Biornimetic hydrogen evolution: MoS_2 nanoparticles as catalyst for hydrogen evolution. *J. Am. Chem. Soc.* **2005**, *127*, 5308–5309. [CrossRef] [PubMed]
5. Dresselhaus, M.S.; Thomas, I.L. Alternative energy technologies. *Nature* **2001**, *414*, 332–337. [CrossRef] [PubMed]
6. Antolini, E. Palladium in fuel cell catalysis. *Energy Environ. Sci.* **2009**, *2*, 915–931. [CrossRef]
7. Tang, H.; Dou, K.P.; Kaun, C.-C.; Kuang, Q.; Yang, S.H. $MoSe_2$ nanosheets and their graphene hybrids: Synthesis, characterization and hydrogen evolution reaction studies. *J. Mater. Chem. A* **2014**, *2*, 360–364. [CrossRef]
8. Chen, Z.B.; Cummins, D.; Reinecke, B.N.; Clark, E.; Sunkara, M.K.; Jaramillo, T.F. Core-shell MoO_3–MoS_2 nanowires for hydrogen evolution: A functional design for electrocatalytic materials. *Nano Lett.* **2011**, *11*, 4168–4175. [CrossRef] [PubMed]
9. Cheng, L.; Huang, W.J.; Gong, Q.F.; Liu, C.H.; Liu, Z.; Li, Y.G.; Dai, H.J. Ultrathin WS_2 nanoflakes as a high-performance electrocatalyst for the hydrogen evolution reaction. *Angew. Chem. Int. Ed.* **2014**, *53*, 7860–7863. [CrossRef] [PubMed]
10. Vrubel, H.; Merki, D.; Hu, X. Hydrogen evolution catalyzed by MoS_3 and MoS_2 particles. *Energy Environ. Sci.* **2012**, *5*, 6136–6144. [CrossRef]
11. Karunadasa, H.I.; Montalvo, E.; Sun, Y.; Majda, M.; Long, J.R.; Chang, C.J. A molecular MoS_2 edge site mimic for catalytic hydrogen generation. *Science* **2012**, *335*, 698–702. [CrossRef] [PubMed]
12. Jaramillo, T.F.; Jørgensen, K.P.; Bonde, J.; Nielsen, J.H.; Horch, S.; Chorkendorff, I. Identification of active edge sites for electrochemical H$_2$ evolution from MoS_2 nanocatalysts. *Science* **2007**, *317*, 100–102. [CrossRef] [PubMed]
13. Bonde, J.; Moses, P.G.; Jaramillo, T.F.; Nørskov, J.K.; Chorkendorff, I. Hydrogen evolution on nano-particulate transition metal sulfides. *Faraday Discuss.* **2009**, *140*, 219–231. [CrossRef]

14. Liu, N.; Yang, L.C.; Wang, S.N.; Zhong, Z.W.; He, S.N.; Yang, X.Y.; Gao, Q.S.; Tang, Y. Ultrathin MoS$_2$ nanosheets growing within an *in-situ*-formed template as efficient electrocatalysts for hydrogen evolution *J. Power Sources* **2015**, *275*, 588–594. [CrossRef]

15. Zhou, T.; Yin, H.; Liu, Y.; Chai, Y.; Zhang, J.; Liu, C. Synthesis, characterization and HDS activity of carbon-containing Ni–Mo sulfide nano-spheres. *Catal. Lett.* **2010**, *134*, 343–350. [CrossRef]

16. Merki, D.; Vrubel, H.; Rovelli, L.; Fierro, S.; Hu, X. Fe, Co, and Ni ions promote the catalytic activity of amorphous molybdenum sulfide films for hydrogen evolution. *Chem. Sci.* **2012**, *3*, 2515–2525. [CrossRef]

17. Jaramillo, T.F.; Bonde, J.; Zhang, J.; Ooi, B.-L.; Andersson, K.; Ulstrup, J.; Chorkendorff, I. Hydrogen evolution on supported incomplete cubane-type [Mo$_3$S$_4$]$^{4+}$ electrocatalysts. *J. Phys. Chem. C* **2008**, *112*, 17492–17498. [CrossRef]

18. Bian, X.J.; Zhu, J.; Liao, L.; Scanlon, M.D.; Ge, P.Y.; Ji, C.; Girault, H.H.; Liu, B.H. Nanocomposite of MoS$_2$ on ordered mesoporous carbon nanospheres: A highly active catalyst for electrochemical hydrogen evolution. *Electrochem. Commun.* **2012**, *22*, 128–132. [CrossRef]

19. Zheng, X.L.; Xu, J.B.; Yan, K.Y.; Wang, H.; Wang, Z.L.; Yang, S.H. Space-confined growth of MoS$_2$ nanosheets within graphite: The layered hybrid of MoS$_2$ and graphene as an active catalyst for hydrogen evolution reaction. *Chem. Mater.* **2014**, *26*, 2344–2353. [CrossRef]

20. Li, Y.G.; Wang, H.L.; Xie, L.M.; Liang, Y.Y.; Hong, G.S.; Dai, H.J. MoS$_2$ nanoparticles grown on graphene: An advanced catalyst for the hydrogen evolution reaction. *J. Am. Chem. Soc.* **2011**, *133*, 7296–7299. [CrossRef] [PubMed]

21. Wang, D.Z.; Pan, Z.; Wu, Z.Z.; Wang, Z.P.; Liu, Z.H. Hydrothermal synthesis of MoS$_2$ nanoflowers as highly efficient hydrogen evolution reaction catalysts. *J. Power Sources* **2014**, *264*, 229–234. [CrossRef]

22. Kong, D.S.; Wang, H.T.; Cha, J.J.; Pasta, M.; Koski, K.J.; Yao, J.; Cui, Y. Synthesis of MoS$_2$ and MoSe$_2$ films with vertically aligned layers. *Nano Lett.* **2013**, *13*, 1341–1347. [CrossRef] [PubMed]

23. Tsai, C.; Pedersen, F.A.; Nørskov, J.K. Tuning the MoS$_2$ edge-site activity for hydrogen evolution via support interactions. *Nano Lett.* **2014**, *14*, 1381–1387. [CrossRef] [PubMed]

24. Lukowski, M.A.; Daniel, A.S.; Meng, F.; Forticaux, A.; Li, L.S.; Jin, S. Enhanced hydrogen evolution catalysis from chemically exfoliated metallic MoS$_2$ nanosheets. *J. Am. Chem. Soc.* **2013**, *135*, 10274–10277. [CrossRef] [PubMed]

25. Shi, J.P.; Ma, D.L.; Han, G.-F.; Zhang, Y.; Ji, Q.Q.; Gao, T.; Sun, J.Y.; Song, X.J.; Li, C.; Zhang, Y.S.; *et al.* Controllable growth and transfer of monolayer MoS$_2$ on Au foils and its potential application in hydrogen evolution reaction. *ACS Nano* **2014** *8*, 10196–10204. [CrossRef] [PubMed]

26. Bissessur, R.; Gallant, D.; Brüning, R. Novel nanocomposite material consisting of poly[oxymethylene-(oxyethylene)] and molybdenum disulfide. *Mater. Chem. Phys.* **2003**, *82*, 316–320. [CrossRef]

27. Lin, B.-Z.; Ding, C.; Xu, B.-H.; Chen, Z.-J.; Chen, Y.-L. Preparation and characterization of polythiophene/molybdenum disulfide intercalation material. *Mater. Res. Bull.* **2009**, *44*, 719–723. [CrossRef]

28. Wu, Z.Z.; Fang, B.Z.; Wang, Z.P.; Wang, C.L.; Liu, Z.H.; Liu, F.Y.; Wang, W.; Alfantazi, A.; Wang, D.Z.; Wilkinson, D.P. MoS$_2$ nanosheets: A designed structure with high active site density for the hydrogen evolution reaction. *ACS Catal.* **2013**, *3*, 2101–2107. [CrossRef]

29. Wang, H.T.; Lu, Z.Y.; Xu, S.C.; Kong, D.S.; Cha, J.J.; Zheng, G.Y.; Hsu, P.-C.; Yan, K.; Bradshaw, D.; Prinz, F.B.; *et al.* Electrochemical tuning of vertically aligned MoS$_2$ nanofilms and its application in improving hydrogen evolution reaction. *Proc. Natl. Acad. Sci. USA* **2013**, *110*, 19701–19706. [CrossRef] [PubMed]

30. Lee, C.G.; Yan, H.G.; Brus, L.E.; Heinz, T.F.; Hone, J.; Ryu, S. Anomalous lattice vibrations of single and few-layer MoS$_2$. *ACS Nano* **2010**, *4*, 2695–2700. [CrossRef] [PubMed]

31. Yu, Y.F.; Huang, S.Y.; Li, Y.P.; Steinmann, S.N.; Yang, W.T.; Cao, L.Y. Layer-dependent electrocatalysis of MoS$_2$ for hydrogen evolution. *Nano Lett.* **2014**, *14*, 553–558. [CrossRef] [PubMed]

32. Peresin, M.S.; Vesterinen, A.-H.; Habibi, Y.; Johansson, L.-S.; Pawlak, J.J.; Nevzorov, A.A.; Rojas, O.J. Crosslinked PVA nanofibers reinforced with cellulose nanocrystals: Water interactions and thermomechanical properties. *J. Appl. Polym. Sci.* **2014**. [CrossRef]

33. Prasanth, S.; Irshad, P.; Raj, D.R.; Vineeshkumar, T.V.; Philip, R.; Sudarsanakumar, C. Nonlinear optical property and fluorescence quenching behavior of PVP capped ZnS nanoparticles co-doped with Mn^{2+} and Sm^{3+}. *J. Lumin.* **2015**, *166*, 167–175. [CrossRef]

34. Popescu, L.M.; Piticescu, R.M.; Petriceanu, M.; Ottaviani, M.F.; Cangiotti, M.; Vasile, E.; Dîrtu, M.M.; Wolff, M.; Garcia, Y.; Schinteie, G.; *et al.* Hydrothermal synthesis of nanostructured hybrids based on iron oxide and branched PEI polymers. Influence of high pressure on structure and morphology. *Mater. Chem. Phys.* **2015**, *161*, 84–95. [CrossRef]

35. Merki, D.; Hu, X. Recent developments of molybdenum and tungsten sulfides as hydrogen evolution catalysts. *Energy Environ. Sci.* **2011**, *4*, 3878–3888. [CrossRef]

36. Chang, Y.-H.; Nikam, R.D.; Lin, C.-T.; Huang, J.-K.; Tseng, C.-C.; Hsu, C.-L.; Cheng, C.-C.; Su, C.-Y.; Li, L.-J.; Chua, D.H.C. Enhanced electrocatalytic activity of MoS_x on TCNQ-treated electrode for hydrogen evolution reaction. *ACS Appl. Mater. Interfaces* **2014**, *6*, 17679–17685. [CrossRef] [PubMed]

37. Xie, J.F.; Zhang, J.J.; Li, S.; Grote, F.; Zhang, X.D.; Zhang, H.; Wang, R.X.; Lei, Y.; Pan, B.C.; Xie, Y. Controllable disorder engineering in oxygen-incorporated MoS_2 ultrathin nanosheets for efficient hydrogen evolution. *J. Am. Chem. Soc.* **2013**, *135*, 17881–17888. [CrossRef] [PubMed]

38. Voiry, D.; Salehi, M.; Silva, R.; Fujita, T.; Chen, M.W.; Asefa, T.; Shenoy, V.B.; Eda, G.; Chhowalla, M. Conducting MoS_2 nanosheets as catalysts for hydrogen evolution reaction. *Nano Lett.* **2013**, *13*, 6222–6227. [CrossRef] [PubMed]

39. Youssef, L.; Alain, D.; Jean-Claude, M. Electrocatalytic hydrogen evolution from molybdenum sulfide-polymer composite films on carbon electrodes. *ACS Appl. Mater. Interfaces* **2015**, *7*, 15866–15875.

40. Heli, H.; Sattarahmady, N.; Jabbari, A.; Moosavi-Movahedi, A.A.; Hakimelahi, G.H.; Tsai, F.-Y. Adsorption of human serum albumin onto glassy carbon surface—Applied to albumin-modified electrode: Mode of protein-ligand interactions. *J. Electroanal. Chem.* **2007**, *610*, 67–74. [CrossRef]

41. Alcaide, F.; Brillas, E.; Cabot, P.-L. EIS analysis of hydroperoxide ion generation in an uncatalyzed oxygen-diffusion cathode. *J. Electroanal. Chem.* **2003**, *547*, 61–73. [CrossRef]

![metals logo] *metals*

MDPI

Article

A Straightforward Route to Tetrachloroauric Acid from Gold Metal and Molecular Chlorine for Nanoparticle Synthesis

Shirin R. King, Juliette Massicot and Andrew M. McDonagh *

School of Mathematical and Physical Sciences, University of Technology Sydney, 15 Broadway,
Ultimo NSW 2007, Australia; shirin.r.king@student.uts.edu.au (S.R.K.); jmassicot@live.fr (J.M.)
* Author to whom correspondence should be addressed; andrew.mcdonagh@uts.edu.au;
  Tel.: +61-2-9514-1035 (ext. 1035).

Academic Editor: Suresh Bhargava
Received: 5 June 2015; Accepted: 12 August 2015; Published: 18 August 2015

Abstract: Aqueous solutions of tetrachloroauric acid of high purity and stability were synthesised using the known reaction of gold metal with chlorine gas. The straightforward procedure developed here allows the resulting solution to be used directly for gold nanoparticle synthesis. The procedure involves bubbling chlorine gas through pure water containing a pellet of gold. The reaction is quantitative and progressed at a satisfactory rate at 50 °C. The gold(III) chloride solutions produced by this method show no evidence of returning to metallic gold over at least twelve months. This procedure also provides a straightforward method to determine the concentration of the resulting solution using the initial mass of gold and volume of water.

Keywords: gold chloride; HAuCl$_4$; chlorine; gold nanoparticles

1. Introduction

Research involving gold nanoparticles has increased significantly over the last two decades, creating with it the need for high purity gold chloride (generally in the form of tetrachloroauric acid) as a starting material for the nanoparticle synthesis. Many synthetic methods utilize aqueous solutions of gold(III) chloride for this purpose. Although gold chloride can be purchased from commercial suppliers, we have found that this can increase laboratory costs significantly, especially when larger quantities of gold nanoparticles are being prepared. Furthermore, solutions of gold(III) chloride can be quite unstable, rapidly returning to metallic gold unless kept under appropriate conditions (e.g., pH < 4 with excess chloride ions) [1]. Here we report a straightforward method to produce highly stable aqueous solutions of gold(III) chloride using gold metal and molecular chlorine as starting materials. The method can be scaled from the millilitre scale to litre scale and requires no expensive reagents.

$$2Au^0_{(s)} + 3Cl_{2\ (g)} + 2HCl_{(aq)} \rightarrow 2HAu^{III}Cl_{4\ (aq)} \tag{1}$$

The reaction of metallic gold with chlorine is shown in Equation (1). This chemistry was applied to industrial processes as early as the 19th century to dissolve gold from gold-bearing ores [2–4], and although chlorination was quickly replaced by cyanidation in the gold mining industry, research into the use of halogens, in particular chlorine, bromine and iodine, has made a resurgence in recent decades due to environmental concerns associated with cyanide use [5,6]. A number of the industrial procedures for ore extraction use chlorine gas injected into the reaction chamber or alternatively water that contains dissolved chlorine gas [7], while other methods employ the reaction between sodium hypochlorite, hydrochloric acid and sodium chloride (Equation (2)) to generate chlorine gas *in situ* [8,9].

$$\text{NaOCl}_{(aq)} + 2\text{HCl}_{(aq)} + \text{NaCl}_{(aq)} \rightarrow 2\text{NaCl}_{(aq)} + \text{Cl}_{2\,(g)} + \text{H}_2\text{O}_{(l)} \quad (2)$$

Of course, *aqua regia* (formed from nitric and hydrochloric acids) has been used for centuries to dissolve gold and form gold(III) chloride, but this process requires cumbersome workup procedures (such as boiling down and repeated hydrochloric acid dilution/concentration cycles to remove nitric acid).

Thus, the method that we report here offers several advantages over the above-mentioned procedures. Importantly, we show that the known reaction of gold with chlorine can be used in a straightforward method to produce high purity and highly stable gold(III) chloride solutions that can be used directly (without further workup procedures) for gold nanoparticle synthesis. In comparison to solutions prepared from commercially available gold(III) chloride, this method offers a significant reduction in expense, ease of concentration determination and is scalable.

2. Experimental Section

2.1. General

Gold pellets (99.99%, ~250–360 mg, AGS Metals, Sydney, Australia), hydrochloric acid (36%; RCI Labscan, Pathumwan Bangkok, Thailand), potassium permanganate (BDH) and sodium thiosulfate pentahydrate (Ajax Chemicals, Sydney, Australia) were all used as received. Ultrapure Milli-Q water (Sartorius, Dandenong South, Australia; 18.2 MΩ cm) was used as the reaction solvent. UV-visible spectra were recorded using an Agilent Technologies Cary 60 UV-Visible spectrophotometer (Mulgrave Victoria, Australia) with 0.1 M aqueous HCl as the solvent. Inductively coupled plasma mass spectrometry was performed with an Agilent Technologies 7500cx series ICP-MS (Mulgrave Victoria, Australia) with sample introduction via a micromist concentric nebuliser (Glass expansion) and a Scott type double pass spray chamber cooled to 2 °C. The sample solution and the spray chamber waste were carried with the aid of a peristaltic pump. ICP-MS extraction lens conditions were selected to maximise the sensitivity of a 1% HNO$_3$:HCl solution containing 1 ng/mL of Li, Co, Y, Ce and Tl. Gold and manganese stock solutions were obtained from Choice Analytical (Thornleigh, Australia). Baseline nitric acid (HNO$_3$) was purchased from Seastar Chemicals Inc. (Sidney, BC, Canada). Calibration standards were prepared in 1% nitric acid. Oleylamine-stabilised gold nanoparticles were prepared by the literature method [10]. The concentration of chlorine in solution was measured using the iodometric titration method [11].

2.2. Synthesis of Tetrachloroauric Acid (HAuCl$_4$)

A laboratory-scale reaction apparatus was assembled as shown in Figure 1. Hydrochloric acid (14 mL, 36%, 0.16 mol) was placed in a pressure-equalizing dropping funnel, and slowly added drop-wise to a side-arm flask containing potassium permanganate (3.0 g, 19 mmol). The resulting chlorine gas (see Equation (3)) was passed into a two-neck round-bottom flask containing 100 mL Milli-Q water and one pellet of gold (~250–360 mg). Undissolved/unreacted chlorine gas was bubbled through a solution of sodium thiosulfate (1.1 g, 7.0 mmol) in 50 mL of water. The reaction mixture was stirred at the selected temperature (25–70 °C) until all of the gold dissolved. The resulting solution of tetrachloroauric acid can be used without further treatment. UV-Vis (λ_{max}, nm [ε, 10^4 M^{-1}·cm^{-1}]): 226 (3.5), 313 (0.54).

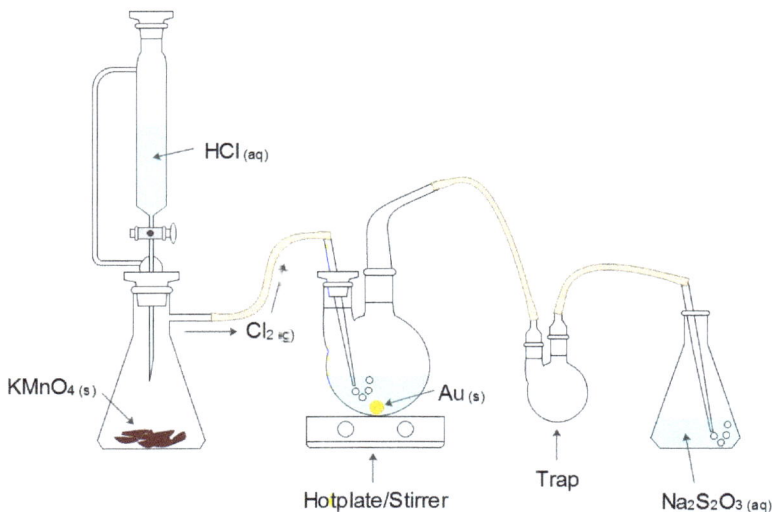

Figure 1. Apparatus for the synthesis of aqueous tetrachloroauric acid.

$$2KMnO_4 \,_{(s)} + 16HCl \,_{(aq)} \rightarrow 5Cl_2 \,_{(g)} + 2KCl \,_{(aq)} + 2MnCl_2 \,_{(aq)} + 8H_2O \,_{(l)} \tag{3}$$

2.3. Measurement of Reaction Times

The rate at which gold dissolved was measured by weighing the gold pellet at regular intervals. The pellet was removed from the reaction vessel using a clean glass spoon, rinsed with acetone and allowed to completely dry, and then weighed before returning to the reaction mixture. For experiments using flattened gold pellets, the pellets were placed between layers of cotton fabric or sheets of polycarbonate and mechanically pressed to a thickness of ~1 mm.

2.4. Synthesis of Citrate-Stabilised Gold Nanoparticles

The citrate method was used to synthesise aqueous gold nanoparticles from the tetrachloroauric acid synthesised as above. All glassware was cleaned with aqua regia, then rinsed and steeped in Milli-Q water prior to use. Aqueous tetrachloroauric acid (6.85 mL, 14.6 mM) was added to a 100 mL volumetric flask and diluted with Milli-Q water to make a 1.00 mM solution. The contents of the volumetric flask were poured into a 250 mL conical flask containing a magnetic stirrer bar, and heated to 90 °C. While stirring vigorously, a freshly prepared solution of tri-sodium citrate (114 mg, 10.0 mL, 38.8 mM) in Milli-Q water was quickly added to the conical flask. The solution was heated for a further 15 min, allowing it to turn a deep burgundy colour, indicating the presence of small gold nanoparticles.

3. Results and Discussion

The oxidation of gold metal to form aqueous solutions of gold(III) chloride was reliably and reproducibly achieved through introduction of chlorine gas to water containing the gold metal. The reaction is quantitative and no manganese-containing by-products (which may have arisen from the chlorine production process) were detected (see Table S1). Thus, the concentration of the gold(III) chloride solutions can be reasonably accurately calculated from the mass of the gold metal and the final volume of water.

The UV-visible spectrum of the resultant gold(III) chloride solution is shown in Figure 2 and displays the expected peaks for a solution at pH 1.05. The absorption bands at 313 and 226 nm are assigned to ligand-to-metal charge transfer [12,13]. It is important to note that the UV-visible spectrum is highly dependent on pH [14]. Therefore, accurate determination of concentrations of aqueous gold(III) chloride solutions using UV-visible spectroscopy requires an accurate determination of pH (and of course knowledge of the molar absorptivity at that particular pH).

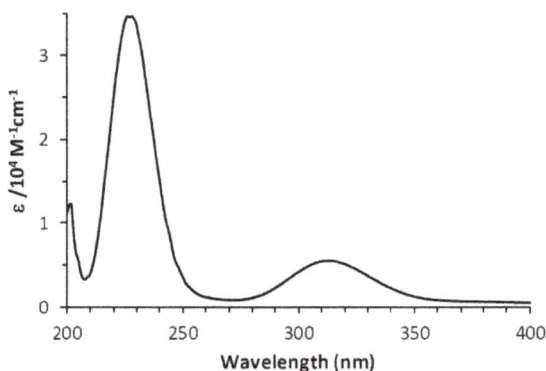

Figure 2. UV-Vis spectrum of $HAuCl_4$ in 0.1 M aqueous HCl.

At room temperature, the reaction proceeded at a moderate rate and the rate of reaction could be significantly increased by increasing the reaction temperature (Figure 3). The rate at 50 °C was significantly greater than that at room temperature and although the rate at 70 °C was greater than that at 50 °C, it was only marginally so. We attribute this to the reduction in solubility of chlorine gas at elevated temperatures [15].

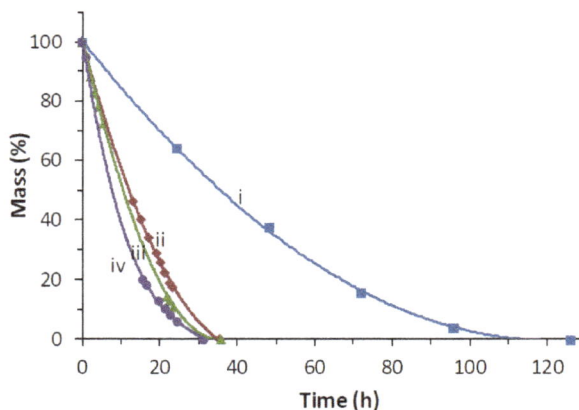

Figure 3. Reaction times at different temperatures, expressed as the percentage remaining of the initial mass of gold as a function of time: (**i**) 281 mg pellet at room temperature; (**ii**) 268 mg pellet at 50 °C; (**iii**) 319 mg pellet at 60 °C; (**iv**) 357 mg pellet at 70 °C. Note: pellets were used as-received from the supplier hence the range of masses used.

The effect of surface area of the gold pellet is also an important factor influencing reaction times. The data shown in Figure 3 were obtained using reasonably spherical (as purchased) gold pellets. Mechanical flattening of the gold pellets to thicknesses <1 mm resulted in significantly shorter reaction times whereby the flattened pellets dissolved in ~12 h at 50 °C (*ca.* 35 h for spherical pellets). However, care needs to be taken to prevent any metallic contamination of the gold during the flattening process. This may be accomplished by placing the gold between layers of cotton fabric or sheets of polycarbonate prior to the mechanical flattening process.

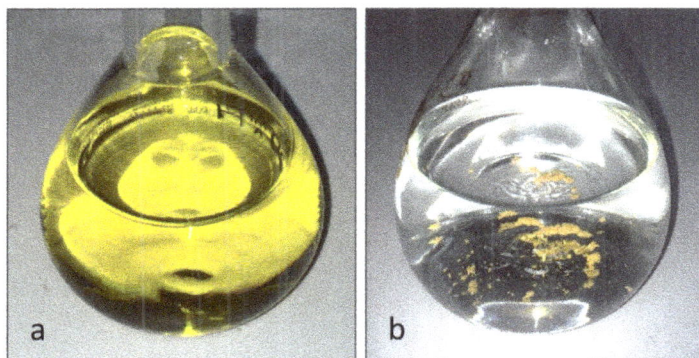

Figure 4. Photographs of gold(III) chloride solutions prepared by: (**a**) chlorine gas method; and (**b**) sodium hypochlorite/hydrochloric acid method (where precipitation occurred ~2 weeks post-synthesis).

The aqueous gold(III) chloride produced by this method is quite stable; no metallic gold precipitation was evident within at least twelve months of synthesis and the UV-visible spectra were invariant over this time (Figure S1). The stabilisation of the gold chloride was achieved without the addition of a source of excess chloride ions, such as sodium chloride, which would have made the gold chloride unusable for some nanoparticle synthesis methods due to the extreme sensitivity of many gold nanoparticles to ionic impurities [16]. The dissolved chlorine concentration of a three month old solution kept in a tightly stoppered flask was measured by iodometric titration to be 8.9×10^{-3} M. This concentration is somewhat lower than the chlorine saturation concentration of 0.08 M at 20 °C [15] and comparable to previous reports of gold chloride solutions with dissolved chlorine [17–19]. The solution pH was 1.6–1.8, where molecular chlorine is expected to be the predominant species [17], and well within the pH range required for gold chloride stability (<pH 4) [1]. Furthermore, the as-prepared gold(III) chloride solution can be directly freeze-dried to produce solid gold(III) chloride (note: the powder is highly hygroscopic, and must therefore be stored under strictly anhydrous conditions). In contrast, gold(III) chloride solutions prepared using sodium hypochlorite as oxidant followed by purification by extraction into diethyl ether, solvent removal and redissolving in pure water or dilute aqueous HCl were found to be significantly less stable over time (see Figure 4).

To test the suitability of the gold(III) chloride solutions for direct use in nanoparticle synthesis, citrate-stabilised aqueous gold nanoparticles were synthesised, which are known to be extremely sensitive to ionic impurities. The formation of the nanoparticles progressed as described in previous literature [20], and remained in a stable colloidal suspension, confirming the high purity of the gold(III) chloride solutions produced by the chlorine gas method. We also used the gold chloride solution as starting material to prepare organic-soluble oleylamine-stabilised gold nanoparticles [10]. This synthesis also proceeded as described in the literature method to yield stable nanoparticles. Scanning electron microscope images and UV-visible spectra of the particles are shown in the Supplementary Information (Figures S2–S4).

4. Conclusions

Aqueous solutions of gold(III) chloride of high purity and stability can be reliably synthesised from gold metal and molecular chlorine via a straightforward procedure. The reaction was found to proceed at a satisfactory rate at 50 °C, and progressed at a faster rate when the gold pellets were flattened compared to the as-purchased spherical beads.

Gold(III) chloride solutions prepared by this method did not precipitate any metallic gold within at least twelve months of synthesis. ICP-MS analysis confirmed the absence of any manganese contamination that may have arisen from the chlorine production process, and the purity of the gold chloride solution was further demonstrated through its direct use in the synthesis of stable aqueous citrate-stabilised as well as oleylamine-stabilised gold nanoparticles. The use of gold metal as the starting material allows for accurate determination of the final gold(III) chloride concentration from the mass of gold and the volume of water.

Acknowledgments: We acknowledge Mr. Alexander Angeloski and Mr. Sanket Aggarwal for laboratory assistance.

Author Contributions: Shirin R. King: sample preparation, data analysis and writing the manuscript. Juliette Massicot: development of the process and data analysis. Andrew M. McDonagh: supervision of the first and second authors and writing the manuscript.

Conflicts of Interest: The authors declare no conflict of interest.

References

1. Nam, K.S.; Jung, B.H.; An, J.W.; Ha, T.J.; Tran, T.; Kim, M.J. Use of Chloride-Hypochlorite Leachants to Recover Gold from Tailing. *Int. J. Miner. Process.* **2008**, *86*, 131–140. [CrossRef]
2. Mellor, J.W. *A Comprehensive Treatise on Inorganic and Theoretical Chemistry*, 2nd ed.; Longmans, Green and Co. Ltd.: London, UK, 1928; Volume 3, pp. 491–618.
3. Konyratbekova, S.S.; Baikonurova, A.; Akcil, A. Non-cyanide Leaching Processes in Gold Hydrometallurgy and Iodine-Iodide Applications: A Review. *Miner. Process. Extr. Metall.: Int. J.* **2014**, *36*, 198–212. [CrossRef]
4. French, A. Process of Obtaining Gold, Silver and Copper from Ores. U.S. Pat. Off. US490193, 17 January 1893.
5. La Brooy, S.R.; Linge, H.G.; Walker, G.S. Review of Gold Extraction from Ores. *Miner. Eng.* **1994**, *7*, 1213–1241. [CrossRef]
6. Syed, S. Recovery of Gold from Secondary Sources—A Review. *Hydrometallurgy* **2012**, *115–116*, 30–51. [CrossRef]
7. Kallmes, J. Improvement in Treating Gold and Silver Ores. U.S. Pat. Off. US138500, 6 May 1873.
8. Puvvada, G.V.K.; Murthy, D.S.R. Selective Precious Metals Leaching from a Chalcopyrite Concentrate Using Chloride/Hypochlorite Media. *Hydrometallurgy* **2000**, *58*, 185–191. [CrossRef]
9. Simpson, C.H. Chlorination Process for Removing Precious Metals from Ore. U.S. Pat. Off. US4439235, 27 March 1984.
10. Choi, H.; Chen, W.T.; Kamat, P.V. Know Thy Nano Neighbor. Plasmonic *versus* Electron Charging Effects of Metal Nanoparticles in Dye-Sensitized Solar Cells. *ACS Nano* **2012**, *6*, 4418–4427. [CrossRef] [PubMed]
11. Skoog, D.A.; West, D.M.; Holler, F.J. *Fundamentals of Analytical Chemistry*, 6th ed.; Saunders College Publishing: Sydney, Australia, 1992; pp. 373–374.
12. Carlin, R.L. *Transition Metal Chemistry*; Marcel Dekker, Inc.: New York, NY, USA, 1965; Volume 1, pp. 245–253.
13. Gangopadhayay, A.K.; Chakravorty, A. Charge Transfer Spectra of some Gold(III) Complexes. *J. Chem. Phys.* **1961**, *35*, 2206–2209. [CrossRef]
14. Peck, J.A.; Tait, C.D.; Swanson, B.I.; Brown, G.E., Jr. Speciation of Aqueous Gold(III) Chlorides from Ultraviolet/Visible Absorption and Raman/Resonance Raman Spectroscopies. *Geochim. Cosmochim. Acta* **1991**, *55*, 671–676. [CrossRef]
15. Alkan, M.; Oktay, M.; Kocakerim, M.M.; Copur, M. Solubility of Chlorine in Aqueous Hydrochloric Acid Solutions. *J. Hazard. Mater.* **2005**, *119*, 13–18. [CrossRef] [PubMed]
16. Turkevich, J. Colloidal Gold. Part II Colour, Coagulation, Adhesion, Alloying and Catalytic Properties. *Gold Bull.* **1985**, *18*, 125–131. [CrossRef]
17. Vinals, J.; Nunez, C.; Herroros, O. Kinetics of the Aqueous Chlorination of Gold in Suspended Particles. *Hydrometallurgy* **1995**, *38*, 125–147. [CrossRef]
18. Diaz, M.A.; Kelsall, G.H.; Welham, N.J. Electrowinning Coupled to Gold Leaching by Electrogenerated Chlorine: I. Au(III)–Au(I)/Au Kinetics in Aqueous Cl_2/Cl^- Electrolytes. *J. Electroanal. Chem.* **1993**, *361*, 25–38. [CrossRef]

19. Sun, T.M.; Yen, W.T. Kinetics of Gold Chloride Adsorption onto Activated Carbon. *Miner. Eng.* **1993**, *6*, 17–29. [CrossRef]
20. Frens, G. Controlled Nucleation for the Regulation of the Particle Size in Monodisperse Gold Suspensions. *Nat. Phys. Sci.* **1973**, *241*, 20–22. [CrossRef]

![metals logo] *metals*

MDPI

Article

Exploring the Possibilities of Biological Fabrication of Gold Nanostructures Using Orange Peel Extract

Laura Castro, María Luisa Blázquez *, Felisa González, Jesús Ángel Muñoz and Antonio Ballester

Department of Materials Science and Metallurgical Engineering, Complutense University of Madrid, Madrid 28040, Spain; lcastror@ucm.es (L.C.); fgonzalezg@ucm.es (F.G.); jamunoz@ucm.es (J.A.M.); ambape@ucm.es (A.B.)

* Author to whom correspondence should be addressed; mlblazquez@ucm.es; Tel.: +34-91-394-4335; Fax: +34-91-394-4357.

Academic Editors: Suresh Bhargava and Rahul Ram
Received: 22 July 2015; Accepted: 6 September 2015; Published: 11 September 2015

Abstract: Development of nanotechnology requires a constant innovation and improvement in many materials. The exploration of natural resources is a promising eco-friendly alternative for physical and chemical methods. In the present work, colloidal gold nanostructures were prepared using orange peel extract as a stabilizing and reducing agent. The initial pH value of the solution and the concentration of the gold precursor had an effect on the formation and morphology of nanoparticles. The method developed is environmentally friendly and allows control of nanoparticles. By controlling the pH and, especially, the gold concentration, we are able to synthesize crystalline gold nanowires using orange peel extract in the absence of a surfactant or polymer to direct nanoparticle growth, and without external seeding. UV-VIS spectroscopy, transmission electron microscopy (TEM), and X-ray diffraction (XRD) were used to characterize the nanoparticles obtained by biosynthesis.

Keywords: biosynthesis; nanoparticles; gold; nanowires; orange peel extract

---

1. Introduction

Nanostructured materials have received considerable attention due to their unique physical and chemical properties and their potential application [1–3]. The synthesis of nanoparticles with a controlled shape and size is one of the most promising research areas. The excellent properties of some materials strongly depend on crystallographic and morphological characteristics [4]. It is well-known that triangular nanoparticles of gold exhibit two characteristic absorption bands referred to as the transverse (out of plane) and longitudinal (in plane) surface plasmon resonance bands. The ability to tune the optical properties of the gold nanotriangles can be very useful in applications such as cancer cell hyperthermia [5] and architectural optical coatings [6]. One-dimensional nanostructures have also unusual anisotropic properties. For example, if metal nanoparticles are smaller than approximately tens of nanometers, electrons should be confined along the diameter of long metallic nanowires. Metallic nanowires are useful to connect different components in nanodevices. Metal nanorods and nanowires are able to absorb and scatter light along the long and the short axis because they have two plasmon bands [7]. These structures have been used as components in flexible electronics [8] and biological or gas sensing applications [9].

Synthesis of metal nanoparticles via chemical and physical methods has been employed in nanotechnology due to their affordability and ease of modulation in functional behavior of nanostructures [10–12]. However, toxic effects of various chemicals and organic solvents used in physical and chemical methods have promoted an increasing interest in biomass as a biosynthetic machinery for the production of metal nanoparticles [13–16]. Biosynthesis of gold nanoparticles has been reported using bacteria [17], yeasts [18], actinomycetes [19], fungi [20], and plants [21]. Although biological methods are regarded as safe, cost-effective, sustainable, and clean processes, they also have some drawbacks in culturing microbes and using biomasses, which are time-consuming and difficult in providing better control over size distribution, shape, and crystallinity [22]. These are the problems that have plagued the biological synthesis approaches.

The use of plants in the recovery of noble metals from ore mines and runoffs is known as phytomining. Compared to the conventional chemical methods, phytomining is a cost-effective, environmentally compatible method [23]. The use of plants and its by-products as sustainable and renewable resources in the synthesis of nanoparticles is more advantageous over prokaryotic microbes, which need expensive methodologies for maintaining microbial cultures and downstream processing [21,24].

Another dimension was added to the "green chemistry" approach for pure metal synthesis with the use of plant broths. Synthesis of noble metal nanoparticles using plant extracts is very cost effective, and therefore can be used as an economic and valuable alternative for the large-scale fabrication of metal nanoparticles. Extracts from plants may act both as reducing and capping agents in nanoparticle synthesis. The bioreduction of metal nanoparticles by combinations of biomolecules found in plant extracts (enzymes, proteins, amino acids, vitamins, polysaccharides, and organic acids, such as citrates) is environmentally benign, despite of its chemical complexity [25,26].

In this work, biosynthesis of gold nanostructures has been investigated using aqueous chloroaurate ions and orange peel extract as a clean technology to recover gold from dilute solutions. Orange peel is a residue obtained after juice extraction in the food industry. Waste orange peel is composed of sugars, cellulose, hemicellulose, pectin and D-limonene [27]. The morphology of the nanoparticles produced could be controlled by varying the initial pH value and the concentration of gold ions in the reaction medium. Triangular nanoplates and nanorods were obtained at low pH and nanospheres at high pH values. Additionally, gold nanowires were produced by increasing the gold ion concentration.

2. Materials and Methods

2.1. Materials

All chemical reagents including chloroauric acid ($HAuCl_4$), sodium hydroxide flakes, and hydrochloric acid (37%) were obtained from Panreac (Barcelona, Spain).

2.2. Synthesis of Gold Nanoparticles

The experimental method consists of mixing aqueous solution of metallic precursor with orange peel extract (9:1) and stirring the mixture. The orange peel extract was previously prepared by boiling the dried orange peel in deionized water. Then, the extract was obtained by filtration using 0.2 μm nylon membrane filters from Whatman (Dassel, Germany). The initial pH of orange peel extract was 5.0. Chloroauric acid ($HAuCl_4$) was used as precursor. Gold nanoparticles were synthesized at room temperature. The influence of solution pH (2, 4, 7, and 9) was investigated in 100 mg/L (0.5 mM Au^{3+}) of metal precursor aqueous solutions. The pH was adjusted by addition of NaOH solution and HCl in Au solutions. The effect of precursor concentration to obtain gold nanowires was tested by using 100 mg/L (0.5 mM Au^{3+}), 250 mg/L (1.25 mM Au^{3+}) and 500 mg/L Au^{3+} (2.5 mM Au^{3+}).

2.3. UV-VIS Absorbance Spectroscopy Studies

The UV-VIS spectra of samples at different pH values were analyzed using a Libra S11 single beam spectrophotometer (Cambridge, UK) operated at a resolution of 5 nm with quartz cells. Blanks of each sample set were prepared with deionized water.

2.4. TEM Measurements

The shape of the nanoparticles was observed by transmission electron microscopy (TEM). Samples of the biosynthesized gold nanoparticles were prepared by placing drops of the product solution onto carbon-coated copper grids and allowing the solvent to evaporate. TEM measurements were performed on a JEOL model JEM-2100 instrument (Tokyo, Japan) operated at an accelerating voltage of 200 kV.

2.5. X-ray Diffraction Analysis

The measurements were carried out by powder X-ray diffraction (XRD) on a Philips X'pert-MPD equipment (El Dorado County, CA, USA) with a Cu anode operating at a wavelength of 1.5406 Å as the radiance source. Samples were placed on off-axis quartz plates (18 mm diameter × 0.5 mm DP cavity). The scanning range was from 10° to 60° 2θ with an angular interval of 0.05° and 4 s counting time. The crystalline phases were identified using standard cards from the International Centre for Diffraction Data (ICDD, Newtown Square, PA, USA) Powder Diffraction File database.

3. Results

The present work was focused on the development of a biosynthetic method to control the production of gold nanostructures using green chemistry. For that, authors propose an efficient exploitation of wastes from food industry production. The color of gold solutions changed from pale yellow to dark blue or pink in 30 min (Figure 1). This change would be an indication of gold reduction by orange peel extract and the formation of nanoparticles. Orange peel extract contains organic acids, amino acids, and proteins, as well as the important presence of saccharides, which provides reduction power for the nanostructures' preparation. In order to complete the study, an attempt to develop a biosynthetic method for the production of gold nanowires using green chemistry was carried out.

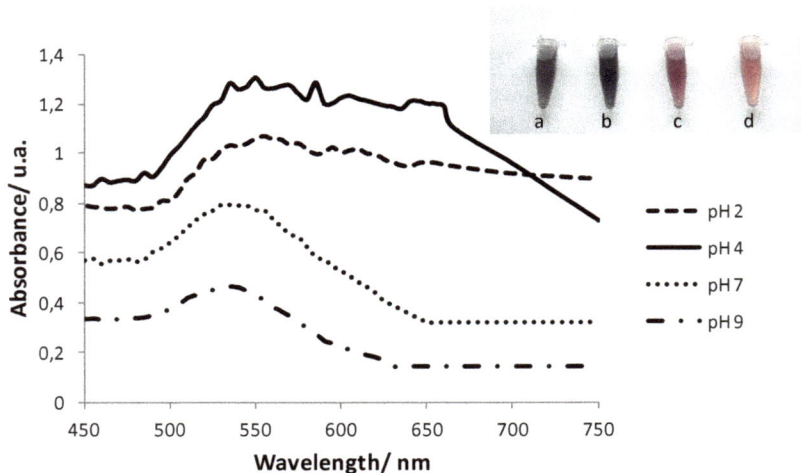

Figure 1. UV-VIS spectra of gold nanoparticles using 100 mg/L $HAuCl_4$ prepared at different initial pH values: pH 2, 4, 7, and 9. The inset shows the solutions corresponding to these spectra: (**a**) pH 2; (**b**) pH 4; (**c**) pH 7; and (**d**) pH 9.

3.1. Biosynthesis of Gold Nanoparticles: Influence of pH

3.1.1. UV-VIS Spectral Study

It is well known that the differences in UV-VIS absorption spectra and color of solution could be dependent on different surrounding mediums and size, shape, and crystallinity of the metal nanoparticles. The color of chloroaurate solutions changed to pink or blue depending on the pH tested, indicating a change in metal oxidation and the formation of gold nanoparticles. As shown in the inset in Figure 1, the color of the colloidal gold solutions is a function of pH.

In all cases, a surface plasmon resonance (SPR) band absorption peak appears centered at approximately 540 nm, which is characteristic of gold nanoparticles. At a longer wavelength, a second band related to aggregates of spherical nanoparticles [28] or anisotropic nanostructures [29] is present in solutions at pH 2 and 4.

Triangular nanoparticles, nanorods, and nanowires of gold display two distinct SPR bands referred to as transverse and longitudinal electron oscillations. The transverse SPR band coincides with the longitudinal SPR band of spherical gold nanoparticles. However, the longitudinal oscillation is very sensitive to the nanoparticle shape. In consequence, slight deviations from spherical shape can lead to dramatic color changes. The longitudinal SPR band is a strong function of the edge length of the triangles [5]. Gold nanorods and nanowires exhibit both a band corresponding to the short axis and another one corresponding to the long axis at a longer wavelength. At pH 2, the absorbance curve is practically flat in the range of 500–825 nm and the intensity is lower than at pH 4. In addition, this spectrum corresponds to the blue sample. In this case, the two peaks of the SPR bands cannot be clearly observed due to the overlap of the longitudinal absorption of the nanorods with different aspect ratios at relevant wavelengths [30].

At pH 7 and 9, the spectra exhibit a unique resonance wavelength at approximately 540 nm associated to the formation of the nanospheres since all their electronic oscillations are equivalent. The band at pH 9, however, is broader indicating the polydispersity of size.

3.1.2. Characterization of Gold Nanoparticles

The morphology of the gold nanoparticles was observed by transmission electron microscopy. Figure 2 shows representative TEM images of the nanoparticles synthesized using orange peel extract at different pH values. TEM observations revealed that gold nanoparticles formed at acidic medium were mainly triangular, polygonal and rod-shaped nanoparticles (Figure 2a,b). The edge length of the rods can reach 120 nm at pH 2. When the pH was increased to 4, the size of the nanoparticles decreased and the length of the triangles' edge and the rods was ~40 nm.

An increase of initial solution pH favored the formation of nanospheres. At pH 7 and 9, spherical nanoparticles were formed (Figure 2c,d). The analysis showed that the average diameter of the nanospheres obtained at pH 7 was about 20 nm. At pH 9, there were nanoparticles with a diameter of 25 nm and smaller nanoparticles of 10 nm of diameter.

EDS microanalysis and XRD of the gold nanoparticles showed that the sample was essentially metallic gold (Figure 3). Elemental gold peaks were found in the EDS study to confirm the nature of the metallic nanoparticles (Figure 3a). The structural properties of gold nanoparticles were confirmed using XRD technique. XRD pattern showed that diffraction peaks corresponded to the diffraction planes (111), (200), (220), (311) and (220) of face-centered cubic gold (Figure 3b).

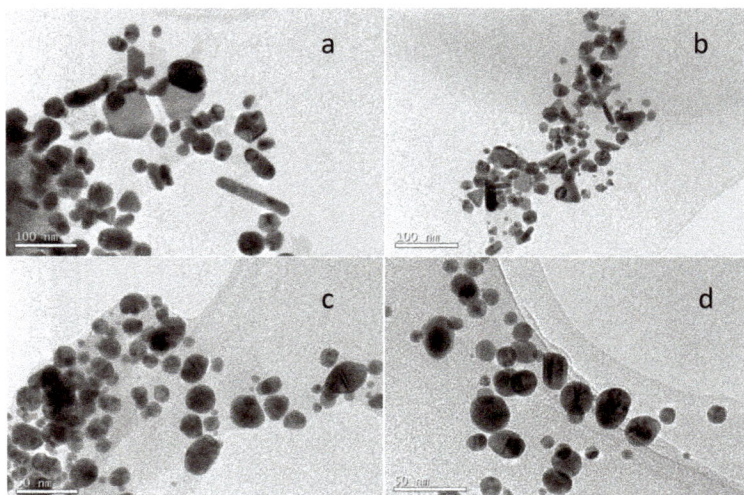

Figure 2. TEM images of gold nanoparticles using 100 mg/L HAuCl$_4$ prepared at different initial pH values: (**a**) pH 2; (**b**) pH 4; (**c**) pH 7; and (**d**) pH 9.

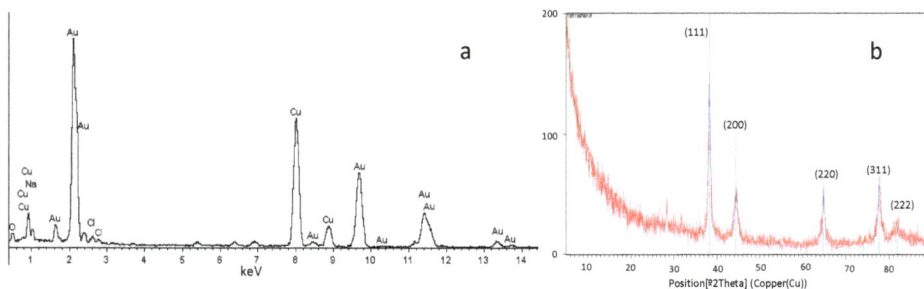

Figure 3. (**a**) EDS spectrum and (**b**) XRD pattern of gold nanoparticles.

3.2. Biosynthesis of Gold Nanowires: Influence of Gold Ion Concentration

It is already known that hydroxide ions play a key role in the production of nanowires. In spite that the capping agent was available, competition between the biomolecules of the biomass and hydroxide ions for gold ions favored the aggregation of nanoparticles due to the lower ability of biomolecules to stabilize the nuclei formed. In turn, nanoparticle aggregation led to the formation of nanowires. This process could be favored not only in the presence of hydroxide ions but also with increasing gold ion concentration. Hence, gold nanowires production was investigated by increasing the concentration of HAuCl$_4$ at high pH value (pH 10).

TEM images of the samples evidenced that gold concentration has an enormous influence in the shape and size of the gold nanoparticles synthesized using orange peel extract. As it is shown in Figure 4a, spherical nanoparticles were obtained at 100 mg/L. There were some particles with an average size of 42 nm and other very tiny nanoparticles of 5 nm. When the initial concentration of gold precursor was 250 mg/L, triangular and polygonal nanoparticles were obtained and their edges can reach up to 110 nm, and were surrounded by spherical nanoparticles with 20 nm of diameter (Figure 4b). When the gold precursor concentration was increased to 500 mg/L, nanowires with an average diameter of 30 nm were formed (Figure 4c). Figure 4d shows the fine nanostructure of one section of the gold nanowires depicted in Figure 4c.

At the lowest initial gold concentration, the amount of capping agent was enough to stabilize the first particles formed in solution.

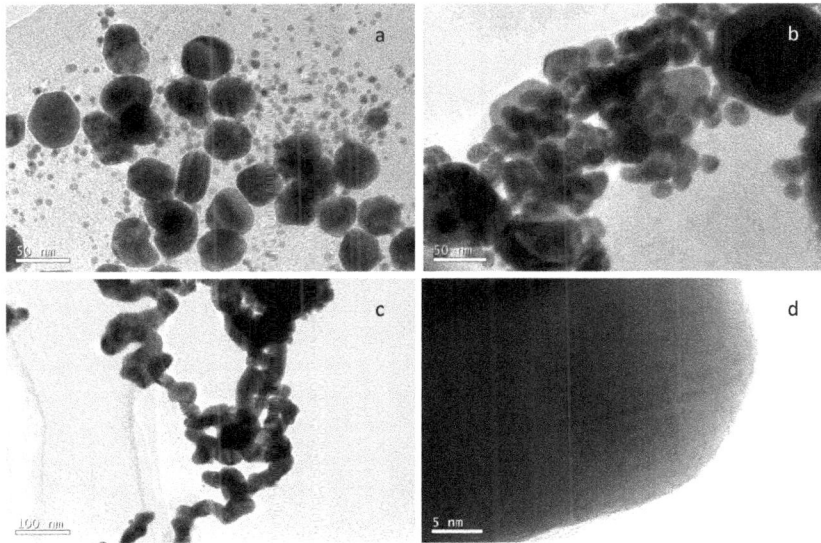

Figure 4. Effect of gold ions concentration. TEM images of gold nanoparticles obtained at pH 10 using different HAuCl$_4$ concentrations: (**a**) 100; (**b**) 250; and (**c**) 500 mg/L; (**d**) detail of nanostructure of the gold nanowires.

Increasing concentration, gold nanoparticles size increased since they were not thermodynamically stable and stuck together. At the highest concentration, gold nanowires were produced. The formation of these nanowires can be related to the Ostwald ripening effect. In Ostwald ripening, smaller particles dissolve preferentially with subsequent crystallization onto larger particles, which involves nucleation and growth processes of larger particles from smaller ones.

4. Discussion

The green synthesis of silver and platinum nanoparticles using orange peel extract have been previously reported [31]. It is found that the pH value influences the morphologies of the nanostructures. The results indicated that the properties of silver and platinum nanoparticles depend strongly on their morphologies and sizes. In general, silver nanospheres with lower diameter posses the higher antimicrobial activity. However, silver nanowires are useful to connect different components in nanodevices. Small and spherical platinum nanoparticles exhibited better catalytic properties.

The biosynthesis of nanoparticles with a controlled shape and size is one of the most promising research areas. Gold nanoparticles are interesting due to their applications in optoelectronic, electronic, and magnetic devices, and because of their use as catalysts and sensors. One- and two-dimensional nanostructures for noble metals are especially attractive.

The initial pH value of the aqueous HAuCl$_4$ solutions for different sets of samples was studied in the synthesis of gold nanoparticles using orange peel extract. At high pH values, are spherical and monodispersed gold nanoparticles were obtained. These nanoparticles could be used in non-toxic carriers for drug and gene delivery applications. With these systems, the gold core imparts stability to the assembly, while the monolayer allows tuning of surface properties such as charge and hydrophobicity. An additional attractive feature of these nanoparticles is their interaction with thiols, providing an effective and selective means of controlled intracellular release [32].

The synthesis of gold nanotriangles under acidic conditions could be produced by slow reduction and crystallization. This leads to the formation of stable multiply twinned particles which evolve into gold nanotriangles due to the shape-directing effect of the constituents of the orange peel. Moreover, the difference in the growth rates of the various crystallographic planes could lead to the changes in the morphology of the nanoparticles. It can also be observed the hexagonal shapes, truncated triangles or icosahedral shapes. These structures with unique and highly anisotropic planer shapes might find application in photonics, optoelectronics, and optical sensing. Applications in cancer cell hyperthermia, but also the effect of different organic solvent vapors like methanol, benzene and acetone on the conductivity of tamarind leaf extract reduced gold nanotriangles, among others, have been investigated [33].

Gold nanorods and nanowires have usually been synthesized by electrochemical reduction with cetyltrimethylammonium bromide (CTAB) and by seed-mediated growth method using a surfactant template. The synthesis of gold nanowires using biological resources has hardly ever been reported. Recently, networked nanowires were synthesized with an extract of *Rhodopseudomonas capsulata* modulating the concentration of $HAuCl_4$ [34] and gold nanowires were also obtained using sugar beet pulp by changing the pH and the gold concentration [30]. Reduction with orange peel extract is a method to produce nanowires in the absence of surfactant or polymer to control nanoparticles growth and without externally added seed crystallites. As mentioned previously, gold is a face centered cubic metal and its surface energies associated with different crystallographic planes are usually different. The free energies decrease in the order $\gamma\{111\} > \gamma\{100\} > \gamma\{110\}$, because of the package and the distinct surface atoms density. Surfactants and capping agents can usually cap planes with $\{100\}$ growth, which might lead to anisotropic growth of gold nanoparticles through $\{111\}$ planes [35]. The effect of the hydroxyl ions in the shape and size of the gold nanoparticles using biomass has been previously reported by our group and the results showed that nanowires are produced in alkaline solutions [30]. The explanation for this probably involves Ostwald ripening and the capping action of the biomolecules. The biomolecules of the orange peel extract were insufficient to stabilize completely gold nuclei. The first gold nanoparticles formed were thermodynamically unstable due to their small size and the insufficient amount of capping agent. The uncovered face of gold nanoparticles could orientate against each other through dipole-dipole interaction of biomolecules and the intermolecular hydrogen bounding interaction of biomolecules and hydroxide ions could act as possible driving forces for the formation of 1D assembly. In addition, the surface energy of larger particles is lower than that of smaller ones [36]. The nanoparticles grew and joined together because of their Brownian motion in the solution, forming wire-like structures. The end of these nanorods may be bounded by $\{111\}$ facets whose interaction with OH^- and with biomolecules is very weak and, thus, reactive to the new atoms formed [37]. The weak complexing agent allows the deposition of newly reduced gold atoms and they sinter together and form larger agglomerates of gold (nanowires).

5. Conclusions

Reduction of tetrachloroaurate with orange peel extract is a simple, conducted at room temperature, efficient, and a clean method to synthesize gold nanostructures from dilute hydrometallurgical solutions. Advantages over other biological methods include the reuse of industry wastes, easy handling and scalability. In addition, one- and two-dimensional nanostructures for noble metals are especially attractive.

The key points for nanowires production are: competition between hydroxide ion and biomolecules for gold ions, shortage of biomolecules to cap the nanoparticles and preferential adsorption of $AuCl_4^-$ on the gold nanoparticles surface.

Acknowledgments: The authors are grateful for the financial support given by the Spanish Ministry of Economy and Competitiveness to fund this work. We acknowledge Juan Luis Bardonado from the Electronic Microscopy Centre of UCM for their technical assistance.

Author Contributions: L. Castro mainly performed the present study and wrote the paper. M.L. Blázquez, F. González, J.A. Muñoz and A. Ballester provided advice and recommendations on the research and polishing of the paper.

Conflicts of Interest: The authors declare no conflict of interest.

References

1. Gao, S.; Li, W.; Cao, R. Palladium-pyridyl catalytic films: A highly active and recyclable catalyst for hydrogenation of styrene under mild conditions. *J. Colloid Interface Sci.* **2015**, *441*, 85–89. [CrossRef] [PubMed]

2. Moghadam, A.H.; Dashtizad, V.; Kaflou A.; Yoozbashizadeh, H.; Ashiri, R. Development of a nanostructured Zr_3Co intermetallic getter powder with enhanced pumping characteristics. *Intermetallics* **2015**, *57*, 51–59. [CrossRef]

3. Jin, Y.; Jia, M. Design and synthesis of nanostructured graphene-SnO_2-polyaniline ternary composite and their excellent supercapacitor performance. *Colloids Surf. A Physicochem. Eng. Asp.* **2015**, *464*, 17–25. [CrossRef]

4. Liz-Marzán, L.M. Nanometals: Formation and color. *Mater. Today* **2004**, *7*, 26–31. [CrossRef]

5. Chandran, S.P.; Chaudhary, M.; Pasricha, R.; Ahmad, A.; Sastry, M. Synthesis of Gold Nanotriangles and Silver Nanoparticles Using Aloevera Plant Extract. *Biotechnol. Prog.* **2006**, *22*, 577–583. [CrossRef] [PubMed]

6. Shankar, S.S.; Rai, A.; Ahmad, A.; Sastry, M. Controlling the Optical Properties of Lemongrass Extract Synthesized Gold Nanotriangles and Potential Application in Infrared-Absorbing Optical Coatings. *Chem. Mater.* **2005**, *17*, 566–572. [CrossRef]

7. Murphy, C.J.; Gole, A.M.; Hunyadi, S.E.; Orendorff, C.J. One-Dimensional Colloidal Gold and Silver Nanostructures. *Inorg. Chem.* **2006**, *45*, 7544–7554. [CrossRef] [PubMed]

8. Vuillaume, D. Molecular-scale electronics. *Comptes Rendus Phys.* **2008**, *9*, 78–94. [CrossRef]

9. Jo, S.Y.; Kang, B.R.; Kim, J.T.; Ra, H.W.; Im, Y.H. The synthesis of single PdAu bimetallic nanowire: Feasibility study for hydrogen sensing. *Nanotechnology* **2010**. [CrossRef] [PubMed]

10. Mikkelsen, K.; Cassidy, B.; Hofstetter, N.; Bergquist, L.; Taylor, A.; Rider, D.A. Block Copolymer Templated Synthesis of Core-Shell PtAu Bimetallic Nanocatalysts for the Methanol Oxidation Reaction. *Chem. Mater.* **2014**, *26*, 6928–6940. [CrossRef]

11. Xiao, A.; Zhou, S.; Zuo, C.; Zhuan, Y.; Ding, X. Electrodeposited porous metal oxide films with interconnected nanoparticles applied as anode of lithium ion battery. *Mater. Res. Bull.* **2014**, *60*, 864–867. [CrossRef]

12. Biswas, A.; Bayer, I.S.; Biris, A.S.; Wang, T.; Dervishi, E.; Faupel, F. Advances in top-down and bottom-up surface nanofabrication: Techniques, applications & future prospects. *Adv. Colloid Interface Sci.* **2012**, *170*, 2–27. [PubMed]

13. Castro, L.; Blázquez, M.L.; González, F.; Muñoz, J.A.; Ballester, A. Extracellular biosynthesis of gold nanoparticles using sugar beet pulp. *Chem. Eng. J.* **2010**, *164*, 92–97. [CrossRef]

14. Castro, L.; Blazquez, M.L.; Munoz, J.A.; Gonzalez, F.; Ballester, A. Biological synthesis of metallic nanoparticles using algae. *IET Nanobiotechnol.* **2013**, *7*, 109–116. [CrossRef] [PubMed]

15. Syed, A.; Saraswati, S.; Kundu, G.C.; Ahmad, A. Biological synthesis of silver nanoparticles using the fungus *Humicola* sp. and evaluation of their cytoxicity using normal and cancer cell lines. *Spectrochim. Acta A Mol. Biomol. Spectrosc.* **2013**, *114C*, 144–147. [CrossRef] [PubMed]

16. Husseiny, M.I.; El-Aziz, M.A.; Badr, Y.; Mahmoud, M.A. Biosynthesis of gold nanoparticles using Pseudomonas aeruginosa. *Spectrochim. Acta A Mol. Biomol. Spectrosc.* **2007**, *67*, 1003–1006. [CrossRef] [PubMed]

17. Rösken, L.; Körsten, S.; Fischer, C.; Schönleber, A.; van Smaalen, S.; Geimer, S.; Wehner, S. Time-dependent growth of crystalline Au^0-nanoparticles in cyanobacteria as self-reproducing bioreactors: 1. *Anabaena* sp. *J. Nanopart. Res.* **2014**, *16*, 1–14. [CrossRef]

18. Agnihotri, M.; Joshi, S.; Kumar, A.R.; Zinjarde, S.; Kulkarni, S. Biosynthesis of gold nanoparticles by the tropical marine yeast Yarrowia lipolytica NCIM 3589. *Mater. Lett.* **2009**, *63*, 1231–1234. [CrossRef]

19. Bansal, V.; Bharde, A.; Ramanathan, R.; Bhargava, S.K. Inorganic materials using "unusual" microorganisms. *Adv. Colloid Interface Sci.* **2012**, *179–182*, 150–168. [CrossRef] [PubMed]

20. Gade, A.; Ingle, A.; Whiteley, C.; Rai, M. Mycogenic metal nanoparticles: Progress and applications. *Biotechnol. Lett.* **2010**, *32*, 593–600. [CrossRef] [PubMed]

21. Iravani, S. Green synthesis of metal nanoparticles using plants. *Green Chem.* **2011**, *13*, 2638–2650. [CrossRef]

22. Grzelczak, M.; Perez-Juste, J.; Mulvaney, P.; Liz-Marzan, L.M. Shape control in gold nanoparticle synthesis. *Chem. Soc. Rev.* **2008**, *37*, 1783–1791. [CrossRef] [PubMed]

23. Narayanan, K.B.; Sakthivel, N. Green synthesis of biogenic metal nanoparticles by terrestrial and aquatic phototrophic and heterotrophic eukaryotes and biocompatible agents. *Adv. Colloid Interface Sci.* **2011**, *169*, 59–79. [CrossRef] [PubMed]

24. Zheng, B.; Kong, T.; Jing, X.; Odoom-Wubah, T.; Li, X.; Sun, D.; Lu, F.; Zheng, Y.; Huang, J.; Li, Q. Plant-mediated synthesis of platinum nanoparticles and its bioreductive mechanism. *J. Colloid Interface Sci.* **2013**, *396*, 138–145. [CrossRef] [PubMed]

25. Gericke, M.; Pinches, A. Biological synthesis of metal nanoparticles. *Hydrometallurgy* **2006**, *83*, 132–140. [CrossRef]

26. Duran, N.; Marcato, P.D.; Duran, M.; Yadav, A.; Gade, A.; Rai, M. Mechanistic aspects in the biogenic synthesis of extracellular metal nanoparticles by peptides, bacteria, fungi, and plants. *Appl. Microbiol. Biotechnol.* **2011**, *90*, 1609–1624. [CrossRef] [PubMed]

27. Balu, A.M.; Budarin, V.; Shuttleworth, P.S.; Pfaltzgraff, L.A.; Waldron, K.; Luque, R.; Clark, J.H. Valorisation of orange peel residues: Waste to biochemicals and nanoporous materials. *ChemSusChem* **2012**, *5*, 1694–1697. [CrossRef] [PubMed]

28. Schwartzberg, A.M.; Grant, C.D.; van Buuren, T.; Zhang, J.Z. Reduction of HAuCl$_4$ by Na$_2$S Revisited: The Case for Au Nanoparticle Aggregates and Against Au$_2$S/Au Core/Shell Particles. *J. Phys. Chem. C* **2007**, *111*, 8892–8901. [CrossRef]

29. Kelly, K.L.; Coronado, E.; Zhao, L.L.; Schatz, G.C. The Optical Properties of Metal Nanoparticles: The Influence of Size, Shape, and Dielectric Environment. *J. Phys. Chem. B* **2002**, *107*, 668–677. [CrossRef]

30. Castro, L.; Blázquez, M.L.; Muñoz, J.A.; González, F.; García-Balboa, C.; Ballester, A. Biosynthesis of gold nanowires using sugar beet pulp. *Process Biochem.* **2011**, *46*, 1076–1082. [CrossRef]

31. Castro, L.; Blázquez, M.L.; González, F.; Muñoz, J.Á.; Ballester, A. Biosynthesis of silver and platinum nanoparticles using orange peel extract: Characterisation and applications. *IET Nanobiotechnol.* **2015**. [CrossRef]

32. Ghosh, P.; Han, G.; De, M.; Kim, C.K.; Rotello, V.M. Gold nanoparticles in delivery applications. *Adv. Drug Deliv. Rev.* **2008**, *60*, 1307–1315. [CrossRef] [PubMed]

33. Ankamwar, B.; Chaudhary, M.; Sastry, M. Gold Nanotriangles Biologically Synthesized using Tamarind Leaf Extract and Potential Application in Vapor Sensing. *Synth. React. Inorg. Met.-Org. Nano-Met. Chem.* **2005**, *35*, 19–26. [CrossRef]

34. He, S.; Zhang, Y.; Guo, Z.; Gu, N. Biological synthesis of gold nanowires using extract of *Rhodopseudomonas capsulata*. *Biotechnol. Prog.* **2008**, *24*, 476–480. [CrossRef] [PubMed]

35. Lin, X.; Wu, M.; Wu, D.; Kuga, S.; Endo, T.; Huang, Y. Platinum nanoparticles using wood nanomaterials: Eco-friendly synthesis, shape control and catalytic activity for *p*-nitrophenol reduction. *Green Chem.* **2011**, *13*, 283–287. [CrossRef]

36. Wang, Y.; Chen, L.Q.; Li, Y.F.; Zhao, X.J.; Peng, L.; Huang, C.Z. A one-pot strategy for biomimetic synthesis and self-assembly of gold nanoparticles. *Nanotechnology* **2010**. [CrossRef] [PubMed]

37. Johnson, C.J.; Dujardin, E.; Davis, S.A.; Murphy, C.J.; Mann, S. Growth and form of gold nanorods prepared by seed-mediated, surfactant-directed synthesis. *J. Mater. Chem.* **2002**, *12*, 1765–1770. [CrossRef]

metals

MDPI

Article

Preparation of Potassium Ferrate from Spent Steel Pickling Liquid

Yu-Ling Wei *, Yu-Shun Wang and Chia-Hung Liu

Department of Environmental Science and Engineering, Tunghai University, Taichung 40704, Taiwan; sonickof2000@hotmail.com (Y.-S.W.), g02340013@thu.edu.tw (C.-H.L.)
* Author to whom correspondence should be addressed; yulin@thu.edu.tw; Tel.: +886-4-2359-1368; Fax: +886-4-2359-6858.

Academic Editors: Suresh Bhargava and Rahul Ram
Received: 31 July 2015; Accepted: 18 September 2015; Published: 24 September 2015

Abstract: Potassium ferrate (K_2FeO_4) is a multi-functional green reagent for water treatment with considerable combined effectiveness in oxidization, disinfection, coagulation, sterilization, adsorption, and deodorization, producing environment friendly Fe(III) end-products during the reactions. This study uses a simple method to lower Fe(VI) preparation cost by recycling iron from a spent steel pickling liquid as an iron source for preparing potassium ferrate with a wet oxidation method. The recycled iron is in powder form of ferrous (93%) and ferric chlorides (7%), as determined by X-ray Absorption Near Edge Spectrum (XANES) simulation. The synthesis method involves three steps, namely, oxidation of ferrous/ferric ions to form ferrate with NaOCl under alkaline conditions, substitution of sodium with potassium to form potassium ferrate, and continuously washing impurities with various organic solvents off the in-house ferrate. Characterization of the in-house product with various instruments, such as scanning electron microscopy (SEM), ultraviolet-visible (UV-Vis), X-ray diffraction (XRD), and X-ray absorption spectroscopy (XAS), proves that product quality and purity are comparative to a commercialized one. Methylene blue (MB) de-colorization tests with in-house potassium ferrate shows that, within 30 min, almost all MB molecules are de-colorized at a Fe/carbon mole ratio of 2/1.

Keywords: potassium ferrate; wet oxidation method; steel pickling liquid; ferrous chloride; ferric chloride; Fe *k*-edge XAS

1. Introduction

1.1. Environmental Application of Potassium Ferrate

Potassium ferrate (K_2FeO_4) is a potent oxidant. Under acidic and alkaline conditions, its respective reduction potentials are 2.20 and 0.700 V, being a potential for replacing traditional oxidants, such as ozone, hypochlorite, permanganate, and others [1]; their respective half-cell reduction potentials in acidic conditions are 2.08, 1.48, and 1 69 V, respectively [2], all less than Fe(VI). Potassium ferrate, other than acting as a powerful oxidant, can be an inorganic coagulant when chemically reduced to $Fe(OH)_3$; it can effectively remove suspended solids, heavy metals, and a variety of contaminants in water [3–7]. Further, using traditional oxidants to treat pollutants/contaminants usually can result in a noteworthy toxic byproduct problem, such as tri-halo-methane and bromates [8]. In contrast, potassium ferrate, as a water treatment agent, is reduced to environment friendly $Fe(OH)_3$ [3,9–13]. Prior to the chlorination process for drinking water, using Fe(VI) as a pre-treatment agent can effectively reduce the formation of hazardous by-products [14,15].

Despite the advantages Fe(VI) can provide, it is expensive for using it to treat pollutants and contaminants. Synthesizing Fe(VI) by using spent steel pickling liquid as an iron source can achieve a dual-win benefit by not only reducing the cost of Fe-source raw chemicals for Fe(VI) synthesis but also recycling spent steel pickling liquid for sustainable spirits. Currently, the annual output rate of spent steel pickling liquid is approximately 150,000 tons in Taiwan [16], and it is considered hazardous due to its high content of various toxic metals that may be a constraint on preparation of Fe(VI) from the pickling liquid.

1.2. Source of Spent Pickling Liquid

Spent pickling liquid is unwantedly produced by the steel industry. Products like steel plates, pipes, and coils always require cleaning with acid to remove their surface impurities before being subjected to further processing. The impurities include black surfaces, iron oxides, and other contaminants. Hydrochloric acid is usually used as the pickling acid for carbon steel products due to its relatively lower price, lower acid consumption rate, and providing a faster pickling process, despite its shortcomings of a larger volatilization rate [17,18].

If spent pickling liquid is treated as waste water, toxic metal present in it is generally removed through different approaches, such as precipitation method, ion exchange, and others. Among them, precipitation as hydroxide is the most often used technology. Although the technology is technically simple, its neutralization step requires a large amount of alkaline- or alkaline earth-based chemicals, and it would be much more praised if spent steel pickling liquid can be recycled.

When hydrochloric acid is used as pickling agent, most iron in spent steel pickling liquid would be present in a form of ferrous chloride through the following chemical reaction [17]:

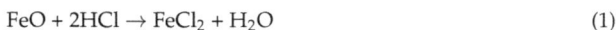

$$FeO + 2HCl \rightarrow FeCl_2 + H_2O \tag{1}$$

1.3. Fe(VI) Preparation with Wet Oxidation Method

Oxidants used for Fe(VI) preparation in previous studies include hypochlorite, Cl_2, and ozone [19]. Among them, ozone is the most powerful [19]. However, this study uses hypochlorite due to its relative ease in laboratory handling. Yet, in our further study, utilization of ozone for Fe(VI) preparation from spent pickling liquid is worthy of study. Theoretically, with the use of hypochlorite, Fe(III) can be oxidized to Na_2FeO_4 in strong basic solution. After adding saturated KOH solution to replace the sodium of Na_2FeO_4, potassium ferrate precipitate can then be formed, separated, washed with organic solvents, and de-watered. Below are the associated chemical reactions [20].

$$Fe^{3+} + 3OH^- \rightarrow Fe(OH)_3 \tag{2}$$

$$2Fe(OH)_3 + 3NaClO + 4NaOH \rightarrow 2Na_2FeO_4 + 3NaCl + 5H_2O \tag{3}$$

$$Na_2FeO_4 + 2KOH \rightarrow K_2FeO_4 + 2NaOH \tag{4}$$

Both the purity and yield of potassium ferrate is affected by the types of oxidants and iron precursors used for its preparation. In general, the use of hypochlorite (OCl^-) can lead to more yield and higher purity, as compared to that using chlorine. As to iron source, ferric nitrate $Fe(NO_3)_3 \cdot 9H_2O$ is more expensive than ferric chloride, which can result in a more rapid synthesis rate and more yield than ferric chloride. As this study intends to recover the spent steel pickling liquid as raw material for potassium ferrate synthesis, there will be no other choice in the aspect of the identity of iron source.

One more thing that needs consideration is what kinds of organic solvents ought to be used for washing crude product potassium ferrate. Organic solvents can barely dissolve potassium ferrate, and their function is to remove the impurities, such as moisture content, KNO_3, KCl, KOH, and others.

During the purification step, *n*-pentane or benzene was used to replace the moisture content in crude potassium ferrate. Methanol or ethanol was employed to dissolve hydroxides, chlorides, nitrates and other impurities. Diethyl ether could speed up the drying process of crude product. The use of inert solvents to remove moisture from crude products is critical. Otherwise, ferrate will be readily reduced to $Fe(OH)_3$, losing its purity grade [21].

For this study, *n*-pentane, rather than benzene, was used due to its lower toxicity and cheaper price. The prices for *n*-pentane and benzene are 80.6 and 96.1 US dollar L^{-1}, respectively. The disadvantage of using *n*-pentane is its lower boiling point (36.1 °C) compared to benzene (80.1 °C); this fact implies a greater loss of *n*-pentane if a pilot-scale or full-scale device is to be set up for this purification process, unless it is a closed process. Methanol was chosen, instead of ethanol, because of its cheaper price and greater effectiveness in removing hydroxides, chlorides, nitrates and other impurities [21]. The price ratio of methanol to ethanol is approximately 1/4 and their respective boiling points are 64.7 and 78.4 °C. Diethyl ether will be used to speed up de-moisture rate of crude product.

1.4. Stability of Fe(VI)

Fe(VI) is very unstable, readily snatching three electrons from neighboring water and being reduced to a more stable form, Fe(III). Solid-phase hexavalent iron is more stable than aqueous iron. It can be stored with negligible reduction reaction for a long time when it is stored in such a way that keeps off the contact of moisture and atmospheric air [22]. When exposed to moisture and atmosphere, it readily reacts with water forming $Fe(OH)_3$, oxygen, and potassium hydroxide. The reduction proceeds as follows [22]:

$$4K_2FeO_4 + 10H_2O \rightarrow 4Fe(OH)_3 + 3O_2 \text{ (g)} + 8KOH \tag{5}$$

1.5. Quantification of Fe(VI)

A few methods can be used for quantitative determination of Fe(VI); they include electrochemical, volumetric, and various spectrophotometric methods [23,24]. Among them, UV-Vis spectral method offers simple steps, rapid experimental work, and accurate outcome [24]. Therefore, this study chooses to use UV-Vis spectral method for determining Fe (VI) concentration and purity. It is briefly described as follows.

Determination of Fe(VI) concentration in water with UV-Vis absorbance is based on Beer's law. Fe(VI) ions in water appears dark purple with an absorbance maximum peak at 505 nm, and its molar absorbance coefficient (ε) was determined to be 1070 M^{-1} cm^{-1} [23]; accordingly, the concentration of Fe(VI) can be calculated using the following equation:

$$\varepsilon = A/BC \tag{6}$$

where the notation "ε" equals 1070 M^{-1} cm^{-1} (molar absorbance coefficient), "A" is UV-Vis absorbance by sample, "B" represents Fe(VI) concentration (M) in the examined sample, and "C" is light path (cm) of the quartz cell used in this study.

For determining Fe(VI) purity of the in-house ferrate prepared in the present study, a proper amount of the product was dissolved in 0.100-L de-ionized water, centrifuged for 10 min at 4000 rounds per minute (centrifuge, Allegra 21/R, Beckman Coulter, Brea, CA, USA), and the top clear solution was measured with a UV-Vis spectrometer (UV-2450, Shimadzu Corporation, Tokyo, Japan). In addition, purity of the product in weight percentage can be calculated using the equation below. It has to be noted that molecular weight of K_2FeO_4 is 198.04 g $mole^{-1}$:

$$K_2FeO_4 \text{ purity} = \frac{A}{1070} \times 0.1 \times \frac{198.04}{\text{weight of sample}} \times 100\% \tag{7}$$

2. Materials and Methods

2.1. Materials

The spent steel pickling liquid was provided by a local steel plant that uses hydrochloric acid to clean its carbon steel surface. It contains 133 g L^{-1} Fe ions and 351 g L^{-1} Cl$^-$, respectively. The chemicals used in this study are summarized in Table 1.

Table 1. Chemicals used in this study.

Chemical	Purity	Producer
Fe(NO$_3$)$_3$·9H$_2$O	98%–101% (reagent grade)	JT Baker, Center Valley, PA, USA
n-C$_5$H$_{12}$	99% (reagent grade)	Merck, Darmstadt, Germany
C$_2$H$_5$OH	96% (reagent grade)	Merck, Germany.
(C$_2$H$_5$)$_2$O	99.5% (reagent grade)	Merck, Germany
FeCl$_2$·4H$_2$O	98% (reagent grade)	Sigma-Aldrich, St. Louis, MO, USA
FeCl$_3$·6H$_2$O	97% (reagent grade)	Sigma-Aldrich, USA
KOH	85% (reagent grade)	Sigma-Aldrich, USA
NaOH	99% (reagent grade)	Sigma-Aldrich, USA
NaOCl solution	6%–14% (reagent grade)	Sigma-Aldrich, USA
K$_2$FeO$_4$	purity ≥ 90%	Sigma-Aldrich, USA
MB	100%	Sigma-Aldrich, USA

2.2. Synthesis Flowchart for K$_2$FeO$_4$

Because the spent pickling liquid used here is very rich in Fe and Cl ions, 133 g L^{-1} and 351 g L^{-1}, respectively, a simple drying process on a 85 °C hot plate can readily vaporize its water content. The solid derived from spent pickling liquid contains FeCl$_2$·4H$_2$O as the major iron species representing approximately 93%, balanced by 7% FeCl$_3$·6H$_2$O on mole basis. The iron speciation was determined by Fe *k*-edge X-ray absorption near edge spectrum (XANES) simulation, and the details will be described below in Section 2.4.

K$_2$FeO$_4$ was synthesized using the derived solid as a raw material and the synthesis flowchart is depicted in Figure 1. Briefly, first, 20 g NaOH and 30 mL 6%–14% NaOCl were mixed for one hour at a stirring speed of 800 rpm, and un-dissolved NaOH solid was filtered off and the filtrate was retained and mixed/reacted (stirring at 800 rpm) for 150 min with 2 g of the derived solid. Second, after the centrifuging and solid/liquid separation, liquid solution was mixed with 20 mL saturated KOH, stirred at 800 rpm for 20 min, centrifuged, and solid/liquid separated. The resulting solid material was added into 10 mL 6.0 M KOH. Third, after a centrifuging process at 3500 rpm for 10 min and solid/liquid separation, the liquid portion was added into an icy-cold 20 mL saturated KOH. Finally, this mixture was filtrated and the resulting crude potassium ferrate was subjected to consecutive washing using each of 0.5 mL *n*-pentane, methyl alcohol, and ethyl ether in approximately 2 min. The solid on filter paper was then dried in a vacuum oven (VO-2000, Pan-Chum Co., Taipei, Taiwan) under a pressure < 30 mm Hg at room temperature for one hour to produce dry, purified K$_2$FeO$_4$ product.

2.3. Crystalline Phase Study with XRD

X-ray diffractometer (XRD, D8 Advance, Bruker AXS, Karlsruhe, Germany) using a copper target was operated at a voltage of 30 kV and a current of 20 mA. X-ray wavelength generated is 1.54056 Å. Sample scanning angle (2θ) is 20°–80° with a scanning rate of 3 deg min^{-1}. Diffraction patterns were identified by comparing them with the database compiled by The Joint Committee on Powder Diffraction Standards (JCPDS).

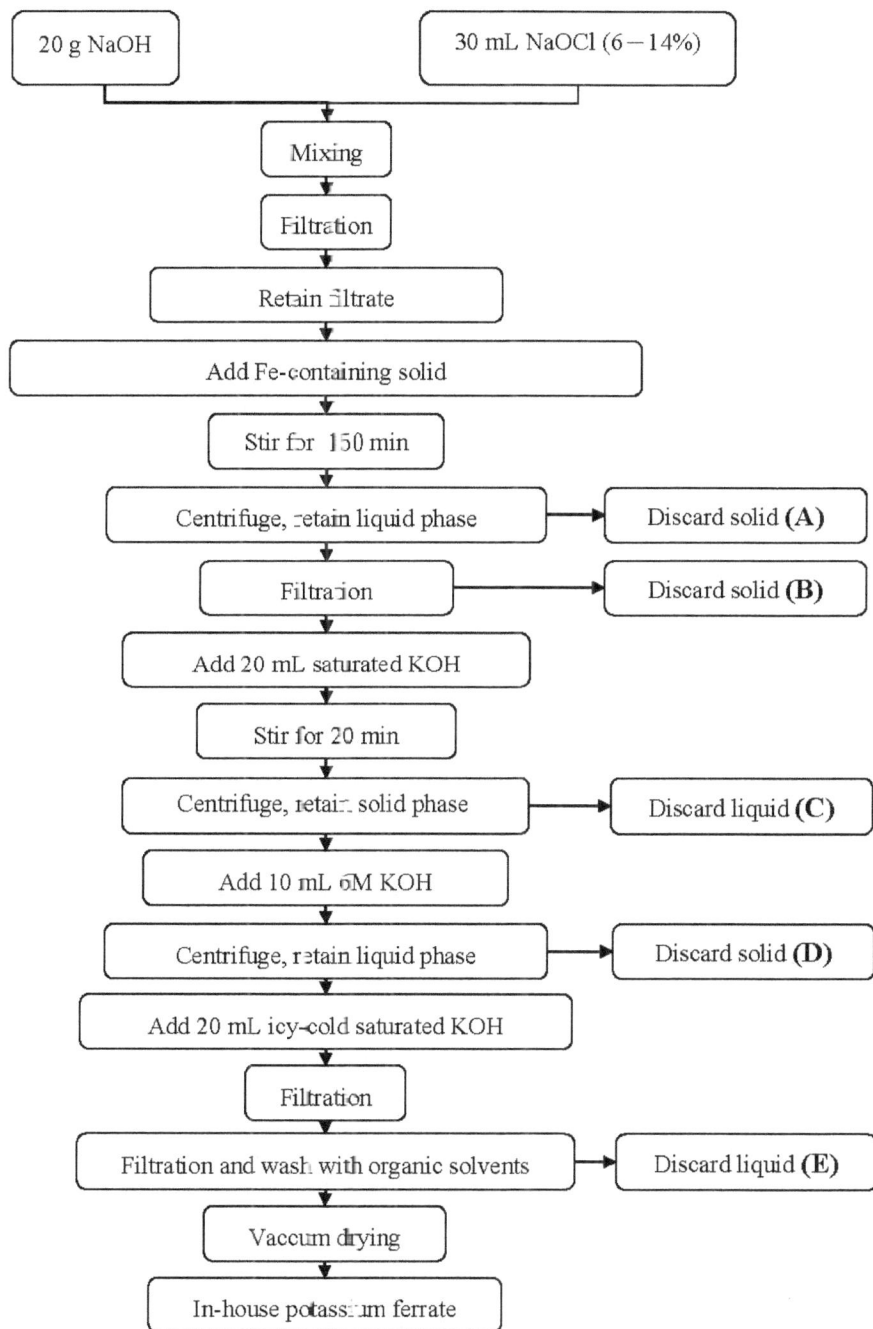

Figure 1. Flowchart for synthesizing in-house potassium ferrate.

2.4. Molecular Study of Potassium Ferrates

X-ray absorption spectroscopy is a non-destructive physical method. Generally, an X-ray absorption spectroscopy (XAS) spectrum can be divided into two parts. The first part is X-ray Absorption Near Edge Spectrum (XANES) for probing electronic properties of target atoms, such as oxidation state and electron occupancy of its *d*-orbitals. The second part is Extended X-ray Absorption Fine Structure (EXAFS) that can be Fourier transformed and simulated to reveal the knowledge of local geometry of coordination atoms, such as atomic type, coordination number (*N*), interatomic distance (*R*), and disorder level of coordination atoms (σ^2).

XAS spectra were recorded on the 16A and Wiggler C (BL-17C) beamlines of National Synchrotron Radiation Center, Hsinchu, Taiwan. During the recording period, its electron storage ring energy was 1.5 GeV, current was 300 mA, and its energy range was 2–14.2 keV. The *k*-edge edge jump of iron is at 7.112 keV. XAS data analysis was carried out using WinXAS 3.1 software [25] (ressler@winxas.de, Thorsten Ressler, Hamburg, Germany). The software can be used to simulate a sample XANES spectrum by linearly combining various standard spectra to quantify component species, based on the least-squares principle from the fingerprints in their XANES spectra [25]. For each standard compound used, two parameters, energy correction and partial mole fraction, are determined. At end of the refining process, standards with a negative mole fraction and/or unreasonable energy shift indicate their absence from the sample [25]. In the present study, XANES spectra from reagent grade $FeCl_2 \cdot 4H_2O$ and $FeCl_3 \cdot 6H_2O$ were recorded and used as standards to simulate the experimental spectrum from dry solid derived from spent pickling liquid.

3. Results and Discussion

3.1. Characteristics of Spent Steel Pickling Liquid

3.1.1. pH and Density of Pickling Liquid

Hydrochloric acid was employed by the steel plant for pickling its steel product. The spent pickling liquid has a pH of <1 and a density of 1.40 kg L^{-3}. The high density is due to its large contents of Fe and Cl.

3.1.2. Composition of Spent Steel Pickling Liquid

Table 2 shows that the pickling liquid contains many kinds of toxic metals with their concentrations being considerably higher than wastewater legal thresholds. The toxic metals include Cr, Bi, Tl, Cd, Co, Cu, Ni, Pb, and Zn. Fe and Cl are the most abundant components, with Mn and Ca (respectively, 1.386 and 1.551 g L^{-1}) being the second most abundant elements. Generally, when Fe content in steel pickling liquid reaches 70–100 g L^{-1}, the liquid is disqualified for further pickling process in the steel industry [17]. The spent pickling liquid (Table 2) has apparently been over-used with an Fe content of 133 g L^{-1}. Because the sum of Fe and Cl, 485 g L^{-1}, accounts for one third of the weight of 1.0 L pickling liquid (*i.e.*, 1.4 kg), only a small amount of heat is required for vaporizing its liquid components, including water and HCl. The pickling liquid was dried on a 85 °C hot plate. Please note that, practically, the HCl can be recovered with a cooling system if desired, but such recovery is not the objective of the present study, rather, at present we are more focusing on investigating the technical feasibility of using the recovered solid iron chloride(s) as a raw material for preparing potassium ferrate. In contrast, direct use of the pickling liquid in potassium ferrate synthesis would alter the required pH value of the solution containing 20 g NaOH and 30 mL NaOCl, as depicted in Figure 1.

Table 2. Chemical compositions of spent steel pickling liquid.

Compositional Element	Content (mg L^{-1})	Spike Recovery [d] (%)
Cr [a]	3.63 ± 0.23	(24 ± 1.2)
Bi [a]	46.48 ± 0.95	(31 ± 4)
Ag [a]	6.86 ± 0.07	(25 ± 3.6)
Al [a]	6.73 ± 0.04	(76 ± 2)
Tl [a]	60.15 ± 0.13	(52 ± 3.1)
Ca [b]	1552 ± 0.3	(97 ± 2.5)
Cd [a]	3.08 ± 0.02	(30 ± 2.4)
Co [a]	5.10 ± 0.036	(31 ± 2.5)
Cu [a]	39.15 ± 0.085	(78 ± 1.4)
Fe [b]	133,000 ± 2160	(89 ± 1.2)
K [a]	7.69 ± 0.00	(111 ± 3.7)
Mg [a]	65.28 ± 0.07	(56 ± 2.6)
Mn [a]	1387 ± 1	(73 ± 0.2)
Na [b]	256.3 ± 1.4	(87 ± 4)
Ni [a]	19.72 ± 0.10	(30 ± 4.1)
Pb [a]	15.08 ± 0.60	(23 ± 2.6)
Zn [b]	188.6 ± 2.1	(91 ± 2.5)
Cl [c]	351,500 ± 3100	(88 ± 3.3)

[a] using an inductively coupled plasma atomic emission spectrometer (ICP-AES, Profile plus, Teledyne Leeman Labs, Hudson, NH, USA) to measure metal concentrations after microwave-assisted digestion; [b] using a flame atomic absorption spectrophotometer (FAAS, Z-6100, Hitachi, Tokyo, Japan) to measure metal concentrations after microwave-assisted digestion; [c] the total amount Cl ion was measured with an ion chromatographer (IC, DX-100, Dionex, Thermal Fisher Scientific, Sunnyvale, CA, USA); [d] the steps are described as follows. First, the pickling liquid was digested with a microwave digester (MWS-4, Berghof Laborprodukte GmbH, Eningen Germany) at 180 °C; metallic ion concentration in the digest was approximately measured with ICP/atomic emission spectrometer or flame atomic absorption spectrometer. Then, exactly 0.5 to 5 times the amount of the approximate ion concentrations were added to the spent steel pickling liquid and ionic concentrations were re-measured after the same digestion process, providing data for the calculation of spike recovery.

3.1.3. Fe Speciation in Solid Raw Material Recycled

As shown in Figure 2, based on XANES simulation, Fe species in the solid raw material recycled by drying the spent steel pickling liquid can be well simulated by approximately summing 93 mol. % ferrous chloride and 7% ferric chloride. The simulative spectrum (dotted curve) and experimental XANES (solid curve) are almost completely overlapped.

As reported in literature, iron sources used in wet oxidation process to synthesize hexavalent iron are mainly ferric chloride and ferric nitrate. Among them, ferric nitrate is preferred due to its characteristics of quick dissolution, rapid chemical reaction with sodium hypochlorite to form hexavalent iron. Oxidation rate of ferric chloride by sodium hypochlorite is relatively slow because chloride ions released by ferric chloride will slow down the rate of sodium hypochlorite dissociation, thus prolonging Fe(VI) synthesis time [12]. However, ferric chloride costs much less than ferric nitrate. Using reagent grade as an example, price ratio among these iron compounds (per unit iron) is approximately 1.0 (ferric chloride):1.2 (ferrous chloride):2.5 (ferric nitrate). It is noteworthy that reagent grade FeCl$_2$·4H$_2$O sells at approximately 120 US dollars kg^{-1} in Taiwan.

A comparison among various studies in K$_2$FeO$_4$ synthesis is presented in Table 3. Included in it are types of raw material (iron source), Fe(VI) product purity, and Fe(VI) yield. The yield is calculated based on the following equation.

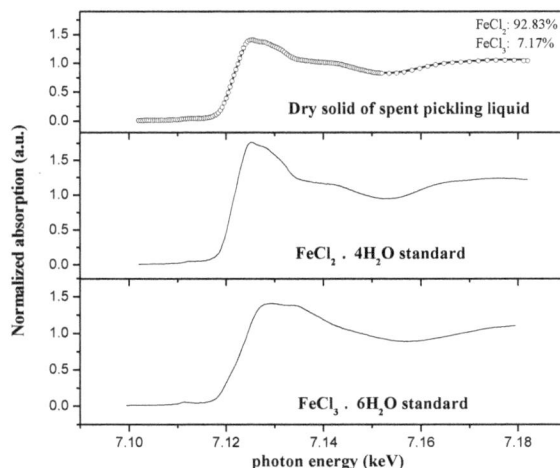

Figure 2. XANES simulation-based Fe speciation in solid raw material recycled from spent steel pickling liquid (dotted curve: simulative XANES spectrum being equivalent to 92.83% ferrous chloride plus 7.17% ferric chloride; solid curve: experimental XANES from dry solid of spent steel pickling liquid).

$$\text{yield (mol. \%)} = \frac{(\text{K}_2\text{FeO}_4 \text{ product weight}) \times (\text{produt purity}) \times (\text{Fe atomic weight})}{(\text{K}_2\text{FeO}_4 \text{ molecular weight}) \times (\text{Fe mass in the solid derived from the pickling liquid})} \qquad (8)$$

Table 3. Purity and yield of potassium ferrate reported in previous studies and this study.

Type of Iron Salt	Oxidizing Agent	Salt Weight (g)	Yield (g)	Purity (wt. %)	Yield (Mol. %)	Literature
Fe(NO$_3$)$_3$·9H$_2$O	NaOCl	25	10.43	63.40	53.9	[23]
FeCl$_3$·6H$_2$O	NaOCl	25	9.64	74.71	39.3	[23]
Fe(NO$_3$)$_3$·9H$_2$O	NaOCl	_ b	_ b	98.5	_ b	[26]
Fe(NO$_3$)$_3$·9H$_2$O	NaOCl	_ b	_ b	90	_ b	[27]
Recycled solid [a]	NaOCl	2	0.254	88	11.7	This study

[a] solid recycled from spent steel pickling liquid; [b] data not reported in both references [25] and [26].

The price of potassium ferrate quite depends on purity. For example, sale price of the Sigma-Aldrich ferrate with ≥90% purity is 99 US dollars g^{-1}, while with purity ≥97%, it used to be sold at approximately 350 US dollars g^{-1}. Although the yield of our in-house ferrate product is only 12.7%, its purity grade is comparative to the Sigma-Aldrich ferrate with ≥90% purity. It is premature to calculate the cost of our in-house ferrate at present because a pilot-plant scale study is required to make such an estimate. However, for sure, the process reported in the present study can save both raw material costs for ferrous chloride (33 US dollar per mole Fe) and ferric chloride (45 US dollar per mole Fe) and local disposal expense for the spent pickling liquid that is currently about 2500 New Taiwan dollars m^{-3} (equivalent to 76.7 US dollars m^{-3} based on the exchange rate of 32.6:1 on 3 September 2015). It is noteworthy that local industrial recycling practitioners recover HCl from the spent pickling liquid, but the HCl solution practically recovered contains only 13 wt. % HCl whose industrial application is very limited.

3.2. Characteristics of In-House and Commercialized Potassium Ferrates

3.2.1. Chemical Compositions of Potassium Ferrates

Chemical compositions of the potassium ferrates were analyzed with ICP-AES after their acidic digestion with microwave assistance. Table 4 shows that both in-house and as-purchase

Sigma-Aldrich potassium ferrates are of high purity with only negligible impurities. Compared with the commercialized one, the in-house product is no less in purity. Most non-iron metals originally present in the pickling liquid have been either washed or precipitated off during the synthesis process.

Figure 3 depicts Fe *k*-edge XANES spectra from in-house (top) and as-purchase Sigma-Aldrich (bottom) potassium ferrates. The height (or area) of pre-edge peak at 7.115 keV is proportional to the ratio Fe(VI)/total Fe. This figure provides evidence that both ferrates are almost identical in pre-edge peak height, implying that they are of similar purity.

Table 4. Chemical compositions of Sigma-Aldrich and in-house potassium ferrates by weight.

Composition	Sigma-Aldrich K$_2$FeO$_4$ Standard		In-House K$_2$FeO$_4$ (This Study)	
	wt. Percentage (%)	Recovery (%)	wt. Percentage (%)	Recovery (%)
Cr [a]	0.11 ± 0	(51 ± 2.5)	0.01 ± 0.0002	(60 ± 2.8)
Bi [a]	0.11 ± 0.0344	(61 ± 1.6)	0.06 ± 0.000196	(51 ± 1.8)
Al [a]	0.01 ± 0.0006	(75 ± 2.6)	0.02 ± 0.00098	(71 ± 2.8)
Ba [a]	0.00 ± 0	(82 ± 2.2)	0.01 ± 0	(88 ± 2.1)
Ca [a]	0.00 ± 0.0004	(73 ± 1.4)	0.02 ± 0.0002	(70 ± 2.6)
Fe [a]	24.99 ± 0.899	(84 ± 2.7)	23.75 ± 0.214	(85 ± 2.1)
K [a]	37.77 ± 0.746	(96 ± 2.5)	38.69 ± 0.697	(94 ± 2.8)
Li [a]	0.20 ± 0.0008	(69 ± 2.3)	0.59 ± 0.00982	(58 ± 1.8)
Mn [a]	0.07 ± 0.002	(80 ± 2.6)	0.10 ± 0.00039	(79 ± 3.1)
Na [a]	0.01 ± 0.0002	(79 ± 1.5)	0.01 ± 0.00039	(77 ± 1.8)
Ni [a]	0.15 ± 0.0656	(69 ± 3.1)	0.04 ± 0.009627	(71 ± 3.0)

[a] analyzed with ICP-AES after microwave-assisted acidic digestion (*n* = 2).

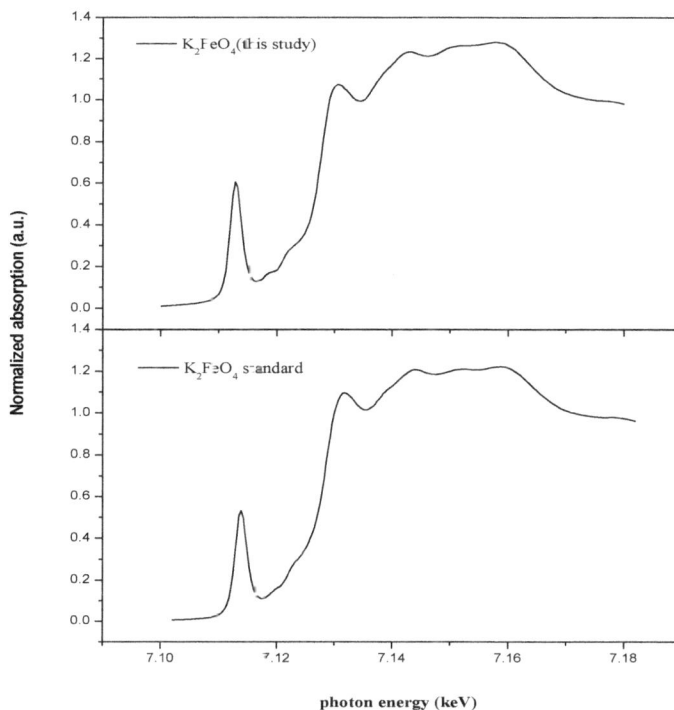

Figure 3. Comparison of Fe *k*-edge XANES of in-house K$_2$FeO$_4$ and Sigma-Aldrich K$_2$FeO$_4$ standard.

3.2.2. UV-Vis Absorption Spectra

Previous literature found that hexavalent iron compounds show a maximum absorbance at wavelength 505 nm [3] and two respective minimum absorbance peaks at 390 nm and 675 nm [15] on UV-Vis spectrum. Figure 4 indicates no difference in UV-Vis absorbance spectrum between our in-house potassium ferrate product and the commercial ferrate that was used as a reference compound in this study.

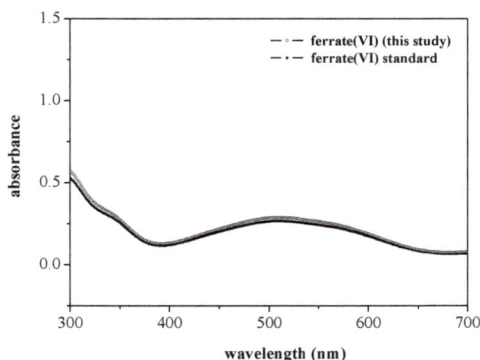

Figure 4. UV-Vis absorbance spectra of in-house and Sigma-Aldrich potassium ferrates at a concentration of 2.88×10^{-4} M.

3.2.3. SEM-EDX Results from Potassium Ferrates

Figures 5 and 6 depict the mapping results from an environmental scanning electron microscope (ESEM-EDX, FEI Quanta 400 F, Hillsboro, OR, USA) spectra of commercialized and in-house potassium ferrates, respectively. Despite the presence of some impurities, such as Pb, Ca, Si, Al, Mg, Na, and Cl, these two ferrate samples are almost identical in EDX spectrum except that Cl peak of in-house product is slightly stronger. Thus, these two ferrates are considered to be not much different in quality.

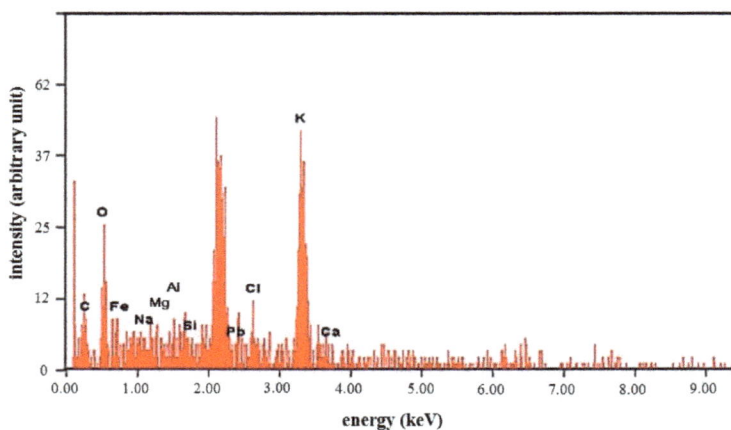

Figure 5. SEM/EDX results from as-purchase potassium ferrate.

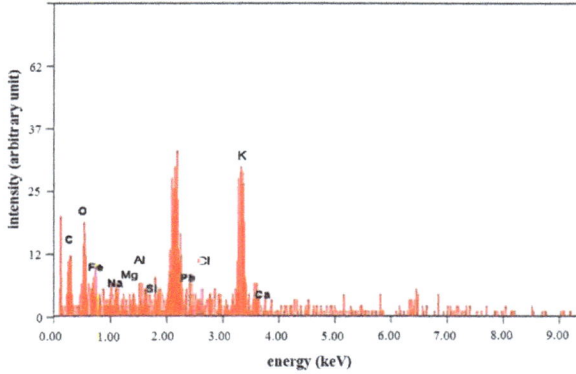

Figure 6. SEM/EDX results from in-house potassium ferrate.

3.2.4. Crystalline Phases of Potassium Ferrates

Figure 7 indicates that XRD patterns of both in-house and commercialized potassium ferrate are consistent with the main patterns belonging to K_2FeO_4 according to The Joint Committee on Powder Diffraction Standards (JCPDS) 25-0652 data file. XRD patterns also indicate that K_2O and FeO(OH) are present as the main impurity in the in-house ferrate product.

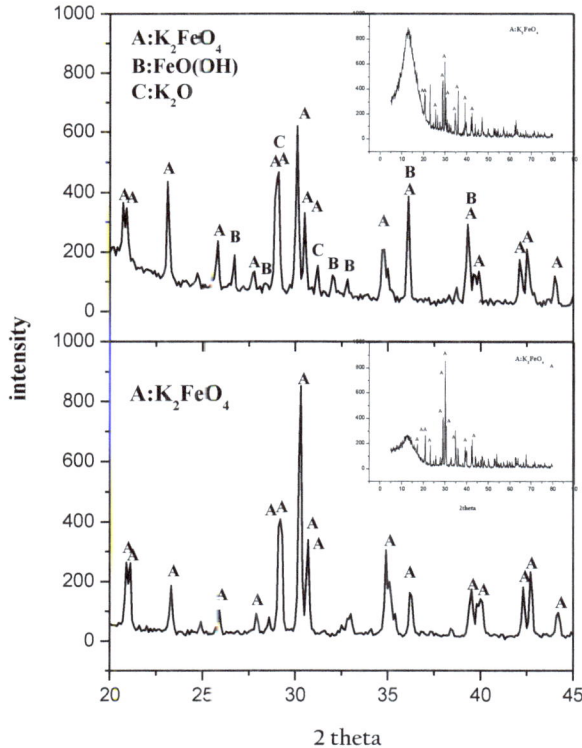

Figure 7. XRD patterns of in-house (top) and as-purchase (bottom) potassium ferrates.

3.2.5. Test of Fe(VI) Stabilization during Storage

Moisture content was recognized to readily reduce Fe(VI) to Fe(III). Thus dry environment is beneficial for Fe(VI) storage. To test storage stability of in-house Fe(VI) product, two identical parts of it were respectively stored in an electronic dryer box (relative humidity: 35%–40%) and in the atmosphere (relative humidity: 67%–70%) for 32 h, and a small amount of each part was retrieved periodically in every 2–3 h interval for analyzing its Fe(VI) purity. The results are presented in Figure 8.

Figure 8 indicates that for a 32 h storage time, only 1.2% chemical reduction occurred to in-house potassium ferrate when stored under 35%–40% relative humidity (RH), while, if stored under the atmosphere of 67%–70% RH, Fe(VI) reduction rate to Fe(III) was 14.5%. Previous study elsewhere has observed an increase in Fe(VI) reduction with increasing RH at room temperature [28]. At a lower humidity (55%–70% RH), Fe(VI) decay was slow, while the decay rate was considerably enhanced at higher humidity such as 90%–95% RH [28]. Formation of $KHCO_3$ at higher RH was suggested to be an important reason for the enhanced Fe(VI) decay rate [28].

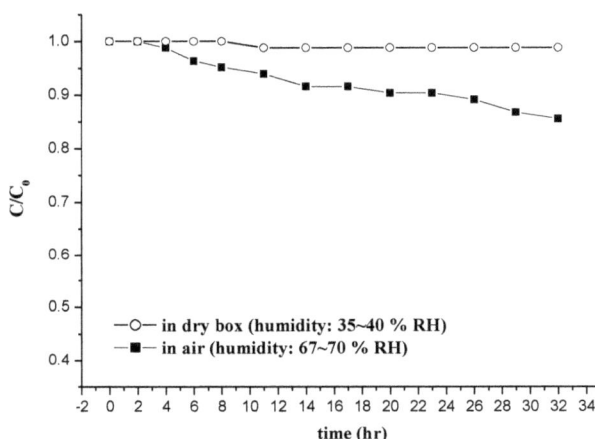

Figure 8. Storage stability of in-house potassium ferrate under different relative humidities.

3.3. Stepwise Loss of Iron in Fe(VI) Synthesis Flowchart

To know which step(s) have lost a large amount of iron for future studies to improve the yield of in-house potassium ferrate, this study performed quantitative analysis of iron in the solid and liquid phases after each solid/liquid separation step of Fe(VI) synthesis. Fe analysis was carried out with the FAAS after microwave-assisted acidic digestion. Table 5 shows that iron was gradually lost along with the synthesis steps. Among them, the centrifugation and filtration steps affect potassium ferrate yield the most. Ideally, if these ion losses can be reduced, yield of potassium ferrate will be improved.

Table 5. Fe loss during wet synthesis of in-house potassium ferrate.

Synthesis Step [b]	Fe Loss [a] (%)
Discarded solid (A)	54.8
Discarded solid (B)	26.5
Discarded liquid (C)	1.49
Discarded solid (D)	8.25
Discarded liquid (E)	0.55

[a] Fe loss $= \frac{\text{(Fe in discarded solid or liquid)}}{\text{total Fe used for synthesis}}$; [b] please refer to Figure 1.

3.4. MB De-Colorization with In-House Potassium Ferrate

To verify how the in-house potassium ferrate performed in treating pollutants, it was employed to de-colorize aqueous MB, a frequently used dye in dying sector. For the de-colorization test, two Fe/C mole ratios, 2/1 and 1/1, were used, given that the oxidation state of carbon of MB molecules will generally be increased from 0 to 4+ while Fe(VI) will be reduced to Fe(III) by a decrease of 3+ oxidation state. Figure 9 depicts time-dependence of methylene blue de-colorization with in-house potassium ferrate, and within 30 min, all MB molecules are de-colorized at a Fe/carbon ratio of 2/1. This fact proves the effectiveness of the in-house product.

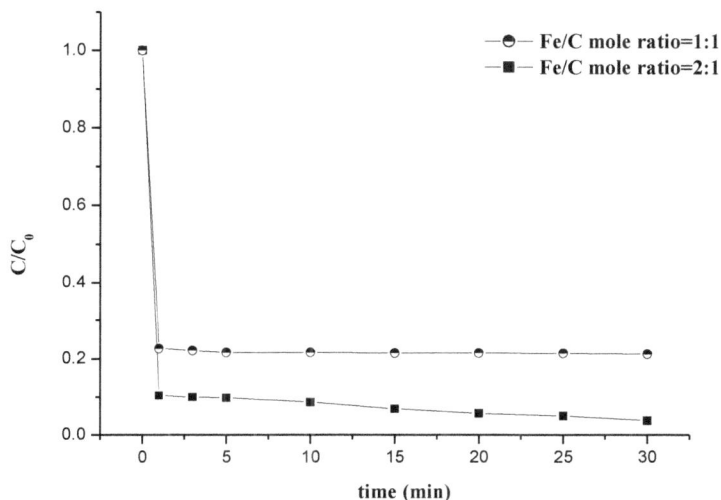

Figure 9. Time-dependence of methylene blue de-colorization by in-house potassium ferrate at two Fe/C mole ratios.

4. Conclusions

A spent hazardous steel pickling liquid was recycled as iron chlorides for synthesizing high-purity in-house potassium ferrate. Below are the conclusions.

First, the spent steel pickling liquid is very rich in iron and chloride ions, being co-existent with various toxic metals that makes the pickling liquid hazardous. Despite the hazardous nature of pickling liquid, after drying as solid material, it was successfully used as a raw material for synthesizing K_2FeO_4. The derived solid material mainly consists of 7% ferric chloride and 93% ferrous chloride, as revealed by Fe k-edge XANES simulation.

Second, as supported by the characterization results from using instruments including SEM-EDX, UV-Vis absorbance spectroscopy, XRD, and XAS, the in-house potassium ferrate product is not inferior to the commercialized ferrate. Varieties of impurity present in the spent steel pickling liquid do not affect the purity of in-house potassium ferrate, compared with the commercialized one.

Third, for a 32 h storage time, only 1.2% decay occurred to in-house potassium ferrate when stored under 35%–40% RH. If stored in an electronic dryer box to effectively cut off contact of atmospheric air, the in-house potassium ferrate can seemingly remain stable without being chemically reduced for a period of time longer than 32 h.

Fourth, the test of methylene blue de-colorization confirms that the in-house potassium ferrate has great oxidation capability to break down pollutants in water.

Finally, for future study, industrial grade chemicals will be used to replace the reagent-grade chemicals for the synthesis of in-house potassium ferrate.

Acknowledgments: This work was sponsored by the Taiwan Ministry of Science and Technology (NSC-102-2221-E-029-001). Fe *k*-edge XAS was recorded on beamlines 17C and 16A of NSRRC, Taiwan.

Author Contributions: Wang and Liu worked together in laboratory under Wei's advice for the project. They also did some literature review. Wei was the principal investigator of the project sponsored by the Taiwan Ministry of Science and Technology and this paper was written by Wei after discussion with Wang and Liu.

Conflicts of Interest: The authors declare no conflict of interest.

References

1. Audette, R.J.; Quail, J.W.; Smith, P.J. Ferrate(VI) ion, a novel oxidizing agent. *Tetrahedron Lett.* **1971**, *3*, 279–282. [CrossRef]
2. Yoon, J.; Lee, Y.; Cho, M.; Kim, J.Y. Chemistry of ferrate (Fe(IV)) in aqueous solution and its application as a green chemistry. *J. Ind. Eng. Chem.* **2004**, *10*, 161–171.
3. Jiang, J.Q.; Lloyd, B. Progress in the development and use of ferrate(VI) salt as an oxidant and coagulant for water and wastewater treatment. *Water Res.* **2002**, *36*, 1397–1408. [CrossRef]
4. Jiang, J.Q. The role of ferrate(VI) in the remediation of emerging micro pollutants: A review. *Desalin. Water Treat.* **2015**, *55*, 828–835. [CrossRef]
5. Sharma, V.K.; Zboril, R.; Varma, R.S. Ferrates: Greener Oxidants with Multimodal Action in Water Treatment Technologies. *Acc. Chem. Res.* **2015**, *48*, 182–191. [CrossRef] [PubMed]
6. Kim, C.; Panditi, V.R.; Gardinali, P.R.; Varma, R.S.; Kim, H.; Sharma, V.K. Ferrate promoted oxidative cleavage of sulfonamides: Kinetics and product formation under acidic conditions. *Chem. Eng. J.* **2015**, *279*, 307–316. [CrossRef]
7. Yates, B.J.; Zboril, R.; Sharma, V.K. Engineering aspects of ferrate in water and wastewater treatment—A review. *J. Environ. Sci. Heal. A* **2014**, *49*, 1603–1614. [CrossRef] [PubMed]
8. Gombos, E. Ferrate treatment for inactivation of bacterial community in municipal secondary effluent. *Bioresour. Technol.* **2012**, *107*, 116–121. [CrossRef] [PubMed]
9. Prucek, R.; Tucek, J.; Kolařĭk J.; Huškováa, I.; Filip, J.; Varma, R.S.; Sharma, V.K.; Zboril, R. Ferrate(VI)-Prompted removal of metals in aqueous media: Mechanistic delineation of enhanced efficiency via metal entrenchment in magnetic oxides. *Environ. Sci. Technl.* **2015**, *49*, 2319–2327. [CrossRef] [PubMed]
10. Gombos, E.; Barkacs, K.; Felfoldi, T.; Vertes, C.; Mako, M.; Palko, G.; Záray, G. Removal of organic matters in wastewater treatment by ferrate(VI) technology. *Microchem. J.* **2013**, *107*, 115–120. [CrossRef]
11. Sharma, V.K. Oxidation of inorganic contaminants by ferrates (VI, V, and IV) kinetics and mechanisms: A review. *J. Environ. Manag.* **2011**, *92*, 1051–1073. [CrossRef] [PubMed]
12. Jiang, J.Q.; Lloyd, B. Preliminary study of ciprofloxacin (cip) removal by potassium ferrate(VI). *Sep. Purif. Technol.* **2013**, *88*, 95–98. [CrossRef]
13. Li, C.; Lib, X.Z.; Grahamc, N.; Gaoa, N.Y. The aqueous degradation of bisphenol A and steroid estrogens by ferrate. *Water Res.* **2008**, *42*, 109–120. [CrossRef] [PubMed]
14. Yang, X.; Guo, W.; Zhang, X.; Chen, F.; Wei, T. Formation of disinfection by-products after pre-oxidation with chlorine dioxide or ferrate. *Water Res.* **2013**, *47*, 5856–5864. [CrossRef] [PubMed]
15. Al-Abdulya, A.; Sharma, V.K. Oxidation of benzothiophene, dibenzothiophene, and methyl-dibenzothiophene by ferrate(VI). *J. Hazard. Mater.* **2015**, *279*, 296–301. [CrossRef] [PubMed]
16. Environmental Protection Administration, R.O.C. Available online: http://www.epa.gov.tw/ (accessed on 21 September 2015).
17. Agrawal, A.; Sahu, K.K. An overview of the recovery of acid from spent acidic solutions from steel and electroplating industries. *J. Hazard. Mater.* **2009**, *171*, 61–75. [CrossRef] [PubMed]
18. Rögener, F.; Buchloh, D.; Reichardt, T.; Schmidt, J.; Knaup, F. Total regeneration of mixed pickling acids from stainless steel production. *Stahl und Eisen* **2009**, *10*, 69–73.
19. Perfiliev, Y.D.; Benko, E.M.; Pankratov, D.A.; Sharma, V.K.; Dedushenko, S.K. Formation of iron(VI) in ozonalysis of iron(III) in alkaline solution. *Inorg. Chim. Acta* **2007**, *360*, 2789–2791. [CrossRef]

20. Thompson, J.E.; Ockerman, L.T.; Schreyer, J.M. Preparation and purification of potassium ferrate(VI). *J. Am. Chem. Soc.* **1951**, *73*, 1379–1381. [CrossRef]

21. Delaude, L.; Laszlo, P. A Novel Oxidizing Reagent Based on Potassium Ferrate(VI). *J. Org. Chem.* **1996**, *61*, 6360–6370. [CrossRef] [PubMed]

22. Sharma, V.K. Potassium ferrate(VI): An environmentally friendly oxidant. *Adv. Environ. Res.* **2002**, *6*, 143–156. [CrossRef]

23. Licht, S.; Naschitz, V.; Halperin, L.; Halperin, N.; Lin, L.; Chen, J.J.; Ghosh, S.; Liu, B. Analysis of ferrate(VI) compounds and super-iron Fe(VI) battery cathodes: FTIR, ICP, titrimetric, XRD, UV-VIS, and electrochemical characterization. *J. Power Sources* **2001**, *101*, 167–176. [CrossRef]

24. Luo, Z.; Strouse, M.; Jiang, J.Q.; Sharma, V.K. Methodologies for the analytical determination of ferrate(VI) A Review. *J. Environ. Sci. Heal. A* **2011**, *46*, 453–460. [CrossRef] [PubMed]

25. Ressler, T. WinXAS: A program for X-ray absorption spectroscopy data analysis under MS-Windows *J. Synchrotron Radiat.* **1998**, *5*, 118–122. [CrossRef] [PubMed]

26. Wang, L.W.; Liu, S.Q.; Zhang, X.Y. Preparation and application of sustained release microcapsules of potassium ferrate(VI) for dinitro butyl phenol (DNBP) wastewater treatment. *J. Hazard. Mater.* **2009**, *169*, 448–453. [CrossRef] [PubMed]

27. Yuan, B.L.; Li, X.Z.; Graham, N. Reaction pathways of dimethyl phthalate degradation in TiO_2–UV–O_2 and TiO_2–UV–Fe(VI) systems. *Chemosphere* **2008**, *72*, 197–204. [CrossRef] [PubMed]

28. Machala, L.; Zboril, R.; Sharma, V.K.; Filip, J.; Jancil, D.; Homonnay, Z. Transformation of solid potassium ferrate(VI) (K_2FeO_4): Mechanism and kinetic effect of air humidity. *Eur. J. Inorg. Chem.* **2009**, *8*, 1060–1067. [CrossRef]

metals

MDPI

Article

The Effect of Grinding and Roasting Conditions on the Selective Leaching of Nd and Dy from NdFeB Magnet Scraps

Ho-Sung Yoon [1], Chul-Joo Kim [1], Kyung Woo Chung [1], Sanghee Jeon [2], Ilhwan Park [2], Kyoungkeun Yoo [2],* and Manis Kumar Jha [3]

[1] Minerals Resource Research Division, Korea Institute of Geoscienece & Mineral Resources (KIGAM), Daejeon 305-350, Korea; hsyoon@kigam.re.kr (H.-S.Y.); cjkim@kigam.re.kr (C.-J.K.); case7@kigam.re.kr (K.W.C.)
[2] Department of Energy & Resources Engineering, Korea Maritime and Ocean University (KMOU), Busan 606-791, Korea; hjeon@kmou.ac.kr (S.J.); ihp2035@naver.com (I.P.)
[3] Metal Extraction & Forming Division, CSIR-National Metallurgical Laboratory, Jamshedpur 831007, India; maniskrjha@gmail.com
* Author to whom correspondence should be addressed; kyoo@kmou.ac.kr; Tel.: +82-51-410-4686; Fax: +82-51-403-4680.

Academic Editor: Suresh Bhargava
Received: 7 June 2015; Accepted: 14 July 2015; Published: 17 July 2015

Abstract: The pretreatment processes consisting of grinding followed by roasting were investigated to improve the selective leaching of Nd and Dy from neodymium-iron-boron (NdFeB) magnet scraps. The peaks of $Nd(OH)_3$ and Fe were observed in XRD results after grinding with NaOH as the amount of water addition increased to 5 cm^3. These results indicate that the components of Nd and Fe in NdFeB magnet could be changed successfully into $Nd(OH)_3$ and Fe, respectively. In the roasting tests using the ground product, with increasing roasting temperature to 500 °C, the peaks of $Nd(OH)_3$ and Fe disappeared while those of Nd_2O_3 and Fe_2O_3 were shown. The peaks of $NdFeO_3$ in the sample roasted at 600 °C were observed in the XRD pattern. Consequently, 94.2%, 93.1%, 1.0% of Nd, Dy, Fe were leached at 400 rpm and 90 °C in 1 $kmol \cdot m^{-3}$ acetic acid solution with 1% pulp density using a sample prepared under the following conditions: 15 in stoichiometric molar ratio of NaOH:Nd, 550 rpm in rotational grinding speed, 5 cm^3 in water addition, 30 min in grinding time, 400 °C and 2 h in roasting temperature and time. The results indicate that the selective leaching of Nd and Dy from NdFeB magnet could be achieved successfully by grinding and then roasting treatments.

Keywords: NdFeB magnet scraps; selective leaching; acetic acid leaching; roasting; planetary ball milling

1. Introduction

Neodymium-iron-boron (NdFeB) magnets were developed in the early 1980s due to the important discovery that modifying Nd_2Fe_{17} produces the ternary compound $Nd_2Fe_{14}B$ [1,2]. The NdFeB magnets could substitute samarium-cobalt magnets because neodymium is more abundant than samarium and the use of iron as transition metal is lower-cost than that of cobalt [1]. The NdFeB magnets have been widely used in various applications such as motors for hard disk and hybrid electric vehicle and magnetic generators for magnetic resonance imaging (MRI) [3–5]. Since most Nd is imported into Korea, there is an increasing demand for recycling process of waste NdFeB magnets.

It has been well known that Nd could be obtained from NdFeB magnet scraps by sulfuric acid leaching followed by precipitation of Nd salts [6–8]. Because this method dissolves iron as well as neodymium, there are disadvantages such as high consumption of sulfuric acid and large amount of wastewater discharged. Many methods have been reported to extract selectively Nd from waste NdFeB magnets such as roasting followed by sulfuric acid leaching [9], selective chlorination of Nd using NH$_4$Cl [4], selective extraction of Nd using molten Mg [5,10] and slag materials [6], and leaching with hydrochloric acid and oxalic acid [11]. These methods require strong acid or should be performed at high temperature over 700 °C. In this study, it was important to change the solid phase of NdFeB magnets for selective recovery of Nd.

High-energy ball milling processes result in repeatedly fracturing and cold welding during collision between balls or ball and inner wall of mill container, so mechanical solid phase reaction could occur during milling [12]. Kim *et al.* reported that they improved the yield of La, Nd and Sm in monazite to around 85% by 120 min-milling with NaOH [12]. Greenberg submitted an application to the US patent regarding Nd recovery where NdFeB magnets are ground mechanically, then leached with acetate, and finally Nd was precipitated selectively as NdF$_3$ [13]. However, the experimental conditions and results were not reported in detail.

This study is aimed at developing the pretreatment process for selective recovery process of Nd and Dy from waste NdFeB magnets. The magnet scraps were ground with NaOH and water, and then roasted under atmosphere conditions. The effects of various conditions such as milling time, milling interval, amount of water and NaOH added, and roasting temperature were investigated. Finally, acetate leaching tests were performed to investigate the selective leaching of Nd.

2. Experimental

2.1. Materials

The NdFeB magnet scraps were obtained from a Korean magnet company, and the chemical composition shows 22.8% Nd and 71.8% Fe as main components and 0.9% B and 0.03% Dy as minor components. The scraps were crushed with a jaw crusher to less than 2 mm before grinding. Pellet-type NaOH (sodium hydroxide) and acetic acid (CH$_3$COOH) are of reagent-grade (Wako Pure Chemicals, Co., Osaka, Japan).

2.2. Grinding and Roasting

The crushed scraps and NaOH were mixed with predetermined stoichiometric ratio, and then 6.4 g of the mixture was introduced into an 80-cm^3 stainless steel jar with 10 stainless steel balls with 15-mm diameter, based on the preliminary tests. The jar was set in the planetary ball mill (Pulverisette 6, Fritsch GmbH, Idar-Oberstein, Germany) and the scraps were ground under the atmosphere conditions with various experimental factors such as adding 0 cm^3 to 5 cm^3 of water, 30 min to 240 min of grinding time, and 15 in stoichiometric ratio of NaOH to Nd. The ground product was washed with deionized-distilled water to remove the remaining NaOH after grinding. After drying at 80 °C for 24 h, the ground powder was roasted at 300 °C to 600 °C for 2 h under the atmosphere conditions.

2.3. Leaching Procedures

The leaching tests of as-received scraps and roasted powder in acetic acid solution were performed in a 500 dm^3 three-necked Pyrex glass reactor using a heating mantle to maintain temperature. The reactor was fitted with an agitator and a reflux condenser. The reflux condenser was inserted in one port to avoid solution loss at high temperatures. In a typical run, 200 dm^3 of solution (1 kmol·m^{-3} acetic acid) was poured into the reactor and allowed to reach the thermal equilibrium (90 °C). Two grams of the samples was then added to the reactor in the experiments, and the agitator was set at 400 rpm. During the experiment, 3 cm^3 of the solution sample was withdrawn periodically at a desired time interval (30−180 min) with a syringe. The sample was filtered with membrane filter and then the filtrate was diluted with 5% HNO$_3$ solution.

2.4. Analytical Methods

The concentrations of Nd, Fe, and Dy were measured by an inductively coupled plasma-atomic emission spectrometry (ICP-AES, OPTIMA 8300DV; PerkinElmer Inc., Waltham, MA, USA). The samples were also characterized by X-ray Diffraction (Smartlab, Rigaku Co., Tokyo, Japan) with Cu target under the following conditions: 30 kV, 200 mA, 20–80°/2θ in angle range, step scanning methods.

3. Results and Discussion

In this study, the pretreatment process consists of grinding and roasting for selective leaching of Nd and Dy from the magnet leaving Fe component as leach residue. The Fe component should be oxidized to ferric ion (Fe^{3+}), since the solubility products (K_{sp}) of $Fe(OH)_3$ and $Fe(OH)_2$ are $10^{-36.51}$ and $10^{-14.18}$, respectively, using standard Gibbs free energy data in Table 1 [14], which means that the solubility of ferric ion (Fe^{3+}) is much lower than that of ferrous ion (Fe^{2+}). The roasting tests were performed to oxidize the Fe, but, since Kim *et al.* reported that the magnet scraps were not oxidized sufficiently at less than 500 °C [15], the grinding tests were conducted to increase the specific surface area of magnet powder and change the solid phase before roasting.

Table 1. Standard Gibbs free energies of species at 25 °C [14].

Species	Fe^{2+}	Fe^{3+}	$Fe(OH)_2$	$Fe(OH)_3$	OH^-
ΔG^0 (kJ/mol)	−91.2	−16.7	−486.6	−696.6	−157.293

The NdFeB magnet scraps were ground with 0 to 5 cm^3 under the following conditions: 15 in stoichiometric molar ratio of NaOH:Nd, 550 rpm in rotational speed, 30 min in grinding time. Figure 1 shows the effect of water addition on the formation of $Nd(OH)_3$ during grinding. When samples were ground without water, XRD peaks show little difference from the as-received sample, while the peaks of Fe and $Nd(OH)_3$ arose as the amount of water addition increased. It has been found that Nd could be oxidized with H_2O as the following equation.

$$2Nd + 6H_2O = 2Nd(OH)_3 + 3H_2 \tag{1}$$

The standard reduction potentials of Nd ($Nd^{3+} + 3e = Nd$) and water ($2H_2O + 2e = H_2 + 2OH^-$) are −2.32 V [16] and −0.83 V [17], respectively, and because the standard electrode potential of Equation (1) is calculated to be 1.49 V, the reaction must be spontaneous. Therefore, 5 cm^3-water addition could accelerate the formation of $Nd(OH)_3$ as shown in Figure 1. However, iron could not be oxidized by water addition, so iron peaks were observed by adding 3 cm^3 or 5 cm^3 water. These results indicate that Nd and Fe of $Nd_2Fe_{14}B$ was changed successfully into $Nd(OH)_3$ and Fe, respectively.

Greenburg proposed the recycling process of NdFeB magnet scraps [13], where Nd was selectively leached after grinding with NaOH. Figure 2 shows XRD patterns of products ground for 30 min and 240 min. However, the intensity of the Fe peak decreased with increasing the grinding time, since Fe remained after 240 min grinding and Fe could be dissolved by acetic acid leaching; therefore, it is difficult to expect selective leaching of Nd. Furthermore, in the preliminary experiments with 30 in stoichiometric molar ratio, the peaks of Fe were observed (data not shown).

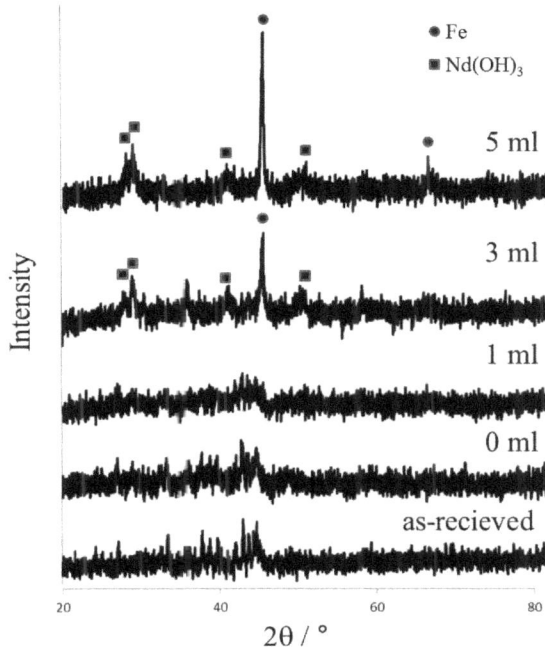

Figure 1. XRD patterns of ground products with water addition (Grinding conditions: 15 in stoichiometric molar ratio, 550 rpm in rotational speed, 30 min in grinding time).

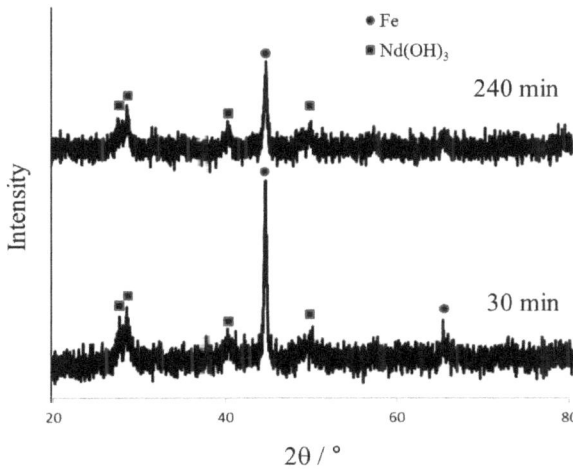

Figure 2. XRD patterns of ground products with grinding time (Grinding conditions: 15 in stoichiometric molar ratio, 550 rpm in rotational speed, 5 cm^3 in water addition).

Therefore, in this study, the magnet scraps were roasted after grinding to oxidize Fe for improving selective leaching efficiency. The ground products were prepared under the conditions such as 15 in stoichiometric molar ratio, 550 rpm in rotational speed, 30 min in total grinding time, and 5 cm^3 in water addition, and then was roasted at 300 °C to 600 °C for 2 h. As shown in Figure 3, the peaks of Fe_2O_3 and Nd_2O_3 as well as $Nd(OH)_3$ and Fe were shown after roasting at 300 °C. After roasting at 400 °C and 500 °C, the peaks of $Nd(OH)_3$ and Fe disappeared, and only peaks of Fe_2O_3 and Nd_2O_3 were observed. The peaks of $NdFeO_3$ emerged after roasting at 600 °C, and the reaction was given as the following equation [9].

$$2Nd + 6H_2O = 2Nd(OH)_3 + 3H_2 \qquad (2)$$

These results indicate that iron could be oxidized at 400 °C and 500 °C, and the components of Nd and Fe were changed into $NdFeO_3$ at 600 °C.

The selective leaching with acetic acid was tested using the roasted product. Figure 4 shows the leaching efficiencies of Nd, Fe, and Dy using product ground under the following conditions: 15 in stoichiometric molar ratio, 550 rpm in rotational speed, 5 cm^3 in water addition, 30 min in grinding time and roasted at 400 °C. The leaching efficiencies of Nd and Dy increased rapidly to more than 90% and remained almost constant after 60 min, and that of Fe remained at less than 1.5%, which indicates that Nd and Dy were selectively leached from scraps. Therefore, the leaching efficiencies measured at 60 min were used for further discussion.

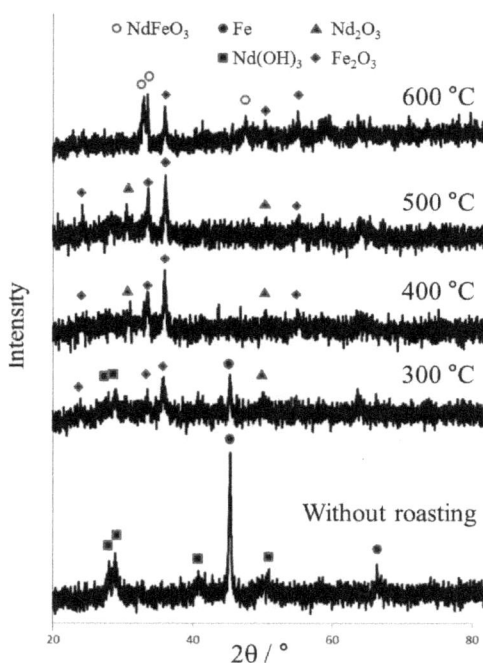

Figure 3. XRD patterns of roasted sample with temperature (Grinding conditions: 15 in stoichiometric molar ratio, 550 rpm in rotational speed, 5 cm^3 in water addition, 30 min in grinding time).

Figure 5 shows the effect of roasting temperature on the leaching efficiencies of Nd, Fe and Dy. The samples were prepared under the grinding conditions: 550 rpm in rotational speed, 5 cm^3 in water addition, 30 min in grinding time and 15 in stoichiometric ratio, and then were leached under the conditions: 1% in pulp density, 1 $kmol \cdot m^{-3}$ in acetic acid, 400 rpm in agitation speed, and 90 °C in temperature. The leaching efficiencies of Nd and Dy decreased rapidly with increasing temperature from 500 °C to 600 °C, and that of Fe decreased and then increased gradually with increasing roasting temperature. The decrease in leaching efficiencies of Nd and Dy would be formation of $NdFeO_3$ shown in Equation (2).

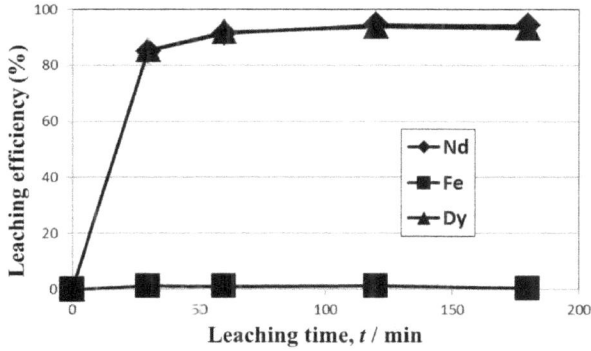

Figure 4. Leaching efficiencies of Nd, Fe, Dy with leaching time (Grinding conditions: 15 in stoichiometric molar ratio, 550 rpm in rotational speed, 5 cm^3 in water addition, 30 min in grinding time, roasting condition; 2 h at 400 °C, leaching conditions: 400 rpm, 1 $kmol \cdot m^{-3}$ acetic acid, 90 °C).

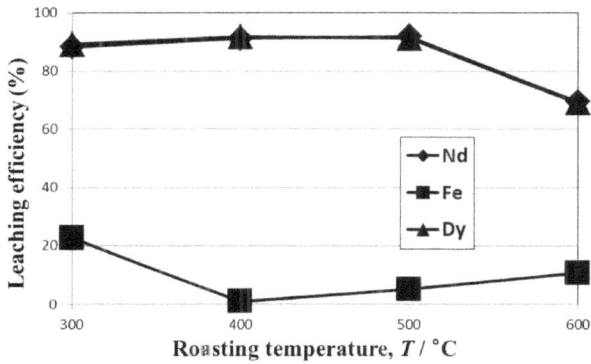

Figure 5. The effect of roasting temperature on the leaching efficiency (Grinding conditions: 550 rpm in rotational speed, 5 cm^3 in water addition, 30 min in grinding time, 15 in stoichiometric molar ratio).

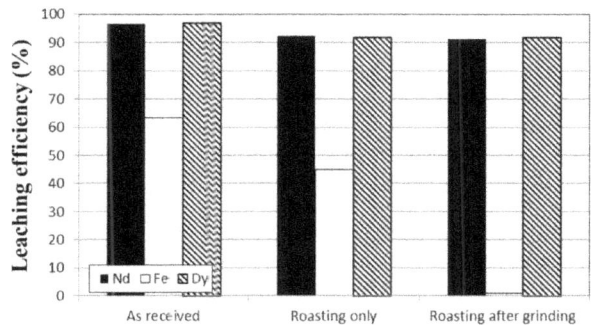

Figure 6. The comparison of leaching efficiencies using samples as as-received, only roasted, and roasted after being ground.

Metals **2015**, *5*, 1306–1314

Figure 6 shows the comparison of leaching efficiencies using samples such as "as-received", "roasting only", which was roasted at 400 °C for 2 h using "as-received" sample, and "roasting after grinding", which was prepared under the grinding conditions: 15 in stoichiometric molar ratio, 550 rpm in rotational speed, 5 cm^3 in water addition, 30 min in grinding time and roasting conditions: 400 °C in temperature and 2 h in roasting time. The leaching efficiency of Fe is calculated to be more than 40% in the cases of "as-received" and "only roasting" samples, and the results indicate that Nd and Dy were not leached selectively from the scraps. The dissolution of Fe would result from insufficient oxidation of Fe as reported by Kim *et al.* [15] However, 30-min grinding with NaOH changes the components of Nd and Fe into Nd(OH)$_3$ and Fe, respectively, which could oxidize Fe easily during roasting. Finally, more than 90% of Nd and Dy could be leached but the leaching efficiency of Fe was 1% at 60 min leaching time, which indicates that selective leaching of Nd and Dy could be achieved by roasting after grinding but not by only roasting or by only grinding. The leaching efficiencies of Nd and Dy increased to 94.2% and 93.1%, respectively, with increasing time to 180 min. These results indicate that the selective leaching of Nd and Dy was achieved successfully.

4. Conclusions

The pretreatment methods consisting of grinding and roasting for the selective leaching of Nd and Dy from NdFeB magnet scraps was investigated for recycling of waste NdFeB magnet. The magnet scraps were ground with NaOH and water, and then roasted under the atmosphere condition to decrease the leaching efficiency of Fe and to accelerate the selectivity of Nd and Dy leaching.

The peaks of Nd(OH)$_3$ and Fe were shown by increasing the amount of water addition in the grinding test. When the ground products were roasted at 400 °C and 500 °C, the peaks of Nd(OH)$_3$ and Fe disappeared while those of Nd$_2$O$_3$ and Fe$_2$O$_3$ were observed in the XRD results. Consequently, 94.2%, 93.1%, 1.0% of Nd, Dy, Fe were leached, respectively, under the following conditions: 15 in stoichiometric molar ratio, 550 rpm in rotational grinding speed, 5 cm^3 in water addition, 30 min in grinding time, 400 °C and 2 h in roasting temperature and time, 90 °C in leaching time, 1 kmol·m^{-3} in acetic acid concentration, 400 rpm in leaching agitation speed, 1% in leaching pulp density, and 180 min in leaching time.

Acknowledgments: The authors express appreciation for support of "R&D Center for Valuable Recycling" (Global-Top R&BD Program) funded by Korea Ministry of Environment (Project No. GT-11-C-01-100-0).

Author Contributions: Yoon and Yoo wrote the paper and contributed to all activities. Kim and Chung performed crushing and roasting samples. Jeon and Park conducted grinding and leaching tests and analyses. Jha contributed to interpretation and discussion of the experimental results.

Conflicts of Interest: The authors declare no conflict of interest.

References

1. Campbell, P. *Permanent Magnet Materials and Their Application*; Cambridge University Press: New York, NY, USA; Melbourne, Australia, 1994; pp. 45–51.
2. Gupta, C.K.; Krishnamurthy, N. *Extractive Metallurgy of Rare Earths*; CRC Press: Boca Raton, FL, USA; London, UK; New York, NY, USA, 2005; pp. 396–411.
3. Lee, K.; Yoo, K.; Yoon, H.S.; Kim, C.J.; Chung, K.W. Demagnetization followed by remagnetization of waste NdFeB magnet for reuse. *Geosystem Eng.* **2013**, *16*, 286–288. [CrossRef]
4. Itoh, M.; Miura, K.; Machida, K. Novel rare earth recovery process on Nd-Fe-B magnet scrap by selective chlorination using NH$_4$Cl. *J. Alloys Compd.* **2009**, *477*, 484–487. [CrossRef]
5. Okabe, T.H.; Takeda, O.; Fukuda, K.; Umetsu, Y. Direct extraction and recovery of neodymium metal from magnet scrap. *Mater. Trans.* **2003**, *44*, 798–801. [CrossRef]
6. Saito, T.; Sato, H.; Ozawa, S.; Yu, J.; Motegi, T. The extraction of Nd from waste Nd-Fe-B alloys by the glass slag method. *J. Alloys Compd.* **2003**, *353*, 189–193. [CrossRef]
7. Lee, C.H.; Chen, Y.J.; Liao, C.H.; Popuri, S.R.; Tsai, S.L. Selective leaching process for neodymium recovery from scrap Nd-Fe-B magnet. *Metall. Mater. Trans. A* **2013**, *44*, 5825–5833. [CrossRef]

8.   Binnemans, K.; Jones, P.T.; Blanpain. B.; Gerven, T.V.; Yang, Y.; Walton, A.; Buchert, M. Recycling of rare earths: A critical review. *J. Clean. Prod.* **2013**, *51*, 1–22. [CrossRef]

9.   Lee, J.C.; Kim, W.B.; Jeong, J.; Yoon, I.J. Extraction of neodymium from Nd-Fe-B magnet scraps by sulfuric acid. *Korean J. Met. Mater.* **1998**, *36*, 967–972.

10.   Xu, Y.; Chumbley, L.S.; Laabs, F.C. Liquid metal extraction of Nd from NdFeB magnet scrap. *J. Mater. Res.* **2000**, *15*, 2296–2304. [CrossRef]

11.   Itakura, T.; Sasai, R.; Itoh, H. Resource recovery from Nd-Fe-B sintered magnet by hydrothermal treatment. *J. Alloys Compd.* **2006**, *408–412*, 1382–1385. [CrossRef]

12.   Kim, W.; Bae, I.; Chae, S.; Shin, H. Mechanochemical decomposition of monazite to assist the extraction of rare earth elements. *J. Alloys Compd.* **2009**, *486*, 610–614. [CrossRef]

13.   Greenberg, B. Neodymium Recovery Process. U.S. Patent 5,362,459, 8 November 1994.

14.   Bard, A.J.; Parsons, R.; Jordan, J. *Standard Potentials in Aqueous Solution*; Marcel Dekker, Inc.: New York, NY. USA; Basel, Switzerland, 1985; pp. 51, 392–394.

15.   Kim, C.J.; Yoon, H.S.; Chung, K.W. The Leaching of Nd Using Dipping Method. KR Patent 10-1147643. 4 May 2012.

16.   Dean, J.A. *Lange's Handbook of Chemistry*; McGraw-Hill, Inc.: New York, NY, USA, 1992; pp. 8–130.

17.   Brett, C.M.A.; Brett, A.M.O. *Electrochemistry, Principles, Methods, and Applications*; Oxford University Press Inc.: New York, NY, USA, 2004; p. 418.

metals

MDPI

Article

Nickel Extraction from Olivine: Effect of Carbonation Pre-Treatment [†]

Rafael M. Santos [1,2,*], Aldo Van Audenaerde [2], Yi Wai Chiang [3], Remus I. Iacobescu [4], Pol Knops [5] and Tom Van Gerven [2]

[1] School of Applied Chemical and Environmental Sciences, Sheridan Institute of Technology, Brampton, ON L6Y 5H9, Canada
[2] Department of Chemical Engineering, KU Leuven, Leuven 3001, Belgium; aldovanaudenaerde@gmail.com (A.V.A.); tom.vangerven@cit.kuleuven.be (T.V.G.)
[3] School of Engineering, University of Guelph, Guelph, ON N1G 2W1, Canada; chiange@uoguelph.ca
[4] Department of Materials Engineering, KU Leuven, Leuven 3001, Belgium; remusion.iacobescu@mtm.kuleuven.be
[5] Innovation Concepts B.V., Twello 7391MG, The Netherlands; knops@innovationconcepts.eu
* Author to whom correspondence should be addressed; rafael.santos@alumni.utoronto.ca; Tel.: +1-905-459-7533 (ext. 5723); Fax: +1-905-874-4321.
† Contents of this paper also appear in the MetSoc of CIM Proceedings of the 7th International Symposium—Hydrometallurgy 2014, Volume II, pp. 755–767.

Academic Editors: Suresh Bhargava and Rahul Ram
Received: 19 August 2015; Accepted: 6 September 2015; Published: 11 September 2015

Abstract: In this work, we explore a novel mineral processing approach using carbon dioxide to promote mineral alterations that lead to improved extractability of nickel from olivine ($(Mg,Fe)_2SiO_4$). The precept is that by altering the morphology and the mineralogy of the ore via mineral carbonation, the comminution requirements and the acid consumption during hydrometallurgical processing can be reduced. Furthermore, carbonation pre-treatment can lead to mineral liberation and concentration of metals in physically separable phases. In a first processing step, olivine is fully carbonated at high CO_2 partial pressures (35 bar) and optimal temperature (200 °C) with the addition of pH buffering agents. This leads to a powdery product containing high carbonate content. The main products of the carbonation reaction include quasi-amorphous colloidal silica, chromium-rich metallic particles, and ferro-magnesite ($(Mg_{1-x},Fe_x)CO_3$). Carbonated olivine was subsequently leached using an array of inorganic and organic acids to test their leaching efficiency. Compared to leaching from untreated olivine, the percentage of nickel extracted from carbonated olivine by acid leaching was significantly increased. It is anticipated that the mineral carbonation pre-treatment approach may also be applicable to other ultrabasic and lateritic ores.

Keywords: nickel; olivine; mineral carbonation; leaching; ferro-magnesite; colloidal silica

1. Introduction

In the last few decades, traditional nickel resources have become scarcer because of ramping global production and growing demand [1]. Nickel is more abundantly present in the Earth's crust than copper and lead, but the availability of high-grade ores is rather limited [2]. The current strong demand for nickel is expected to carry into the future, and the scarcity of high-grade recoverable ores will inevitably call for the exploitation of low-grade ores as a source for nickel. Therefore, increasingly more research is underway investigating the feasibility of recovering nickel from low-grade ores [3–5]. Magnesium-iron silicates, minerals that are widely distributed on the Earth's crust and that contain relatively dilute, yet considerable amounts, of nickel arise as one possible, yet challenging, opportunity.

The main objective of this study was to investigate the possibility of processing olivine ($(Mg,Fe)_2SiO_4$) for the production of nickel. Olivine is solid-solution of iron- and magnesium-silicates containing relatively small amounts of nickel and chromium, and is the precursor of weathered lateritic ores. Olivine (also known as dunite, an ore containing at least 90% olivine [6]) is abundantly present in the Earth's upper mantle [7], and intrudes in some locations into the Earth's crust, most notably in the Fjordane Complex of Norway, which contains the largest ore body (approx. two billion metric tons) under commercial exploitation [6]. The use of olivine as a nickel source could, thus, possibly solve the scarcity problem of high-grade ores. Due to its small nickel content, conventional extraction and recovery methods (e.g., high pressure acid leaching, agitation leaching or heap leaching [3]) are not viable, as reagent and processing costs become too high [8]. In this work a novel approach was investigated, whereby the mineral is first carbonated in a pre-treatment step before the nickel is extracted by leaching. Carbonation may allow for an easier recovery due to a better accessibility of the nickel during leaching as a result of morphological and mineralogical changes. Recently, considerable research has focused on the carbonation of olivine and other alkaline silicates as an option for sustainable carbon dioxide sequestration [9,10]. In the present work, however, CO_2 is utilized primarily as a processing agent; such an approach can be termed "carbon utilization" (more specifically turning CO_2 from a waste into an acid).

Nickel is able to replace magnesium in olivine's magnesium silicate matrix forming a magnesium-nickel silicate ($(Mg,Ni)_2SiO_4$) called liebenbergite or nickel-olivine. This replacement is possible due to certain similarities of nickel and magnesium in the silicate structure. Their ionic radii are similar (Mg = 0.66 Å; Ni = 0.69 Å), their valences are the same (Mg^{2+}, Ni^{2+}), and they both belong to the same orthorhombic system [11]. The amount of nickel in olivine is variable and depends on the ore's origin, varying between <0.1 and 0.5 wt. % Ni [12]. These concentrations are rather low compared to the grade of nickel deposits presently used in industrial processes, which ranges from 0.7 to 2.7 wt. % Ni [11].

Olivine is highly susceptible to weathering processes and alterations by hydrothermal fluids. These alteration reactions involve hydration, silicification, oxidation, and carbonation; common alteration products are serpentine, chlorite, amphibole, carbonates, iron oxides, and talc [11]. The fact that olivine is highly susceptible to weathering also makes it suitable for intensified carbonation. Due to this suitability and its high abundance, olivine has been the subject of intensive research for carbon dioxide sequestration using mineral carbonation, whereby the formation of stable magnesium carbonates act as carbon sinks [13–17].

Carbonating olivine converts the silicates (mainly forsterite (Mg_2SiO_4) and fayalite (Fe_2SiO_4) [6]) into carbonates and silica. This reaction is exothermic and is, thus, thermodynamically favored. The reaction mechanism contains three main steps: the dissolution of CO_2 in the aqueous solution to form carbonic acid; the dissolution of magnesium in the aqueous solution, and the precipitation of magnesium carbonate. The overall reaction schemes for carbonation of forsterite and fayalite are given in Equations (1) and (2):

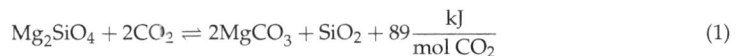

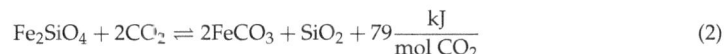

$$Mg_2SiO_4 + 2CO_2 \rightleftharpoons 2MgCO_3 + SiO_2 + 89\frac{kJ}{mol\ CO_2} \tag{1}$$

$$Fe_2SiO_4 + 2CO_2 \rightleftharpoons 2FeCO_3 + SiO_2 + 79\frac{kJ}{mol\ CO_2} \tag{2}$$

The formed magnesite ($MgCO_3$) and siderite ($FeCO_3$), as well as the residual silica (SiO_2), are thermodynamically stable products that are environmentally friendly. These reaction products can, thus, be readily disposed of in the environment or reutilized as commercial products.

The focus of this work was to investigate the leaching behavior of carbonated olivine. When carbonated olivine is leached, the acid will have to dissolve a carbonate structure instead of a silicate structure. These reactions can be seen in Equations (3) and (4):

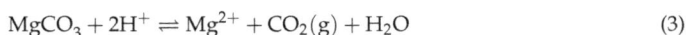

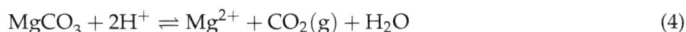

$$MgCO_3 + 2H^+ \rightleftharpoons Mg^{2+} + CO_2(g) + H_2O \tag{3}$$

$$MgCO_3 + 2H^+ \rightleftharpoons Mg^{2+} + CO_2(g) + H_2O \tag{4}$$

Through these alterations of the olivine mineral, which may increase specific surface area, nickel might become more accessible to leaching. Secondly, the C–O (360 kJ/mol) bonds are weaker than their Si–O (466 kJ/mol) counterparts [18], which can lead to an easier leaching of the carbonated olivine compared to natural olivine.

This paper reports the results of a series of tests that aimed to: (i) find the optimal carbonation conditions that maximize the desired mineral and morphological alterations; (ii) characterize the carbonated products with a focus on the fate of nickel; (iii) compare the leaching performance of an array of organic and inorganic acids, and assess the efficiency and extent of nickel extraction from carbonated olivine compared to natural olivine; and (iv) provide the proof-of-concept of using carbonation as a pre-treatment step for nickel recovery from low-grade silicate ores and elucidate directions for future research.

2. Experimental Section

2.1. Olivine Characterization

Olivine was supplied by Eurogrit B.V. (a subsidiary of Sibelco, Antwerp, Belgium) and originated from Åheim, Norway. The material obtained, classified as GL30, had the following properties described by the supplier: sub-angular to angular shape, pale green color, hardness of 6.5 to 7 Mohs, specific density of 3.25 kg/dm^3, and a grain size between 0.063 and 0.125 mm. The olivine was milled before any further use to increase the reactivity of the material to carbonation and leaching by increasing the specific surface area. The milling was performed using a centrifugal mill (Retsch ZM100, Haan, Germany) operated at 1400 rpm with an 80 μm sieve mesh. After milling, a total of 86 vol. % of the material had a particle size below 80 μm, and the average mean diameter D{4,3}, determined by Laser Diffraction Analysis (LDA, Malvern Mastersizer 3000, Worcestershire, UK), was equal to 34.8 μm. The particle size distribution is shown in Figure S1. The morphology of the particles was imaged by Scanning Electron Microscopy (SEM, Philips XL30 FEG, Eindhoven, The Netherlands), and is shown in Figure S2. For SEM analysis, particles were gold-coated and mounted on conductive carbon tape.

The material was extensively analyzed to obtain the chemical and mineralogical composition. Table 1 presents the elemental composition results obtained by digestion followed by Inductively-Coupled Plasma Mass Spectrometry (ICP-MS, Thermo Electron X Series, Waltham, MA, USA) analysis; Co, Mg, Mn, and Si content were determined by Wavelength Dispersive X-ray Fluorescence (XRF, Panalytical PW2400, Almelo, The Netherlands).

The mineralogy of the fresh olivine was analyzed by powder X-ray Diffraction (XRD, Philips PW1830, Almelo, The Netherlands) with quantification by Rietveld refinement; the diffractogram is shown in Figure 1. As can be expected, the material contains mostly forsterite (84.5 wt. %; in fact ferroan-forsterite, which is forsterite with iron substitution) as well as a smaller amount of fayalite (2.5 wt. %). Other minor components present include some hydrated silicates (clinochlore ((Mg,Fe^{2+})$_5$Al(Si$_3$Al)O$_{10}$(OH)$_8$, 2.1 wt. %), lizardite (Mg$_3$Si$_2$O$_5$(OH)$_4$, 2.7 wt. %), talc (Mg$_3$Si$_4$O$_{10}$(OH)$_2$, 0.5 wt. %), and tirodite (Na(Na,Mn^{2+})(Mg$_4$,Fe^{2+})Si$_8$O$_{22}$(OH)$_2$, 3.1 wt. %)), carbonates (magnesian calcite (Ca$_{0.85}$Mg$_{0.15}$CO$_3$, 1.0 wt. %), and magnesite (MgCO$_3$, 0.2 wt. %)), magnesium (hydr)oxides (periclase (MgO, 0.1 wt. %), and brucite (Mg(OH)$_2$, 0.7 wt. %)), chromite (FeCr$_2$O$_4$, 1.1 wt. %) and quartz (SiO$_2$, 0.2 wt. %).

Table 1. Elemental composition of fresh olivine, in decreasing order, determined by ICP-MS (Al, Ca, Cr, Fe, Ni) and XRF * (Co, Mg, Mn, Si).

Element	Mass %
Mg	27.2
Si	20.7
Fe	3.7
Ni	0.27
Cr	0.24
Al	0.17
Ca	0.17
Mn	0.09
Co	0.02

* XRF was used for Si and Mg analysis as it is a more accurate method for determination of these elements. Data for Co and Mn is not available by ICP-MS, so XRF data is presented; it should be noted that due to their low concentration, these data are to be considered semi-quantitative.

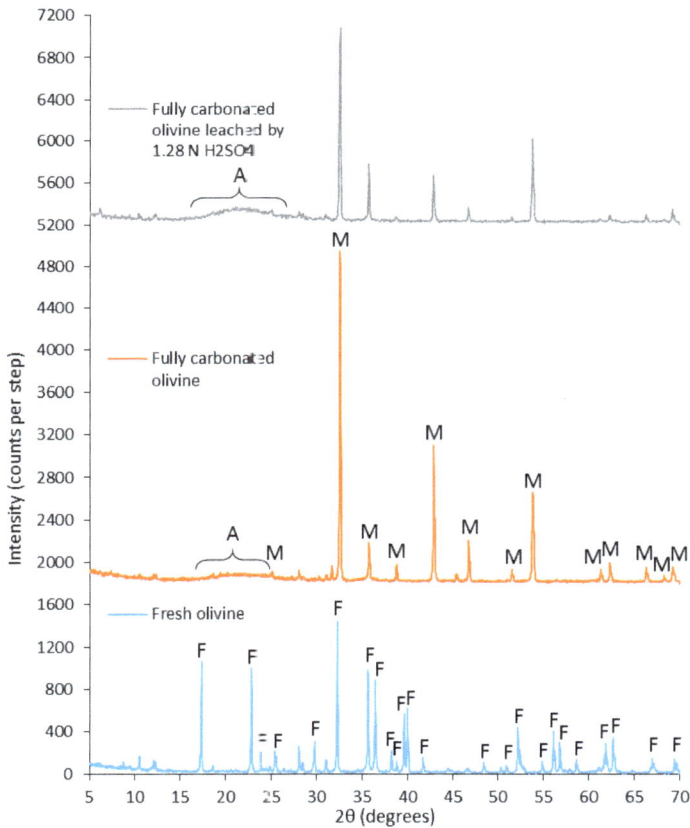

Figure 1. XRD diffractograms of fresh olivine, fully carbonated olivine and fully carbonated olivine after leaching in 1.28 N H_2SO_4 for 24 h; major mineral peaks are indicated: F = ferroan-forsterite $((Mg,Fe)_2SiO_4)$; M = magnesite/ferro-magnesite $((Mg_{1-x},Fe_x)CO_3)$; A = quasi-amorphous phase.

The olivine was also analyzed with a Jeol Hyperprobe JXA-8530F Field Emission Gun Electron Probe Micro-Analyzer (FEG EPMA, Akishima, Japan), equipped with five wavelength dispersive spectrometers, to map the concentration of each element within the particles. The EPMA was capable of detecting elements down to a concentration of 100 ppm and map them down to a spatial resolution of 0.1 μm. A small representative surface area (80 × 100 μm) of a polished sample (pelletized and embedded in resin) was fully mapped to give the distribution of elements in the material. The EPMA was operated at 15 kV, a probe current of 100 nA, and dwell time of 30 ms per 0.3 × 0.3 μm pixel. Both peak and background were measured under these conditions. Nickel was found to be dispersed in the material, as can be seen in Figure 2. This would indicate that it replaces magnesium in the magnesium silicate structure to form a magnesium-nickel silicate (($Mg,Ni)_2SiO_4$). There are also small particles that are highly concentrated (shown as white) in nickel, chromium and iron. Figure S3 shows the elemental distribution of other elements (Al, C, Ca, Co, Cr, Fe, Mg, Mn, Si). Figure S4 helps to visualize that nickel-rich regions exist; in some, nickel is associated with iron (cyan color in composite map), and in some nickel is not associated with iron nor chromium (green color in composite map). In the case of chromium, it is present mainly in select regions, and those regions are highly concentrated in iron as well (suggestive of chromite), but not in nickel.

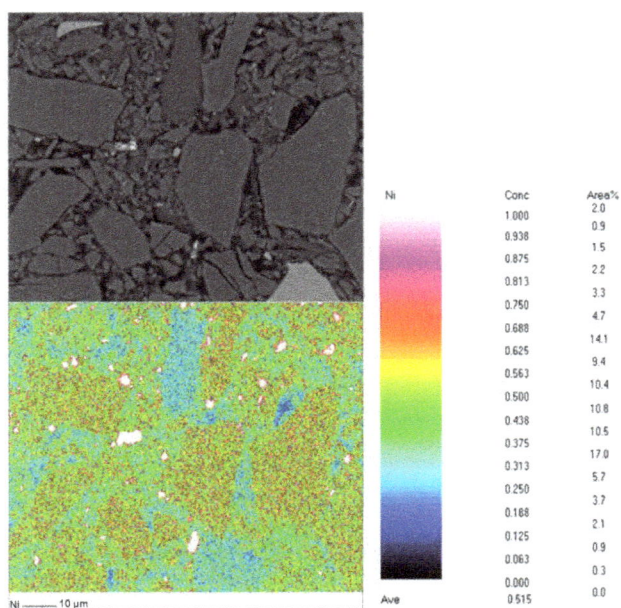

Figure 2. Fresh olivine backscattered scanning electron image (**top**) and EPMA mapping of nickel concentrations (**bottom**); concentration scale is relative to max/min levels.

2.2. Carbonation

Carbonation experiments were conducted in a Büchi Ecoclave continuously-stirred tank reactor (CSTR, Uster, Switzerland). The reactor has a volume of 1.1 liters and is capable of withstanding pressures up to 60 bar and temperatures up to 250 °C. Carbon dioxide gas (99.5% purity) was continuously injected from a compressed cylinder. It should be noted that for industrial implementation, gases with lower CO_2 purity (e.g., combustion flue gases) may be used for mineral carbonation so long as the desired CO_2 partial pressure is met by gas compression. All experiments in this study were conducted with 35 bar CO_2 partial pressure; steam made up the balance pressure up to 55 bar total, depending on the temperature. The reactor was equipped with a Rushton turbine stirrer and a baffle to ensure adequate mixing of the reactor contents; 1000 rpm stirring rate was used. The liquid volume in the reactor was kept constant at 800 mL.

The experimental parameters varied are detailed in Table 2; these were temperature, solids loading, residence time, and additive concentrations. Increasing the temperature influences the equilibrium constants. An increase in the dissociation constants of carbonic acid leads to a decrease in pH (higher acidity) and an increase in both bicarbonate and carbonate ion concentrations; this enhances the dissolution of magnesium as well as the precipitation of magnesium carbonate (under suitable pH, *i.e.,* not excessively acidic). These effects are counteracted by an increase of Henry's constant, which leads to a lower solubility of CO_2 in the solution. Lastly, a decrease of the solubility product of magnesium carbonate stimulates its precipitation. These opposing effects indicate that an optimal temperature exists. Increasing the solids loading in the reaction process has been reported to increase the extent of carbonation due to an increase in particle-particle collisions that remove passivating layers and increase the surface area available for carbonation [19]. The use of additives aims at enhancing the dissolution of magnesium, the dissociation of carbonic acid, and/or the precipitation of magnesium carbonate. Sodium chloride (NaCl) and sodium bicarbonate ($NaHCO_3$) were tested as carbonation enhancing additives as suggested by Chen *et al.* [20].

After completion, the reacted slurry content was filtered to recover the liquid and solid portions; solids were dried at 105 °C for 24 h. Most experiments were conducted in duplicate, and data presented are average values.

Table 2. Parameter values used in the carbonation experiments.

Parameters	Tested Values
Temperature	150–185–200 °C
Solids loading	50–100–200 g/800 mL
Residence time	4–24–48–72 h
NaCl concentration	0–1–2 M
$NaHCO_3$ concentration	0–0.64–2.5 M

2.3. Leaching

Leaching experiments were conducted by atmospheric agitation methodology. A certain amount of olivine (typically two grams), either fresh or carbonated, was added to plastic flasks together with 100 mL of a solution containing various concentrations of a certain acid. The flasks were shaken at 25 °C for the desired reaction time (typically 24 h). When finished, solids and liquids were separated using a centrifuge. The supernatant liquids and the dried solids were further analyzed. Leaching experiments were conducted in duplicate, and data presented are average values. A low leaching temperature was used as this study's main aim was to investigate mineralogical effects on chemical equilibrium rather than leaching kinetics. Low temperature leaching (*i.e.,* ambient) is typical in heap leaching operations [3].

For the leaching with inorganic acids, sulfuric acid (H_2SO_4), nitric acid (HNO_3) and hydrochloric acid (HCl) were chosen. A preliminary test was performed using a large variety of organic acids: citric acid ($HOC(COOH)(CH_2COOH)_2$), oxalic acid (HOOCCOOH), succinic acid ($HOOC(CH_2)_2COOH$), lactic acid ($CH_3CH(OH)COOH$), acetic acid (CH_3COOH), formic acid (HCOOH) and butyric acid (C_3H_7COOH). These organic acids were chosen as they are reportedly produced by microorganisms utilized in bioleaching of silicate minerals [21,22]. Based on preliminary experimental results (Figure S5) the three most promising organic acids, citric acid, formic acid, and lactic acid were selected for further use in the experiments discussed hereon

2.4. Analytical Methods

The concentrations of soluble elements in aqueous solutions were determined by ICP-MS. The mineralogical, morphological, and microstructural properties of carbonated solids were characterized by XRD, SEM, nitrogen adsorption (BET, Micromeritics TriStar 3000, Norcross, GA, USA), LDA, and EPMA. The CO_2 uptake of the carbonated solids was determined by thermogravimetric analysis (TGA, TA Instruments Q500, New Castle, DE, USA), conducted in duplicate. The weight loss between 250 and 900 °C was attributed to the decomposition of carbonates (XRD results suggest minimal formation of hydration products that could interfere in this range, and there is good agreement between quantitative XRD and TGA determination of magnesite content (Figure S6)). The maximal theoretical CO_2 uptake ($M_{CO_2,max}$) of natural olivine, 0.521 g, CO_2/g, olivine, was estimated based on its magnesium and iron content. Extent of carbonation (ξ) is expressed as the percentage ratio of actual to maximal uptake values: $\xi = M_{CO_2,actual} / M_{CO_2,max}$.

3. Results and Discussion

3.1. Influence of Carbonation Parameters

The dependencies of the temperature, residence time, solids loading, and NaCl and $NaHCO_3$ concentrations on the carbonation extent are given in Figure 3.

The influence of the reactor temperature on the carbonation is shown in Figure 3a. The extent of carbonation increases with increasing temperature between 150 °C and 200 °C, both for 4 h and 24 h residence times. This is due to the increase in both acid dissociation constants of carbonic acid, which contributes to magnesium silicate dissolution, as well as the decrease in the solubility product of magnesium carbonate, which promotes magnesium carbonate precipitation. These two effects are mutually beneficial, since as more magnesium precipitates as carbonate, more magnesium can leach from the silicate, propagating the reaction. Increasing the temperature also increases Henry's constant for the dissolution of CO_2 in the water, which can have a negative impact on the carbonation [20], but this was not observed here. O'Connor *et al.* [23] found that these counteracting temperature effects lead to an optimal olivine carbonation temperature of 185 °C. In our experiments, no maximum was reached between 150 °C and 200 °C. The difference in results can be explained because O'Connor *et al.* [23] use other parameter values in their experiments; most importantly, they operated at CO_2 pressures of 150 bar, whereas our experiments operated at 35 bar. At higher CO_2 pressure, the solubility limit of CO_2 will be reached at a lower temperature.

The extent of carbonation increases linearly with an increase in residence time, as can be seen in Figure 3b. There seems to be an initially fast carbonation rate due to parts of the olivine that are more easily carbonated (fines, particle surfaces, and more reactive minerals (e.g., periclase, brucite)), after which the carbonation continues linearly with time. Due to this linear increase with time, there is either no limitation by the formation of a passivating layer, or the passivating layer is broken down sufficiently by particle collisions. This was confirmed by SEM analysis of partially carbonated olivine. Figure S7 shows that the residual silica and precipitated magnesite form separate particles, rather than forming a passivating layer around unreacted olivine. More discussion on this is presented in the Section 3.2. The residence times used in this study are relatively long, which was necessary because of the relatively low CO_2 partial pressure utilized (35 bar), as restricted by the reactor's pressure rating. Higher CO_2 partial pressures should accelerate the processes, from the order of days to the order of hours, as indicated by other studies conducted at higher pressures (e.g., 139 atm [14]) and modeling work [13].

As can be seen from Figure 3c, increasing the solids loading greatly enhances carbonation. A solids loading increase from 50 g (5.9 wt. %) to 200 g (20 wt. %) almost doubles the carbonation extent for both the 24 h and 72 h experiments at 200 °C with the addition of 1 M NaCl. These results confirm previous results from Béarat *et al.* [19] who also noticed a substantial increase in carbonation, proportional to (wt. %)$^{1/3}$, when increasing the solids loading from 5 to 20 wt. %. The higher amount of solids in the reactor will lead to more collisions of the olivine particles, promoting the removal of passivating layers and the breakage of unreacted particles. Julcour *et al.* [16] emphasized the importance of attrition/exfoliation, conducting olivine carbonation reactor in a stirred bead mill and achieving 80% conversion in 24 h at 180 °C, 800 rpm and 20 bar CO_2.

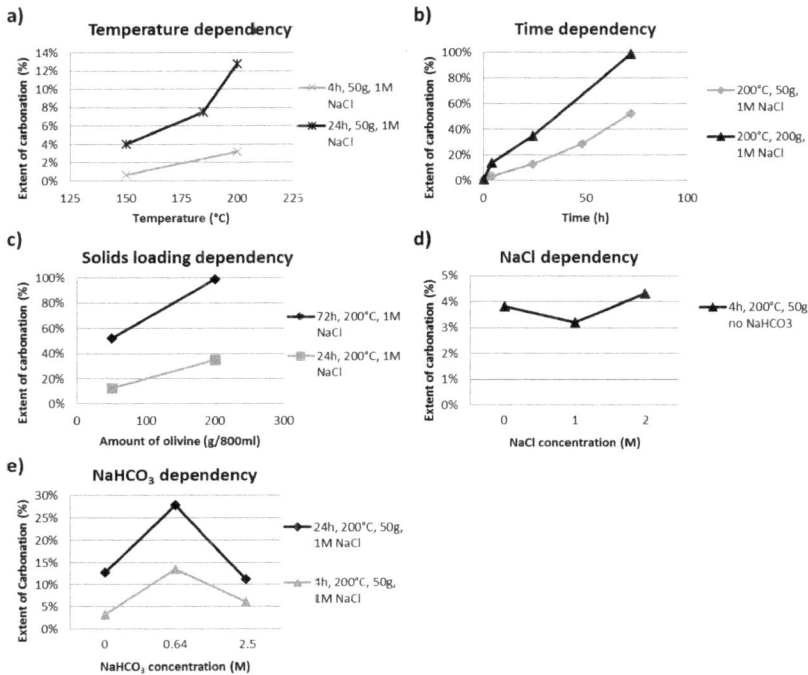

Figure 3. Influence of carbonation process parameters (temperature (**a**); residence time (**b**); solids loading (**c**); NaCl concentration (**d**); and $NaHCO_3$ concentration (**e**)) on extent of olivine carbonation; Table S1 provides detailed data values and statistics on replicates.

The addition of NaCl does not seem to enhance the extent of carbonation. As can be seen in Figure 3d, using one or two molar solutions of NaCl has a very limited impact on the extent of carbonation. O'Connor *et al.* [23] proposed the addition of both 1 M NaCl and 0.64 M $NaHCO_3$, although they also remarked that the addition of sodium bicarbonate has a much larger impact than sodium chloride on the carbonation extent. It can be concluded that, in view of minimizing processing cost or complexity, NaCl addition can be omitted. However, ionic strength can play a role in surface charges and particle aggregation, and should, thus, be investigated in view of product properties such as particle size distribution, specific surface area, and mineral separation. An economical source of saline solution, if desired, would be seawater.

The addition of $NaHCO_3$, on the other hand, has a substantial impact on the carbonation reaction. As can be seen from Figure 3e, the extent of carbonation is highest when 0.64 M of $NaHCO_3$ is added. The substantial impact of $NaHCO_3$ on the carbonation is due to its dissolution into Na^+ and HCO_3^- ions. Chen *et al.* [20] state that adding sodium bicarbonate reduces the concentration of magnesium ions required to exceed the solubility product for magnesium carbonate and, thus, promotes the precipitation of magnesite. They explain that this is because in solutions with a large amount of $NaHCO_3$, the concentration of CO_3^{2-} is inversely proportional to CO_2 pressure and proportional to the square of the concentration of $NaHCO_3$. A reversal of this effect occurs at higher concentrations of $NaHCO_3$ possibly because the solution pH increases excessively, slowing the dissolution of the silicate minerals.

3.2. Characterization of Fully-Carbonated Olivine

Full conversion (0.515 g, CO_2/g, olivine = 99.0% $\pm$ 3.9%) of olivine was achieved by carbonating it for 72 h at 200 °C and 35 bar CO_2 partial pressure with the addition of 1 M NaCl and using a solids loading of 200 g/800 mL. This fully carbonated olivine was the only carbonated material used in the acid leaching experiments presented in Section 3.3. The chemical and mineralogical compositions as well as the microstructural characteristics of the carbonated olivine are very important for interpretation of the leaching results.

The chemical composition of the fully carbonated olivine was determined by digestion followed by ICP-MS (for Al, Ca, Cr, Fe, and Ni) and by XRF (for Co, Mg, Mn, and Si). The obtained results, in decreasing order, were: 18.1 wt. % Mg; 13.8 wt. % Si; 2.5 wt. % Fe; 0.19 wt. % Ni; 0.17 wt. % Cr; 0.13 wt. % Ca; 0.11 wt. % Al; 0.06 wt. % Mn; 0.01 wt. % Co. The respective weight percentages are lower compared to fresh olivine (Table 1) due to the conversion of magnesium silicate to magnesium carbonate; the addition of CO_2 increases the total mass of the olivine by roughly 50%, thus reducing the elemental concentrations.

The mineralogy of the fully-carbonated olivine was analyzed using XRD; the diffractogram is shown in Figure 1. The magnesium silicate (ferroan-forsterite) that was predominant in fresh olivine (84.5 wt. %) is now converted nearly completely (0.5 wt. % remaining) to magnesium carbonate (magnesite ($MgCO_3$), 87.0 wt. %). Periclase and brucite are absent, since they easily carbonate and are also converted to magnesite. The various other minerals that were found in fresh olivine are still present in the carbonated olivine, as they either did not react during carbonation or were formed as hydrated by-products of the reaction: clinochlore (3.7 wt. %), talc (3.1 wt. %), lizardite (2.3 wt. %), tirodite (1.4 wt. %), fayalite (0.9 wt. %), quartz (0.9 wt. %), magnesian calcite (0.3 wt. %), and chromite (0.2 wt. %). Hydrated silicates are known to require dehydroxylation to enable accelerated carbonation [24], and fayalite is known to require elevated pressures to convert into siderite [25].

The particle size distribution of the fully-carbonated olivine can be seen in Figure 4. The average particle size is considerably lower after carbonation as 90 vol. % of the carbonated olivine has a particle size below 42 μm. The BET specific surface area increased from 0.49 m^2/g to 16.1 m^2/g. This confirms that carbonation can act as a substitute to more intense comminution of fresh olivine. Two distinctive peaks can clearly be identified in the particle size distribution, with one peak around 5 μm and one peak around 30 μm. The SEM and EDX analyses presented in Figure 5 provide insight into their occurrence. The carbonated olivine consists of small particles that are clusters of small spheres, and larger crystalline particles. The small particles are analyzed to be silica (SiO_2) rich (Figure 5c), whereas the crystalline particles are primarily magnesium carbonate ($MgCO_3$), seemingly in solid-solution with iron carbonate ($FeCO_3$) (Figure 5b). The appearance of clustered silica particles is likely due to the aggregation of smaller polymerized silica particles. Surface silica will polymerize setting free a water molecule. The polymerized silica will break off from the surface of the olivine particle forming small silica particles free in solution. These small particles can either grow by further condensation or by aggregating together forming the clusters that are shown in Figure 5c. This reaction mechanism was proposed for the preparation of silica from olivine by Lieftink and Geus [26] and was confirmed by Lazaro *et al.* [27].

The fully carbonated olivine was analyzed using EPMA to map the concentration of each element within the particles. The concentrations of chromium, iron, magnesium and nickel are shown in Figure 6. Chromium is concentrated in iron-rich particles, in agreement with chromite composition. Iron distribution largely coincides with that of magnesium, confirming the solid-solution carbonate formation ((Mg_{1-x},Fe_x)CO_3). Nickel appears to be more dispersed throughout the material, although the outline of the carbonate particles can be seen in its map, which would suggest a preference for the carbonate phase. Highly concentrated nickel is also found in a few small particles and the few chromium-rich particles.

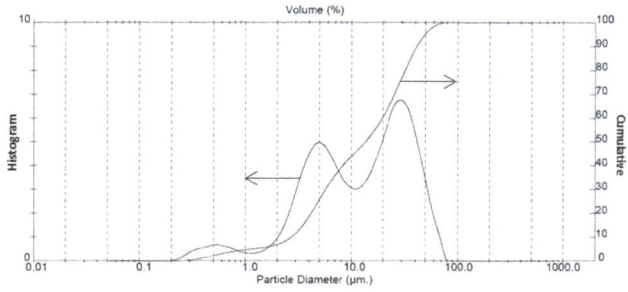

Figure 4. Particle size distribution of fully carbonated olivine, determined by LDA.

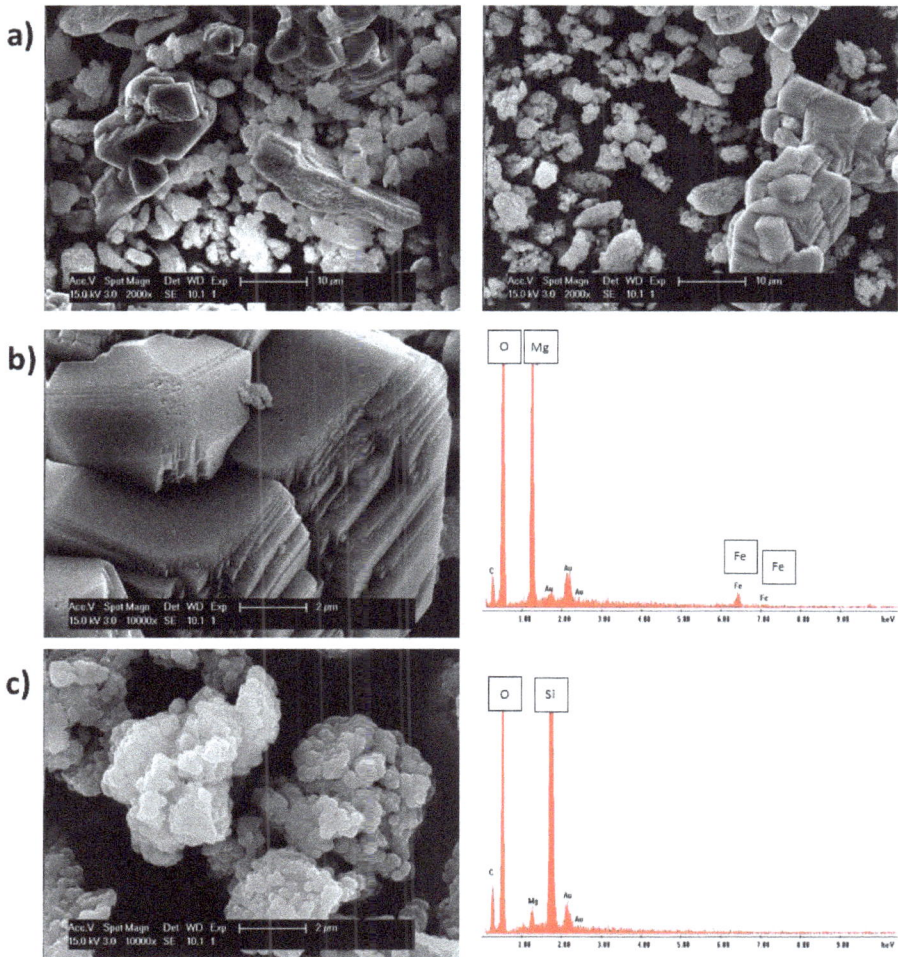

Figure 5. SEM and EDX analyses of individual fully carbonated olivine particles: (**a**) fully carbonated olivine at low magnification; (**b**) ferro-magnesite crystal; (**c**) colloidal silica cluster. Note that Au and C signals are also attributable to gold coating and carbon tape.

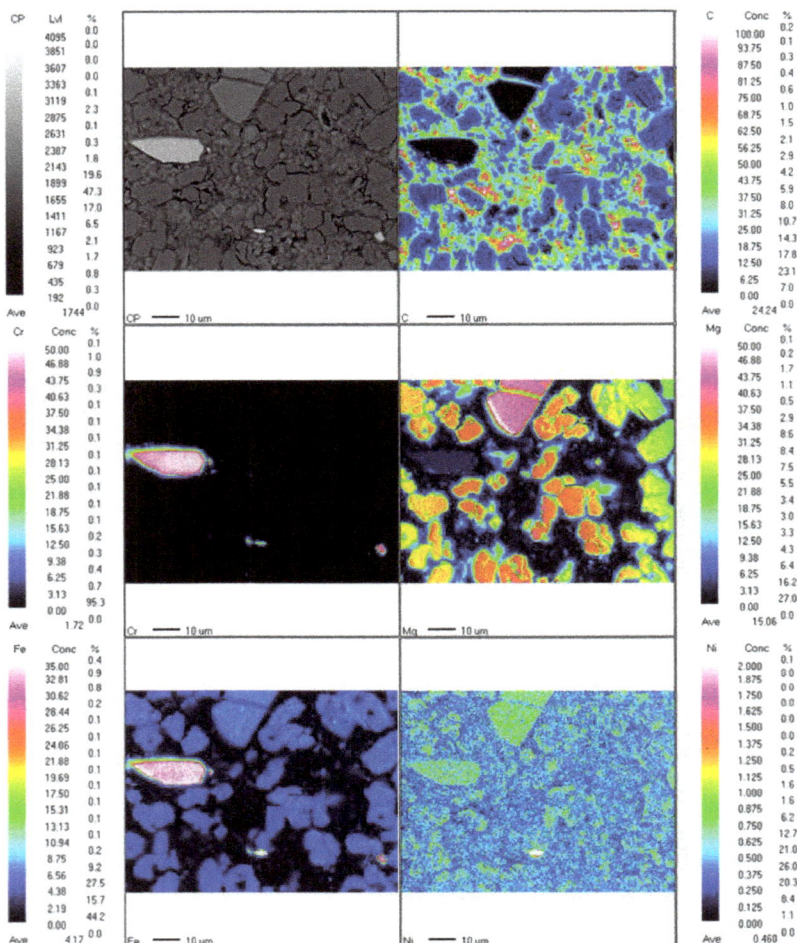

Figure 6. Backscattered scanning electron image and elemental concentration mapping (carbon, chromium, magnesium, iron and nickel) of fully carbonated olivine, captured by EPMA; concentration scale is relative to max/min levels; color gradient in Mg map is due to curvature artifact that occurs at these relatively low magnifications.

Additional elemental distributions are mapped in Figure 7 and Figure S8, taken at slightly lower magnification of another area of the embedded sample. In Figure 7 it is seen that silicon is present in regions poor in magnesium and iron, which supports the EDX results (Figure 5) in that silicon forms distinct particles separate from the carbonate phase. The few calcium-rich particles have magnesium, silicon, and aluminum co-present (Figure 7), and low levels of carbon (Figure S8), which could indicate a calcium-magnesium-aluminum silicate originally present in the olivine or formed during the reaction. A possible natural analogue would be alumoåkermanite (($Ca,Na)_2(Al,Mg,Fe^{2+})(Si_2O_7)$) [28].

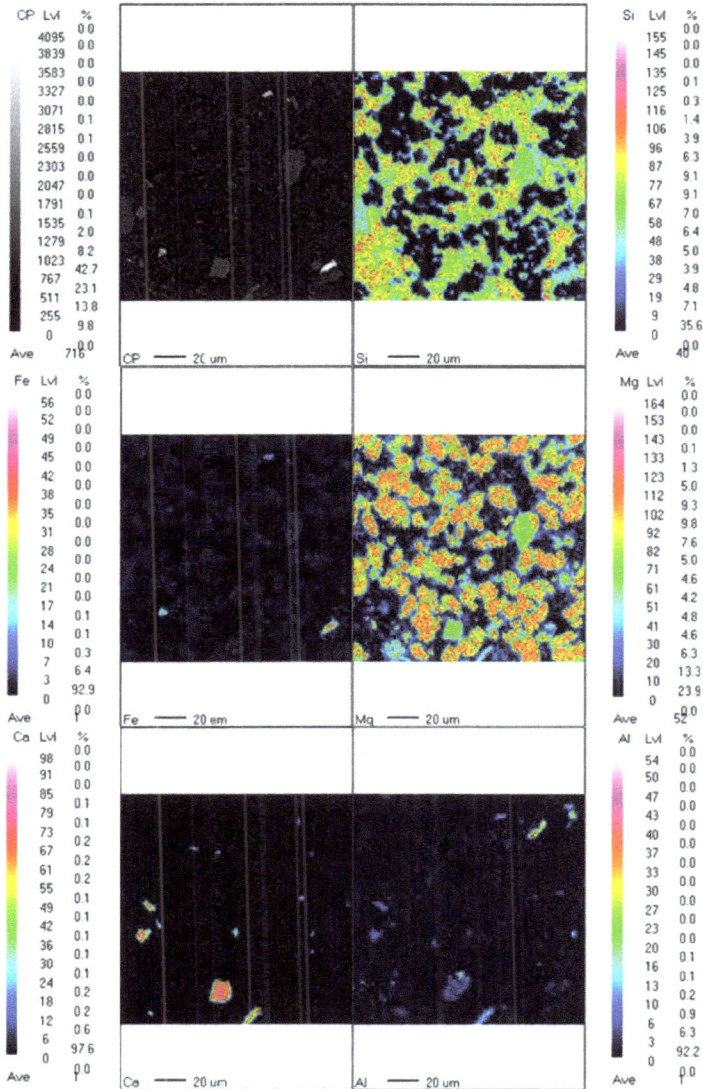

Figure 7. Backscattered scanning electron image and elemental concentration mapping (silicon, iron, magnesium, calcium and aluminum) of fully carbonated olivine, captured by EPMA; color scale indicates max/min levels.

Spot EPMA analysis on certain regions of the polished sample is shown in Figure 8, with their chemical compositions given in Table 3. The four analyzed spot areas are very distinctive. Area 001 is a chromium- and iron-concentrated chromite particle. Area 002 contains mainly a combination of magnesium and silicon, meaning it is a rare grain of unreacted ferroan-forsterite. Area 003 contains magnesium, iron and carbon, indicative of ferro-magnesite ((Mg_{1-x},Fe_x)CO_3). Area 004 is a polymerized silica cluster, as it contains high concentrations of silicon, low concentrations of magnesium and carbon that originates from the embedding resin that penetrates the gaps between the small agglomerated silica particles. These results reaffirm the EDX analysis presented in Figure 5, indicating that carbonate and silica phases (as well as unreacted mineral grains) form distinct particles in the product powder. This means that these particles may be separable by physical means, in view of producing high-value product streams.

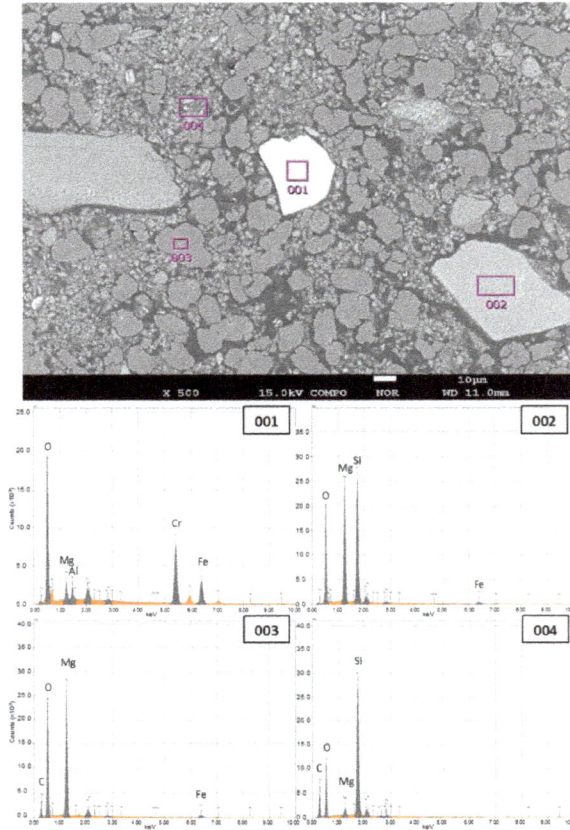

Figure 8. Backscattered scanning electron image of fully carbonated olivine and EPMA scanning of selected areas; mounted sample was coated with platinum and palladium for the analysis.

Table 3. Chemical composition (%), determined by EPMA, of spot areas in Figure 8; carbon as elemental C and other elements as oxides.

Chemical Formula	001	002	003	004
C	0.5	1.2	19.8	37.7
MgO	4.8	35.3	68.6	2.6
SiO_2	0	57.6	0	59.8
Al_2O_3	3.3	0	0	0
Cr_2O_3	55.5	5.8	0	0
Fe_2O_3	35.9	0	11.6	0

3.3. Acid Leaching of Fresh and Carbonated Olivine

3.3.1. Inorganic Acids

Limited research mentions the acid leaching of nickel from olivine. Our own experiments have two main objectives. The first objective is to look at the leaching behavior of various inorganic and organic acids for olivine. The second objective is to investigate the impact of carbonation as a pre-treatment step to leaching.

The metal extractions for the leaching from fresh and carbonated olivine with the three inorganic acids tested can be seen in Figure 9. Sanemasa *et al.* [29] found that there is no preferential leaching of the silicate structure in olivine of one metal over the other. Our results, both for fresh as well as carbonated olivine, confirm these findings: for each acid, at the various concentrations, there is no preferential leaching of magnesium, iron or nickel. In all cases, chromium leaching was minimal, which confirms that it is located in different particles than the ones containing nickel, and that those chromium-containing particles do not undergo dissolution during leaching, while the ones containing nickel do.

Terry *et al.* [30] noted that olivine dissolves congruently, meaning a complete breakdown of the silicate structure to give, percentage-wise, the same amount of silica and metal cation leached. The leaching results for fresh olivine in Figure 9 do not entirely confirm these findings. It seems that less silica is solubilized than would be expected based on a congruent dissolution. This indicates that dissolved silica will precipitate, possibly by forming a silica gel. Notably, the results for carbonated olivine (Figure 9b,d,f) show substantially less leaching of silica compared to fresh olivine. This occurs because the metal silicates in the fresh olivine are transformed to metal carbonates in carbonated olivine. Carbon dioxide will now be released instead of silica during the acid leaching. The polymerized silica in the carbonated olivine does not significantly partake in the dissolution reactions of the acid attack.

Increasing the acid concentration enhances the leaching of most elements. The leaching of Mg, Fe and Ni from fresh olivine increased from below 10% at 0.02 N to above 60% at 2.56 N for all inorganic acids tested. This impact was even more apparent for the leaching of carbonated olivine with HCl and HNO_3, where Mg, Fe, and Ni extractions increased from below 10% at 0.02 N to nearly 100% at 2.56 N. Most notably, the leaching of nickel from carbonated olivine is substantially better than from fresh olivine when using hydrochloric and nitric acids. A concentration of 2.56 N of either acid leaches, respectively, 100% and 91% of nickel from carbonated olivine. For fresh olivine these values are only 66% and 64%, respectively.

In terms of the differences between the three acids, all three inorganic acids show similar leaching behavior of fresh olivine for the same acid normalities. Between 60% and 70% of Mg, Fe and Ni is leached with 2.56 N of either HCl, H_2SO_4 or HNO_3. At intermediate normalities, sulfuric acid performed better; Sanemasa *et al.* [29] explain that sulfate is better at stabilizing the metal cations than the chloride anions. For carbonated olivine, however, there is a clear and opposite distinction between the leaching behaviors of nitric and hydrochloric acid compared to sulfuric acid. H_2SO_4 leaches substantially less Mg, Fe and Ni than HCl and HNO_3. Whereas H_2SO_4 leaches only about 30% Mg, Fe and Ni at 2.56 N, for HCl and HNO_3 this is equal to about 90% to 100%. Additional leaching tests, discussed later on, were performed to investigate this phenomenon.

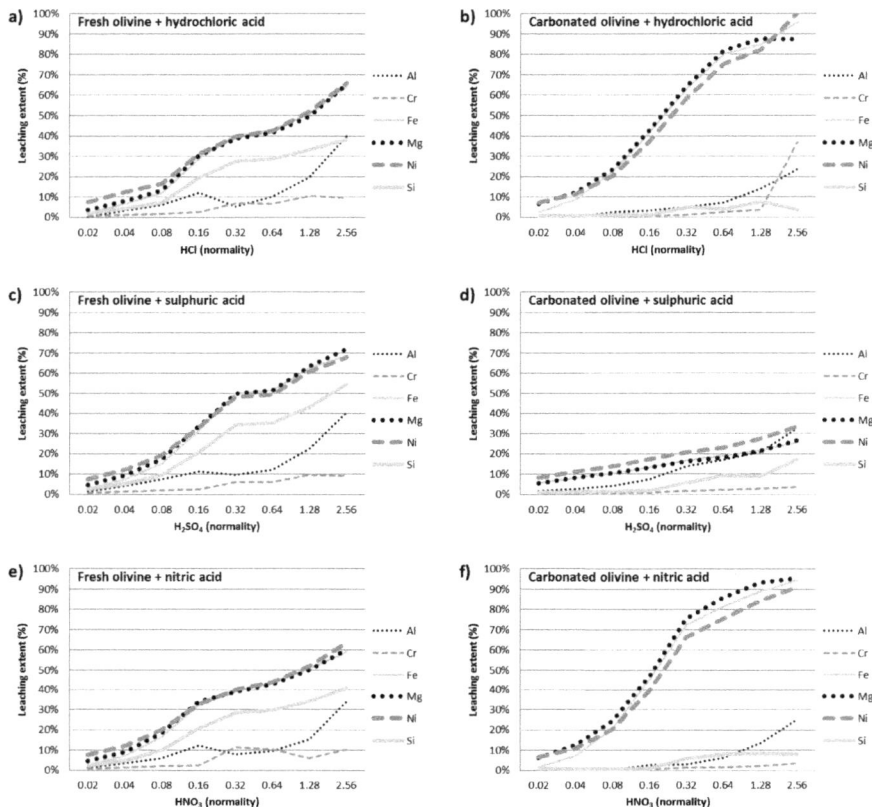

Figure 9. Leaching extent from fresh (**a,c,e**) and carbonated (**b,d,f**) olivine with hydrochloric (**a,b**), sulfuric (**c,d**) and nitric (**e,f**) acids; Tables S2 and S3 provide statistical data on replicates.

The leaching enhancement ratio of nickel from carbonated olivine compared to fresh olivine can be seen in Table 4; the enhancement ratio (φ) is calculated as the percent of nickel leached from carbonated olivine ($\chi_{Ni,carb}$) over the percent of nickel leached from fresh olivine ($\chi_{Ni,fresh}$): $\varphi = \chi_{Ni,carb} / \chi_{Ni,fresh}$. For HCl and HNO$_3$, the leaching ratio increases with increasing acid concentration, peaking at 0.64 N, where carbonated olivine leaches, respectively, 1.77 and 1.72 times better than fresh olivine. The slight decrease at even higher concentrations can be attributed to the fact that the leaching of carbonated olivine is approaching completion, while fresh olivine still benefits significantly from higher acid concentrations (Figure 9). For H$_2$SO$_4$, the leaching of carbonated olivine compared to fresh olivine follows the opposite path, decreasing with increasing acid concentrations. This indicates that the inhibiting effect of H$_2$SO$_4$ is dependent on the sulfate ion concentration. The leaching ratio for H$_2$SO$_4$ reaches a minimum at 0.32 N with only 0.42 times the percentage of nickel extracted from carbonated olivine compared to fresh olivine.

Table 4. Nickel leaching enhancement ratios of carbonated olivine over fresh olivine for inorganic and organic acids.

Leaching Ratio φ (Carbonated/Fresh)	HCl	HNO$_3$	H$_2$SO$_4$	Leaching Ratio φ (Carbonated/Fresh)	Lactic	Citric	Formic
0.02 N	0.99	0.85	1.11	0.25 N	1.01	1.15	0.46
0.04 N	0.93	0.92	0.92	0.5 N	1.18	1.25	0.46
0.08 N	1.23	1.01	0.72	1.0 N	1.38	1.16	0.47
0.16 N	1.20	1.22	0.52	2.0 N	1.48	1.00	0.57
0.32 N	1.47	1.65	0.42	4.0 N *	1.35	-	0.68
0.64 N	1.77	1.72	0.47				
1.28 N	1.58	1.62	0.45				
2.56 N	1.52	1.43	0.49				

* exceeds citric acid solubility.

3.3.2. Organic Acids

The metal extractions from fresh and carbonated olivine by the three organic acids tested are shown in Figure 10. Some similar conclusions can be drawn as for inorganic acids. There is also essentially no preferential dissolution of Mg, Fe and Ni both from fresh and carbonated olivine. Although there is also less silica in solution as would be expected based on a congruent dissolution of the fresh olivine silicate structure, this difference is less than for inorganic acids. Silica is, thus, less prone to precipitation in the presence of organic acids, but is equally insoluble when already precipitated as colloidal silica in the case of carbonated olivine.

The leaching enhancement ratio for lactic acid increases with increasing acid concentrations, reaching a maximum of 1.48 at 2 N (Table 4). Contrary to lactic and formic acids, leaching with citric acid remains constant (for fresh olivine), or decreases slightly (for carbonated olivine) with increasing acid concentration (Figure 10). The enhancement ratio remains above one up to 1 N, with a maximum of 1.25 at 0.5 N (Table 4). Citrate ions act as chelating agents and will form strong metal-ligand complexes that enhance leaching. It appears that due to these highly soluble complexes, citric acid already reaches its maximum leaching potential at low concentrations. Tzeferis *et al.* [31] also found that increasing the citric acid concentration from 0.5 M to 1.5 M did not increase the nickel or iron extraction from nickeliferous ores at low pulp densities. At higher pulp densities, the percentages of nickel and iron leached at 0.5 M were substantially lower than for low pulp densities, and they did increase when the citric acid concentration was increased to 1.5 M. This confirms the idea that citric acid has an intrinsic maximum for the leaching of metals or specific ores, which, once reached, will not increase further when increasing the citric acid concentration.

Comparing the leaching results from fresh olivine with carbonated olivine, it can be seen that there is also a large discrepancy among the organic acids (Table 4). Nickel leaching by citric and lactic acids is enhanced when olivine is carbonated (*i.e.*, enhancement ratios > 1). Leaching by formic acid, however, just like with sulfuric acid, experiences a considerable decrease (−32% to −54%) in nickel extraction from carbonated olivine compared to fresh olivine.

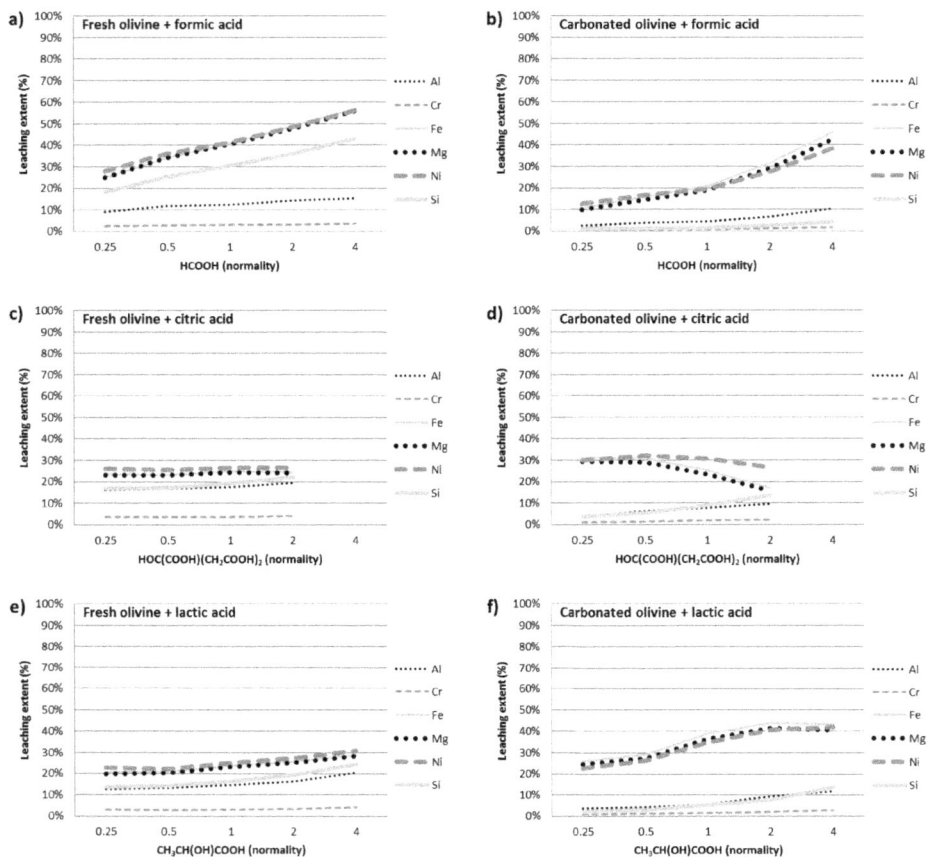

Figure 10. Leaching extent from fresh (**a**,**c**,**e**) and carbonated (**b**,**d**,**f**) olivine with formic (**a**,**b**), citric (**c**,**d**), and lactic (**e**,**f**) acids; Tables S4 and S5 provides statistical data on replicates.

3.3.3. Sulfuric Acid Leaching Investigation

Aforementioned results for fresh olivine show a limited leaching of nickel, magnesium and iron (the cationic components of ferroan-forsterite) with H_2SO_4 compared to HCl and HNO_3. Terry *et al.* [30] noted that sulfate ions form stronger metal cation-acid anion complexes than chloride ions, which would lead to an increased reactivity with sulfuric acid. For carbonated olivine, however, this is completely reversed and thus does not follow the existing theory. One possibility is that the sulfate ions form insoluble compounds with the components of carbonated olivine (carbonate and silica phases), passivating the particles. To test this hypothesis, leaching tests were conducted using sulfuric acid and mineral mixtures. The mineral mixtures consisted of one or more of the following components: pure magnesite ($MgCO_3$), pure fumed (amorphous) silica (SiO_2), fresh olivine and fully carbonated olivine. If the carbonation products (magnesite and silica) had unexpected behavior in contact with sulfuric acid, these tests would help uncover these effects. Since the pure phases did not contain nickel or iron, leaching data for magnesium was collected.

Figure 11 presents the data on magnesium leaching from the pure components (Figure 11a) and from the mineral mixtures (Figure 11c). Figure 11a presents the raw leaching data (expressed as g, Mg/100 mL) and the data expressed as a percentage of the theoretical maximum leaching extent (based on Mg content in the mineral: 0.577 g, Mg/100 mL for pure $MgCO_3$, 0.543 g, Mg/100 mL for fresh olivine, and 0.362 g, Mg/100 mL for fully carbonated olivine). It is found that pure $MgCO_3$ leaches completely in 1.28 N H_2SO_4, so there is no negative effect of sulfate anions, and no precipitation of insoluble compound. In the case of olivine, leaching is much more extensive in the case of fresh olivine (77%) compared to fully carbonated olivine (20%). This confirms the effects seen in previous results (Figure 9). Figure 11b presents kinetic data on the leaching of fresh and fully carbonated olivine by 1.28 N H_2SO_4. Leaching of the latter is slower and stalls after two hours, while leaching of the former continuously increases over time, although it is also relatively slow (leaching after 2 h is less than a third that after 24 h).

Figure 11c presents the leaching results of four mineral mixtures. For each mixture, data is presented in raw format (g, Mg/100 mL) and as a percentage of the theoretical maximum leaching based on the mixture's composition and the leaching results obtained for the singular components (Figure 11a). The first mixture contains only the pure minerals, and shows that SiO_2 does not prevent leaching of $MgCO_3$, as the leaching extent reaches 97% of the predicted value (*i.e.*, within experimental uncertainty). The second mixture shows that $MgCO_3$ does not alter the leaching of fresh olivine, as the leaching extent is equal to the predicted value; hence, no insoluble precipitate forms. Likewise, the third mixture shows that SiO_2 does not alter the leaching of fresh olivine (the value greater than 100% is within experimental uncertainty); again, no insoluble precipitate forms. Finally, the last mixture shows that carbonated olivine still leaches poorly and that leaching of $MgCO_3$ is not affected by the presence of carbonated olivine, since the leaching extent is approximately equal to the calculated prediction (97%). This last mixture result shows that no component of carbonated olivine prevents the leaching of pure $MgCO_3$, although the ferro-magnesite present in carbonated olivine is affected.

To better understand what happens with the sulfuric acid-leached carbonated olivine, the leaching residue was characterized by XRD, and results are shown in Figure 1. The diffraction pattern of the leaching residue is very similar to that of the pre-leaching fully carbonated olivine. No additional peaks form after leaching, which indicates that no crystalline precipitates form. Additionally, all significant peaks present in the pre-leaching mineral (attributable to magnesite) are still present after leaching, meaning that magnesite leaching is not extensive, as the magnesium leaching data indicated (Figure 11a). The main difference seen between the two diffractograms is that the leached residue has a larger "bump", which is attributable to a quasi-amorphous phase. Since this bump is near the theoretical location for crystalline silica (*i.e.*, quartz), it is possible to infer that it represents the colloidal silica content of the material (and hence is also present in the pre-leached mineral). The reason why the bump grows after leaching, is that the crystalline content is reduced, due to partial dissolution of magnesite. These XRD results suggest that no precipitate forms after leaching, either crystalline or quasi-amorphous in nature.

Based on leaching and XRD results, it appears that the only explanation for the poor leaching results of fully carbonated olivine in sulfuric acid is that the magnesite present in carbonated olivine is a solid-solution of $MgCO_3$ and $FeCO_3$. Since pure $MgCO_3$ leaches adequately in sulfuric acid, and amorphous silica does not affect leaching, it could be that the iron content of the ferro-magnesite helps to passivate the mineral particles after an initial limited leaching extent. Further research is needed to characterize this mechanism.

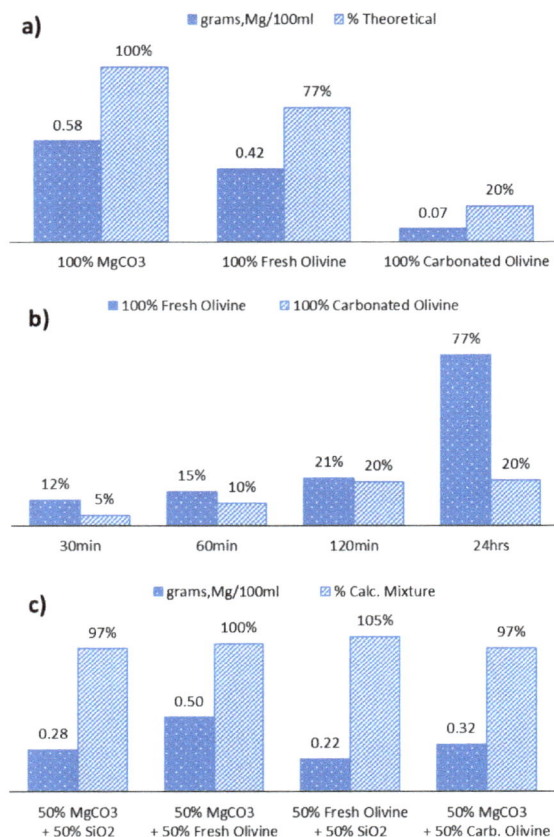

Figure 11. (**a**) leaching extent (in 1.28 N H_2SO_4, 2 g/100 mL, 24 h) from pure magnesium carbonate, fresh olivine and carbonated olivine; (**b**) leaching extent (in 1.28 N H_2SO_4, 2 g/100 mL) from fresh olivine and carbonated olivine as a function of time; (**c**) leaching extent (in 1.28 N H_2SO_4, 2 g/100 mL, 24 h) from mixtures of pure magnesium carbonate, pure silica, fresh olivine and carbonated olivine.

4. Conclusions

The increasing demand and diminishing availability of raw materials requires us to look beyond conventional sources. In the future, the importance of low-grade ores and waste streams as a source for raw materials will only increase. The objective of this work was to look at the extraction of nickel from a low-grade silicate ore, namely olivine. This was achieved by combining conventional acid leaching with a pre-treatment step in which the olivine underwent mineral carbonation. It is anticipated that the mineral carbonation pre-treatment approach may also be applicable to other ultrabasic and lateritic ores.

In a first processing step, olivine was fully carbonated at high CO_2 partial pressures (35 bar) and optimal temperature (200 °C) with the addition of pH buffering agents. Although substantial research has looked into the carbonation of olivine, reported extents of carbonation are usually lower than those achieved in this work (*i.e.*, full carbonation). The carbonation increased linearly with time, indicating that carbonation is not limited by the formation of a passivating silica layer under the processing conditions used. This was confirmed by SEM analysis of partially carbonated olivine, showing that after carbonation distinct crystalline magnesium carbonate particles and clusters of nano-silica are formed. High solids loading and mixing rate appear to enhance the carbonation reaction substantially due to the olivine particles being eroded by increased particle collisions. Using electron probe micro analysis it was possible to map the distribution of both major (C, Mg, Si) and minor (Al, Ca, Cr, Fe, Ni) elemental components in this material. The main products of the carbonation reaction included quasi-amorphous colloidal silica, chromium-rich metallic particles, and ferro-magnesite.

The second stage of this work looked at the extraction of nickel and other metals by leaching fresh as well as carbonated olivine with an array of inorganic and organic acids to test their leaching efficiency. Compared to leaching from untreated olivine, the percentage of nickel extracted from carbonated olivine by acid leaching was significantly increased. For example, using 2.6 N HCl and HNO$_3$, 100% and 91% of nickel was respectively leached from carbonated olivine. This compares to only 66% and 64% nickel leached from untreated olivine using the same acids. Similar trends were observed with the organic acids used, where the leaching enhancement reached a factor of 1.25 using 0.5 N citric acid, and a factor of 1.48 using 2 N lactic acid. It was found that two acids, sulfuric and formic, are unsuitable for leaching of carbonated olivine.

Looking at future developments, it should be pointed out that in the present work the metal extraction was performed after carbonation. Selective metal recovery of the olivine during carbonation might substantially enhance the leaching and reduce extractant consumption. During carbonation, the metals dissolve due to the acidic aqueous solution as a result of carbonic acid formation. The residual solids could then be separated at high temperatures and pressures prior to exiting the reactor, producing a purified silica stream and a concentrated metal liquor or metal precipitate. Another option is to separate the final products (*i.e.*, silica-rich clusters, carbonate crystals, and metallic particles) prior to leaching, and leach only the metal-rich fraction; this may reduce the amount of sequestered CO$_2$ liberation. Lastly, re-use of any CO$_2$ released during acidification can contribute to lowering processing costs associated with CO$_2$ concentration from industrial emission sources. These processes are presently in development using the proprietary reactor technology of Innovation Concepts B.V. called the "CO$_2$ Energy Reactor" [13]. This reactor makes use of a Gravity Pressure Vessel (GPV) that supports hydrostatically built supercritical pressures, runs in autothermal regime by recycling the exothermic carbonation heat, and operates under turbulent three-phase plug flow configuration. This reactor is expected to allow faster carbonation conversion and more economical processing.

Acknowledgments: The KU Leuven Industrial Research Fund (IOF) is gratefully acknowledged for funding the Knowledge Platform on Sustainable Materialization of Residues from Thermal Processes into Products (SMaRT-Pro$_2$) in which this work was performed. Rafael Santos is thankful for the PGS-D support from the Natural Sciences and Engineering Research Council of Canada (NSERC), and for postdoctoral financial support (PDM) from the KU Leuven Special Research Fund (BOF). The EPMA work was made possible due to the support of the Hercules Foundation (project ZW09-09). The KU Leuven Department of Earth and Environmental Sciences is acknowledged for the use of XRD and LDA equipment.

Author Contributions: Rafael M. Santos conceptualized and managed the research, co-supervised the Master student, characterized materials and samples, and co-wrote the paper. Aldo Van Audenaerde performed experiments, analyzed the data and co-wrote the paper. Yi Wai Chiang co-supervised the Master student, performed BET analyses and co-wrote the paper. Remus I. Iacobescu performed EPMA analyses and co-wrote the paper. Pol Knops conceptualized the research and provided industrial advice. Tom Van Gerven supervised the personnel, provided funding and managed the laboratory facilities.

References

1. Goldie, R.J. The shortage of nickel. Prospectors & Developers Association of Canada 2004 Convention. Available online: http://www.webcitation.org/query?url=http%3A%2F%2Fwww.pdac.ca%2Fdocs%2Fdefault-source%2Fpublications---papers-presentations---conventions%2Ftechprgm-goldie.pdf%3Fsfvrsn%3D4&date=2015-07-26 (accessed on 26 July 2015).
2. Kerfoot, D.G.E. Nickel. In *Ullmann's Encyclopedia of Industrial Chemistry*; Wiley-VCH Verlag GmbH & Co. KGaA: Weinheim, Germany, 2012; Volume 27, pp. 37–101.

3. McDonald, R.G.; Whittington, B.I. Atmospheric acid leaching of nickel laterites review—Part I: sulphuric acid technologies. *Hydrometallurgy* **2008**, *91*, 35–55. [CrossRef]

4. Zhai, Y.-C.; Mu, W.-N.; Liu, Y.; Xu, Q. A green process for recovering nickel from nickeliferous laterite ores. *Trans. Nonferrous Metals Soc. China* **2010**, *20*, s65–s70. [CrossRef]

5. Chen, N.; Cao, Z.-F.; Zhong, H.; Fan, F.; Qiu, P.; Wang, M.-M. A novel approach for recovery of nickel and iron from nickel laterite ore. *Metall. Res. Technol.* **2015**, *112*, 306. [CrossRef]

6. Harben, P.W.; Smith, C., Jr. Olivine. In *Industrial Minerals & Rocks: Commodities, Markets, and Uses*, 7th ed.; Kogel, J.E., Trivedi, N.C., Barker, J.M., Krukowski, S.T., Eds.; Society for Mining, Metallurgy and Exploration, Inc.: Littleton, CO, USA, 2006; pp. 679–684.

7. Smyth, J.R.; Frost, D.J.; Nestola, F.; Holl, C.M.; Bromiley, G. Olivine hydration in the deep upper mantle: Effects of temperature and silica activity. *Geophys. Res. Lett.* **2006**. [CrossRef]

8. Agatzini-Leonardou, S.; Dimaki, D. Heap Leaching of Poor Nickel Laterites by Sulphuric Acid at Ambient Temperature. In *Hydrometallurgy '94*; Springer: Dordrecht, The Netherlands, 1994; pp. 193–208.

9. Bodor, M.; Santos, R.M.; van Gerven, T.; Vlad, M. Recent developments and perspectives on the treatment of industrial wastes by mineral carbonation—A review. *Cent. Eur. J. Eng.* **2013**, *3*, 566–584. [CrossRef]

10. Sanna, A.; Hall, M.R.; Maroto-Valer, M. Post-processing pathways in carbon capture and storage by mineral carbonation (CCSM) towards the introduction of carbon neutral materials. *Energy Environ. Sci.* **2012**, *5*, 7781–7796. [CrossRef]

11. Ahmad, W. Nickel laterites—Fundamentals of chemistry, mineralogy, weathering processes, formation, and exploration. VALE Inco-VITSL. Available online: http://www.scribd.com/doc/83755762/Nickel-Laterites-Vale-Inco-Oct-2008 (accessed on 7 July 2012).

12. Thompson, J.F.H.; Barnes, S.J.; Dyke, J.M. The distribution of nickel and iron between olivine and magmatic sulfides in some natural assemblages. *Can. Miner.* **1984**, *22*, 55–66.

13. Santos, R.M.; Verbeeck, W.; Knops, P.; Rijnsburger, K.; Pontikes, Y.; van Gerven, T. Integrated mineral carbonation reactor technology for sustainable carbon dioxide sequestration: "CO$_2$ energy reactor". *Energy Proc.* **2013**, *37*, 5884–5891. [CrossRef]

14. Gadikota, G.; Matter, J.; Kelemen, P.; Park, A.-H.A. Chemical and morphological changes during olivine carbonation for CO$_2$ storage in the presence of NaCl and NaHCO$_3$. *Phys. Chem. Chem. Phys.* **2014**, *16*, 4679–4693. [CrossRef] [PubMed]

15. Gadikota, G.; Swanson, E.J.; Zhao, H.; Park, A.-H.A. Experimental design and data analysis for accurate estimation of reaction kinetics and conversion for carbon mineralization. *Ind. Eng. Chem. Res.* **2014**, *53*, 6664–6676. [CrossRef]

16. Julcour, C.; Bourgeois, F.; Bonfils, B.; Benhamed, I.; Guyot, F.; Bodénan, F.; Petiot, C.; Gaucher, É.C. Development of an attrition-leaching hybrid process for direct aqueous mineral carbonation. *Chem. Eng. J.* **2015**, *262*, 716–726. [CrossRef]

17. Rigopoulos, I.; Petallidou, K.C.; Vasiliades, M.A.; Delimitis, A.; Ioannou, I.; Efstathiou, A.M.; Kyratsi, T. Carbon dioxide storage in olivine basalts: Effect of ball milling process. *Powder Technol.* **2015**, *273*, 220–229. [CrossRef]

18. Oxtoby, D.W.; Freeman, W.A.; Block, T.F. *Chemistry: Science of Change*, 4th ed.; Brooks/Cole: Belmont, NY, USA, 2002.

19. Béarat, H.; McKelvy, M.J.; Chizmeshya, A.V.G.; Gormley, D.; Nunez, R.; Carpenter, R.W.; Squires, K.; Wolf, G.H. Carbon sequestration via aqueous olivine mineral carbonation: Role of passivating layer formation. *Environ. Sci. Technol.* **2006**, *40*, 4802–4808. [CrossRef] [PubMed]

20. Chen, Z.-Y.; O'Connor, W.K.; Gerdemann, S.J. Chemistry of aqueous mineral carbonation for carbon sequestration and explanation of experimental results. *Environ. Prog.* **2006**, *25*, 161–166. [CrossRef]

21. De Windt, L.; Devillers, P. Modeling the degradation of Portland cement pastes by biogenic organic acids. *Cem. Concr. Res.* **2010**, *40*, 1165–1174. [CrossRef]

22. Chiang, Y.W.; Santos, R.M.; van Audenaerde, A.; Monballiu, A.; van Gerven, T.; Meesschaert, B. Chemoorganotrophic Bioleaching of Olivine for Nickel Recovery. *Minerals* **2014**, *4*, 553–564. [CrossRef]

23. O'Connor, W.K.; Dahlin, D.C.; Rush, G.E.; Gerdemann, S.J.; Penner, L.R.; Nilsen, D.N. Aqueous mineral carbonation—Mineral availability, pretreatment, reaction parametrics, and process studies. National Energy Technology Laboratory, US DOE. Available online: http://www.webcitation. org/query?url=https%3A%2F%2Fwww.netl.doe.gov%2FFile%2520Library%2FResearch%2FCoal% 2FNETLAlbanyAqueousMineralCarbonation.pdf&date=2015-07-26 (accessed on 26 July 2015).

24. Larachi, F.; Daldoul, I.; Beaudoin, G. Fixation of CO_2 by chrysotile in low-pressure dry and moist carbonation: *Ex-situ* and *in-situ* characterizations. *Geochim. Cosmochim. Acta* **2010**, *74*, 3051–3075. [CrossRef]

25. Qafoku, O.; Kovarik, L.; Kukkadapu, R.K.; Ilton, E.S.; Arey, B.W.; Tucek, J.; Felmy, A.R. Fayalite dissolution and siderite formation in water-saturated supercritical CO_2. *Chem. Geol.* **2012**, *332–333*, 124–135. [CrossRef]

26. Lieftink, D.J.; Geus, J.W. The preparation of silica from the olivine process and its possible use as a catalyst support. *J. Geochem. Explor.* **1998**, *62*, 1–3. [CrossRef]

27. Lazaro, A.; Brouwers, H.J.H.; Quercia, G.; Geus, J.W. The properties of amorphous nano-silica synthesized by the dissolution of olivine. *Chem. Eng. J.* **2012**, *211–212*, 112–121. [CrossRef]

28. Wiedenmann, D.; Zaitsev, A.N.; Britvin, S.N.; Krivovichev, S.V.; Keller, J. Alumoåkermanite, $(Ca,Na)_2(Al,Mg,Fe^{2+})(Si_2O_7)$, a new mineral from the active carbonatite-nephelinite-phonolite volcano Oldoinyo Lengai, northern Tanzania. *Miner. Mag.* **2009**, *73*, 373–384. [CrossRef]

29. Sanemasa, I.; Yoshida, M.; Ozawa, T. The dissolution of olivine in aqueous solution of inorganic acids. *Bull. Chem. Soc. Jpn.* **1972**, *45*, 1741–1746. [CrossRef]

30. Terry, B. The acid decomposition of silicate minerals part I. Reactivities and modes of dissolution of silicates. *Hydrometallurgy* **1983**, *10*, 135–150. [CrossRef]

31. Tzeferis, P.G.; Agatzini-Leonardou, S. Leaching of nickel and iron from Greek non-sulphide nickeliferous ores by organic acids. *Hydrometallurgy* **1994**, *36*, 345–360. [CrossRef]

metals

MDPI

Article

Direct Aqueous Mineral Carbonation of Waste Slate Using Ammonium Salt Solutions

Hwanju Jo [1], Ho Young Jo [2],*, Sunwon Rha [2] and Pyeong-Koo Lee [1]

[1] Korea Institute of Geoscience and Mineral Resources, Daejeon 34132, Korea; chohwanju@kigam.re.kr (H.J.); pklee@kigam.re.kr (P.-K.L.)

[2] Department of Earth and Environmental Sciences, Korea University, Seoul 02841, Korea; sunwon671@korea.ac.kr

* Author to whom correspondence should be addressed; hyjo@korea.ac.kr; Tel.: +82-2-3290-3179; +82-2-3290-3189.

Academic Editors: Suresh Bhargava, Mark Pownceby and Rahul Ram
Received: 18 October 2015; Accepted: 15 December 2015; Published: 18 December 2015

Abstract: The carbonation of asbestos-containing waste slate using a direct aqueous mineral carbonation method was evaluated. Leaching and carbonation tests were conducted on asbestos-containing waste slate using ammonium salt (CH_3COONH_4, NH_4NO_3, and NH_4HSO_4) solutions at various concentrations. The CH_3COONH_4 solution had the highest Ca-leaching efficiency (17%–35%) and the NH_4HSO_4 solution had the highest Mg-leaching efficiency (7%–24%) at various solid dosages and solvent concentrations. The $CaCO_3$ content of the reacted materials based on thermogravimetric analysis (TGA) was approximately 10%–17% higher than that of the as-received material for the 1 M CH_3COONH_4 and the 1 M NH_4HSO_4 solutions. The carbonates were precipitated on the surface of chrysotile, which was contained in the waste slate reacted with CO_2. These results imply that CO_2 can be sequestered by a direct aqueous mineral carbonation using waste slate.

Keywords: CO_2 sequestration; mineral carbonation; waste slate; ammonium salts; asbestos

1. Introduction

CO_2 mineral carbonation is a method to permanently sequester CO_2 as a form of carbonate minerals with or without aqueous phase. For the CO_2 mineral carbonation process with aqueous phase, CO_2 is reacted with raw materials containing alkaline earth metals (mostly Ca and Mg) in aqueous solutions (*i.e.*, direct aqueous mineral carbonation) or is reacted with alkaline earth metals leached from raw materials in aqueous solutions (*i.e.*, indirect aqueous mineral carbonation) to form carbonate minerals [1,2].

The mineral carbonation to sequester CO_2 has been extensively studied due to its following advantages: (1) non-requirement for any underground geological storage sites; (2) permanent CO_2 sequestration without long-term monitoring; (3) potential immobilization of toxic elements contained in raw materials; and (4) beneficial use of produced carbonates. However, the major challenge of the mineral carbonation method is to enhance the leaching capacity of alkaline earth metals from raw materials, which is a main factor affecting rate and degree of mineral carbonation [3,4]. The leaching processes of alkaline earth metals from raw materials are generally expensive due to the need for increasing temperature and pressure, acid or base solutions, and grinding raw materials.

Previous researches regarding aqueous mineral carbonation have mainly focused on enhancing the leaching efficiency of alkaline earth metals by pre-treating raw materials before carbonation processes [4–8]. Natural alkaline materials and alkaline industrial wastes have been extensively evaluated for the mineral carbonation, e.g., [8–19]. Natural alkaline materials are relatively abundant compared with alkaline industrial wastes but their use requires pretreatment to enhance the leaching

Metals **2015**, *5*, 2413–2427

of the alkaline earth metal due to their strong chemical stability. Even if industrial wastes have more limited availability, the alkaline earth metals can be relatively easily leached from the alkaline industrial byproducts because of their chemical instability [20].

Asbestos-containing waste slate is one of the potential alkaline industrial wastes to sequester CO_2 due to its high content of Ca and Mg. Asbestos-containing waste slate is mostly comprised of cement, which is a Ca source, and chrysotile, which is an Mg source. Chrysotile, which is considered as a carcinogenic material, is the fibrous magnesium silicate mineral ($Mg_3(Si_2O_5)(OH)_4$) in the serpentine group. Asbestos-containing slate was used as a roofing material in South Korea during 1960 and 1970s due to its high insulating capacity, but subsequently has not been legally allowed for use due to the toxicity of chrysotile, which is a form of asbestos. Currently, the asbestos containing waste slate is mostly disposed of in government certified landfills without reuse in South Korea.

Due to the health risks such as asbestosis, lung cancer, and mesothelioma, a number of studies suggested safe disposal schemes of asbestos containing materials. Most treatment methods of asbestos-containing materials focused on morphological alteration of asbestos using chemical and thermal treatment. Chemical treatment methods of asbestos-containing material using oxalic acid [21], Na-oxalate and Na-acetate [22], sulfuric acid [23], and hydrogen peroxide [24] could alter significantly asbestos material. Thermal treatment methods of asbestos using microwave [25–27] and microwave air plasma [28] could also alter effectively asbestos to non-hazardous form. However, thermal treatment requires vast amount of energy. Gualtieri and Tartaglia [29] reported that temperature higher than 1000 °C is required to entire transformation of asbestos to non-hazardous silicate glass phase. Yoshikawa *et al.* [27] also showed that asbestos was completely transformed to non-hazardous form at the temperature >1000 °C with microwave treatment.

The main objective of this study was to evaluate the feasibility of CO_2 sequestration using waste slate by a direct aqueous mineral carbonation. Ammonium salt (CH_3COONH_4, NH_4Cl, NH_4NO_3, and NH_4HSO_4) solutions were used as solvents for a direct aqueous mineral carbonation of asbestos-containing waste slate. The leaching and carbonation behaviors of asbestos-containing waste slate were investigated at room temperature and atmospheric pressure conditions. For the carbonation tests, CO_2 was injected into the mixture of waste slate and ammonium salt solution. Morphological alteration of asbestos-containing waste slate after carbonation was also evaluated.

2. Experimental Section

2.1. Materials

Waste roofing slate panels were collected from an abandoned house in South Korea. The waste slate was broken using a hand hammer to get particle size less than 0.425 mm. The slate particles were used in the leaching and carbonation tests. CH_3COONH_4, NH_4NO_3, and NH_4HSO_4 (Sigma-Aldrich Co., St. Louis, Mo, USA) solutions were used as solvents. Deionized (DI) water was used to prepare the ammonium salt solutions.

2.2. Leaching Tests

The leaching test was performed with various slate solid dosages ranging between 20 and 150 g/L in a 50 mL polypropylene copolymer centrifuge tube using an ammonium salt solution for four hours. 1–10 g of oven-dried slate was added into a 50 mL polypropylene copolymer centrifuge tube with 1, 2, and 4 M solutions. The mixture was shaken at room temperature and atmospheric pressure using a water bath shaker and the pH and electrical conductivity (EC) were then measured. The mixture was centrifuged at 5000 rpm for 30 min (VS-550i, Vision Scientific CO., Daejeon, Korea) and was then filtered using a 0.2 μm filter (ADVANTEC®, Advantec MFS, Inc. Tokyo, Japan). All of the leaching tests were conducted a single time.

2.3. Carbonation Tests

For the carbonation test, the waste slate and 1 M CH_3COONH_4 solution or 1 M NH_4HSO_4 solution at various solid dosages ranged from 20 to 150 g/L were mixed thoroughly for four hours in a 500 mL Elrenmeyer flask (Dongsung Scientific, Busan, Korea). The initial pH of the mixtures of the waste slate and 1 M CH_3COONH_4 solution or 1 M NH_4HSO_4 solution were about 9.4 and 9.2, respectively. A 15 vol. % CO_2 gas mixture with 85 vol. % N_2 was then injected at a flow rate of 200 mL/min into the slurry of waste slate and 1 M CH_3COONH_4 solution or 1 M NH_4HSO_4 solution at various solid dosages in the 500 mL Erlenmeyer flask. The 15 vol. % CO_2 gas mixture was chosen because the flue gas from coal fired power plants in South Korea generally contains 15 vol. % CO_2. During the carbonation tests, the slurry was mixed using a magnetic stirrer at a stirring rate of approximately 300 rpm. The procedure of the carbonation test used in this study is similar to that used in Jo *et al.* [8]. The carbonation test was performed at a room temperature and atmospheric pressure. The CO_2 injection was stopped at a pH of around 7.5 for preventing $CaCO_3$ dissolution [30]. All of the carbonation tests were conducted a single time.

2.4. Chemical Analysis and Material Characterization

The cation concentrations of filtrates from the leaching and carbonation tests were analyzed by inductivity coupled plasma-atomic emission spectroscopy (ICP-AES, OPTIMA 3000XL, Perkin Elmer, Wellesley, MA, USA). The elemental composition of the as-received material was determined by using microwave digestion. 0.5 g of the waste slate raw material with 10 mL of HNO_3 was digested using microwave digestion for 10 min at 175 °C and the leachate was then analyzed by ICP-AES (Perkin Elmer, Wellesley, MA, USA).

An X-ray diffractometer (Xpert MPD, Phillips, Almelo, The Netherlands) and a field emission scanning electron microscope (FE-SEM: S-4300, Hitachi, Tokyo, Japan) equipped with energy dispersive X-ray spectroscopy (EDX: Ex-20, Horiba, Kyoto, Japan) were used to characterize the as-received and reacted materials. The thermogravimetric analysis (TGA) (SDTG-60H, Shimadzu Corp., Kyoto, Japan) was performed on the reactant and products to determine the calcium carbonate ($CaCO_3$) content.

3. Results and Discussion

3.1. Raw Material

The as-received waste slate contained mainly Ca (25.2%). (Table 1). Table 1 shows selected elemental composition determined by using microwave digestion. The as-received waste slate mainly consists of calcite, chrysotile, and Ca-Mg-Al silicates (Figure 1). The $CaCO_3$ content of the waste slate determined using the results of TGA [8], as shown in Figure 2 was approximately 29.8%. The $CaCO_3$ may have been formed by natural carbonation because the slate had been exposed to atmosphere for a very long period of more than 20 years. It is well known that cement based materials exposed to atmosphere can absorb CO_2 during the service life [31].

Table 1. Chemical and mineralogical characteristics of as-received waste slate sample.

Materials	Composition (wt. %)					Minerals (XRD)
	Ca	Mg	Al	Fe	Si	
Waste slate	25.2	2.2	1.4	1.3	1.4	Calcite, Chrysotile, and Ca-Mg-Al-Si-oxide minerals

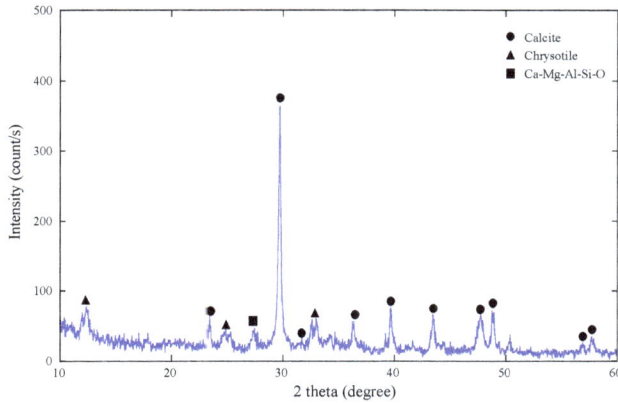

Figure 1. X-ray diffraction (XRD) patterns of as-received waste slate.

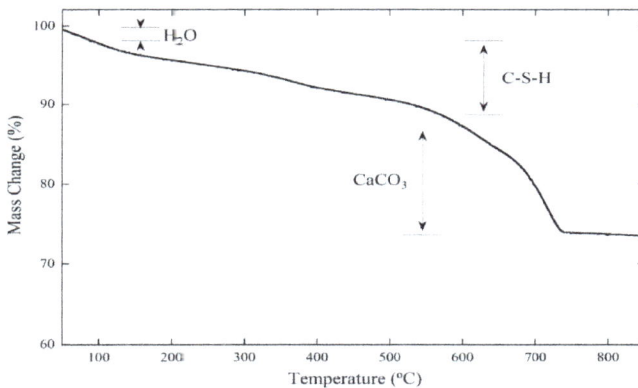

Figure 2. Thermogravimetric analysis (TGA) results of as-received waste slate. "H_2O", "C–S–H", and "$CaCO_3$" indicate the mass change occurred by the decomposition of H_2O, C–S–H, and $CaCO_3$, respectively.

3.2. Leaching Behaviors

The Ca- and Mg-leaching efficiencies after leaching tests as a function of solid dosage (20, 50, 100, 150, and 200 g/L) and solvent concentration (1, 2, and 4 M) are shown in Figures 3 and 4. The leaching efficiency was determined by following equation:

$$\text{Metal leaching efficiency}(\%) = \frac{M_{\text{Me-leached}}(g)}{M_t(g) \times C_{\text{Meo}}(\%)} \times 100 \quad (1)$$

where $M_{\text{Me-leached}}$ is the mass (g) of the metal in the leachate obtained after the leaching test, M_t is the total mass (g) of the material used in the leaching test, C_{MeO} is the metal content of the material determined by the total elemental analysis.

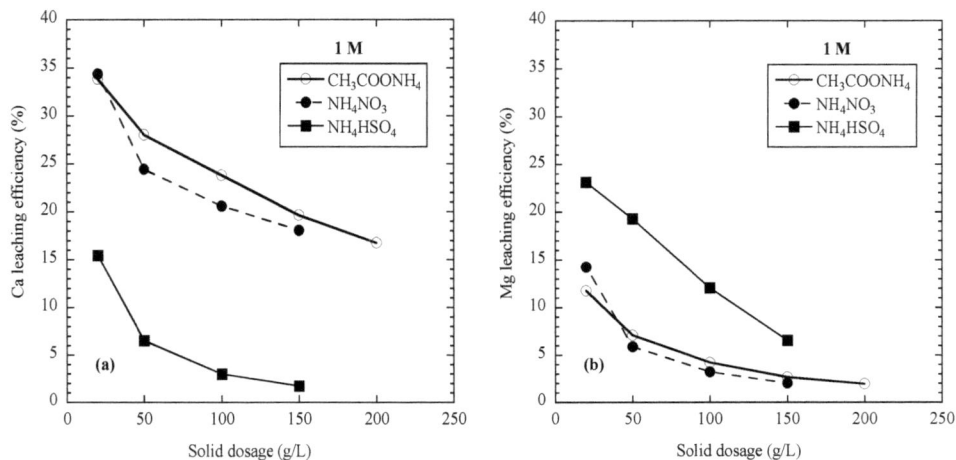

Figure 3. Ca and Mg-leaching efficiencies as a function of the solid dosage with the 1.0 M ammonium solution for waste slate samples.

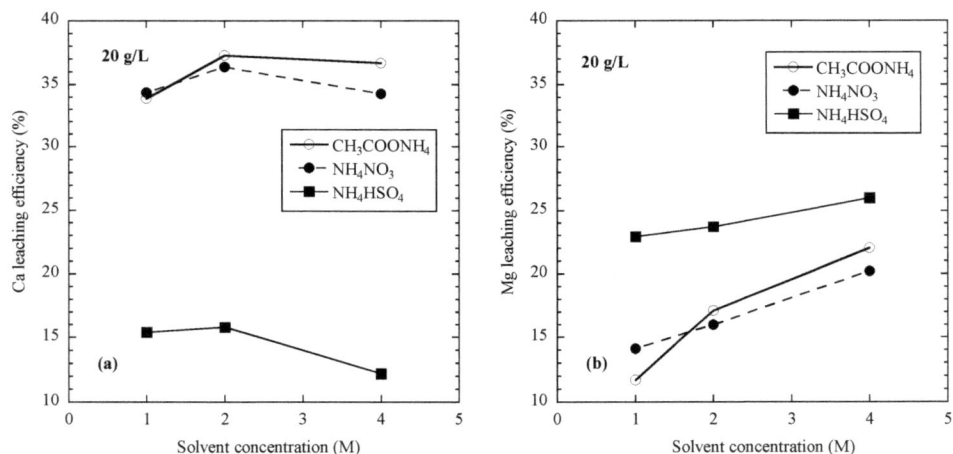

Figure 4. Ca and Mg-leaching efficiencies as a function of solvent concentration at the solid dosage of 20 g/L for waste slate samples.

For all solvent conditions, the metal leaching efficiencies were decreased when the solid dosage increased from 20 to 200 g/L (Figures 3 and 4). Among the 1 M ammonium salts solutions, the CH_3COONH_4 and NH_4NO_3 solutions had higher Ca-leaching efficiencies than did the NH_4HSO_4 solution, regardless of solid dosage. Jo *et al.* [8] reported that 1 M NH_4NO_3 and CH_3COONH_4 solutions could effectively dissolve Ca ions from waste cement. The NH_4HSO_4 solution may have had a lower Ca-leaching efficiency than the NH_4NO_3 and CH_3COONH_4 solutions due to the lower Ca selectivity or the precipitation of gypsum ($CaSO_4$) during the leaching step (Figure 5). In contrast, the NH_4HSO_4 solution had the highest Mg-leaching efficiency among all the solvents. Wang and Maroto-Valer [6] reported that 1.4 M NH_4HSO_4 solution leached 100% of Mg and 98% of Fe, but only 13.6% of Si, from serpentine at 100 °C. However, the lower Mg-leaching efficiency (<24%) obtained in this study was probably due to the lower temperature and solvent concentration. Nevertheless, Mg ions were leached dominantly from waste slate in the NH_4HSO_4 solution.

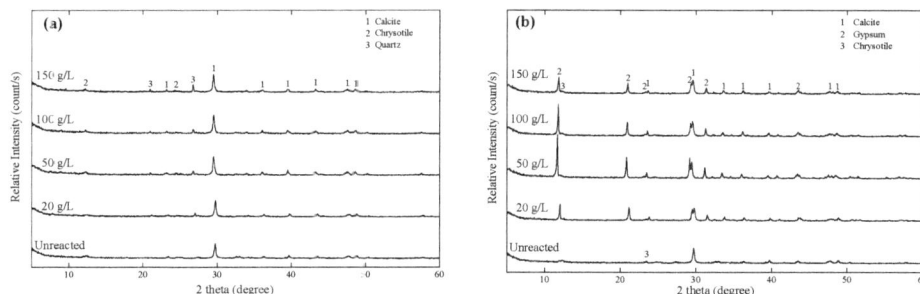

Figure 5. X-ray diffraction (XRD) patterns of as-received waste slate (Unreacted) and reacted waste slate samples obtained from leaching tests using the (**a**) 1 M CH_3COONH_4 and (**b**) 1 M NH_4HSO_4 solutions at various solid dosages (20, 50, 100, and 150 g/L).

For the 20 g/L of the solid dosage, the CH_3COONH_4 and NH_4NO_3 solutions had relatively high Ca-leaching efficiencies (~37%) at high solvent concentrations (>2 M) (Figure 4). However, the CH_3COONH_4 and NH_4NO_3 solutions had a slightly lower Ca-leaching efficiency (34%) at low solvent concentration (1 M). The NH_4HSO_4 solution had the lowest Ca-leaching efficiency at all solvent concentrations. On the contrary, the NH_4HSO_4 solution had the highest Mg-leaching efficiency (18%–20%) at all solvent concentrations. The Mg-leaching efficiencies of both CH_3COONH_4 and NH_4NO_3 solutions were lower than that of the NH_4HSO_4 solution but increased with increasing the solvent concentration. The Mg-leaching efficiency of NH_4HSO_4 was slightly affected by the solvent concentration, suggesting that the CH_3COONH_4 and NH_4HSO_4 solutions were the most efficient solvents for Ca and Mg leaching from waste slate, respectively.

The XRD patterns on the unreacted and reacted waste samples are shown in Figure 5. For the 1 M CH_3COONH_4 solution, the chrysotile peak intensity was slightly decreased, regardless of the solid dosage (Figure 5). Even though Ca ions were dissolved into the leachate, the calcite peak intensity barely changed after the leaching tests, possibly indicating that Ca-silicates in the waste slate was a main Ca source in the slurry. The quartz peak identified after the leaching tests was further possible evidence for the dissolution of Ca-silicates.

For the 1 M NH_4HSO_4 solution, gypsum ($CaSO_4 \cdot 2H_2O$) was identified at all solid dosages, suggesting that gypsum was precipitated during the leaching process. The gypsum peak intensity was higher at the solid dosage of 50 g/L and decreased with increasing the solid dosage. The calcite peak intensity also barely changed after the leaching test using the 1 M NH_4HSO_4 solution. In addition, the chrysotile peak was still observed after the leaching tests. The Mg-leaching efficiency (~20%) also confirmed that chrysotile, which was an Mg source, was not fully decomposed after the leaching tests using the CH_3COONH_4, NH_4NO_3, and NH_4HSO_4 solutions.

3.3. Carbonation Behaviors

The carbonation tests were carried out on suspensions from the leaching test using 1 M CH_3COONH_4 and 1 M NH_4HSO_4 solutions. The 1 M CH_3COONH_4 and NH_4HSO_4 solutions were selected due to their high metal leaching efficiency and selectivity in the leaching test. The 1 M CH_3COONH_4 solution had a high Ca-leaching efficiency and Ca selectivity and the 1 M solution NH_4HSO_4 showed high Mg-leaching efficiency and Mg selectivity (Figures 3 and 4).

The cation concentrations of filtrates obtained from the leaching and carbonation tests are shown in Table 2. After the leaching tests, the pH of mixtures using 1 M CH_3COONH_4 and 1 M

NH_4HSO_4 solutions ranged between 9.0 and 9.8, which was suitable for mineral carbonation. For the CH_3COONH_4 solution, the Ca concentration decreased significantly after carbonation test due to the precipitation of $CaCO_3$. For the NH_4HSO_4 solution, however, the Ca concentration was not greatly changed after the carbonation test. In contrast, the Mg concentration increased after the carbonation test for both CH_3COONH_4 and NH_4HSO_4 solutions, probably because Mg ions were released from the solid phase during the CO_2 injection. Dissolution of Mg ion is favorable at pH < 8.0 [32]. The Mg ion might dissolve further during carbonation because CO_2 injection was stopped when pH of the suspension reached about 7.5. The Si concentration decreased after the carbonation for both CH_3COONH_4 and NH_4HSO_4 solution, probably due to the precipitation of SiO_2 from dissolved Ca-Mg-Al-Si-O complex during the CO_2 injection.

Table 2. Results of chemical analysis for filtrates obtained from leaching and carbonation tests.

Solvent	Solid dosage (g/L)	pH Leaching	Cation Concentration (mg/L)							
			Ca		Mg		Fe		Si	
			Leaching	Carbonation	Leaching	Carbonation	Leaching	Carbonation	Leaching	Carbonation
1 M CH$_3$COONH$_4$	20	9.0	1707.2	375.2	51.3	136.8	0.1	0.9	44.3	21.6
	50	9.4	3533.2	652.3	77.5	345.9	0.2	0.6	41.3	32.1
	100	9.7	6002.2	1955.0	91.9	784.1	0.2	0.5	39.9	24.7
	150	9.8	7429.2	2597.0	87.4	1200	0.7	0.4	37.0	30.2
1 M NH$_4$HSO$_4$	20	9.0	778.1	709.9	100.9	154.4	1.0	0.6	40.5	16.4
	50	9.2	822.9	797.5	210.3	342.7	0.1	0.4	50.3	24.0
	100	9.7	755.9	751.9	263.3	824.1	0.4	1.2	52.5	16.1
	150	9.8	655.1	852.7	214.3	1234	0.3	0.5	48.9	26.3

The carbonation efficiency was determined using the Ca concentration of the filtrates obtained from the leaching and carbonation tests using the following equation:

$$\text{Ca carbonation efficiency } (\%) = \frac{M_{\text{Ca-leached}}(g) - M_{\text{Ca-carbonation}}(g)}{M_t(g) \times \frac{C_{\text{Ca}}(\%)}{100}} \times 100(\%) \qquad (2)$$

where $M_{\text{Ca-leached}}$ is the mass (g) of Ca in the filtrate obtained from the leaching test, $M_{\text{carbonation}}$ is the mass (g) of Ca in the filtrate obtained from the carbonation test, M_t is the total mass (g) of the waste slate used in the tests, and C_{Ca} is the calcium content (%) of the waste slate determined by total elemental analysis (Table 1).

The carbonation efficiency is shown in Figure 6. The 1 M CH_3COONH_4 solution had a higher Ca carbonation efficiency than the 1 M NH_4HSO_4 solution, regardless of the solid dosage. The Ca carbonation efficiencies for both 1 M CH_3COONH_4 and 1 M NH_4HSO_4 solutions decreased as the solid dosage increased, which was comparable to the Ca-leaching efficiency.

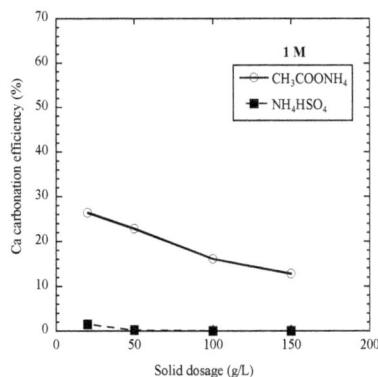

Figure 6. Ca carbonation efficiencies as a function of the solid dosage with the 1.0 M CH_3COONH_4 and 1.0 M NH_4HSO_4 solutions for waste slate samples.

3.4. Characteristics of Reacted Asbestos-Containing Waste Slate

The $CaCO_3$ content of the reacted materials determined using results of TGA is shown in Figure 7. The $CaCO_3$ content of the reacted material was approximately 10%–17% higher than that of the as-received material, except for the 1 M NH_4HSO_4 solution, at a solid dosage of 100 g/L. These results suggest that a maximum CO_2 sequestration capacity of ~0.048 kg CO_2/kg waste slate was obtained at a solid dosage of 20 g/L using the 1 M CH_3COONH_4 solution. The $CaCO_3$ content of the reacted material was slightly decreased with increasing the solid dosage from 20 to 150 g/L, except the solid dosage of 100 g/L, with the 1 M NH_4HSO_4 solution. The $CaCO_3$ content in the 1 M CH_3COONH_4 solution was approximately 10% higher than that in the 1 M NH_4HSO_4 solution. On the other hand, the carbonation efficiency in the 1 M CH_3COONH_4 solution based on Ca concentrations in the filtrate was approximately 20% higher than that in the 1 M NH_4HSO_4 solution. These results suggest that Ca ions obtained during the leaching process and Ca ions obtained by dissolving further from waste slate might be precipitated during the CO_2 injection.

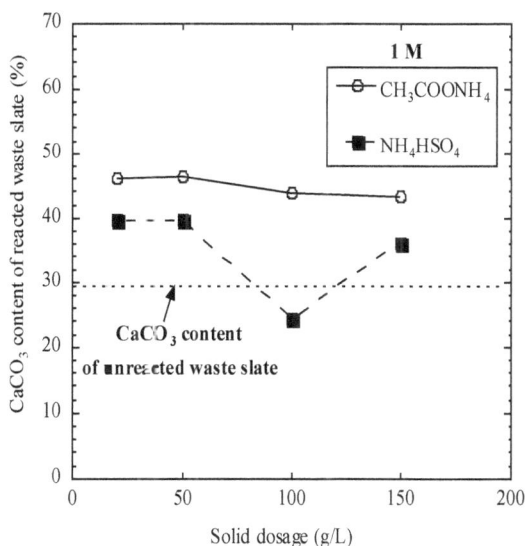

Figure 7. $CaCO_3$ content of reacted waste slate samples determined using TGA as a function of the solid dosage with the 1.0 M CH_3COONH_4 and NH_4HSO_4 solutions.

After the carbonation tests using 1 M CH_3COONH_4 and 1 M NH_4HSO_4 solutions, the calcite peak intensity increased at all the solid dosages (Figure 8). For both 1 M CH_3COONH_4 and 1 M NH_4HSO_4 solutions, a chrysotile peak was still observed in the reacted materials after the carbonation tests. For the 1 M CH_3COONH_4 and 1 M NH_4HSO_4 solutions, almost similar XRD results were obtained, except for the formation of gypsum ($CaSO_4 \cdot 2H_2O$) for the 1 M NH_4HSO_4 solution (Figure 8).

The images of SEM analysis on the as-received and reacted materials obtained after the carbonation tests using 1 M CH_3COONH_4 and 1 M NH_4HSO_4 solutions are shown in Figures 9 and 10, respectively. No carbonates were observed on the surface of needle-like chrysotile in the as-received waste slate, as shown in Figure 9a, even though carbonates were formed by natural carbonation due to the long atmospheric exposure of the slate panels. In addition, the SEM images confirmed that the chrysotile originally contained in the waste slate was not fully decomposed after the carbonation tests.

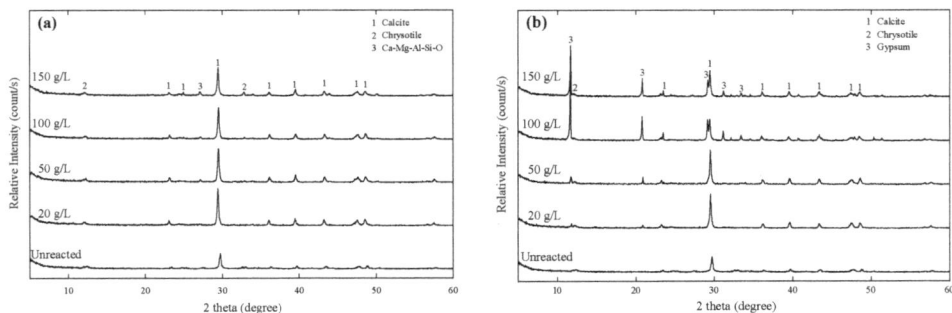

Figure 8. X-ray diffraction (XRD) patterns of unreacted waste slate (Unreacted) and reacted waste slate samples obtained from carbonation tests using (**a**) 1 M CH_3COONH_4 and (**b**) 1 M NH_4HSO_4 solutions at various solid dosages (20, 50, 100, and 150 g/L).

Mg ions from chrysotile can be dissolved at high pressure (>3 MPa) and temperature (>600 °C) conditions or moderate temperature (>70 °C) with very strong acid solutions (e.g., hydrochloric acid, oxalic acid, and sulfuric acid), e.g., [33,34]. However, cubic calcite was precipitated on the surface of needle-like chrysotile at all solid dosages for both 1 M CH_3COONH_4 and 1 M NH_4HSO_4 solutions after the carbonation tests (Figures 9 and 10). Health impact of asbestos is mainly caused by its needle-like morphological characteristics. The mobility of needle-like chrysotile coated by carbonates might be decreased, which probably mitigated the chrysotile's toxicity by changing its morphology. The carbonates also changed chemical composition of asbestos surface. Aust *et al.* [35] reported that the asbestos toxicity is strongly related to surface chemical composition of respirable elongated mineral particles (e.g., Fe associated with the fibers). The carbonates coated chrysotile may inhibit the Fe-related reaction between chrysotile and biological molecules [35]. These results suggest that the aqueous mineral carbonation of asbestos-containing waste slate may reduce toxicity of asbestos.

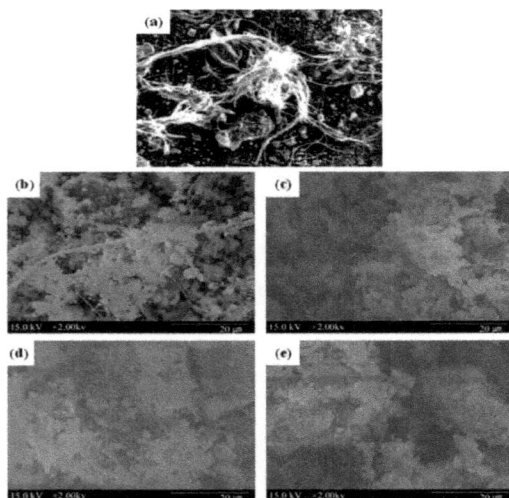

Figure 9. Scanning electron microscope (SEM) images of (**a**) unreacted waste slate and reacted waste slate samples obtained from carbonation tests using the 1 M CH_3COONH_4 solution at solid dosages of (**b**) 20 g/L, (**c**) 50 g/L, (**d**) 100 g/L, and (**e**) 150 g/L.

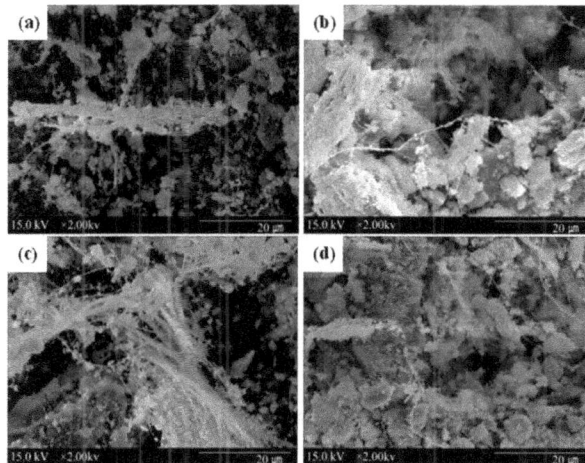

Figure 10. Scanning electron microscope (SEM) images of reacted waste slate samples obtained from carbonation tests using the 1 M NH_4HSO_4 solution at solid dosages of (**a**) 20 g/L, (**b**) 50 g/L, (**c**) 100 g/L, and (**d**) 150 g/L.

4. Conclusions

Leaching and carbonation tests on asbestos-containing waste slate using ammonium salt (CH_3COONH_4, NH_4NO_3, and NH_4HSO_4) solutions were conducted at room temperature and atmospheric pressure conditions. The Ca and Mg-leaching efficiencies increased with decreasing the solid dosage and increasing the solvent concentration. The CH_3COONH_4 solution had the highest Ca-leaching efficiency (17%–35%) and the NH_4HSO_4 solution had the highest Mg-leaching efficiency (7%–24%). The NH_4HSO_4 solution had the lowest Ca-leaching efficiency, probably due to precipitation of gypsum.

The carbonation efficiency determined using the Ca concentrations of the filtrate obtained before and after the carbonation tests was correlated to the Ca-leaching efficiency. The carbonation efficiency in the CH_3COONH_4 solution ranged between 14 and 27%. The carbonation efficiency was lower in the NH_4HSO_4 solution due to its lower Ca-leaching efficiency. However, the carbonation efficiency based on the $CaCO_3$ content of the reacted materials was approximately 10%–17% higher than the carbonation efficiency determined using the Ca concentration of the leachate for both the 1 M CH_3COONH_4 and the 1 M NH_4HSO_4 solutions because further Ca^{2+} ions were dissolved from the waste slate during the carbonation tests. These results suggest that a maximum CO_2 sequestration capacity of ~0.07 kg CO_2/kg waste slate was obtained at a solid dosage of 50 g/L using the 1 M CH_3COONH_4 solution. The cost of CH_3COONH_4 in this direct aqueous carbonation procedure using 1 M CH_3COONH_4 solution and a solid dosage of 50 g/L can be estimated to be about US$4000 to sequester one ton of CO_2 assuming that the price of CH_3COONH_4 is US$200/ton, which is too expensive. Thus, further study on reducing the cost of chemical is necessary.

The carbonates were precipitated on the surface of chrysotile, which was contained in the reacted materials obtained after the carbonation tests. However, no carbonates were observed on the surface of chrysotile in the as-received waste slate, even though natural carbonation had been occurred in the slate. Consequently, the aqueous mineral carbonation changed the morphology and the surface composition of the needle-like chrysotile by coating it with carbonates. The results of this study suggest that the direct aqueous mineral carbonation of waste slate can be used to sequester CO_2 and to reduce the human body risk of asbestos-containing waste slate. However, the results should be verified by conducting multiple tests because a single test for leaching and carbonation was conducted in this study. In addition, further studies are necessary to investigate the human body risk of the solid particles obtained after the carbonation tests and the potential effectiveness of more appropriate conditions to further mitigate the toxicity of asbestos-containing waste slate.

Acknowledgments: This research was supported by the Basic Research Project (Study on the mineral carbonation of the alkaline industrial products, GP2015-009) of the Korea Institute of Geoscience and Mineral Resources (KIGAM) funded by the Ministry of Science, ICT and Future Planning and also supported by the Basic Science Research Program through the National Research Foundation of Korea (NRF) funded by the Ministry of Education (2012R1A1B3002473).

Author Contributions: Ho Young Jo designed and managed the research and co-wrote and edited the paper. Hwanju Jo characterized the materials, interpreted the data, and co-wrote the paper. Sunnwon Rha conducted leaching and carbonation tests and characterized the materials. Pyeong-Koo Lee conceptualized the experiment and analyzed interpreted the data.

Conflicts of Interest: The authors declare no conflict of interest.

References

1. Lackner, K.S.; Wendt, C.H.; Butt, D.P.; Joyce, B.L.; Sharp, D.H. Carbon dioxide disposal in carbonate minerals. *Energy* **1995**, *20*, 1153–1170. [CrossRef]
2. Sipilä, J.; Teir, S.; Zevenhoven, R. *Carbon dioxide sequestration by mineral carbonation: Literature review update 2005–2007*; Report VT; Abo Akademi University: Turku, Finland, 2008.
3. Intergovernmental Panel on Climate Change. Climate change 2007: Mitigation of climate change: Contribution of working group III to the fourth assessment report of the intergovernmental panel on climate change. Cambridge University Press: Cambridge, NY, USA, 2007.
4. Jo, H.; Jo, H.Y.; Jang, Y.N. Effect of leaching solutions on carbonation of cementitious materials in aqueous solutions. *Environ. Technol.* **2012**, *33*, 1391–1401. [CrossRef] [PubMed]
5. Kodama, S.; Nishimoto, T.; Yamamoto, N.; Yogo, K.; Yamada, K. Development of a new pH-swing CO_2 mineralization process with a recyclable reaction solution. *Energy* **2008**, *33*, 776–784. [CrossRef]
6. Wang, X.; Maroto-Valer, M.M. Dissolution of serpentine using recyclable ammonium salts for CO_2 mineral carbonation. *Fuel 90*, 1229–1237. [CrossRef]
7. Eloneva, S.; Said, A.; Fogelholm, C.J.; Zevenhoven, R. Preliminary assessment of a method utilizing carbon dioxide and steelmaking slags to produce precipitated calcium carbonate. *Appl. Energy* **2012**, *90*, 329–334. [CrossRef]
8. Jo, H.; Park, S.H.; Jang, Y.N.; Chae, S.C.; Lee, P.K.; Jo, H.Y. Metal leaching and indirect mineral carbonation of waste cement material using ammonium salt solutions. *Chem. Eng. J.* **2012**, *254*, 313–323. [CrossRef]
9. Park, A.H.A.; Fan, L.S. CO_2 mineral sequestration: Physically activated dissolution of serpentine and pH swing process. *Chem. Eng. Sci.* **2004**, *59*, 5241–5247. [CrossRef]
10. Huijgen, W.J.J.; Witkamp, G.J.; Comans, R.N.J. Mineral CO_2 sequestration by steel slag carbonation. *Environ. Sci. Technol.* **2005**, *39*, 9676–9682. [CrossRef] [PubMed]
11. Katsuyama, Y.; Yamasaki, A.; Iizuka, A.; Fujii, M.; Kumagai, K.; Yanagisawa, Y. Development of a process for producing high-purity calcium carbonate ($CaCO_3$) from waste cement using pressurized CO_2. *Environ. Prog.* **2005**, *24*, 162–170. [CrossRef]
12. Baciocchi, R.; Polettini, A.; Pomi, R.; Prigiobbe, V.; Zedwitz, V.N.V.; Steinfeld, A. CO_2 sequestration by direct gas-solid carbonation of air pollution control (APC) residues. *Energy Fuels* **2006**, *20*, 1933–1940. [CrossRef]
13. Teir, S.; Kuusik, R.; Fogelholm, C.J.; Zevenhoven, R. Production of magnesium carbonates from serpentinite for long-term storage of CO_2. *Int. J. Miner. Process.* **2007**, *85*, 1–15. [CrossRef]
14. Bonenfant, D.; Kharoune, L.; Sauve, S.; Hausler, R.; Niquette, P.; Mimeault, M.; Kharoune, M. CO_2 Sequestration potential of steel slags at ambient pressure and temperature. *Ind. Eng. Chem. Res.* **2008**, *47*, 7610–7616. [CrossRef]
15. Huntzinger, D.N.; Gierke, J.S.; Sutter, L.L.; Kawatra, S.K.; Eisele, T.C. Mineral carbonation for carbon sequestration in cement kiln dust from waste piles. *J. Hazard. Mater.* **2009**, *168*, 31–37. [CrossRef] [PubMed]
16. Baciocchi, R.; Costa, G.; Bartolomeo, E.D.; Polettini, A.; Pomi, R. Carbonation of stainless steel slag as a process for CO2 storage and slag valorization. *Waste Biomass Valorization* **2010**, *1*, 467–477. [CrossRef]

17. Kashef-Haghighi, S.; Ghoshal, S. CO₂ Sequestration in concrete through accelerated carbonation curing in a flow-through reactor. *Ind. Eng. Chem. Res.* **2010**, *49*, 1143–1149. [CrossRef]

18. Wang, X.; Maroto-Valer, M.M. Integration of CO₂ capture and mineral carbonation by using recyclable ammonium salts. *ChemSusChem* **4**, 1291–1300. [CrossRef] [PubMed]

19. Baciocchi, R.; Costa, G.; Gianfilippo, M.D.; Polettini, A.; Pomi, R.; Stramazzo, A. Thin-film *versus* slurry-phase carbonation of steel slag: CO₂ uptake and effects on mineralogy. *J. Hazard. Mater.* **2015**, *283*, 302–313. [CrossRef] [PubMed]

20. Kirchofer, A.; Becker, A.; Brandt, A.; Wilcox, J. CO₂ mitigation potential of mineral carbonation with industrial alkalinity sources in the United States. *Environ. Sci. Technol.*, **2013**, *47*, 7548–7554.

21. Jang, Y.N.; Chae, S.C.; Lee, M.K.; Won, H.I.; Ryu, K.W. Method of removing asbestos from asbestos-containing materials by 99% through low temperature heat treatment. US9079055, 14 July 2014.

22. Gadikota, G.; Natali, C.; Boschi, C.; Park, A.H.A. Morphological changes during enhanced carbonation of asbestos containing material and its comparison to magnesium silicate minerals. *J. Hazard. Mater.* **2014**, *264*, 42–52. [CrossRef] [PubMed]

23. Nam, S.N.; Jeong, S.; Lim, H. Thermochemical destruction of asbestos-containing roofing slate and the feasibility of using recycled waste sulfuric acid. *J. Hazard. Mater.* **2014**, *265*, 151–157. [CrossRef] [PubMed]

24. Pacella, A.; Fantauzzi, M.; Turci, F.; Cremisini, C.; Montereali, M.R.; Nardi, E.; Atzei, D.; Rossi, A.; Andreozzi, G.B. Surface alteration mechanism and topochemistry of iron in tremolite asbestos: A step toward understanding the potential hazard of amphibole asbestos. *Chem. Geol.* **2015**, *405*, 28–38. [CrossRef]

25. Leonelli, C.; Veronesi, P.; Boccaccini, D.N.; Rivasi, M.R.; Barbieri, L.; Andreola, F.; Lancellotti, I.; Rabitti, D.; Pellacani, G.C. Microwave thermal inertisation of asbestos containing waste and its recycling in traditional ceramics. *J. Hazard. Mater.* **2006**, *135*, 149–155. [CrossRef] [PubMed]

26. Horikoshi, S.; Sumi, T.; Ito, S.; Dillert, R.; Kashimura, K.; Yoshikawa, N.; Sato, M.; Shinohara, N. Microwave-driven asbestos treatment and its scale-up for use after natural disasters. *Environ. Sci. Technol.* **2014**, *48*, 6882–6890. [CrossRef] [PubMed]

27. Yoshikawa, N.; Kashimura, K.; Hashiguchi, M.; Sato, M.; Horikoshi, S.; Mitani, T.; Shinohara, N. Detoxification mechanism of asbestos materials by microwave treatment. *J. Hazard. Mater.* **2015**, *284*, 201–206. [CrossRef] [PubMed]

28. Averroes, A.; Sekiguchi, H.; Sakamoto, K. Treatment of airborne asbestos and asbestos-like microfiber particles using atmospheric microwave air plasma. *J. Hazard. Mater.* **2011**, *195*, 405–413. [CrossRef] [PubMed]

29. Gualtieri, A.F.; Tartaglia, A. Thermal decomposition of asbestos and recycling in traditional ceramics. *J. Eur. Ceram. Soc.* **2000**, *20*, 1409–1418. [CrossRef]

30. Stumm, W.; Morgan, J.J. Aquatic chemistry: Chemical equilibria and rates in natural water, 3rd ed.Wiley-Interscience publications: New York, NY, USA, 1996.

31. Pade, C.; Guimaraes, M. The CO₂ uptake of concrete in a 100 year perspective. *Cem. Concr. Res.* **2007**, *37*, 1348–1356. [CrossRef]

32. Back, M.; Kuehn, M.; Stanjek, H.; Peiffer, S. Reactivity of alkaline lignite fly ashes towards CO₂ in water. *Environ. Sci. Technol.* **2008**, *42*, 4520–4526. [CrossRef] [PubMed]

33. Lee, M.G.; Ryu, K.W.; Jang, Y.N.; Kim, W.; Bang, J.H. Effect of oxalic acid on heat pretreatment for serpentine carbonation. *Mater. Trans.* **2011**, *52*, 235–238. [CrossRef]

34. Ryu, K.W.; Jang, Y.N.; Lee, M.G. Enhancement of chrysotile carbonation in alkali solution. *Mater. Trans.* **2012**, *53*, 1349–1352. [CrossRef]

35. Aust, A.E.; Cook, P.M.; Dodson, R.F. Morphological and chemical mechanisms of elongated mineral particle toxicities. *J. Toxic. Environ. Health Part* **2011**, *14*, 40–75. [CrossRef] [PubMed]

metals

MDPI

Article

Chalcopyrite Dissolution at 650 mV and 750 mV in the Presence of Pyrite

Yubiao Li [1,2,*], Gujie Qian [2], Jun Li [2] and Andrea R. Gerson [2,3,*]

[1] School of Resources and Environmental Engineering, Wuhan University of Technology, Wuhan 430070, Hubei, China

[2] Minerals and Materials Science & Technology, Mawson Institute, University of South Australia, Mawson Lakes, SA 5095, Australia; gujie.qian@unisa.edu.au (G.Q.); jun.li@unisa.edu.au (J.L.)

[3] Blue Minerals Consultancy, 13 Mill Terrace, Middleton, SA 5213, Australia

* Authors to whom correspondence should be addressed; yubiao.li@whut.edu.cn (Y.L.); andrea.gerson@bigpond.com (A.R.G.); Tel.: +86-18702778563 (Y.L.); +61-4-22112516 (A.R.G.); Fax: +86-27-87882128 (Y.L.); +61-8-83025545 (A.R.G.).

Academic Editor: Suresh Bhargava
Received: 6 August 2015; Accepted: 25 August 2015; Published: 28 August 2015

Abstract: The dissolution of chalcopyrite in association with pyrite in mine waste results in the severe environmental issue of acid and metalliferous drainage (AMD). To better understand chalcopyrite dissolution, and the impact of chalcopyrite's galvanic interaction with pyrite, chalcopyrite dissolution has been examined at 75 °C, pH 1.0, in the presence of quartz (as an inert mineral) and pyrite. The presence of pyrite increased the chalcopyrite dissolution rate by more than five times at E_h of 650 mV (SHE) (Cu recovery 2.5 *cf.* 12% over 132 days) due to galvanic interaction between chalcopyrite and pyrite. Dissolution of Cu and Fe was stoichiometric and no pyrite dissolved. Although the chalcopyrite dissolution rate at 750 mV (SHE) was approximately four-fold greater (Cu recovery of 45% within 132 days) as compared to at 650 mV in the presence of pyrite, the galvanic interaction between chalcopyrite and pyrite was negligible. Approximately all of the sulfur from the leached chalcopyrite was converted to S^0 at 750 mV, regardless of the presence of pyrite. At this E_h approximately 60% of the sulfur associated with pyrite dissolution was oxidised to S^0 and the remaining 40% was released in soluble forms, e.g., SO_4^{2-}.

Keywords: chalcopyrite; pyrite; galvanic interaction; dissolution

1. Introduction

Chalcopyrite ($CuFeS_2$) is the most abundant copper containing mineral on the planet, accounting for approximately 70% of the total copper resources [1]. Although copper is predominantly extracted using pyrometallurgical processing, much attention has recently been paid to the development of an effective hydrometallurgical extraction methodology [2–8] due to the potential for improved economics and reduced environmental impact, especially for ore of low copper grade.

The most common reaction (Equation (1)) used to describe [4,6,9–11] chalcopyrite dissolution is its oxidation by Fe^{3+}.

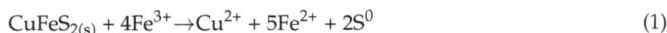

$$CuFeS_{2(s)} + 4Fe^{3+} \rightarrow Cu^{2+} + 5Fe^{2+} + 2S^0 \tag{1}$$

Equation (1) can be considered to be the sum of anodic and cathodic half-cell reactions as shown in Equations (2) and (3) [12–15].

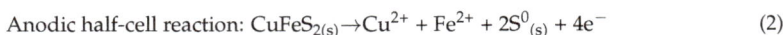

$$\text{Anodic half-cell reaction: } CuFeS_{2(s)} \rightarrow Cu^{2+} + Fe^{2+} + 2S^0_{(s)} + 4e^- \tag{2}$$

$$\text{Cathodic half-cell reaction: } 4Fe^{3+} + 4e^- \rightarrow 4Fe^{2+} \quad\quad (3)$$

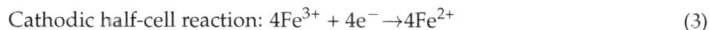

Chalcopyrite is frequently associated with pyrite (FeS_2), the predominant contributor to the serious environmental issue of acid and metalliferous drainage (AMD). However, to date the role of chalcopyrite in AMD is not well understood. In the absence of pyrite, both anodic and cathodic reactions occur on the chalcopyrite surface. The slow dissolution rate has been proposed to be due to a slow cathodic half-cell reaction on the chalcopyrite surface [11,16]. In the presence of pyrite, the cathodic half-cell reaction occurs on the pyrite surface as the rest potential of pyrite is greater (660 mV SHE) than that of chalcopyrite (560 mV SHE), forming a galvanic cell with electron transfer from the lower rest potential material (anode) to the material with the greater rest potential (cathode) material [3]. It has been reported that when chalcopyrite is in intimate contact with pyrite, the dissolution rate of chalcopyrite is increased significantly as compared chalcopyrite in isolation [11,14,17,18]. On the other hand, the dissolution rate of chalcopyrite is significantly decreased in the presence of sulfide minerals with lower rest potential such as sphalerite and galena [10].

Li *et al.* [19] have investigated the effect of E_h on chalcopyrite dissolution and found that the chalcopyrite dissolution rate at 750 mV is significantly greater than that at 650 mV. Further study has shown that the chacopyrite dissolution rate at 650 mV is increased in the presence of added soluble iron [20]. However, the effect of pyrite at these two E_h conditions (*i.e.*, 650 and 750 mV SHE) is not yet well understood. Hence, better understanding of chalcopyrite dissolution behaviour, in the presence of pyrite as a function E_h is of geochemical and industrial importance for minimising AMD and optimising Cu extraction during hydrometallurgical processing, respectively [21–23].

2. Experimental Section

2.1. Mineral Characterisation

The chalcopyrite sample used in this study was from Sonora, Mexico. The pyrite originated from Huanzala, Peru and was purchased from Ward's Scientific. Quartz, added as an inert material in this study, originated from Bolivia, NSW, Australia and was purchased from Geo Discoveries (Gosford, NSW, Australia). A chunk of each mineral, *i.e.*, chalcopyrite, pyrite and quartz, was crushed and subsequently rod milled to obtain a size fraction of 38–75 μm via wet sieving. The resulting sample was then sonicated to remove clinging fines, and dried at 70 °C in an oven purged with N_2. These samples were then placed into a plastic tube which was sealed after being filled with N_2 gas to minimise surface oxidation by air. All the samples were stored in a freezer prior to use. BET analysis (University of South Australia, Adelaide, Australia) indicated a surface area of 0.24 ± 0.04 $m^2 \cdot g^{-1}$ for chalcopyrite and 0.39 ± 0.04 $m^2 \cdot g^{-1}$ for pyrite. As quartz was not dissolved in the dissolution experiment, the BET was not performed.

X-ray powder diffraction Rietveld analysis (University of South Australia, Adelaide, Australia) indicated that the chalcopyrite sample contained 92 wt. % chalcopyrite, 2 wt. % quartz, 1 wt. % pyrite, 1 wt. % sphalerite and another 4 wt. % amorphous (or unidentified) component(s). This is consistent with the ICP-AES measurement (Table 1) suggesting a stoichiometry of 95.9 wt. % $Cu_{0.96}Fe_{0.99}S_{2.00}$ [24]. A stoichiometry for the pyrite samples of 94.3 wt. % $Fe_{1.91}S_{2.00}$ was also indicated (Table 1). This is in agreement with Rietveld XRD analysis of the pyrite which indicated the only phase present to be pyrite with approximately 4 wt. % amorphous. SiO_2, was determined to be of more than 99 wt. % purity by XRD analysis.

Table 1. Elemental composition of chalcopyrite and pyrite samples.

Element (wt. %)	S	Cu	Fe	Si	Ca	Bi	Zn	Pb	As	Al	Co	K	Mg
Chalcopyrite	34.1	32.5	29.3	0.9	0.8	0.3	0.5	0.3	0.06	0.04	0.1	0.02	0.02
Pyrite	51.5	0.01	42.8	-	0.01	-	0.01	0.004	0.003	0.01	-	0.006	-

Note: "-" less than detection limit (Si, Mg 0.003 wt. %, Bi, Co 0.001 wt. %).

2.2. Dissolution Conditions

Glass vessels (1.2 L) and dissolution cells (PTFE, 60 mL) were used for the flow-through dissolution experiments. The 5-port lid of the glass vessel provided access to the dissolution liquor and was used to house a thermometer, Teflon impeller, high temperature E_h probe, hydrogen peroxide (H_2O_2) inlet and a reflux condenser. The four blade Teflon impeller was driven by a digitally controlled stirrer with a constant agitation speed of 500 rpm. Heating was provided by a thermostatically controlled silicone oil bath. In addition to the glass vessels, a peristaltic pump (20 rpm·min^{-1}) was used to recycle the dissolution solution between each vessel and the PTFE cell (60 mL) through connecting tubes. The PTFE cells were hosted and heated in an adjacent oven to the oil bath. Both the reaction vessels and the PTFE cells were maintained at 75 °C.

A total 10 g of sample (38 < x < 75 μm) was placed into the bottom of the flow-through cell. For the reference experiment 4 g of chalcopyrite and 6 g of quartz were mixed uniformly. In another cell 4 g of chalcopyrite and 6 g of pyrite were mixed uniformly to investigate chalcopyrite dissolution in the presence of pyrite. A solution flow rate of 8 L·d^{-1} was used. The leach liquor was controlled to have E_h of either 650 or 750 mV (SHE) to within ±10 mV, using 7.5 wt. % H_2O_2 solution, and an auto titrator (EUTECH pH 200 series (Thermo Scientific, Singapore) with a Master Flex pump (John Morris Scientific, Chatswood, Australia). The leach solution pH was checked during sampling and maintained at pH 1.0 by addition of 5 M sulfuric acid.

In order to better understand the surface S species on the leached chalcopyrite polished chalcopyrite and pyrite samples (8 × 8 × 5 mm, using 600 then 1200 grit silicon carbide paper) were used. The presence of pyrite in the dissolved residue containing the mixture of pyrite and chalcopyrite would perturb the interpretation of S species upon chalcopyrite surface. The reference experiment used one chalcopyrite slab and the chalcopyrite-pyrite galvanic couples was examined using one chalcopyrite and one pyrite slab tied together using cable ties. In both cases these samples were leached in 750 mV leach liquor at 75 °C and pH 1.0. The leach system that was used is described in [25] and in both instances a stirring speed of 500 rpm was applied. It has been shown that approximately 30% Cu was leached from powdered (38–75 μm) chalcopyrite at 30 h upon being subjected to this degree of agitation. Hence, it is reasonable to suggest that surface species differences due to galvanic interaction would be apparent after 30 h.

2.3. Bulk and Surface Analyses

During the dissolution process, 10 mL of the reaction solution was sampled periodically and filtered using a 0.22 μm membrane. The liquor was analysed for soluble Fe, Cu and S analysis using inductively coupled plasma—atomic emission spectroscopy (ICP-AES) by Analytical Chemistry Sector at Rio Tinto Technology and Innovation in Bundoora, Australia. Experimental uncertainty of the ICP-AES measurements was estimated to be around 1%. The reaction solution in the glass vessel was replenished upon each sampling with a fresh 10 mL of lixiviant to maintain the original volume.

At the end of each experiment, a portion of the leached residue was rinsed three times using pH 1.0 perchloric acid to remove any surface adsorbed species from the lixiviant. These solids were immediately frozen using liquid N_2 to minimise post-dissolution oxidation. These frozen samples were subsequently used for surface-sensitive X-ray photoelectron spectroscopy (XPS) analysis. The remaining solids were washed with distilled water and dried in the oven at 35 °C for XRD and SEM analysis.

A SEM (Phillips XL30 field emission SEM, Adelaide Microscopy, Adelaide, Australia) equipped with energy dispersive spectrometers (EDS) was employed to record images, using both backscattered (BSE) and secondary electron (SE) modes, and elemental quantification. XRD measurements, on selected residues, were measured using a Bruker D4 ENDEAVOR (University of Adelaide, Adelaide, Australia) with Co Kα1 = 1.78897 Å at 35 kV and 40 mA. A computer program DIFFRA.EVA (Version 3.0.0.9) from Bruker in conjunction with the crystallography open database (REV 30738 2011.11.2) was applied to identify mineral phases while quantitative phase analysis of the XRD data was performed using the Rietveld method with the aid of DIFFRAC.SUIT TOPAS (Version 4.2, Bruker, Melbourne, Australia).

The XPS instrument used was a Kratos Axis Ultra instrument (University of South Australia, Adelaide, Australia). The X-ray source was a monochromatic aluminium cathode running at 225 W with a characteristic energy of 1486.6 eV. Pass energies of 160 and 20 eV were used for survey and high-resolution scans, respectively. The charge neutraliser was utilised to compensate for surface static charging resulting from electron emission. The area of analysis (Iris aperture) was a 0.3 × 0.7 mm slot; the analysis depth was less than 15 nm. The analysis vacuum was 4×10^{-9} Torr. To minimise the sublimation of elemental sulfur (S^0), the samples were sample stage was cooled using liquid N_2 (150 K). Details of the XPS data analysis methodology are described in [24].

3. Results

3.1. Dissolution

Figure 1a indicates no significant difference in the extracted Cu concentrations in the presence of quartz or pyrite within 132 days, the duration of the experiment, when E_h was controlled at 750 mV. Within the initial 40 days, the Cu extracted at 750 mV increased in a manner similar to that at 650 mV in the presence of pyrite; approximately 6% of the Cu was extracted during this stage. Subsequently, the concentration of the Cu in the 750 mV leach liquor increased significantly, considerably more than that in the 650 mV leach liquor, regardless of the presence of quartz or pyrite. Approximately 45% of the Cu was released within 132 days when the E_h was 750 mV. In contrast, the Cu concentration increased slowly from 40 days (approximately 6%) within 132 days (approximately 12% Cu) for the dissolution at 650 mV in the presence of pyrite. When chalcopyrite was dissolved at 650 mV in the presence of quartz, less than 2.5% Cu was released within 132 days, indicating that the presence of pyrite resulted in a significantly increase in Cu dissolution rate at 650 mV, due most likely to the galvanic interaction between chalcopyrite and pyrite.

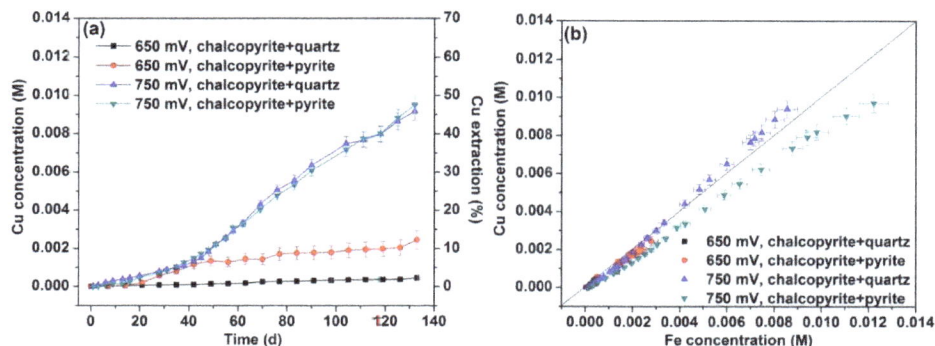

Figure 1. Leach data resulting from the flow-through experiments conducted at 75 °C, pH 1.0, 650 mV and 750 mV: (**a**) Cu concentration and % extraction *versus* time; (**b**) Cu concentration against Fe concentration.

The Cu: Fe ratio was slightly greater than 1 when chalcopyrite was dissolved with quartz at 750 mV (Figure 1b). This ratio increased gradually, but was still close to 1, in the later dissolution stage when both Cu and Fe concentrations increased, indicating Cu to Fe stoichiometric dissolution of chalcopyrite. In contrast, when chalcopyrite was dissolved at 750 mV in the presence of pyrite, the Cu to Fe ratio was slightly less than 1 from the beginning of dissolution, and continued to decrease to approximately 0.8 at the end of the experiment (*i.e.*, 132 days), indicating the dissolution of Fe from pyrite. However, the Cu to Fe ratio in the leach liquor was approximately 1 during the entire experiment duration when E_h was controlled at 650 mV, regardless of the presence of quartz or pyrite, indicating stoichiometric Cu–Fe dissolution of chalcopyrite.

3.2. XRD and SEM Analyses

Figure 2 shows the XRD patterns collected from the leach residues subjected to 750 mV, 75 °C and pH 1.0. Quantitative Rietveld analysis indicated a leach residue phase composition of 71 wt. % quartz, 19 wt. % chalcopyrite and 10 wt. % S^0 for the chalcopyrite-quartz leach system (Figure 2a) and 47 wt. % pyrite, 25 wt. % chalcopyrite and 28 wt. % S^0 for the chalcopyrite-pyrite system (Figure 2b). No other phases were identified. XRD patterns collected from the leach residues at 650 mV are not shown as most of the residues were almost identical to the feed (less than 12% of the chalcopyrite was dissolved, Figure 1a).

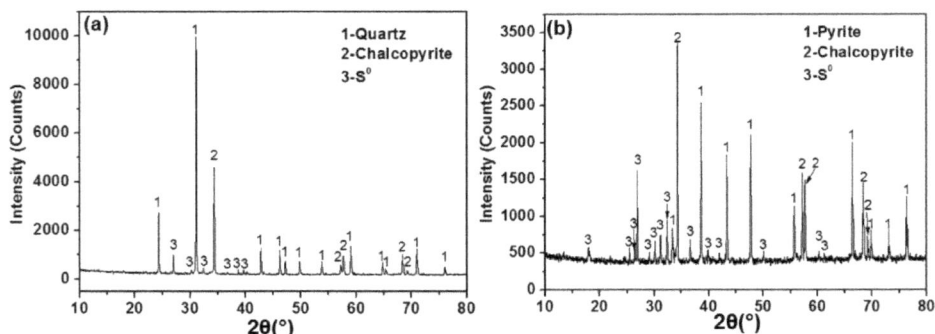

Figure 2. XRD patterns from the chalcopyrite leach residue subjected to 750 mV, 75 °C and pH 1.0: (**a**) chalcopyrite and quartz; (**b**) chalcopyrite and pyrite.

If the quantity of quartz (6 g) in the residues from the chalcopyrite-quartz system is assumed to be unchanged 1.607 g chalcopyrite and 0.845 g of S^0 are calculated to be present in the leach residue (Table 2). This weight of S^0 converts into 2.423 g of leached chalcopyrite. Hence, the total weight of chalcopyrite (remaining and dissolved) was 4.03 g which is nearly identical to that added initially (4 g). This suggests that approximately all of the sulfur from the chalcopyrite was oxidised to S^0, or alternately that no $SO_4{}^{2-}$ was produced during chalcopyrite dissolution controlled at 750 mV, pH 1.0 and 75 °C (in the presence of inert quartz).

Table 2. Initial and final weights of mineral phases for the flow-through leach experiments. It is assumed that no quartz dissolution occurs.

	Remaining (Dissolved) Quartz	Remaining (Dissolved) Chalcopyrite	Remaining (Dissolved) Pyrite	S^0 Formed
4 g chalcopyrite + 6 g quartz	6 (0) g	1.607 (2.423) g	N.A.	0.845 g
4 g chalcopyrite + 6 g pyrite	N.A.	1.607 (2.423) g	3.021 (2.979) g	1.800 g

As shown in Figure 1a, Cu was dissolved in a similar manner in the presence of either quartz or pyrite at 750 mV, indicating an identical (within experimental error) quantity of chalcopyrite remaining in these two residues, *i.e.*, 1.607 g (equating to 25 wt. % of the leach residue in the presence of pyrite). Hence, the remaining total residue, and the amounts of pyrite and S^0 formed are calculated to be 6.428 g, 3.021 g and 1.800 g, respectively. If it is assumed that the leached chalcopyrite (*i.e.*, 2.423 g) produces 0.845 g of S^0 in the chalcopyrite-pyrite system, 0.955 g of S^0 must result from pyrite oxidation.

This suggests that approximately 60% S (corresponding to 1.791 g of pyrite) of the leached pyrite (2.979 g) was oxidised to S^0 while 40% S (corresponding to 1.188 g of pyrite) from the pyrite was oxidised to soluble sulfur species, e.g., SO_4^{2-}.

Figure 3 shows SEM images, using two imaging modes, SE and BSE (e.g., Figure 3a,b) of the leach residues collected at 750 mV, 75 °C and pH 1.0. As indicated in Figure 3b, a portion of the chalcopyrite particles in the chalcopyrite-quartz leach system were oxidised to S^0, with the chalcopyrite particles being surrounded/armoured by S^0 (Figure 3c). In addition some chalcopyrite particles appeared unoxidised, presenting a smooth surface (Figure 3a,b).

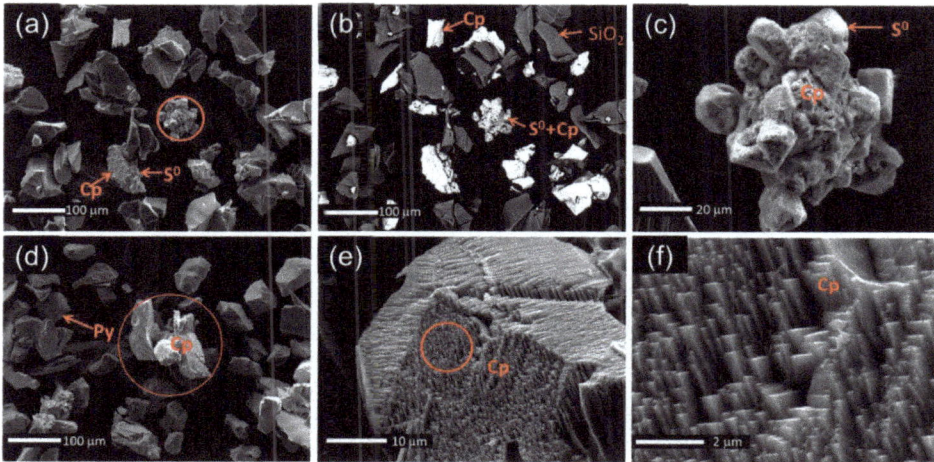

Figure 3. SEM images of leach residues collected at the end of the experiment, 132 days, (750 mV, 75 °C and pH 1.0 shown in Figure 1): (a) full view of chalcopyrite dissolved with quartz (SE image); (b) full view of chalcopyrite dissolved with quartz (BSE image); (c) selected area from (a); (d) full view of chalcopyrite dissolved with pyrite (SE image); (e) selected area from (d); (f) selected area from (e). Cp—chalcopyrite, S^0—elemental sulfur, SiO_2—quartz, Py—pyrite.

In contrast, when pyrite was present, the morphology of the oxidised chalcopyrite was significantly changed from that with in the chalcopyrite-quartz system. Figure 3d shows that both chalcopyrite and pyrite were present in the residue and Figure 3e indicates that the chalcopyrite surface was S^0-free. In addition, Figure 3e,f suggest that chalcopyrite dissolution behaviour may be different on different crystal faces. However, the interpretation of these images in terms of leach behaviour is not yet clear and will form a next step in our study.

3.3. XPS Analyses

XPS has been employed in order to understand the evolution of the uppermost surface species. As the presence of pyrite in the dissolved residue may convolute the interpretation of XPS S 2p species on the chalcopyrite surface, chalcopyrite slab surfaces leached in the glass vessel (instead of particles in the flow-through cell) were used.

Figure 4a shows that a significant amount of S^0 and S_n^{2-} (polysulfide) formed on the leached chalcopyrite surface in galvanic interaction with pyrite, as compared to that leached in isolation (Figure 4b). Figure 4c provides the difference spectra of Figure 4a minus Figure 4b. Figure 4d provides the fitting of the S 2p spectrum shown in Figure 4a. The presence of SO_4^{2-} on the chalcopyrite surface leached in isolation (Figure 4b) might be due to the slow leach rate. As water was used as the lubricating agent during the polishing process, the chalcopryite surface may react with water to form

SO_4^{2-}. But when chalcopyrite was contacting with pyrite, the increased leach would be expected to result in complete dissolution of SO_4^{2-} (Figure 4a).

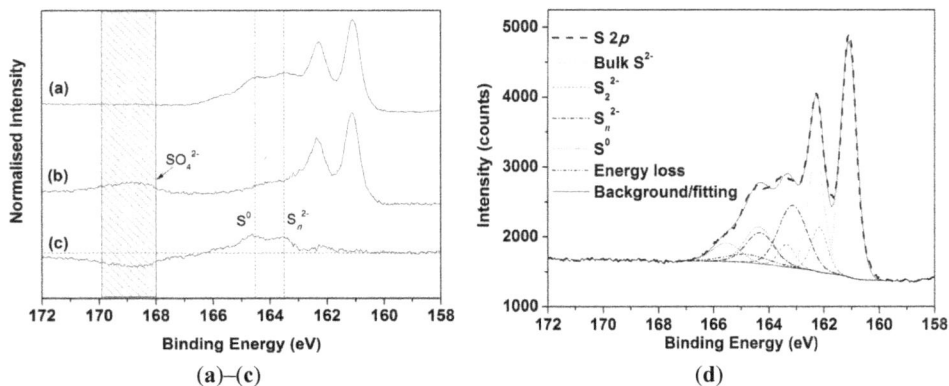

Figure 4. S 2p spectra collected from the chalcopyrite slabs dissolved at 750 mV, 75 °C and pH 1.0 for 30 h: (**a**) chalcopyrite from galvanic cell; (**b**) chalcopyrite only; (**c**) (**a**) minus (**b**); (**d**) fitting of (**a**).

Table 3 provides the quantification of S species on the surfaces of the chalcopyrite slabs. Both S_n^{2-} and S^0 are present in relatively greater quantities (as % of total S) where pyrite was in contact with the chalcopyrite as compared to the chalcopyrite only system, indicating enhanced sulfur oxidation occurred on the chalcopyrite slab surface involved in the galvanic cell.

However, these observations are apparently inconsistent with the dissolution data provided in Figure 1a which shows no enhancement in the Cu extraction rate in the galvanic chalcopyrite-pyrite system at 750 mV. In the flow-through dissolution experiment considerably greater Fe^{2+} is released from the chalcoyrite powder particles, as compared to the system used to leach the slab samples. This will be rapidly oxidised to Fe^{3+} which will dominate the subsequent dissolution (Equation (1)) at 750 mV. It is also worth considering that the absolute amount of surface S is not known. Hence although the relative amounts of oxidised S species (as % S) are greater on the chalcopyrite surfaces in contract with pyrite (Table 3), the total S on these surfaces may actually be smaller as compared to those in contact with quartz. If this is the case so the absolute concentrations of the oxidised S species may also be smaller.

Table 3. S species (% S) on the leached chalcopyrite slab surfaces (30 h).

S Species	Binding Energy (eV)	FWHM * (eV)	Chalcopyrite Only	Chalcopyrite (Galvanic Cell)
Surface S^{2-}	160.6–160.8	0.5–0.6	3	0
Bulk S^{2-}	161.1	0.5–0.6	51	49
S_2^{2-}	161.8–162.2	0.6–0.7	5	7
S_n^{2-}	162.8–163.4	1.2–1.5	22	28
S^0	163.9–164.3	1.2–1.5	2	13
SO_4^{2-}	168.7–169.1	1.2–1.5	14	0
Energy loss	164.8–165.2	1.8–2.0	3	3

* FWHM—full width at half maximum. For the estimation of FWHM values refer to [24,26].

No significant difference was observed between the O 1s spectra collected from chalcopyrite slab surface with and without pyrite present. Only chemisorbed H_2O and electrical isolated H_2O were observed in the O 1s spectra collected at these two surfaces, with no O^{2-} or OH^-/SO observed (Figure 5).

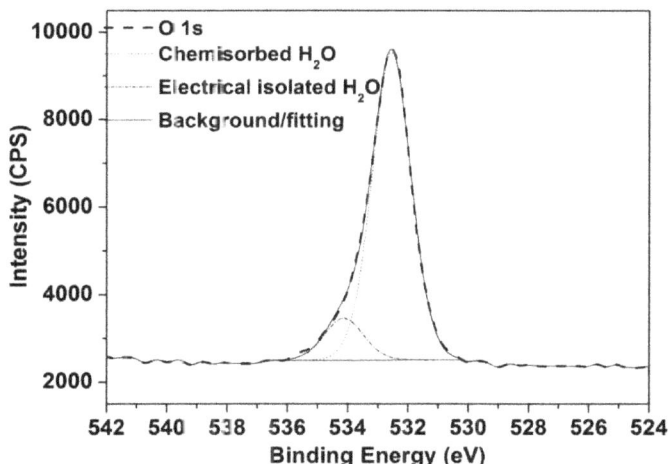

Figure 5. Fitted O 1s spectrum collected from the chalcopyrite slab in contact with pyrite (galvanic cell) leached at 750 mV, 75 °C and pH 1.0 for 30 h.

The strongest peak in the Fe $2p_{3/2}$ XPS spectra is observed at around 708 eV and another weak peak was also observed at 710–712 eV. This indicates that Fe(III)–S was the major Fe containing species with a component of Fe–O/OH also present [3,26–31] (Figure 6a). The Cu 2p spectrum (Figure 6b) shows that the oxidation stat of Cu in/on chalcopyrite was +1 [32–35].

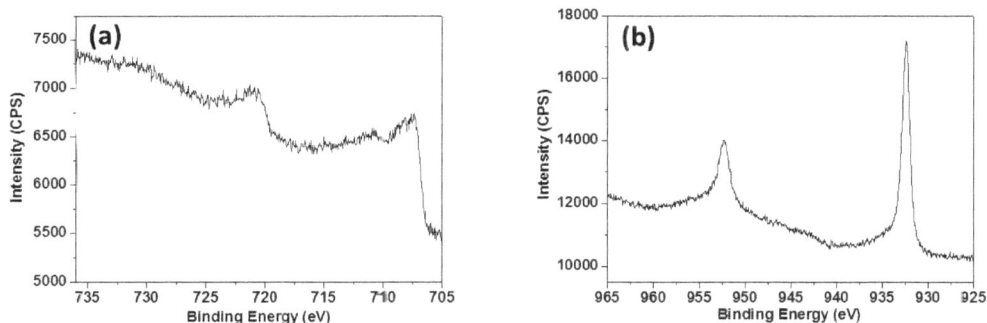

Figure 6. XPS spectra collected from the chalcopyrite surface in contact with pyrite (galvanic cell) leached for 30 h at 750 mV, 75 °C and pH 1.0: (**a**) Fe 2p; (**b**) Cu 2p.

4. Discussion

Equation (1) indicates that chalcopyrite can be oxidised by Fe^{3+} to produce S^0. As indicated by the Pourbaix diagram of the $CuFeS_2$–H_2O system [36], pH less than 4 and E_h greater than 580 mV (SHE) is required for effective Cu extraction from chalcopyrite. A greater Cu dissolution rate was observed at greater E_h (750 *cf.* 650 mV, Figure 1a), in accord with our previous chalcopyrite batch dissolution study [19].

The total Cu extracted from chalcopyrite at 650 mV increased significantly across the 132 days experimental duration when pyrite, as compared to quartz, was present (Figure 1a, 12% *cf.* 2.5%). Based on the ICP results (Figure 1a), the average Cu dissolution rate across 132 days at 650 mV is calculated to be 3.36×10^{-6} M·d^{-1} in the absence of pyrite and 1.86×10^{-5} M·d^{-1} when pyrite is present, indicating a fivefold increase of the Cu dissolution rate due to galvanic interaction. The mean chalcopyrite dissolution rate across 132 days at 750 mV is approximately four times greater than that at 650 mV in the presence of pyrite, *i.e.*, 7.11×10^{-5} M·d^{-1} and 7.35×10^{-5} M·d^{-1} for the chalcopyrite-quartz and chalcopyrite-pyrite systems respectively. This suggests that E_h plays a vital role in chalcopyrite dissolution rate-control. The similar Cu dissolution rates at 750 mV, in the presence and absence of pyrite, indicate that the galvanic interaction between chalcopyrite and pyrite is negligible.

Our findings are consistent with those of Tshilombo [37] who, using electrochemical techniques, reported that pyrite did not dissolve significantly at low solution potentials and under these conditions the leach rate of chalcopyrite was enhanced due to the presence of the cathodic sites on pyrite, *i.e.*, increased cathodic surface area. However, at greater solution potentials the galvanic interaction stopped and pyrite oxidation was apparent at redox potentials above 700 mV (SHE) [37].When Fe^{2+} is released from chalcopyrite or pyrite particles into solution, Fe^{3+} is available as the E_h is controlled and a portion of Fe^{2+} is oxidised rapidly by the addition of H$_2$O$_2$. The reduction of Fe^{3+} occurs on the pyrite surface in preference to the chalcopyrite surface when pyrite is present and contacting with chalcopyrite. However, when the solution E_h (e.g., 750 mV) is greater than the rest potential of pyrite, the reduction of Fe^{3+} on both pyrite and chalcopyrite is possible, resulting in the dissolution from both chalcopyrite and pyrite as indicated by Figure 1b. Therefore at 750 mV leaching of both chalcopyrite and pyrite is likely to take place. The dissolution of pyrite therefore increases the Fe concentration, resulting in a decreased Cu to Fe ratio as shown in Figure 1b. However, stoichiometric Cu to Fe dissolution of chalcopyrite at 650 mV is expected (Figure 1b), regardless of the presence of pyrite, as this E_h is less than the rest potential of pyrite but greater than that of chalcopyrite [38] and little dissolution of pyrite occurs.

These findings are of significance in providing better understanding of the factors affecting dissolution rate of mixed chalcopyrite and pyrite systems in AMD systems within the most often observed E_h range of 500–800 mV (SHE) [39–41]. In particular the findings suggest that metalliferous Cu-containing drainage may be of importance in this E_h range and that acidic drainage from pyrite dissolution may be partially inhibited if the E_h remains below pyrite's rest potential. In addition, this study also reveals that the galvanic effect between chalcopyrite and pyrite (or other sulfides) in the hydrometallurgical processing of chalcopyrite can be managed via controlling the solution E_h.

5. Conclusions

Chalcopyrite leaching has been conducted at 75 °C and pH 1.0. At E_h of 650 mV, the presence of pyrite, as compared to the same amount of the inert mineral of quartz, increases chalcopyrite dissolution rate by more than five times due to the effective galvanic interaction between chalcopyrite and pyrite. In addition, chalcopyrite leaching of Cu and Fe was found to be stoichiometric suggesting insignificant pyrite dissolution.

The dissolution rate at 650 mV is significantly less than that at 750 mV, regardless of the presence of pyrite or quartz. At 750 mV the galvanic effect derived from pyrite is negligible as this E_h is greater than the rest potentials of both chalcopyrite and pyrite so that both minerals dissolve.

XRD and SEM results indicate that almost 100% S of the dissolved chalcopyrite is oxidised to S^0, regardless of the presence of pyrite (at 750 mV). When pyrite contacts with chalcopyrite at 750 mV, around 60% of the S due to pyrite leaching is oxidised to S^0 and the remaining 40% is released as soluble species, e.g., SO$_4$$^{2-}$. Therefore, a reasonable E_h should be selected if galvanic interaction is expected in chalcopyrite dissolution.

Acknowledgments: Financial support from Rio Tinto and Australian Research Council via the ARC-Linkage Project "Solution and surface speciation evolution during chalcopyrite leaching" (LP110200326) is gratefully acknowledged.

Author Contributions: Yubiao Li and Andrea Gerson wrote the paper and contributed to all the activities related to this paper. Gujie Qian cooperated in conducting the dissolution experiment and data analysis. Jun Li designed the dissolution experiment and contributed to all the discussion.

Conflicts of Interest: The authors declare no conflict of interest.

References

1. Córdoba, E.M.; Muñoz, J.A.; Blázquez, M.L.; González, F.; Ballester, A. Leaching of chalcopyrite with ferric ion. Part I: General aspects. *Hydrometallurgy* **2008**, *93*, 81–87. [CrossRef]
2. Córdoba, E.M.; Muñoz, J.A.; Blázquez, M.L.; González, F.; Ballester, A. Leaching of chalcopyrite with ferric ion. Part III: Effect of redox potential on the silver-catalyzed process. *Hydrometallurgy* **2008**, *93*, 97–105. [CrossRef]
3. Li, Y.; Kawashima, N.; Li, J.; Chandra, A.P.; Gerson, A.R. A review of the structure, and fundamental mechanisms and kinetics of the leaching of chalcopyrite. *Adv. Colloid Interface Sci.* **2013**, *197–198*, 1–32. [CrossRef] [PubMed]
4. Watling, H.R. Chalcopyrite hydrometallurgy at atmospheric pressure: 1. Review of acidic sulfate, sulfate-chloride and sulfate-nitrate process options. *Hydrometallurgy* **2013**, *140*, 163–180. [CrossRef]
5. Munoz-Castillo, B.P. Reaction Mechanisms in the Acid Ferric-Sulfate Leaching of Chalcopyrite. Ph.D. Thesis, University of Utah, Salt Lake City, UT, USA, 1977.
6. Hirato, T.; Majima, H.; Awakura, Y. The leaching of chalcopyrite with ferric sulfate. *Metall. Mater. Trans. B* **1987**, *18*, 489–496. [CrossRef]
7. Klauber, C. A critical review of the surface chemistry of acidic ferric sulphate dissolution of chalcopyrite with regards to hindered dissolution. *Int. J. Miner. Process.* **2008**, *86*, 1–17. [CrossRef]
8. Watling, H.R. The bioleaching of sulphide minerals with emphasis on copper sulphides—A review. *Hydrometallurgy* **2006**, *84*, 81–108. [CrossRef]
9. Hiroyoshi, N.; Miki, H.; Hirajima, T.; Tsunekawa, M. Enhancement of chalcopyrite leaching by ferrous ions in acidic ferric sulfate solutions. *Hydrometallurgy* **2001**, *60*, 185–197. [CrossRef]
10. Dutrizac, J.E. Elemental sulphur formation during the ferric sulphate leaching of chalcopyrite. *Can. Metall. Quart.* **1989**, *28*, 337–344. [CrossRef]
11. Dixon, D.G.; Mayne, D.D.; Baxter, K.G. Galvanox (TM)—A novel galvanically-assisted atmospheric leaching technology for copper concentrates. *Can. Metall. Quart.* **2008**, *47*, 327–336. [CrossRef]
12. Liu. Q.; Li, H.; Zhou, L. Galvanic interactions between metal sulfide minerals in a flowing system: Implications for mines environmental restoration. *Appl. Geochem.* **2008**, *23*, 2316–2323. [CrossRef]
13. Koleini, S.M.J.; Jafarian, M.; Abdollahy, M.; Aghazadeh, V. Galvanic leaching of chalcopyrite in atmospheric pressure and sulfate media: Kinetic and surface studies. *Ind. Eng. Chem. Res.* **2010**, *49*, 5997–6002. [CrossRef]
14. Koleini, S.M.J.; Aghazadeh, V.; Sandström, A. Acidic sulphate leaching of chalcopyrite concentrates in presence of pyrite. *Miner. Eng.* **2011**, *24*. 381–386. [CrossRef]
15. Misra, M.; Fuerstenau, M.C. Chalcopyrite leaching at moderate temperature and ambient pressure in the presence of nanosize silica. *Miner. Eng.* **2005**, *18*, 293–297. [CrossRef]
16. Tshilombo, A.F.; Petersen, J.; Dixon, D.G. The influence of applied potentials and temperature on the electrochemical response of chalcopyrite during bacterial leaching. *Miner. Eng.* **2002**, *15*, 809–813. [CrossRef]
17. Berry, V.K.; Murr, L.E.; Hiskey, J.B. Galvanic interaction between chalcopyrite and pyrite during bacterial leaching of low-grade waste. *Hydrometallurgy* **1978**, *3*, 309–326. [CrossRef]
18. Mehta, A.P.; Murr, L.E. Kinetic study of sulfide leaching by galvanic interaction between chalcopyrite, pyrite, and sphalerite in the presence of T. ferrooxidans (30 °C) and a thermophilic microorganism (55 °C). *Biotechnol. Bioeng.* **1982**, *24*, 919–940. [CrossRef] [PubMed]
19. Li, Y.; Qian, G.; Li, J.; Gerson, A. The rate controlling parameters in the hydrometallurgical leaching of chalcopyrite. In Proceedings of ALTA 2014 Nickel-Cobalt-Copper Sessions, Perth, Australia, 24–31 May 2014; pp. 399–410.
20. Li, Y.; Qian, G.; Li, J.; Gerson, A.R. Kinetics and roles of solution and surface species of chalcopyrite dissolution at 650 mV. *Geochim. Cosmochim. Acta* **2015**, *161*, 188–202. [CrossRef]
21. Li, J.; Kawashima, N.; Kaplun, K.; Absolon, V.J.; Gerson, A.R. Chalcopyrite leaching: The rate controlling factors. *Geochim. Cosmochim. Acta* **2010**, *74*, 2881–2893. [CrossRef]
22. Kaplun, K.; Li, J.; Kawashima, N.; Gerson, A.R. Cu and Fe chalcopyrite leach activation energies and the effect of added Fe^{3+}. *Geochim. Cosmochim. Acta* **2011**, *75*, 5865–5878. [CrossRef]

23. Kimball, B.E.; Rimstidt, J.D.; Brantley, S.L. Chalcopyrite dissolution rate laws. *Appl. Geochem.* **2010**, *25*, 972–983. [CrossRef]
24. Li, Y.; Chandra, A.P.; Gerson, A.R. Scanning photoelectron microscopy studies of freshly fractured chalcopyrite exposed to O_2 and H_2O. *Geochim. Cosmochim. Acta* **2014**, *133*, 372–386. [CrossRef]
25. Qian, G.; Li, J.; Li, Y.; Gerson, A.R. Probing the effect of aqueous impurities on the leaching of chalcopyrite under controlled conditions. *Hydrometallurgy* **2014**, *149*, 195–209. [CrossRef]
26. Acres, R.G.; Harmer, S.L.; Beattie, D.A. Synchrotron XPS, NEXAFS, and ToF-SIMS studies of solution exposed chalcopyrite and heterogeneous chalcopyrite with pyrite. *Miner. Eng.* **2010**, *23*, 928–936. [CrossRef]
27. Buckley, A.N.; Woods, R. An X-ray photoelectron spectroscopic study of the oxidation of chalcopyrite. *Aust. J. Chem.* **1984**, *37*, 2403–2413. [CrossRef]
28. Smart, R.S.C. Surface layers in base metal sulphide flotation. *Miner. Eng.* **1991**, *4*, 891–909. [CrossRef]
29. Fairthorne, G.; Fornasiero, D.; Ralston, J. Effect of oxidation on the collectorless flotation of chalcopyrite. *Int. J. Miner. Process.* **1997**, *49*, 31–48. [CrossRef]
30. Brion, D. Photoelectron spectroscopic study of the surface degradation of pyrite (FeS_2), chalcopyrite ($CuFeS_2$), sphalerite (ZnS), and galena (PbS) in air and water. *Appl. Surf. Sci.* **1980**, *5*, 133–152. [CrossRef]
31. Acres, R.G.; Harmer, S.L.; Shui, H.W.; Chen, C.H.; Beattie, D.A. Synchrotron scanning photoemission microscopy of homogeneous and heterogeneous metal sulfide minerals. *J. Synchrotron Radiat.* **2011**, *18*, 649–657. [CrossRef] [PubMed]
32. Yin, Q.; Kelsall, G.H.; Vaughan, D.J.; England, K.E.R. Atmospheric and electrochemical oxidation of the surface of chalcopyrite ($CuFeS_2$). *Geochim. Cosmochim. Acta* **1995**, *59*, 1091–1100. [CrossRef]
33. Harmer, S.L.; Pratt, A.R.; Nesbitt, H.W.; Fleet, M.E. Reconstruction of fracture surfaces on bornite. *Can. Mineral.* **2005**, *43*, 1619–1630. [CrossRef]
34. Pearce, C.I.; Pattrick, R.A.D.; Vaughan, D.J.; Henderson, C.M.B.; van der Laan, G. Copper oxidation state in chalcopyrite: Mixed Cu d^9 and d^{10} characteristics. *Geochim. Cosmochim. Acta* **2006**, *70*, 4635–4642. [CrossRef]
35. Klauber, C.; Parker, A.; van Bronswijk, W.; Watling, H. Sulphur speciation of leached chalcopyrite surfaces as determined by X-ray photoelectron spectroscopy. *Int. J. Miner. Process.* **2001**, *62*, 65–94. [CrossRef]
36. Garrels, R.M.; Christ, C.L. *Solutions, Minerals, and Equilibria*; Harper & Row: New York, NY, USA, 1965.
37. Tshilombo, A.F. Mechanism and Kinetics of Chalcopyrite Passivation and Depassivation during Ferric and Microbial Leaching. Ph.D. Thesis, University of Brithish Columbia, Vancouver, BC, Canada, December 2004.
38. Payant, R.; Rosenblum, F.; Nesset, J.E.; Finch, J.A. The self-heating of sulfides: Galvanic effects. *Miner. Eng.* **2012**, *26*, 57–63. [CrossRef]
39. Costa, M.; Martins, M.; Jesus, C.; Duarte, J. Treatment of acid mine drainage by sulphate-reducing bacteria using low cost matrices. *Water Air Soil Pollut.* **2008**, *189*, 149–162. [CrossRef]
40. Borkowski, A.; Parafiniuk, J.; Wolicka, D.; Kowalczyk, P. Geomicrobiology of acid mine drainage in the weathering zone of pyrite-bearing schists in the rudawy janowickie mountains (Poland). *Geol. Q.* **2013**, *57*, 601–612. [CrossRef]
41. Azzali, E.; Marescotti, P.; Frau, F.; Dinelli, E.; Carbone, C.; Capitani, G.; Lucchetti, G. Mineralogical and chemical variations of ochreous precipitates from acid sulphate waters (asw) at the Roşia Montană gold mine (Romania). *Environ. Earth Sci.* **2014**, *72*, 3567–3584. [CrossRef]

![metals logo] *metals*

MDPI

Review

Hydrometallurgical Recovery of Precious Metals and Removal of Hazardous Metals Using Persimmon Tannin and Persimmon Wastes

Katsutoshi Inoue [1,*], Manju Gurung [1], Ying Xiong [1], Hidetaka Kawakita [1,†], Keisuke Ohto [1,†] and Shafiq Alam [2,†]

1   Department of Applied Chemistry, Faculty of Science & Engineering, Saga University, Honjo-Machi 1, Saga 840-8502, Japan; grgmanju@gmail.com (M.G.); xiongying_1977@hotmail.com (Y.X.); kawakita@cc.saga-u.ac.jp (H.K.); ohtok@cc.saga-u.ac.jp (K.O.)
2   Department of Chemical and Biological Engineering, College of Engineering, University of Saskatchewan, 57 Campus Drive, Saskatoon, SK S7N 5A9, Canada; shafiq.alam@usask.ca
*   Author to whom correspondence should be addressed; inoueka@cc.saga-u.ac.jp; Tel.: +81-90-5290-7575; Fax: +81-952-288548.
†   These authors contributed equally to this work.

Academic Editors: Suresh Bhargava, Mark Pownceby and Rahul Ram
Received: 3 June 2015; Accepted: 29 September 2015; Published: 23 October 2015

Abstract: Novel and environmentally benign adsorbents were prepared via a simple sulfuric acid treatment process using the wastes of astringent persimmon, a type of biomass waste, along with persimmon tannin extract which is currently employed for the tanning of leather and as natural dyes and paints. The effectiveness of these new biosorbents was exemplified with regards to hydrometallurgical and environmental engineering applications for the adsorptive removal of uranium and thorium from rare earths, cesium from other alkaline metals such as sodium, hexa-valent chromium from zinc as well as adsorptive recovery of gold from chloride media. Furthermore, reductive coagulation of gold from chloride media for the direct recovery of metallic gold and adsorptive recovery of palladium and platinum using chemically modified persimmon tannin extract were studied.

Keywords: adsorption; precious metals; radioactive elements; chromium(VI); persimmon tannin; biomass wastes

---

1. Introduction

Within the last decades, increasing attention has been paid towards energy saving hydrometallurgical processes for the recovery of valuable metals and the removal of hazardous materials. In hydrometallurgical processing, a leach liquor is subjected to metal separation and purification, which includes solvent extraction, ion exchange, precipitation, membrane filtration, and so forth. However, all the aforementioned methods suffer from significant disadvantages, which include incomplete metal recovery, high capital costs, high reagent and/or energy requirements and generation of toxic sludge or other waste products that demand additional disposal considerations, hence increasing the overall cost. Due to their high efficiency, flexibility, applicability over a wide range of metal concentration, and simple sludge free operation for the effective recovery of analyte, adsorption techniques demonstrate great promise for the recovery of metal values from aquatic environment. From the technical and economical points of view, in recent years, much attention has been focused on low cost adsorbents prepared from various biomass wastes as they are relatively inexpensive, environmentally benign, biodegradable and renewable. In such biomass wastes, the chemical components which exhibit effective adsorption behaviors include several polysaccharides such as pectic acid, alginic acid, chitin and cellulose, polyphenol compounds such as lignin and tannin, and proteins. Additionally, such adsorption behaviors of the original components can be significantly enhanced using simple chemical treatments or chemical modification processes. For example, although the adsorption of chitin for metal ions is poor, that of chitosan which can be produced by simple hydration reaction using concentrated sodium hydroxide solution is attractive particularly with regards to hydrometallurgical processing [1]. Furthermore, although cellulose itself exhibits negligible adsorption for nearly all metal ions, it exhibits extraordinary high and selective adsorption for gold as a consequence of a simple concentrated sulfuric acid treatment [2].

Concerning tannin compounds, it is well known that they possess ion exchange and complexation characteristics towards metal ions; that is, the potential of tannin rich biomass materials for the recovery of heavy and precious metals has been reported by some authors [3,4].

Persimmon is a popular fruit in East Asian countries such as China, Korea and Japan. There are two kinds of persimmon fruits, sweet persimmon and astringent persimmon, which differ by their content of water soluble persimmon tannin. Although the former can be eaten, as they are similar to apples, the latter are rich in the water soluble persimmon tannin which generates a very astringent taste and, are thus unsuitable for direct consumption. However, such astringent persimmon can be edible by peeling and drying in the open air for a few weeks during the winter wherein the water soluble persimmon tannin is transformed by a spontaneous condensation reaction into water insoluble polymerized tannin and its astringent taste is diminished. Here, persimmon tannin is a kind of polyphenol compound possessing a complicated chemical structure as shown in Scheme 1 containing large amounts of catechol and pyrogallol functional groups [5]. Persimmon extract, the juice of astringent persimmon, contains large quantities of persimmon tannin and has been traditionally employed in various applications as natural paints, dyes, tanning agent for leather tanning, a coagulating agent for proteins, *etc.* During the preparation of dried persimmon fruits and astringent persimmon extract, large quantities of peels and juice residue, or biomass wastes, are generated. These wastes still contain large amounts of persimmon tannin. In our recent works, several novel types of environmentally benign adsorption gels were prepared from the powder of the astringent persimmon juice and also from the above-mentioned persimmon wastes as potential agents for the recovery of valuable metals and the removal of hazardous metals. In the present paper, these works are summarized and briefly reviewed from the viewpoints of hydrometallurgy and environmental engineering.

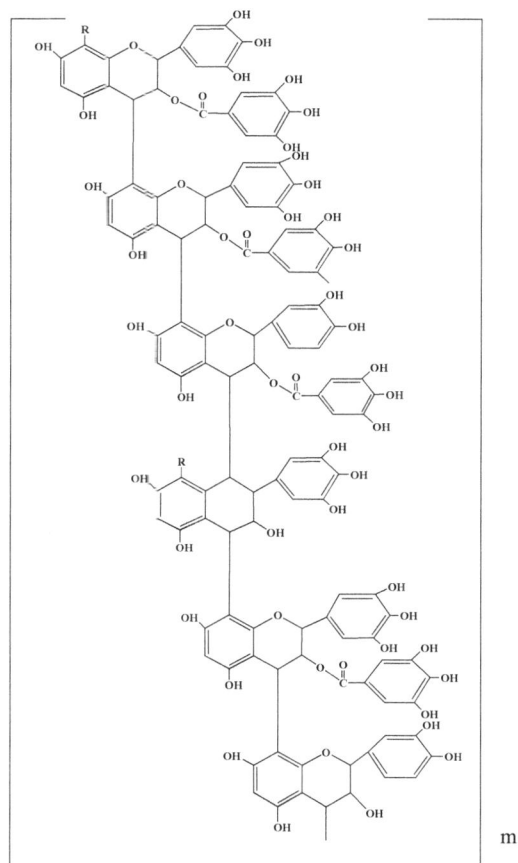

Scheme 1. Schematic structure of persimmon tannin where m stands for the number of the unit.

2. Adsorptive Removal of Uranium and Thorium

After the oil crisis in the 1970s, the recovery of uranium from sea water received a considerable amount of attention in Japan. From such interest, Sakaguchi and Nakajima found that some tannin compounds such as Chinese gallotannin (tannic acid) immobilized on cellulosic matrices such as cellulose powder are effective for the recovery of uranium from sea water [3]. They further discovered that persimmon extract exhibits an extremely high affinity for uranium, and subsequently prepared an adsorption gel for the recovery of uranium by crosslinking it using glutaraldehyde [6]. They attempted to employ this gel not only for the recovery of uranium from sea water but also for the removal of trace amounts of uranium from uranium milling wastewater. Environmental pollution by radioactive elements such as uranium and thorium is not limited to uranium mines and nuclear facilities. Rare earth ores, monazite, bastnaesite and xenotime, contain some quantities of these elements, which are removed in the mineral dressing stage at mine sites. However, unfavorable treatment and management of tailings of these ores may bring about serious environmental contamination. For instance, in the early 1980s a serious environmental pollution problem took place at a rare earth plant at Ipoh, Malaysia. As seen from this case, selective removal of uranium and thorium from rare earths has been required in order to avoid such environmental problems.

From such a viewpoint, studies were conducted on the selective removal of uranium(VI) and thorium(IV) from lutetium(III) (the heaviest rare earth exhibiting similar chemical behaviors with these actinide elements) using three kinds of adsorbents [7,8]. These are as follows: (1) A persimmon peel gel (PP gel) prepared from persimmon peel waste generated in the production of dried persimmon fruits by boiling in concentrated sulfuric acid for the dehydration condensation reaction for crosslinking between polyphenols and polysaccharides such as cellulose; (2) a gallic acid resin prepared by interacting gallic acid with a commercially available weak base anion exchange resin (DIAION WA30, Mitsubishi Chemical Corp., Tokyo, Japan); and (3) a kakishibu resin, which was prepared by impregnating persimmon extract (kakishibu) into a high porous resin without any functional groups (DIAION HP 20, Mitsubishi Chemical Corp., Tokyo, Japan), kindly donated by Tomiyama Corp., Kyoto, Japan.

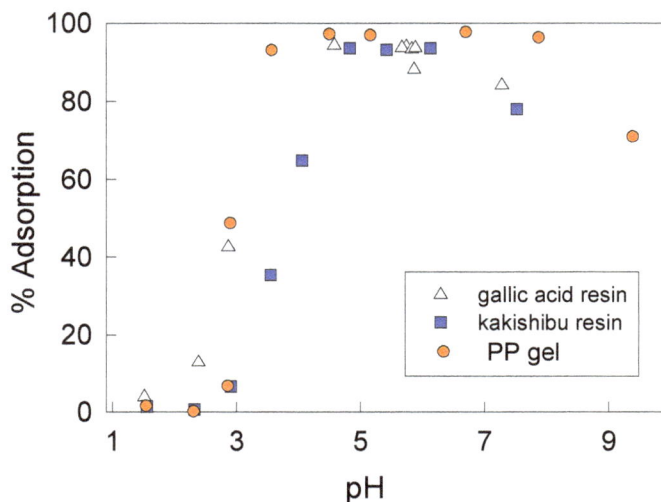

Figure 1. Percentage adsorption of uranium(VI) at varying equilibrium pH on PP gel, gallic acid resin and kakishibu resin.

Figure 1 shows the pH dependency of % adsorption of uranium(VI) on the PP gel, the gallic acid resin and the kakishibu resin for comparison. For all adsorbents, the adsorption increases with increasing pH, indicating that adsorption takes place according to a cation exchange mechanism with the phenolic hydroxyl groups of the persimmon tannin releasing hydrogen ions as suggested by Sakaguchi and Nakajima [6] (a similar adsorption mechanism will be discussed in Section 4 (chromium(VI) adsorption) by the reaction equation shown in Scheme 2b) wherein nearly quantitative adsorption is achieved at pH values greater than 4.5. The adsorption on the PP gel appears to be nearly equal to that on the gallic acid resin and higher than that of the kakishibu resin. On the other hand, the adsorption behavior for thorium(IV) was found to be nearly equal for all of the three adsorbents while the adsorption of lutetium(III) on the PP gel was found to be higher than that on other two adsorbents.

Figure 2 shows the comparison of the % adsorption of uranium(VI), thorium(IV) and lutetium(III) on the PP gel to compare the order of selectivity among these three metal ions. As seen from this figure, it is as follows: Th(IV) > U(VI) > Lu(III). That is, uranium(VI) and thorium(IV) are selectively adsorbed on the gel over lutetium(III), supporting the feasibility of selective adsorptive removal of the radioactive elements, uranium(VI) and thorium(IV), contained in rare earth ores using this gel.

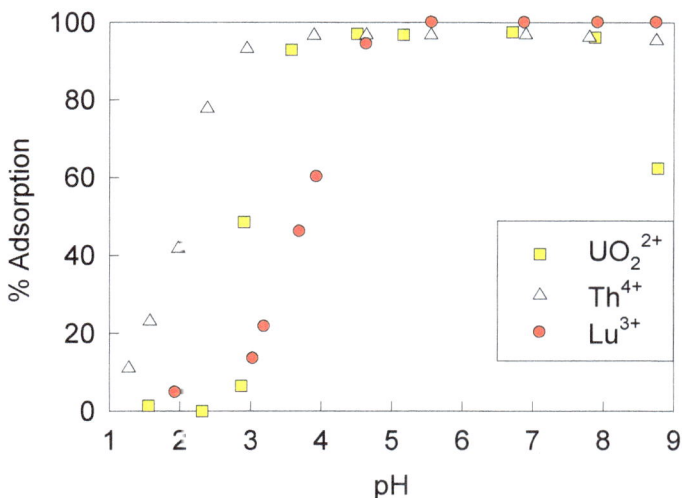

Figure 2. Percentage Adsorption of uranium(VI), thorium(IV) and lutetium(III) at varying equilibrium pH on PP gel [8] (with permission for reuse from TMS).

From the study of the adsorption isotherms of uranium(VI), thorium(IV) and lutetium(III) on the PP gel, it was found that the adsorption appears to exhibit the Langmuir type adsorption; *i.e.*, it increases with increasing metal concentration and tends to approach constant values corresponding to each metal ion, from which maximum adsorption capacities were evaluated for uranium(VI), thorium(IV) and lutetium(III) as follows, respectively: 1.29, 0.49 and 0.28 mol·kg^{-1}-dry gel. In contrast to the selectivity order, uranium(VI) is much more greatly adsorbed than thorium(IV) and lutetium(III). Compared with the adsorption capacities of commercially available chelating resins, that of uranium(VI) on the PP gel is very high as in the case of the persimmon tannin gel reported by Sakaguchi and Nakajima [6].

Based on experimental results of the above-mentioned batch wise adsorption tests, mutual separation between uranium(VI) and lutetium(III) was tested using a small glass column packed with PP gel and model solution containing 0.21 mmol·dm^{-3} uranium(VI) and 0.29 mmol·dm^{-3} lutetium(III) at pH = 3.81.

Figure 3 shows the breakthrough profiles of these two metal ions; *i.e.*, a plot of the relative concentration of these ions in the effluent (their outlet concentrations) compared with their feed concentrations against contact time after the initiation of the flow. As seen from this figure, breakthrough of lutetium(III) began at the contact time before 10 h just after the initiation of the feed while that of uranium(VI) began at 20 h, suggesting that the mutual separation of uranium(VI) from lutetium(III) can be successfully achieved using the column packed with the PP gel.

Figure 4, on the other hand, shows the elution profiles for these two metal ions from the column after eluting the loaded column with a 1 mol·dm^{-3} hydrochloric acid solution. As seen from this figure, uranium(VI) was eluted at a concentration greater than 25 times that of the feed solution for the breakthrough test, though a small amount of lutetium(III) coexisted as a contaminant in the eluted solution.

Further, from the results of the cycle test, it was apparent that repeating the adsorption of uranium(VI) followed by elution did not compromise the adsorption capacity of the PP gel for uranium(VI) even after 10 cycles, further supporting its feasibility for practical application.

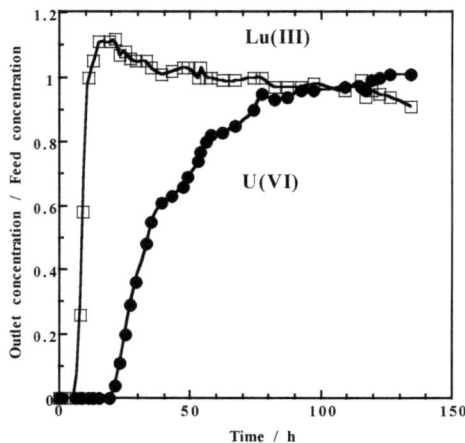

Figure 3. Breakthrough profile of uranium(VI) and lutetium(III) from the column packed with persimmon peel (PP) gel.

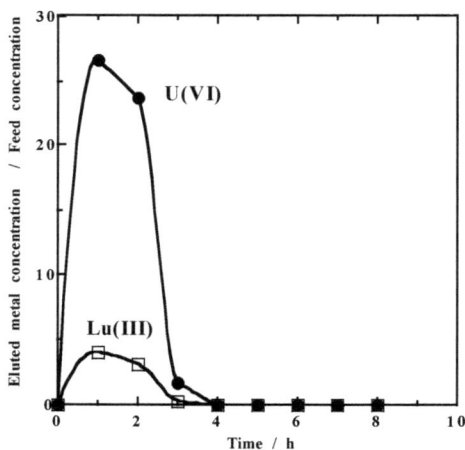

Figure 4. Elution profiles of uranium(VI) and lutetium(III) from the PP gel packed column after breakthrough using 1 mol·dm^{-3} hydrochloric acid solution.

3. Adsorptive Removal of Cesium

Cesium (^{137}Cs) is one of the fission products of nuclear fuels contained in radioactive wastes. Due to its long half life (30 years) and bio-toxicity, its separation and removal from aquatic environments has always attracted special attention, and this has become much more important in recent days after the nuclear plant accident at Fukushima, Japan.

Studies on the removal of cesium(I) ion from aqueous solution have been focused largely on adsorption and ion-exchange methods [9–12]. Natural and synthetic zeolites, clay minerals and synthetic organic or inorganic ion-exchangers have been employed for large scale separation of ^{137}Cs from low and intermediate-level radioactive waste effluents or contaminated water [9,10,13,14]. However, the majority of these materials appear to be uneconomical as a consequence of their high operational cost and poor performance. It has been reported that the phenolic resins such as Duolite

S-30 (phenolic), and Duolite CS-100 (phenolic-carboxylic), can selectively remove cesium(I) from alkaline solutions [15,16]. The phenolic OH groups are, however, ionizable and only effective at high pH. Because of their comparatively lower pKa *versus* the monohydroxy phenols, the polyhydroxyphenols are ionizable and can function as metal chelators at relatively lower pH than the monohydroxyphenols. Based on such findings, we anticipated that bioadsorbents having polyphenolic moieties would exhibit adsorption selectivity towards cesium(I) ions at approximately neutral pH. From this perspective, our focus was towards developing highly effective adsorption materials for cesium(I) removal using a persimmon extract rich in persimmon tannin [17,18]. That is, the adsorbent in the present case, which is abbreviated as CPT (cross-linked persimmon tannin) gel hereafter, was prepared from commercially available dried persimmon tannin extract powder (Persimmon-Kaki Technology Development Co. Ltd., Jincheng, China) also by means of crosslinking in boiling concentrated in sulfuric acid at 100 °C.

Figure 5 shows the effect of initial pH on the adsorption behavior of CPT gel towards cesium(I) and sodium(I) ions. It is clear from this figure that the adsorption of these metal ions is also greatly affected by pH similar to the adsorption of uranium(VI), thorium(IV) and lutetium(III) as described in the preceding section. Although cesium(I) is negligibly adsorbed on this adsorbent in acidic region, its adsorption increases with increasing pH, achieving near quantitative removal at pH higher than 4.5 with moderate adsorption of sodium(I) ion (around 40%), which supports the applicability of this adsorbent for the selective removal of cesium(I) at around neutral pH.

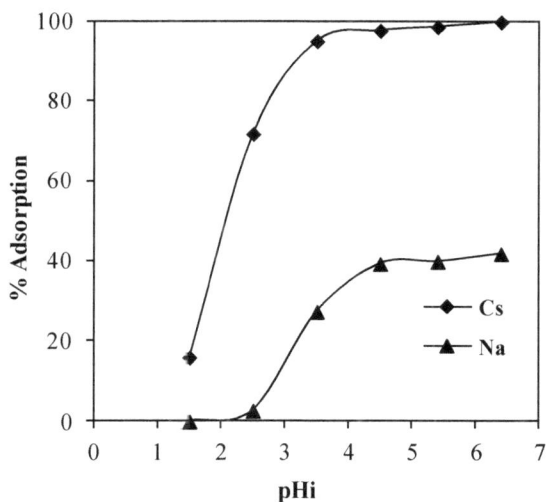

Figure 5. Percentage adsorption of cesium(I) and sodium(I) ions on crosslinked persimmon tannin (CPT) gel at varying initial pH [17] (with permission for reuse from Elsevier B.V.).

Figure 6 shows the isotherm plots for the adsorption of cesium(I) (plot of amount of adsorption (*q*) *vs.* equilibrium concentration (Ce)) as well as sodium(I) on CPT gel. The plots indicate the typical Langmuir's monolayer type of adsorption. From the Langmuir plot using these data, the maximum adsorption capacities for cesium(I) and sodium(I) on this adsorbent were evaluated as 1.33 and 0.45 mol·kg^{-1}, respectively.

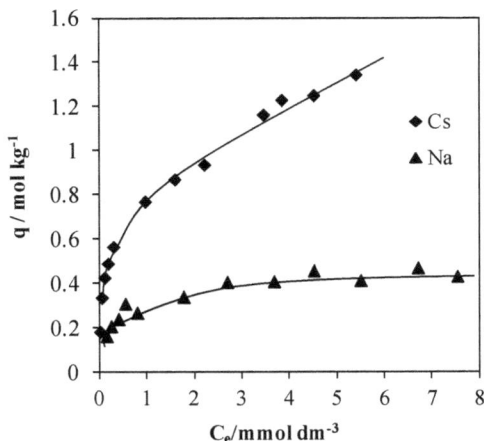

Figure 6. Adsorption isotherms of cesium(I) and sodium(I) ions on CPT gel [17] (with permission for reuse from Elsevier B.V.).

A comparative evaluation of the maximum adsorption capacity of this adsorbent towards cesium(I) ion was carried out with those of different adsorbents reported in literatures as listed in Table 1. From this table, it is deduced that the present adsorbent containing polyphenolic groups as ionizable and coordinating functional group have significant adsorption potential for cesium(I) removal from aqueous solutions compared with other adsorbents.

Table 1. Cesium(I) ion adsorption capacity comparison of the CPT gel and other adsorbents.

Adsorbent	Adsorption Capacity $(mol \cdot kg^{-1})$	pH of the Solution	Reference
Zeolite A	1.57	6.0	[10]
Phosphate-modified montmorillonite	0.70	5.0	[9]
Ammonium molybdophosphate-polyacrylonitrile	0.61	6.5	[19]
Copper ferrocyanide functionalized mesoporous silica	0.12	7.8	[9]
Prussian blue	0.09	7.8	[9]
Natural clinoptilolite	0.37	6.5	[13]
Brewery waste biomass	0.07	4.0	[12]
Arca shell biomass	0.03	5.5	[11]
Sulfuric acid crosslinked *Pseudochoricystis ellipsoidea*	1.36	6.5	[20]
CPT gel	1.34	6.5	[17]

Taking into account the remarkable selectivity and adsorption capacity of CPT gel towards cesium(I) over sodium(I), practical applicability of this adsorbent for selective adsorption, preconcentration and removal/recovery of cesium(I) from the mixture of both the ions was studied by continuous column experiment.

Figure 7 shows the breakthrough profiles of cesium(I) and sodium(I) ions using a small glass column packed with CPT gel. In this figure, relative concentrations at outlet to the feed concentration (C_t/C_i) are plotted against the bed volume (B.V.) which is defined as the ratio of the total volume of the feed solution passed though the column to the volume of the adsorbent packed in the column. As expected from the results of batch wise experiment, breakthrough of sodium(I) took place after a few hours while that of cesium(I) began to take place after 100 h (996 B.V.), suggesting satisfactory separation between these two metal ions.

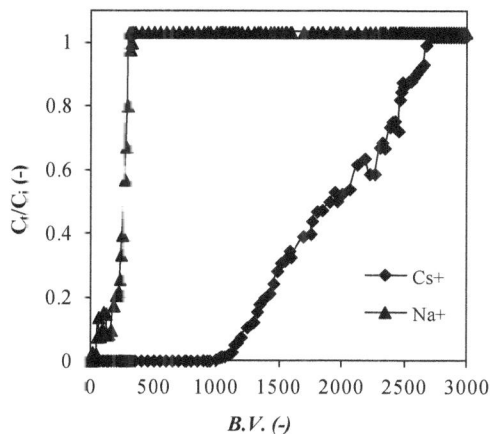

Figure 7. Breakthrough profiles of cesium(I) and sodium(I) ions from the column packed with CPT gel [17] (with permission for reuse from Elsevier B.V.).

After the adsorption bed was saturated, elution test was carried out by using 1 mol·dm^{-3} hydrochloric acid solution in order to recover the adsorbed cesium(I) in concentrated form. It was found that the elution was quite fast and the concentration of cesium(I) in the elute was much higher than that of the feed solution, *i.e.*, the preconcentration factor (=outlet concentration of the column/feed concentration) for cesium(I) was evaluated as 140. Adsorptive separation of cesium(I) with high preconcentration factor and quantitative elution in a short period of time suggest that the CPT gel can be used repeatedly and effectively for the separation and removal of cesium(I) from the mixture of sodium(I) ions.

4. Adsorptive Removal of Chromium(VI)

Chromium exists as chromium(III) and chromium(VI) in aqueous medium. Chromium(VI) is 500 times more toxic than chromium(III). Its toxicity includes lung cancer as well as kidney, liver and gastric damage [21]. The maximum level of chromium(VI) permitted in waste water is 0.5 mg·dm^{-3}. Industrial effluents from plating industries, leather tanning and so on sometimes contain higher levels of chromium(VI) than the permitted level [22]. It is essential for such industries to treat their effluents to reduce the chromium(VI) content to the acceptable level. In conventional chromium plating, chromium is plated as a thin layer on zinc-coated steel plate. This generates post treatment wastewater containing a high concentration of zinc(II) and a small amount of chromium(VI). In the present work, persimmon waste generated in the production of persimmon extract as the residue has been tested for the removal of chromium(VI), taking account of chromium(VI)-containing effluents from the plating industry [23]. That is, the adsorbent was prepared from the persimmon extract residue also by boiling in concentrated sulfuric acid at high temperature for cross-linking in this case. This adsorbent is abbreviated as CPW (crosslinked persimmon waste) gel, hereafter.

Chromium(VI) exist in aqueous solution as $HCrO_4^-$, H_2CrO_4, $HCr_2O_7^-$ and CrO_4^{2-} depending on the pH of the solution and the total chromium concentration.

Figure 8 shows the % adsorption of several metal ions as a function of equilibrium pH on CPW gel. This figure shows a high selectivity of this adsorbent for chromium(VI) at pH < 3. It also shows that the % adsorption of all metal ions investigated in this experiment, except for chromium(VI), was found to increase with increasing pH in the pH range of 1–5 whereas opposite behavior was observed for chromium(VI). That is, the % adsorption of chromium(VI) increased with increasing pH in the pH range of 1–3 and then decreased with further increases in pH. This contrasting behavior for chromium(VI) adsorption compared to the other metal ions can be attributed to a different adsorption mechanism shown in Scheme 2 [4]. As stated earlier, Cr(VI) exists in aqueous medium as $HCrO_4^-$, H_2CrO_4, $HCr_2O_7^-$ and CrO_4^{2-}, among which the $HCrO_4^-$ is the dominant species up to pH 5 [24].

Scheme 2a shows the adsorption mechanism of the oxo anionic form of chromium(VI) on CPW gel from an acidic solution in terms of an esterification reaction, which suggests that a high concentration of H^+ ion enhances chromium(VI) adsorption whereas a high concentration of OH^- ion suppresses chromium(VI) adsorption, thus accounting for the decrease in the % adsorption of chromium(VI) with increasing the pH of the solution at pH higher than 3. The reason for decreasing % adsorption with decreasing pH of the solution at pH less than 3 is due to the reduction of chromium(VI) to chromium(III) in acidic solution as will be discussed later.

On the other hand, the adsorption of other metal ions, except for chromium(VI), which exist as cationic species in aqueous solution, are adsorbed on CPW gel according to a cation exchange mechanism as shown in Scheme 2b, which accounts for the increase in the % adsorption with increasing pH of the solution.

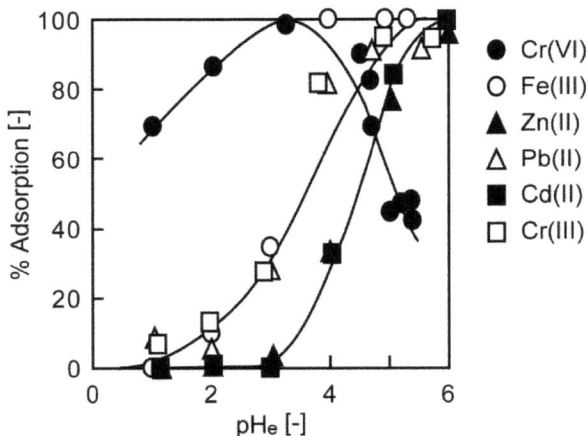

Figure 8. Effect of equilibrium pH on the adsorption of different metal ions on CPW gel [23] (with permission for reuse from Elsevier B.V.).

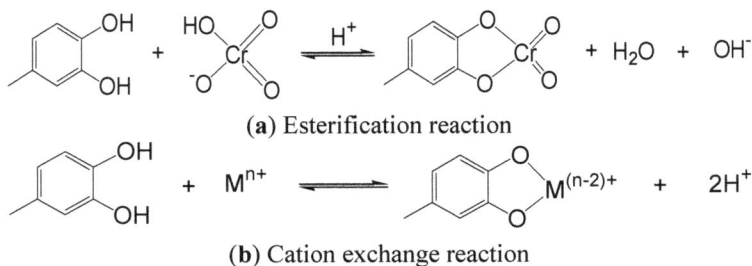

(**a**) Esterification reaction

(**b**) Cation exchange reaction

Scheme 2. The binding mechanism of (**a**) chromium(VI) and (**b**) other metals to the hydroxyl groups of catechol units on the cross-linked persimmon gel.

Batch wise kinetic experiments were carried out in order to elucidate the equilibrium time and mechanism of the adsorption reaction of Cr(VI).

The time variation of % adsorption of total chromium at varying pH is shown in Figure 9, which reveals that the chromium adsorption monotonously increases with increasing contact time until reaching constant values corresponding to pH values, except for the case at pH = 1. On the other hand, at pH = 1, after the adsorption rapidly increases at the initial stage similar to the cases at other pH values, it slightly decreases and tends to approach a constant value. This phenomenon was observed also in the adsorption on the crosslinked grape waste gel and it was concluded to be attributed to the adsorption coupled reduction mechanism of chromium(VI) to chromium(III) which is not adsorbed at low pH as seen from Figure 8 [25].

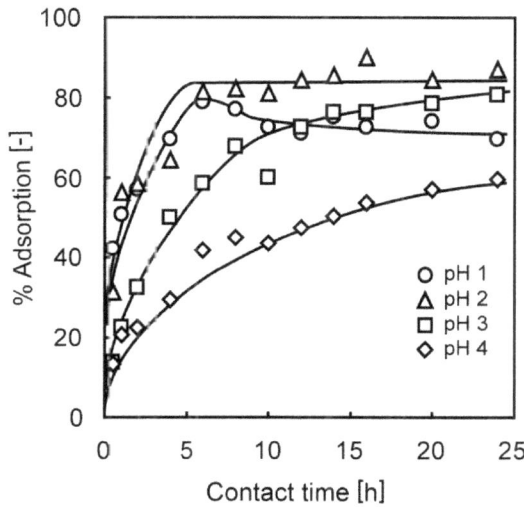

Figure 9. Effect of contact time on the adsorption of total chromium on cross-linked persimmon waste (CPW) gel from chromium(VI) solution at different pH values [23] (with permission for reuse from Elsevier B.V.).

In order to elucidate the actual mechanism of chromium(VI) adsorption by CPW gel, the rate of change of both chromium(VI) and chromium(III) concentrations in solution were measured at pH = 1 as shown in Figure 10. The result shows that chromium(VI) ions were completely removed from the aqueous solution at this pH and chromium(III) ions were generated which were not present in the original solution. This result demonstrates that there is not only an adsorption mechanism of chromium(VI) but also a reduction mechanism taking place in this pH range. That is, the oxo anionic species of chromium(VI) that is adsorbed on the adsorbent by the esterification reaction as described earlier is reduced to chromium(III) by the electron rich polyphenolic aromatic ring that supplies electrons for the reduction reaction as described by the following reactions:

$$CrO_4^{2-} + 8H^+ + 3e^- \rightarrow Cr^{3+} + 4H_2O \tag{1}$$

After reduction, the adsorbed chromium is released into the aqueous solution in the form of chromium(III) ions, thus increasing the concentration of chromium(III) in the solution with contact time.

Isotherm studies were carried out to investigate the effect of metal concentration on the adsorption of chromium(VI) on the CPW gel. Figure 11 shows the experimental isotherm plots for chromium(VI) adsorption by the CPW gel at different pH values ranging from 1 up to 4. From the result in this figure, the adsorption of chromium(VI) at all the pH values tested appears to take place according to the Langmuir model, that is, the amount of adsorption increases with increasing chromium(VI) concentration at low concentration while it tends to approach constant values corresponding to each pH at high concentration in the aqueous solution.

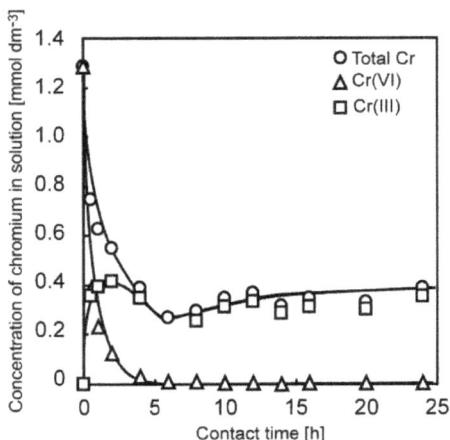

Figure 10. Time variation of the concentrations of chromium(VI) and chromium(III) as well as total chromium in the aqueous solution at pH = 1 during adsorption [23] (with permission for reuse from Elsevier B.V.).

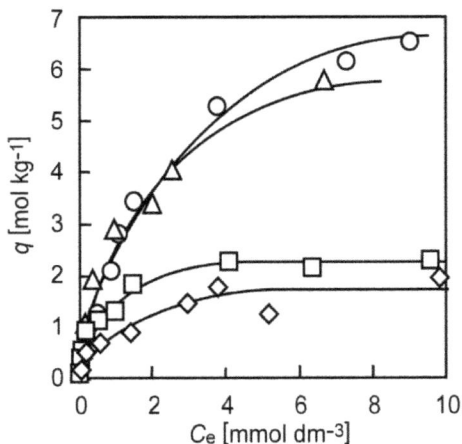

Figure 11. Adsorption isotherms of chromium(VI) on CPW gel [23] (with permission for reuse from Elsevier B.V.). ○ pH 1, △ pH 2, □ pH 3, ◊ pH 4.

From the Langmuir plots of these experimental results, the maximum adsorption capacities of chromium(VI) on the CPW gel were evaluated as 7.18, 6.11, 2.38 and 1.96 $mol \cdot kg^{-1}$ at pH = 1, 2, 3 and 4, respectively. This result demonstrates the increasing maximum adsorption capacity with decreasing pH, indicating that chromium(VI) removal by the CPW gel is greatly influenced by hydrogen ion.

Table 2 shows the maximum adsorption capacities reported for various adsorbents for comparison, which shows that the polyphenol-containing adsorbents such as mimosa tannin (condensed tannin) [4] and persimmon tannin [26] have higher adsorption capacities than other biomasses adsorbents such as grape gel [25], chitosan [27,23] and sugarcane bagasse [29]. It is noteworthy that the CPW gel containing polyphenol groups investigated in this study has the highest adsorption capacity among these, and it is also much higher than those of other tannin-containing gels.

Table 2. The maximum adsorption capacity for chromium ion(VI) for different adsorbents.

Adsorbent	pH	Temperature [K]	Maximum Adsorption Capacity (mol kg^{-1})	Reference
Mimoza tannin	2.0	303	5.52	[4]
Persimmon tannin	3.0	303	5.27	[26]
Quaternary chitosan	4.5	298	0.58	[27]
Sugarcane bagasse	3.0	298	1.97	[29]
Ocimun americanum	1.5	300	1.60	[30]
Crosslinked grape gel	4	303	1.91	[25]
Carbonized wheat straw	2	303	1.67	[31]
Carbonized barley straw	3	303	1.68	[31]
CPW gel	1	303	7.18	[23]

Similar to the cases of other metals, mutual separation of chromium(VI) from other metals such as zinc(II), for example, was tested on the basis of the results of the batch wise experiments mentioned earlier.

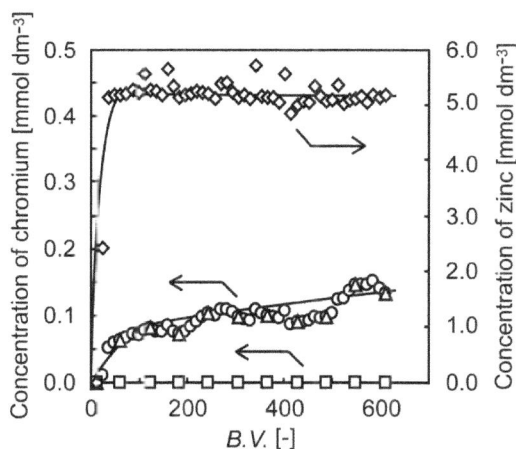

Figure 12. Breakthrough profiles of various chromium and zinc species from the column packed with CPW gel at pH = 1.0 [23] (with permission for reuse from Elsevier B.V.). ○: Total Cr △: Cr(III) □: Cr(VI) ◇: Zn(II).

Figures 12 and 13 show the breakthrough profiles of zinc(II), total chromium and chromium(VI) at pH = 1 and 4, respectively, for a column packed with the CPW gel using an initial feed solution containing 0.5 mmol·dm^{-3} chromium(VI) and 5 mmol·dm^{-3} zinc(II). From these figures, it is seen that breakthrough of zinc(II) took place just after the initiation of the feed whereas that of chromium(VI) occurred later, suggesting that mutual separation between zinc(II) and chromium(VI) is feasible using the CPW gel. The breakthrough of zinc(II) ion just after the start of feeding is due to the non-adsorbing nature of this ion at this pH on the CPW gel. Although some amount of total chromium was found to have leaked from the column from the start of the feeding at pH = 1, no chromium(VI) was detected in the outlet solution; *i.e.*, the total chromium detected in the outlet solution is concluded to be chromium(III) reduced from chromium(VI) in the feed solution by the aid of the CPW gel as described earlier. Consequently, it is apparent that small or trace amounts of chromium(VI) can be completely removed from an excess concentration of zinc(II) by using the column packed with the CPW gel, indicating that it can be applicable for the treatment of effluent generating from chromium plating industries which contain high concentration of zinc(II) but low concentration of chromium(VI).

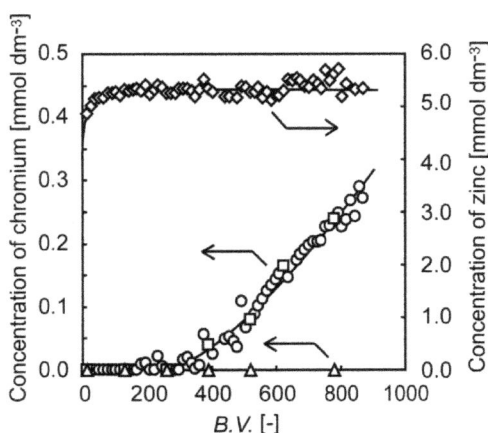

Figure 13. Breakthrough profiles of various chromium and zinc species from the column packed with CPW gel at pH = 4.0 [23] (with permission for reuse from Elsevier B.V.). ○: Total Cr △: Cr(III) □: Cr(VI) ◇: Zn(II).

5. Recovery of Gold from Acidic Chloride Media by Means of Coagulation Using Persimmon Tannin Extract Liquor

Nowadays, precious metals such as gold, silver and platinum group metals (PGM) are in extensive use not only as traditional jewelry materials but also as useful components in a variety of well-known advanced applications such as electric and electronic devices, catalysts, and medical instruments. Since these metals are limited resources existing only in small amounts on the earth, they should be effectively recovered from various wastes for recycling and reuse purposes. From an economical point of view, the recovery process should be such that the precious metals are highly selectively separated from base metals such as iron, copper, and zinc which often coexist with the precious metals in disproportionate amounts.

Due to the high price of precious metals, they have been recovered from various wastes for many years. According to the classical refining method of precious metals, feed materials including various wastes are leached using aqua regia, where gold, platinum and palladium are dissolved leaving other precious metals like silver as well as base metals in leach residue. From the leach liquor, gold is recovered at first as gold sponge by means of the reduction using various reducing agents like ferrous sulfate, followed by the recovery of platinum and palladium using ammonium chloride. However, because such conventional methods are tedious and require a significant amount of energy as well as long processing times, it is currently being replaced by new methods which involve the complete dissolution of feed materials using chlorine containing hydrochloric acid except for silver, followed by solvent extraction and ion exchange or distillation for selective separation into high purity of each precious metal. In this dissolution process, chlorine is dissociated into hydrochloric acid and hypochlorous acid which functions as oxidation agent and is converted into hydrochloric acid. Consequently, the mutual separation among precious metals by means of solvent extraction and ion exchange is carried out from hydrochloric acid. Solvent extraction reagents as well as diluents and ion exchange resins, synthetic organic materials, are not environmentally benign taking account of their dissolution in water and the treatments after their use. On such viewpoints, we attempted to employ persimmon tannin extract (kakishibu), a natural material, as a coagulating agent for the refining of gold while, as mentioned earlier, persimmon tannin extract has been employed as a coagulating agent for proteins for a long time [32]. In the present work, persimmon tannin extract liquid was kindly donated by Tomiyama Corp., Kyoto, Japan, and employed by diluting with water into 10 (V/V) % aqueous solution.

As shown in Figure 14, the minor addition of the persimmon tannin extract into hydrochloric acid solution containing gold(III) ions caused an observable color change of the solution, which gradually became turbid. In the course of time, such dark turbid materials were coagulated and sank down on the bottom of vessel. After the filtration of the coagulated materials, the filter cake was observed by using an optical microscope and by means of X-ray diffraction (XRD) analysis.

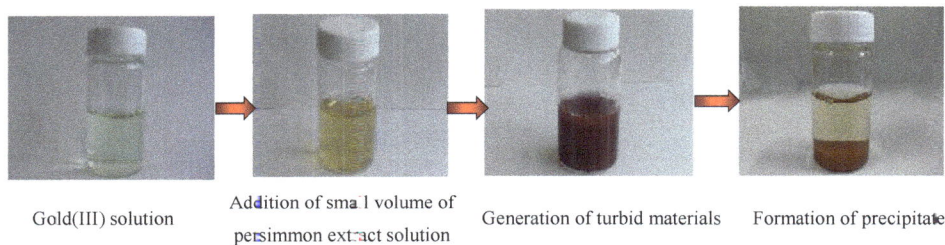

Gold(III) solution   Addition of small volume of persimmon extract solution   Generation of turbid materials   Formation of precipitate

Figure 14. Change in the hydrochloric acid solution containing small amount of gold(III) after the addition of trace volume of persimmon extract liquid [32] (with permission for reuse from The Society of Chemical Engineers of Japan).

Figure 15. Image of the filter cake of the precipitate generated by the addition of persimmon extract liquor into the gold(III) solution observed by optical microscope (×800).

Figure 15 shows the image of the filter cake observed by an optical microscope. Aggregates of brilliant yellow fine particles are observed in the filter cake.

From the observation by the X-ray diffraction pattern of the filter cake, 4 sharp peaks were confirmed at 2θ = 38.2, 44.4, 64.6 and 77.5 degree, surely suggesting the existence of metallic gold.

From these observations, it can be concluded that gold(III) ion was reduced into elemental gold (gold(0)) and coagulated into the aggregates by the aid of persimmon tannin extract.

Similar coagulation tests were carried out for hydrochloric acid solutions containing other metal ions. Figure 16 shows the plots of % recovery as coagulated precipitates against hydrochloric acid concentration for the individual solutions of platinum(IV), palladium(II), copper(II), zinc(II), and iron(III) as well as gold(III).

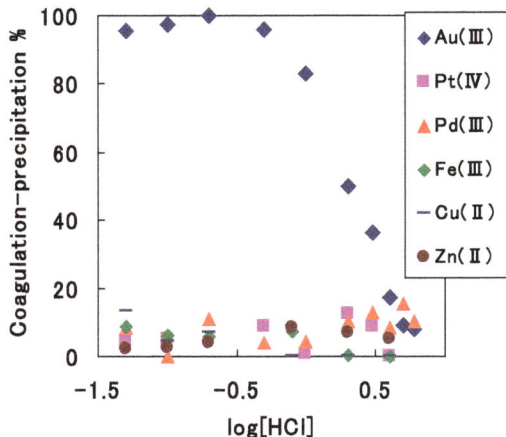

Figure 16. Percentage recovery of various metal ions by means of coagulation using persimmon extract liquor from varying concentration (mol·dm^{-3}) of hydrochloric acid solution [32] (with permission for reuse from The Society of Chemical Engineers of Japan).

As seen from this figure, the recovery of other metal ions is negligible, suggesting that only gold is much selectively recovered by this method using persimmon tannin extract. The high selectivity to gold is attributable to the higher oxidation reduction potential (ORP) of gold(III) ion than other metal ions.

The reduction of gold(III) ions to metallic gold (gold(0)) is inferred to take place according to the following reactions where P stands for the polymer matrices of persimmon tannin.

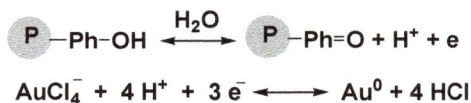

$$\text{P}-\text{Ph}-\text{OH} \;\underset{}{\overset{H_2O}{\rightleftharpoons}}\; \text{P}-\text{Ph}{=}\text{O} + \text{H}^+ + \text{e}$$

$$\text{AuCl}_4^- + 4\,\text{H}^+ + 3\,\text{e}^- \rightleftharpoons \text{Au}^0 + 4\,\text{HCl}$$

Figure 17 shows the time variation of the % recovery of gold from 0.1 mol·dm^{-3} hydrochloric acid solution at varying temperatures. As seen from this figure, although it takes very long time to reach the quantitative recovery at 293 and 303 K, such kinetic behavior of the gold recovery can be much improved by elevating temperature.

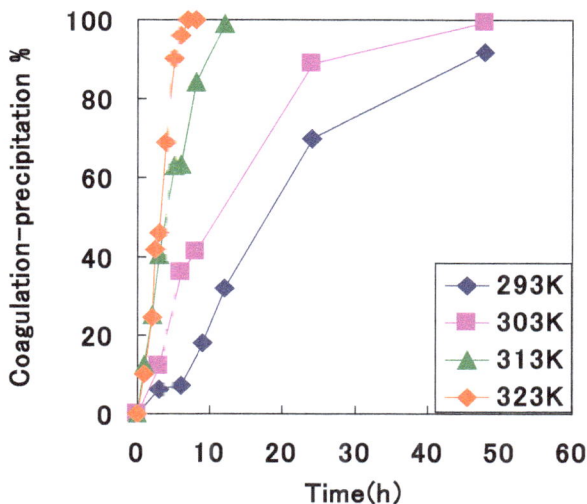

Figure 17. Percentage recovery of gold(III) by means of coagulation using persimmon extract liquor from 0.1 mol·dm^{-3} hydrochloric acid solution at varying temperatures [32] (with permission for reuse from The Society of Chemical Engineers of Japan).

6. Adsorptive Recovery of Gold from Acidic Chloride Media Using Gels of Dried Persimmon Tannin Extract and Waste of Persimmon Peel

As mentioned in the preceding section, the liquid of persimmon tannin extract exhibits a strong reduction behavior for gold. On the basis of such observation, we carried out the adsorptive recovery of gold using a CPT gel [33] which was employed for the adsorptive removal of cesium(I) as described in the former section and also using the gel prepared from persimmon peel waste generated in the production of dried persimmon fruit still containing large quantities of persimmon tannin [34].

Figures 18 and 19 show the % adsorption of gold(III), platinum(IV) and palladium(II) as well as some base metals such as copper(II) from varying concentration of hydrochloric acid solution on CPT gel and crude persimmon tannin extract (PT) powder for comparison, respectively.

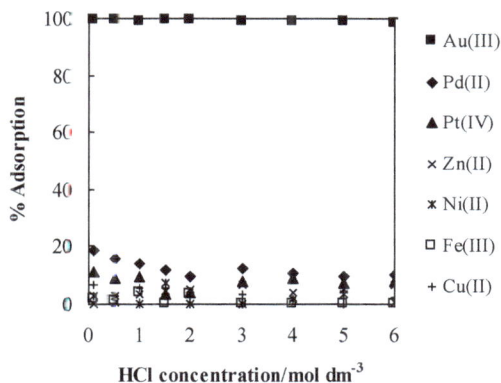

Figure 18. Percentage adsorption of various metal ions on CPT gel as a function of concentration of hydrochloric acid [33] (with permission for reuse from Elsevier B.V.).

Quantitative adsorption of gold(III) was achieved on CPT gel over the whole concentration range of hydrochloric acid while the adsorption of gold(III) on crude PT powder decreased significantly in higher acid concentration regions (greater than 3.0 mol·dm^{-3}). Although both adsorbents exhibited about 10%–20% adsorption for platinum(IV) and palladium(II), the adsorption of base metals was practically negligible under the present experimental conditions. The remarkably high selectivity of CPT gel for gold(III) over base metals as well as platinum(IV) and palladium(II) in a wide concentration range of hydrochloric acid is immensely useful for selective and quantitative recovery of gold(III) from acidic chloride media.

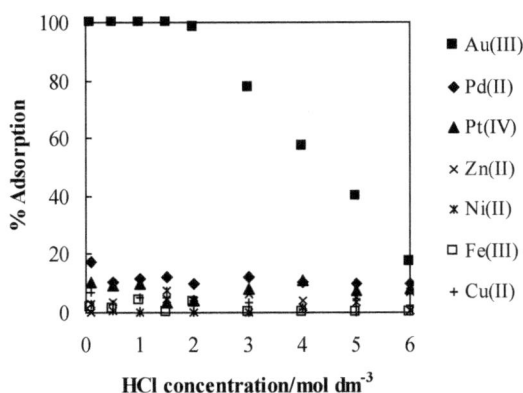

Figure 19. Percentage adsorption of various metal ions on crude PT powder as a function of concentration of hydrochloric acid [33] (with permission for reuse from Elsevier B.V.).

Figure 20 shows the adsorption isotherms of gold(III) on CPT gel and crude PT powder. As seen from this figure, it is clear that the amount of gold(III) adsorption on both adsorbents increased with increasing metal concentration of the test solution and tended to approach the constant value. That is, the gold(III) adsorption capacity of both adsorbents tended to approach a constant value at around 4.5 mol·kg^{-1}. Interestingly, as can be seen in this figure, gold(III) adsorption capacity of both adsorbents again increased, after tending to approach the constant value, with further increase in gold(III) concentration in the test solution, resulting a typical BET type adsorption isotherm based on multilayer adsorption model. Consequently, the gold uptake capacity of CPT gel reached as high as 7.7 mol·kg^{-1} (=1.52 kg-gold·kg^{-1}-dry adsorbent) and that of crude PT powder reached 5.8 mol·kg^{-1} (=1.14 kg-gold·kg^{-1}-dry adsorbent), respectively, suggesting that greater quantities of gold(III) relative to the dry weight of these adsorbents were adsorbed. From this result, it is clear that CPT gel is more effective than the crude PT powder for uptake of gold(III) from hydrochloric acid solutions. The higher uptake of CPT gel than crude PT powder towards gold(III) may be attributable to the improvement of structure or surface morphology of the polymer matrices of the gel through crosslinking related to the adsorption coupled reduction mechanism. That is, it was found in our previous work that polysaccharides such as cellulose which have no functional groups for binding metal ions exhibit very high affinity only for gold to give rise to metallic gold particles by special reduction reaction after the crosslinking using concentrated sulfuric acid [2]. In the case of CPT gel, in addition to the reductive adsorption of persimmon tannin itself, the crosslinked polysaccharides coexisting in CPT gel may exhibit the additional reductive adsorption for gold(III).

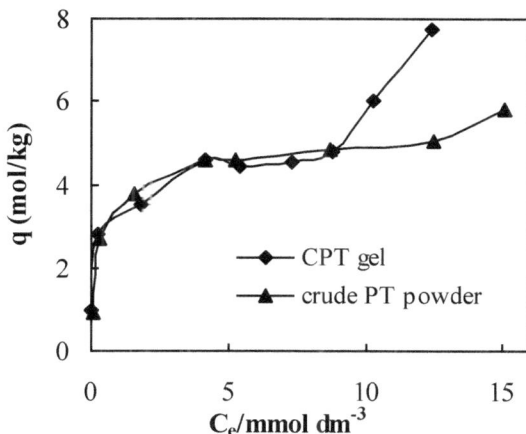

Figure 20. Adsorption isotherms of gold(III) on CPT gel and crude persimmon extract tannin (PT) powder from 0.1 mol·dm^{-3} hydrochloric acid solution [33] (with permission for reuse from Elsevier B.V.).

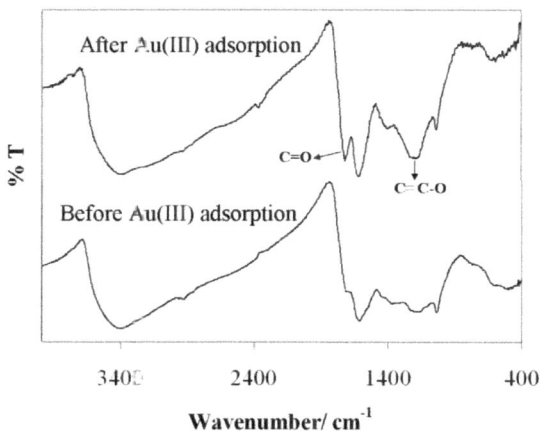

Figure 21. FT-IR spectra of CPT gel before and after the adsorption of gold(III) [33] (with permission for reuse from Elsevier B.V.).

The above-mentioned mechanism of adsorption of gold(III) followed by reduction to elemental gold was supported by the observation of FT-IR spectra before and after the adsorption of Au(III) on CPT gel and crude PT powder as presented in Figures 21 and 22, respectively. In the FT-IR spectrum after gold adsorption, the intensity of the band at 1709 cm^{-1} assigned for quinine type C=O stretching was increased. It is interesting to note that the band at 1185 cm^{-1} attributable to C=C–O asymmetrical stretching was also increased in intensity. These results support the fact that oxidation of phenolic hydroxyl groups has actually taken place as mentioned earlier during the adsorption of gold(III) on phenol-rich tannin matrix.

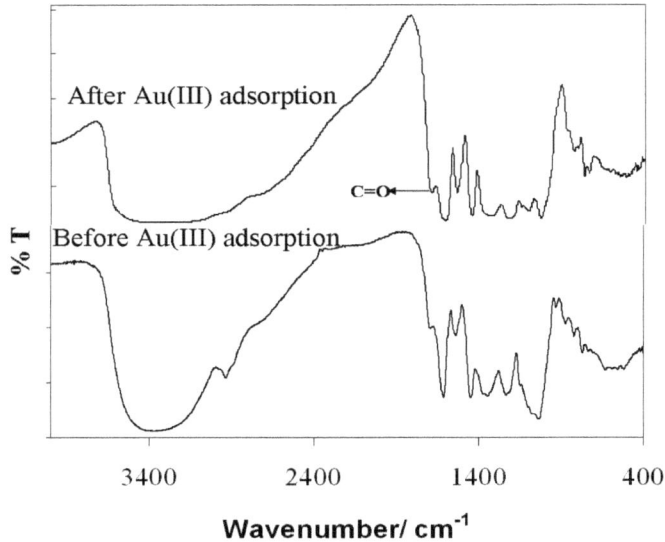

Figure 22. FT-IR spectra of crude PT powder before and after the adsorption of gold(III) [33] (with permission for reuse from Elsevier B.V.).

Once the adsorption occurs, reduction of gold(III) to elemental gold also takes place. The elemental gold is aggregated and released from the surface of the adsorbents creating active vacant sites for further adsorption, leading to an apparent uptake of higher amounts of gold(III). Consequently, it is not unreasonable to consider that the unique BET type of adsorption behaviors of CPT gel and crude PT powder towards gold(III) may be attributable to the adsorption-reduction cycles as such.

The details of the adsorption kinetics of gold(III) on both the adsorbents were studied by varying the solution temperature from 293 to 323 K and the results are presented in Figure 23 for CPT gel as an example. Although similar kinetic behavior was observed for crude PT powder, it was relatively faster than that observed for the CPT gel at the initial stage of adsorption. However, it also required longer times to reach equilibrium. This result is attributable to the high aqueous solubility of crude PT powder since tannin molecules of low molecular weight existing in crude PT powder are dissolved in the solution at the initial stage of the contact. Because the interaction of these dissolved molecules with gold(III) ion in solution is the homogeneous reaction, it takes place more rapidly than does the heterogeneous reaction taking place in the case of gold(III) adsorption on solid surface of CPT gel particles.

The adsorption rates for both adsorbents at different temperatures were analyzed in terms of the pseudo-first-order kinetic model according to the following equation:

$$\ln\left(1 - \frac{q_t}{q_e}\right) = -k_1 t \tag{2}$$

where, q_e and q_t are the amount of adsorbed metal (mol·kg^{-1}) at equilibrium and any time, respectively, and k_1 (h^{-1}) is the pseudo-first-order rate constant. The fitting of this rate expression was checked by linear plot of $\ln\{1-(q_t/q_e)\}$ *vs.* t at four different temperatures. It was confirmed that the plots were lying on proportional straight lines with the correlation coefficient greater than 0.96.

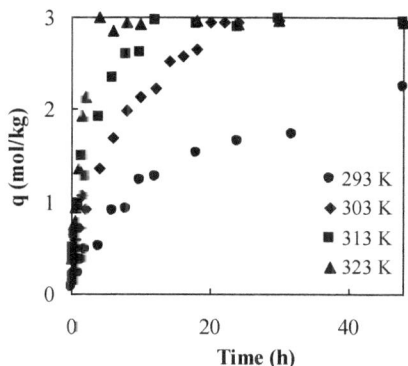

Figure 23. Time variation of the adsorption of gold(III) on CPT gel at different temperatures [33] (with permission for reuse from Elsevier B.V.).

The relationships between the evaluated pseudo-first-order rate constant and temperature were rearranged according to the Arrehnius equation as shown below:

$$k_1 = A\, e^{-Ea/RT} \tag{3}$$

where k_1 is rate constant, E_a is activation energy, T is absolute temperature, A is Arrehenius constant and R is gas constant.

The results are shown in Figure 24 for CPT gel as an example. From the slopes of the straight line in this figure, the apparent activation energy, E_a, was evaluated as 73 kJ/mol for CPT gel. Similarly, that for crude PT powder was evaluated as 54 kJ/mol.

Similar experimental work was carried out using the adsorption gel prepared by the same method from persimmon peel waste generated in the production of dried persimmon fruit [34]. Such waste still contains large quantities of persimmon tannin. This gel is abbreviated as PP (persimmon peel) gel, hereafter. Figure 25 shows % adsorption of various metal ions including precious metals such as gold(III), platinum(IV) and palladium(II) as well as base metals such as copper(II) and iron(III) on PP gel from varying concentration of hydrochloric acid containing individual metal ions. Similar to the adsorption on CPT gel, selective adsorption nearly only to gold(III) was observed also in this case.

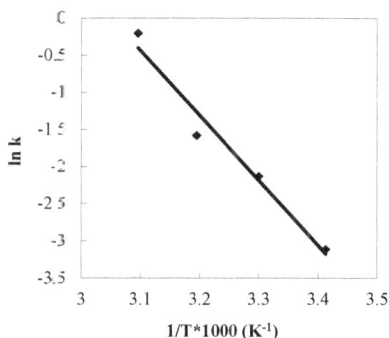

Figure 24. Arrhenius plots for the pseudo-first-order rate constants for CPT gel where the straight line is expressed by the linear equation, $Y = -8.57x + 26.7$, with the correlation factor $(R^2) = 0.9671$ [33] (with permission for reuse from Elsevier B.V.).

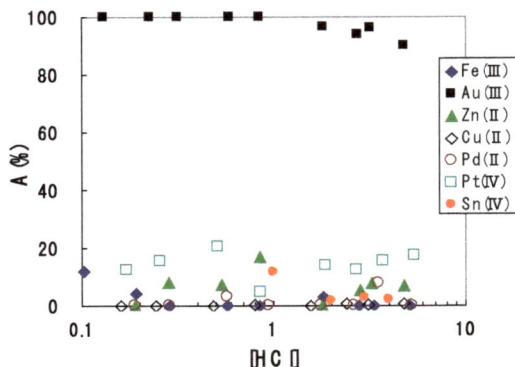

Figure 25. Percentage adsorption of some metal ions on PP gel from varying concentration (mol·dm^{-3}) of hydrochloric acid [34] (with permission for reuse from Elsevier B.V.).

Figure 26 shows the adsorption isotherm of gold(III) on PP gel from 1 mol·dm^{-3} hydrochloric acid solution as well as those on the adsorption gels prepared from peels of lemon and grape by the same method for comparison.

Similar to the case of CPT gel, typical BET type adsorption isotherms based on multilayer adsorption model were also observed for these adsorption gels. Extraordinary high amounts of gold(III) adsorption are noteworthy on the adsorbents of PP gel and lemon gel in particular; *i.e.*, the maximum adsorption amount observed for PP gel is as high as around 9 mol·kg^{-1} (=1.77 kg-gold·kg^{-1}-dry adsorbent), suggesting that considerably greater quantity of gold(III) was adsorbed also on this gel relative to the dry weight of the adsorbent.

Also in this case, it was confirmed by XRD analysis that the adsorbed gold(III) existed as metallic gold on the adsorbent similar to the case of CPT gel. Figure 27 shows the image of the filter cake observed by optical microscope, after the adsorption of gold(III) on PP gel followed by filtration. It is evident from this figure that the brilliant gold particles surely exist in the filter cake separated from the gel particles.

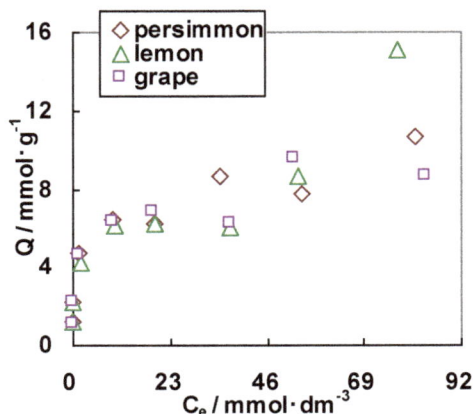

Figure 26. Adsorption isotherms of gold(III) from 1 mol·dm^{-3} hydrochloric acid solution on the adsorption gels prepared from peels of persimmon, lemon and grape treated in boiling concentrated sulfuric acid [34] (with permission for reuse from Elsevier B.V.).

Figure 27. Image of filter cake after the adsorption of gold(III) on PP gel observed by an optical microscope. Brilliant yellow is the particle of metallic gold while black materials are particles of PP gel.

It may be inferred that the surfaces of CPT and PP gels as well as PT powder function as catalysts for the reductive adsorption reaction of gold(III) as follows: (1) adsorption of gold(III) ion on the surface; (2) reduction of gold(III) ion into metallic gold (gold(0)); (3) release of nanoparticles of metallic gold from the surface and (4) formation of aggregates of nanoparticles of metallic gold. The extraordinary high selectivity and high adsorption capacity for gold(III) may be attributable to this mechanism. In the step (1), the adsorption takes place by the interaction of anionic tetrachloro-complex of gold(III), $AuCl_4{}^-$, with phenolic hydroxyl groups of persimmon tannin molecules, which is followed by the reduction of the step (2). Here, the phenolic hydroxyl groups are oxidized into quinone groups, which are again converted into phenolic hydroxyl groups by the aid of hydrogen ions and again take place the adsorption followed by the reduction reaction. Consequently, such reductive adsorption takes place only under acidic conditions.

7. Chemical Modification of Persimmon Tannin Extract and Persimmon Waste and the Adsorption Behaviors of the Modified Gels

Although, as mentioned earlier, CPT and PP gels were found to exhibit extraordinary high selectivity for gold(III), we further attempted to enhance the affinity also for platinum(IV) and palladium [35–40]. It is easy to chemically modify persimmon tannin extract and persimmon waste including persimmon peel by immobilizing a variety of functional groups which exhibit special affinities for some metal ions onto their surface similar to other biomass wastes [41,42]. In our research works, persimmon waste was chemically modified by immobilizing functional group of dimethylamine (DMA) to prepare the adsorption gel functioning as a typical weak base type of anion exchange material and investigated its adsorption behavior for some precious and base metals from hydrochloric acid solutions [35,36]. Further, a variety of adsorption gels were prepared from persimmon tannin extract powder by the chemical modifications immobilizing other functional groups such as tertiary amine where the immobilized tertiary amine groups function as quaternary ammonium compounds (QA) [37], a typical strong base, as well as tetraethylenepentamine (TEPA) [38], glycidyltrimethyl ammonium (GTA) [40], aminoguanidine (AG) [40], and bisthiourea (BTU) [41]. These are abbreviated as DMA-PW gel and QAPT, BTU-PT, AG-PT, TEPA-PT and GTA-PT gels, respectively, hereafter.

DMA-PW gel was prepared according to Scheme 3. Powder of PW, the same feed material employed for the adsorptive removal of chromium(VI) mentioned earlier, was interacted with para- formaldehyde, 1,4-dioxane, acetic acid, phosphoric acid and hydrochloric acid to prepare the chloromethylated intermediate product. Here, the polymer matrices of PT are not only chloromethylated but also crosslinked by the aid of paraformaldehyde under acidic condition. It was then interacted with dimethylamine to prepare the final product. The abundance of dimethylamine groups in the DMA-PW gel thus prepared was evaluated as 2.09 mol·kg^{-1} dry gel.

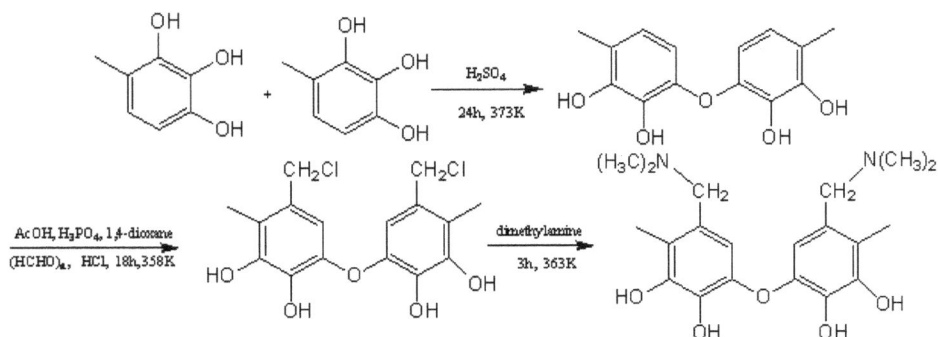

Scheme 3. Synthetic route of dimethylamine modified persimmon waste (DMA-PW) gel.

Figure 28 shows % adsorption of various metal ions on DMA-PW gel from varying concentration of hydrochloric acid solution. Although only gold(III) is selectively adsorbed on CPT and PP gels to any considerable extent as mentioned earlier, adsorption is observed not only for gold(III) but also for platinum(IV) and palladium(II) while no affinity was observed for base metals such as copper(II), iron(III), nickel(II), and zinc(II) in the case of DMA-PW. This result indicates that the precious metals can be selectively recovered from any other coexisting base metal ions in hydrochloric acid medium using the DMA-PW gel. Additionally, the % adsorption of the precious metals decreased with increasing hydrochloric acid concentration though nearly quantitative adsorption was observed for gold(III) over the whole concentration region of hydrochloric acid similar to the cases of CPT and PP gels.

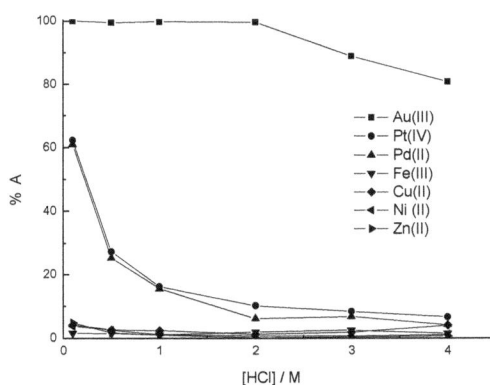

Figure 28. Percentage adsorption of various metal ions on DMA-PW gel from varying concentration (mol·dm^{-3}) of hydrochloric acid solution [36] (with permission for reuse from Elsevier B.V.).

In acidic chloride media, tertiary amine group of DMA-PW gel is protonated as follows:

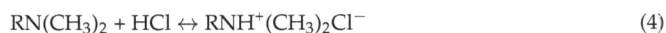

$$RN(CH_3)_2 + HCl \leftrightarrow RNH^+(CH_3)_2Cl^- \tag{4}$$

The protonated DMA-PW gel is interacted with anionic chloride complexes of gold(III), palladium(II) and platinum(IV), $AuCl_4^-$, $PtCl_6^{2-}$ and $PdCl_4^{2-}$, to form ion pair complexes on the gel surface as follows.

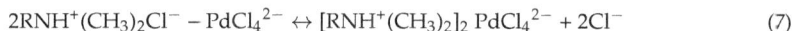

$$RNH^+(CH_3)_2Cl^- + AuCl_4^- \leftrightarrow [RNH^+(CH_3)_2]AuCl_4^- + Cl^- \tag{5}$$

$$2RNH^+(CH_3)_2Cl^- - PtCl_6^{2-} \leftrightarrow [RNH^+(CH_3)_2]_2\,PtCl_6^{2-} + 2Cl^- \tag{6}$$

$$2RNH^+(CH_3)_2Cl^- - PdCl_4^{2-} \leftrightarrow [RNH^+(CH_3)_2]_2\,PdCl_4^{2-} + 2Cl^- \tag{7}$$

The decrease in the amount of adsorption of these metal ions with increasing hydrochloric acid concentration as shown in Figure 28 can be reasonably interpreted by the above-described adsorption reaction equations. However, the adsorbed gold(III) ion is further reduced by the aid of the functional groups of polyphenols of persimmon tannin into metallic gold similar to the cases of CPT and PP gels.

Figure 29 shows the time variations of amount of the adsorption of gold(III) on CPW and DMA-PW gels for comparison, which clearly indicates that equilibrium is attained within 5 h in the case of DMA-PW whereas it takes much longer in the case of the CPW gel; that is, as a consequence of the chemical modification immobilizing the dimethylamine functional groups, the adsorption kinetics of gold(III) was significantly enhanced by the interactions between the protonated amine groups and anionic chloro-complex of gold(III).

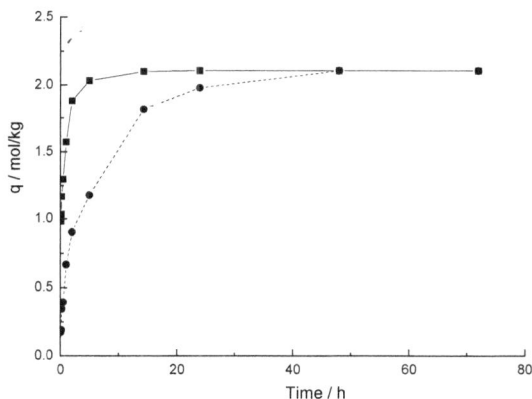

Figure 29. Time variation of the amount of gold(III) on CPW (●)gel and DMA-PW (■)gel from $0.1\ mol\cdot dm^{-3}$ hydrochloric acid solution [36] (with permission for reuse from Elsevier B.V.).

From the studies of adsorption isotherms of gold(III), platinum(IV) and palladium(II) on DMA-PW gel, it was found that all of these are satisfactorily fitted with the Langmuir's adsorption isotherm and the maximum adsorption capacities were evaluated as 5.63, 0.26 and 0.42 $mol\cdot kg^{-1}$, respectively, while that of gold(III) on CPW gel was evaluated as 4.94 $mol\cdot kg^{-1}$. Compared with the adsorption of platinum(IV) and palladium(II), the extraordinary higher adsorption capacities of gold(III) are noteworthy, which is attributable to the reduction of the adsorbed gold(III) ion into metallic gold (gold(0)) by the polyphenolic functional groups of persimmon tannin as mentioned earlier. In addition, the adsorption capacity of gold(III) itself is also enhanced by this chemical modification.

On the basis of the above-mentioned results of the batch wise experimental works, an adsorption tests was conducted for the recovery of precious metals from actual industrial effluent sample, which is an acidic chloride solution dissolving various metal scraps generated in a precious metals industry. Since the CPW gel has exhibited high selectivity and capacity only for gold, while the DMA-PW has exhibited a high selectivity for palladium and platinum, the flow experiment was conducted using a column packed with PW gel at first for the separation of gold(III) from platinum(IV) and palladium(II), connected in series with the second column packed with DMA-PW gel for recovering platinum(IV) and palladium(II). Here, the acid concentration was around 1 $mol \cdot dm^{-3}$ and the concentrations of precious metals and some base metals in the actual test solution were as follows (in $mg \cdot dm^{-3}$): Cu 3360, Zn 1040, Fe 760, Au 100, Pd 6.3, and Pt 4.1.

Figures 30a and 31a show the breakthrough profiles from the first column packed with the CPW gel and from the second column packed with DMA-PW gel, respectively. As seen from these figures, gold(III) is selectively separated from platinum(IV), palladium(II) and base metals at the first column while platinum(IV) and palladium(II) are separated from all base metals at the second column. All precious metals adsorbed in these columns were effectively eluted using a mixture of 0.1 $mol \cdot dm^{-3}$ thiourea in 1 $mol \cdot dm^{-3}$ hydrochloric acid solution as shown in Figures 30b and 31b. The % recovery of gold(III), platinum(IV) and palladium(II) were 94.2%, 92.9% and 96.1%, respectively. Almost complete recovery of the loaded precious metal ions including the metallic gold from the loaded gel using acidic thiourea solution verifies the easy regeneration of the gel for repeated use. These findings are highly encouraging in terms of consistent recovery of precious metals from actual complex mixtures.

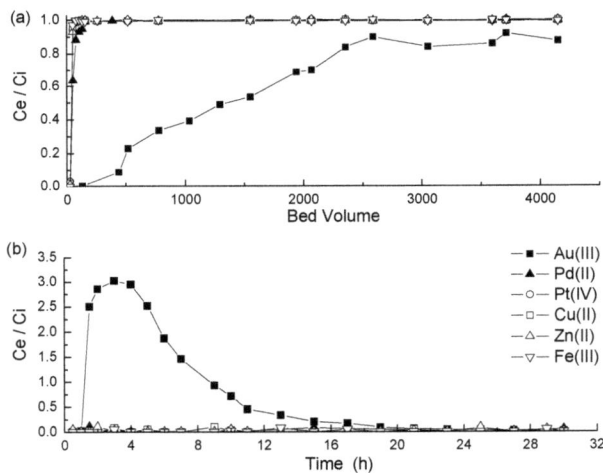

Figure 30. (**a**) Breakthrough profiles of various metal ions contained in the sample solution of the actual effluent of the precious metals industry from the first column column packed with CPW gel; (**b**) Elution profiles of various metal ions from the first column using aqueous mixture of 0.1 $mol \cdot dm^{-3}$ thiourea in 1 $mol \cdot dm^{-3}$ hydrochloric acid solution [36] (with permission for reuse from Elsevier B.V.).

The chemical structures of QAPT, BTU-PT, AG-PT, TEPA-PT and GTA-PT gels are shown in Scheme 4.

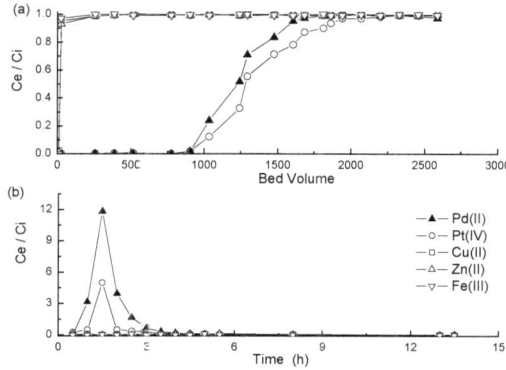

Figure 31. (a) Breakthrough profiles of various metal ions from the second column packed with DMA-PW gel; (b) Elution profiles of various metal ions from the second column using aqueous mixture of 0.1 mol·dm^{-3} thiourea in 1 mol·dm^{-3} hydrochloric acid solution [36] (with permission for reuse from Elsevier B.V.).

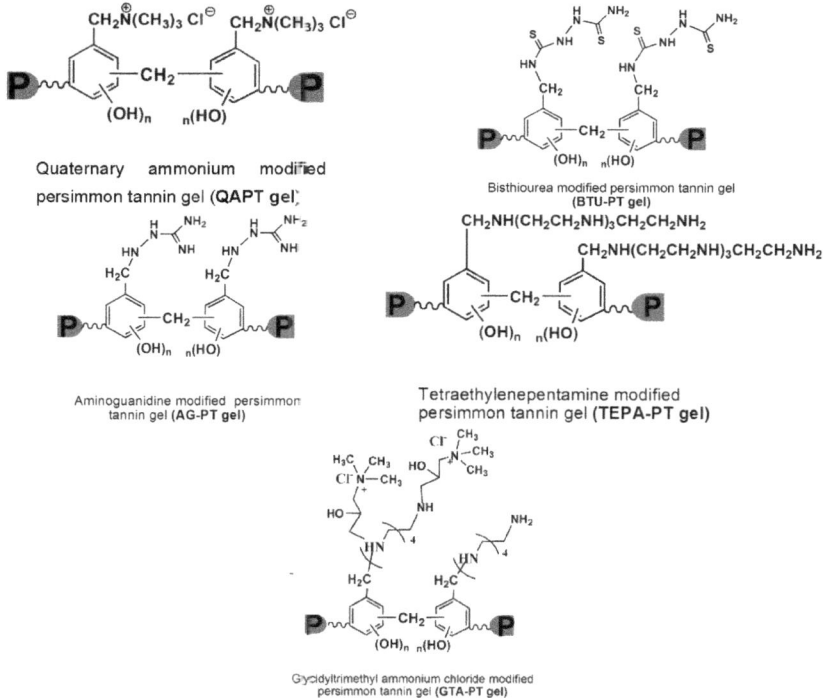

Scheme 4. Chemical structures of the gels prepared from persimmon tannin extract powder via chemical modification immobilizing functional groups of quaternary ammonium (QA), bisthiourea (BTU), aminoguanidine (AG), tetraethylenepentamine (TEPA) and glycidyltrimethyl ammonium (GTA) where P stands for persimmon tannin moiety.

Figure 32 shows the % adsorption of various metal ions at varying concentrations of hydrochloric acid on these chemically modified gels for comparison. As seen from this figure, although platinum(IV), palladium(II) and gold(III) in particular are adsorbed on all of these modified gels, only negligible adsorption takes place for the base metals. Among these gels, it is noticeable that QAPT gel exhibits preferable adsorption behavior compared with other modified gels not only for gold(III) but also for platinum(IV) and palladium(II) over the relatively wide concentration range of hydrochloric acid. That is, the adsorption of gold(III) on QAPT gel is nearly quantitative over the whole concentration region of hydrochloric acid. Additionally, nearly quantitative adsorption of palladium(II) and platinum(IV) was achieved in the low concentration range (0.1–1.5 mol·dm^{-3}) of hydrochloric acid. From the comparison of this figure with Figure 28, it is evident that the adsorption of palladium(II) and platinum(IV) on QAPT gel is much higher than that on DMA-PW gel. Both of tertiary amine functional groups of DMA-PW gel protonated in acidic solutions and quaternary ammonium functional groups of QAPT containing positive charges regardless of pH interact with negatively charged chloro-complexes of precious metals by electrostatic interactions. However, because the basicity of QAPT gel is stronger than that of MDA-PW gel, the interaction by the former is stronger than that by the latter. Furthermore, because precious metals such as gold(III), platinum(IV) and palladium(II) give rise to much more stable chloro-complexes with chloride ion, much more negatively charged complexes, than base metals such as iron(III), zinc(II) and copper(II), these precious metals are much more preferentially adsorbed on these modified gels, which may be inferred to be the origin of the high selectivity to precious metals over base metals.

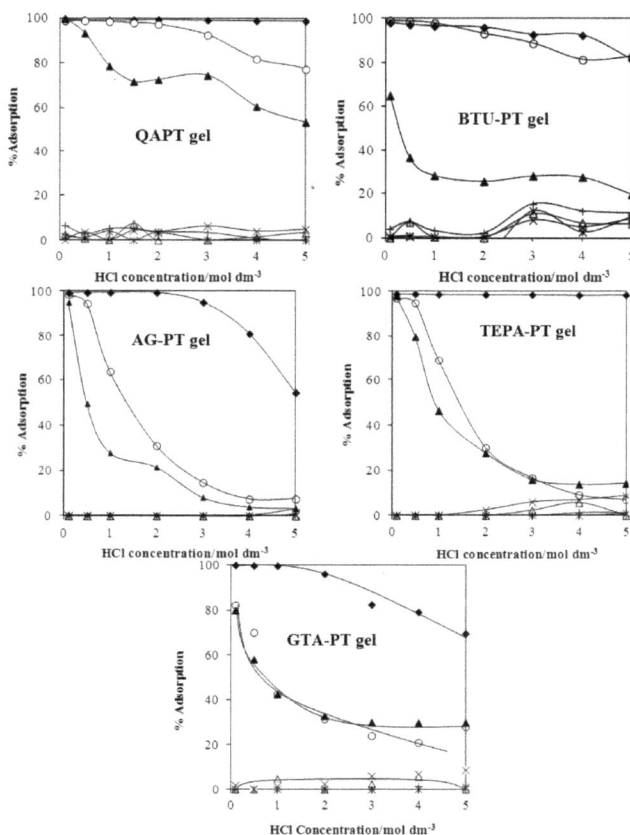

Figure 32. Percentage adsorption of various metal ions on some chemically modified PT gel from varying concentration of hydrochloric acid. ◆ Au(III), ○ Pd(II), ▲ Pt(IV), | Cu(II), × Zn(II), △ Ni(II), * Fe(III).

The results acquired from the adsorption isotherm studies of gold(III), palladium(II), and platinum(IV) on these modified gels suggested that they were adequately fitted with the Langmuir type adsorption isotherms. The maximum adsorption capacities of these precious metals on these modified gels as well as the DMA-PW gel evaluated according to the Langmuir's equation are listed in Table 3 together with those on other adsorbents for comparison.

Table 3. The maximum adsorption capacities of gold(III), platinum(IV) and palladium(II) on various adsorbents.

Adsorbent	Maximum Adsorption Capacity (mol·kg⁻¹)			HCl Concentration (mol·dm⁻³) or pH	Reference
	Au(III)	Pt(IV)	Pd(II)		
Persimmon extract tannin (PT) powder	5.39	-	-	0.1	[33]
Crosslinked persimmon tannin (CPT)	7.7	-	-	0.1	[33]
Crosslinked persimmon waste (CPW)	4.95	-	-	0.1	[35,36]
Dimethylamine modified persimmon waste (DMA-PW)	5.53	0.28	0.42	0.1	[35,36]
Quaternary ammonium modified persimmon tannin (QAPT)	4.16	0.52	0.84	0.1	[37]
Tetraethylenepentamine- modified persimmon tannin (TEPA-PT)	5.93	1.48	1.76	0.1	[38]
Glycidyltrimethyl ammonium chloride modified persimmon tannin (GTA-PT)	3.30	1.00	1.67	0.1	[39]
Aminoguanidine modified persimmon tannin (AG-PT)	8.90	1.00	2.00	0.1	[40]
Bisthiourea modified persimmon tannin (BTU-PT)	5.22	0.70	1.72	0.1	[41]
Quaternary ammonium modified microalgal residue		1.33	2.95	0.1	[43]
Crosslinked microalgal residue	3.25	0.15	0.25	0.1	[44]
Crosslinked chitosan		1.6	2.1	0.01	[1]
Collagen fiber immobilized bayberry tannin		0.495	0.80	pH = 5.6	[45]
Lysine modified cross-linked chitosan	0.55	0.66	1.03	pH = 1 for Pt; pH = 2 for Pd, and Au	[46]
Glycine modified cross-linked chitosan	0.86	0.62	1.13	pH = 2	[47]
PEI-modified corynebacterium glutamicum		1.66		0.1	[48]
Ethylenediamine modified chitosan nanoparticle		0.87	1.30	pH = 2	[49]
Thiourea-modified chitosan microsphere		0.66	1.06	pH = 2	[50]
Duolite GT-73	0.58	0.26		pH = 2	[51]
Polyallylamine modified *Escherichia coli* biomass			2.50	pH = 3	[52]
Dimethylamine modified waste paper	4.50	0.90	2.10	1	[42]

Among these modified persimmon tannin gels, the high adsorption capacities of TEPA-PT, GTA-PT, AG-PT and BTU-PT not only for gold(III) but also for platinum(IV) and palladium(II) are noteworthy though the concentration range of hydrochloric acid effective for the high adsorption is limited. Also for these modified persimmon tannin gels, the exceptionally high adsorption capacities for gold(III) are attributable to the combination of adsorption and reduction into metallic gold particles by the aid of polyphenolic groups of persimmon tannin, which was confirmed by the observations using optical microscope and XRD.

The adsorption kinetics of gold(III), palladium(II) and platinum(IV) on the QAPT gel were studied in detail and compared with that of gold(III) on crude PT powder. It was observed that equilibrium was attained within 2 h for palladium(II) and platinum(IV) for the adsorption on QAPT gel, whereas it was attained within 6 h for gold(III) adsorption. On the other hand, the crude PT powder requires 24 h to reach equilibrium of gold(III) adsorption under the same condition. The rapid kinetics of gold (III) adsorption with the QAPT gel is attributable to the effect of abundant quaternary ammonium groups.

which concentrate the reactant species, $AuCl_4^-$, on the gel surface by anion exchange adsorption, enhancing the reduction rate of gold(III).

8. Conclusions

Persimmon tannin extract rich in polyphenol compounds has been traditionally employed for various purposes such as tanning of leather, natural dyes and paints as well as coagulating agents for proteins in East Asian countries. By effectively using special characteristics of persimmon tannin extract and persimmon wastes which have not been utilized to date, new separation technologies were developed using adsorption gel prepared from these feed materials by a simple treatment in boiling sulfuric acid in the authors' previous research works. These are as follows.

(1) Adsorptive separation of trace concentration of uranium(VI) and thorium(III) from rare earths(III), which can be expected to be used for resolving the environmental problems of rare earth mines.

(2) Adsorptive removal of cesium(I) from other alkaline metals such as sodium, which can be expected to be useful for the remediation of water resources or wetlands that have been polluted by the accidents of atomic energy facilities or atomic energy wastes.

(3) Adsorptive removal of chromium(VI) from other metal ions such as zinc(II), which can be expected to be useful for the treatment of spent chromium plating solutions.

(4) Highly selective adsorptive recovery of gold(III) from acidic chloride media in the form of fine particles of metallic gold in which gold(III) ions are reduced to metallic gold by the catalytic reduction mechanism functioned by hydroxyl groups contained in these gels.

Additionally, such reduction reaction for gold(III) was also observed in the precipitation using persimmon extract liquid itself as a coagulating agent. These interesting behaviors exhibited by persimmon extract and persimmon wastes can be expected to be effectively used for the recovery of gold from various waste including e-wastes.

Furthermore, it was also found that these persimmon tannin extracts and persimmon wastes can be easily chemically modified by immobilizing a variety of functional groups such as quaternary ammonium compounds to improve the adsorption behaviors for platinum(IV) and palladium(II).

Acknowledgments: The authors are greatly indebted for the financial support for the study of cesium(I) removal by Grant-in-Aid for Challenging Exploratory Research by Japan Society for the Promotion of Science (JSPS), for that of gold recovery by the Industrial Technology Research Grant Program in 2006 by the New Energy and the Industrial Technology Development Organization (NEDO), Japan, and for those of the recovery of chromium(VI) and precious metals by the Grant-in-Aid for Scientific Research about Establishing a Sound Material-Cycle Society by the Ministry of Environment of Japanese Government. We are also indebted to Hiroshi Murakami, Kumiko Kajiyama, Rie Yamauchi and Hisashi Nakagawa for their experimental works.

Author Contributions: Manju Gurung contributed to the works related to Sections 3, 6 and 7. Ying Xiong contributed to the works related to Sections 6 and 7. All of other authors contributed equally to this work.

Conflicts of Interest: The authors declare no conflict of interest.

References

1. Inoue, K.; Baba, Y.; Yoshizuka, K. Adsorption of metal ions on chitosan and crosslinked copper(II)-complexed chitosan. *Bull. Chem. Soc. Jpn.* **1993**, *66*, 2915–2921. [CrossRef]

2. Pangeni, B.; Paudyal, H.; Abe, M.; Inoue, K.; Kawakita, H.; Ohto, K.; Adhikari, B.B.; Alam, S. Selective recovery of gold using some cross-linked polysaccharide gels. *Green Chem.* **2012**, *14*, 1917–1927. [CrossRef]

3. Sakaguchi, T.; Nakajima, A. Recovery of uranium from seawater by immobilized tannin. *Sep. Sci. Technol.* **1987**, *22*, 1609–1623. [CrossRef]

4. Nakano, Y.; Takeshita, K.; Tsutsumi, T. Adsorption mechanism of hexavalent chromium by redox within condensed-tannin gel. *Water Res.* **2001**, *35*, 496–500. [CrossRef]

5. Matsuo, T.; Ito, S. The chemical structure of Kaki-tannin from immature fruit of the persimmon (*Diospyros kaki* L.). *Agric. Biol. Chem.* **1978**, *42*, 1637–1643. [CrossRef]

6. Sakaguchi, T.; Nakajima, A. Accumulation of uranium by immobilized persimmon tannin. *Sep. Sci. Technol.* **1994**, *29*, 205–221. [CrossRef]

7. Inoue, K.; Kawakita, H.; Ohto, K.; Oshima, T.; Murakami, H. Adsorptive removal of uranium and thorium with a crosslinked persimmon peel gel. *J. Radioanal. Nucl. Chem.* **2006**, *267*, 435–442. [CrossRef]

8. Inoue, K.; Alam, S. Refining and mutual separation of rare earths using biomass wastes. *JOM* **2013**, *65*, 1341–1347. [CrossRef]

9. Sangvanich, T.; Sukwarotwat, V.; Wiacek, R.J.; Grudzien, R.M.; Fryxell, G.E.; Addleman, R.S.; Timchalk, C.; Yantasee, W. Selective capture of cesium and thallium from natural waters and simulated wastes with copper ferrocyanide functionalized mesoporous silica. *J. Hazard. Mater.* **2010**, *182*, 225–231. [CrossRef] [PubMed]

10. El-Kamash, A.M. Evaluation of zeolite A for the sorptive removal of Cs^+ and Sr^{2+} ions from aqueous solutions using batch and fixed bed column operations. *J. Hazard. Mater.* **2008**, *151*, 432–445. [CrossRef] [PubMed]

11. Dahiya, S.; Tripathi, R.M.; Hedge, A.G. Biosorption of heavy metals and radionuclide from aqueous solutions by pre-treated arca shell biomass. *J. Hazard. Mater.* **2008**, *150*, 376–386. [CrossRef] [PubMed]

12. Chen, C.; Wang, J. Removal of Pb^{2+}, Ag^+, Cs^+, and Sr^{2+} from aqueous solution by brewery's waste biomass. *J. Hazard. Mater.* **2008**, *151*, 65–70. [CrossRef] [PubMed]

13. Smičiklas, I.; Dimović, S.; Plećaš. I. Removal of Cs^+, Sr^{2+} and Co^{2+} from aqueous solutions by adsorption on natural clinoptilolite. *Appl. Clay Sci.* **2007**, *35*, 139–144. [CrossRef]

14. Bhattacharyya, K.G.; Gupta, S.S. Adsorption of a few heavy metals on natural and modified kaolinite and montmorillonite: A review. *Adv. Colloid Interface Sci.* **2008**, *140*, 114–131. [CrossRef] [PubMed]

15. Samanta, S.K.; Ramaswamy, M. Studies on cesium uptake by phenolic resins. *Sep. Sci. Technol.* **1992**, *27*, 255–267. [CrossRef]

16. Dumont, N.; Favre-Réguillon, A.; Dunjic, B.; Lemaire, M. Extraction of cesium from an alkaline leaching solution of spent catalysts using an ion-exchange column. *Sep. Sci. Technol.* **1996**, *31*, 1001–1010. [CrossRef]

17. Gurung, M.; Adhikari, B.B.; Alam, S.; Kawakita, H.; Ohto, K.; Inoue, K.; Harada, H. Adsorptive removal of Cs(I) from aqueous solution using polyphenols enriched biomass-based adsorbents. *Chem. Eng. J.* **2013**, *231*, 113–120. [CrossRef]

18. Pangeni, B.; Paudyal, H.; Inoue, K.; Ohto, K.; Kawakita, H.; Alam, S. Preparation of natural cation exchanger from persimmon waste and its application for the removal of cesium from water. *Chem. Eng. J.* **2014**, *242*, 109–116. [CrossRef]

19. Park, Y.; Lee, Y.-C.; Shin, W.S.; Choi, S.-J. Removal of cobalt, strontium and cesium from radioactive laundry wastewater by ammonium molybdophosphate-polyacrylonitrile (AMP-PAN). *Chem. Eng. J.* **2010**, *162*, 685–695. [CrossRef]

20. Inoue, K.; Gurung, M.; Adhikari, B.B.; Alam, S.; Kawakita, H.; Ohto, K.; Murata, M.; Atsumi, K. Adsorptive removal of cesium using bio-fuel extraction microalgal waste. *J. Hazard. Mater.* **2014**, *271*, 196–201. [CrossRef] [PubMed]

21. Mohan, D.; Singh, K.P.; Singh, V.K. Removal of hexavalent chromium from aqueous solution using low-cost activated carbons derived from agricultural waste materials and activated carbon fabric cloth. *Ind. Eng. Chem. Res.* **2005**, *44*, 1027–1042. [CrossRef]

22. Tarley, C.R.T.; Arruda, M.A.Z. Biosorption of heavy metals using rice milling by-products. Characterization and application for removal of metals from aqueous effluents. *Chemosphere* **2004**, *54*, 987–995. [CrossRef] [PubMed]

23. Inoue, K.; Paudyal, H.; Nakagawa, H.; Kawakita, H.; Ohto, K. Selective adsorption of chromium(VI) from zinc(II) and other metal ions using persimmon waste gel. *Hydrometallurgy* **2010**, *104*, 123–128. [CrossRef]

24. Park, D.; Yun, Y.S.; Lee, H.; Park, J.M. Advanced kinetic model of the Cr(VI) removal by biomaterials at various pHs and temperatures. *Bioresour. Technol.* **2008**, *99*, 1141–1147. [CrossRef] [PubMed]

25. Chand, R.; Narimura, K.; Kawakita. H.; Ohto, K.; Watari, T.; Inoue, K. Grape waste as a biosorbent for removing Cr(VI) from aqueous solution. *J. Hazard. Mater.* **2009**, *163*, 245–250. [CrossRef] [PubMed]

26. Nakajima, A.; Baba, Y. Mechanism of hexavalent chromium adsorption by persimmon tannin gel. *Water Res.* **2004**, *38*, 2859–2864. [CrossRef] [PubMed]

27. Spinelli, V.A.; Lalanjeira, M.C.M.; Favere, V.T. Preparation and characterization of quaternary chitosan salt: Adsorption equilibrium of chromium(VI) ion. *React. Funct. Polym.* **2004**, *61*, 347–352. [CrossRef]

28. Tan, T.; He, X.; Du, W. Adsorption behavior of metal ions on imprinted chitosan resin. *J. Chem. Technol. Biotechnol.* **2001**, *76*, 191–195.

29. Wartelle, L.H.; Marshall, W.E. Chromate ion adsorption by agricultural by-products modified with dimethyloldihydroxyethylene urea and choline chloride. *Water Res.* **2005**, *39*, 2869–2871. [CrossRef] [PubMed]

30. Levankumar, L.; Muthukumaran, V.; Gobinath, M.B. Batch adsorption and kinetics of chromium(VI) removal from aqueous solutions by *Ocimum americanum* L. seed pods. *J. Hazard. Mater.* **2009**, *161*, 709–713. [CrossRef] [PubMed]

31. Chand, R.; Watari, T.; Inoue, K.; Torikai, T.; Yada, M. Evaluation of wheat straw and barley straw carbon for Cr(VI) adsorption. *Sep. Purif. Technol.* **2009**, *65*, 331–336. [CrossRef]

32. Kawakita, H.; Ohto, K.; Harada, H.; Inoue, K.; Parajuli, D.; Yamauchi, R. Recovery of gold by means of aggregation and precipitation using astringent persimmon extract. *Kagaku Kogaku Ronbunshu* **2008**, *34*, 230–233. [CrossRef]

33. Gurung, M.; Adhikari, B.B.; Kawakita, H.; Ohto, K.; Inoue, K.; Alam, S. Recovery of Au(III) by using low cost adsorbent prepared from persimmon tannin extract. *Chem. Eng. J.* **2011**, *174*, 556–563. [CrossRef]

34. Parajuli, D.; Kawakita, H.; Inoue, K.; Ohto, K.; Kajiyama, K. Persimmon peel gel for the selective recovery of gold. *Hydrometallurgy* **2007**, *87*, 133–139. [CrossRef]

35. Xiong, Y.; Adhikari, C.R.; Kawakita, H.; Ohto, K.; Harada, H.; Inoue, K. Recovery of precious metals by selective adsorption on dimethylamine-modified persimmon peel. *Waste Biomass Valoriz.* **2010**, *1*, 339–345. [CrossRef]

36. Xiong, Y.; Adhikari, C.R.; Kawakita, H.; Ohto, K.; Inoue, K.; Harada, H. Selective recovery of precious metals by persimmon waste chemically modified with dimethylamine. *Bioresour. Technol.* **2009**, *100*, 4083–4089. [CrossRef] [PubMed]

37. Gurung, M.; Adhikari, B.B.; Khunathai, K.; Kawakita, H.; Ohto, K.; Harada, H.; Inoue, K. Quaternary amine modified persimmon tannin gel: An efficient adsorbent for the recovery of precious metals from hydrochloric acid media. *Sep. Sci. Technol.* **2011**, *46*, 2250–2259. [CrossRef]

38. Gurung, M.; Adhikari, B.B.; Alam, S.; Kawakita, H.; Ohto, K.; Inoue, K. Persimmon tannin-based new sorption material for resource recycling and recovery of precious metals. *Chem. Eng. J.* **2013**, *228*, 405–414. [CrossRef]

39. Gurung, M.; Adhikari, B.B.; Inoue, K.; Kawakita, H.; Ohto, K.; Alam, S. Adsorptive recovery of palladium and platinum from acidic chloride media using chemically modified persimmon tannin. In Proceedings of TMS 2016, Nashville, TN, USA, 14–18 February 2016. Submitted.

40. Gurung, M.; Adhikari, B.B.; Morisada, S.; Kawakita, H.; Ohto, K.; Inoue, K.; Alam, S. *N*-aminoguanidine modified persimmon tannin: A new sustainable material for selective adsorption, preconcentration and recovery of precious metals from acidic chloride solution. *Bioresour. Technol.* **2013**, *129*, 108–117. [CrossRef] [PubMed]

41. Gurung, M.; Adhikari, B.B.; Kawakita, H.; Ohto, K.; Inoue, K.; Alam, S. Selective recovery of precious metals from acidic leach liquor of circuit boards of spent mobile phones using chemically modified persimmon tannin gel. *Ind. Eng. Chem. Res.* **2012**, *51*, 11901–11913. [CrossRef]

42. Adhikari, C.R.; Parajuli, D.; Kawakita, H.; Inoue, K.; Ohto, K.; Harada, H. Dimethylamine-modified waste paper for the recovery of precious metals. *Environ. Sci. Technol.* **2008**, *42*, 5486–5491. [CrossRef] [PubMed]

43. Khunathai, K.; Inoue, K.; Ohto, K.; Kawakita, H.; Kurata, M.; Atsumi, K.; Fukuda, H.; Alam, S. Adsorptive recovery of palladium(II) and platinum(IV) on the chemically modified-microalgal residue. *Solvent Extr. Ion Exch.* **2013**, *31*, 320–334. [CrossRef]

44. Khunathai, K.; Xiong, Y.; Biswas, B.K.; Adhikari, B.B.; Kawakita, H.; Ohto, K.; Inoue, K.; Kato, H.; Kurata, M.; Atsumi, K. Selective recovery of gold by simultaneous adsorption-reduction using microalgal residues generated from biofuel conversion processes. *J. Chem. Technol. Biotechnol.* **2012**, *87*, 393–401. [CrossRef]

45. Wang, R.; Liao, X.; Shi, B. Adsorption behaviors of Pt(IV) and Pd(II) on collagen fiber immobilized bayberry tannin. *Ind. Eng. Chem. Res.* **2005**, *44*, 4221–4226. [CrossRef]

46. Fujiwara, K.; Ramesh, A.; Maki, T.; Hasegawa, H.; Ueda, K. Adsorption of platinum(IV), palladium(II) and gold(III) from aqueous solutions on L-lysine modified crosslinked chitosan resin. *J. Hazard. Mater.* **2007**, *146*, 39–50. [CrossRef] [PubMed]

47. Ramesh, A.; Hasegawa, H.; Sugimoto, W.; Maki, T.; Ueda, K. Adsorption of gold(III), platinum(IV) and palladium(II) onto glycine modified crosslinked chitosan resin. *Bioresour. Technol.* **2008**, *99*, 3801–3809. [CrossRef] [PubMed]

48. Won, S.W.; Park, J.; Mao, J.; Yun, Y.-S. Utilization of PEI-modified *Corneybacterium glutamikum* biomass for the recovery of Pd(II) in hydrochloric solution. *Bioresour. Technol.* **2011**, *102*, 3888–3893. [CrossRef] [PubMed]

49. Zhou, L.; Xu, J.; Liang, X.; Liu, Z. Adsorption of platinum(IV) and palladium(II) from aqueous solutions by magnetic crosslinking chitosan nanoparticles modified with ethylenediamine. *J. Hazard. Mater.* **2010**, *182*, 518–524. [CrossRef] [PubMed]

50. Zhou, L.; Liu, J.; Liu, Z. Adsorption of platinum(IV) and palladium(II) from aqueous solution by thiourea-modified chitosan microspheres. *J. Hazard. Mater.* **2009**, *172*, 439–446. [CrossRef] [PubMed]

51. Iglesias, M.; Antico, E.; Salvado, V. Recovery of palladium(II) and gold(III) from diluted liquors using the resin Duolite GT-73. *Anal. Chim. Acta* **1999**, *381*, 61–67. [CrossRef]

52. Park, J.; Won, S.W.; Mao, J.; Kwak, I.S.; Yun, Y.-S. Recovery of Pd(II) from hydrochloric solution using polyallylamine hydrochloride-modified *Escherichida coli* biomass. *J. Hazard. Mater.* **2010**, *181*, 794–800. [CrossRef] [PubMed]

metals

MDPI

Article

Copper Recovery from Polluted Soils Using Acidic Washing and Bioelectrochemical Systems

Karin Karlfeldt Fedje [1,2,*], Oskar Modin [1] and Ann-Margret Strömvall [1]

[1]   Water Environment Technology, Department of Civil and Environmental Engineering, Chalmers University of Technology, Gothenburg SE-412 96, Sweden; oskar.modin@chalmers.se (O.M.); ann-margret.stromvall@chalmers.se (A.-M.S.)

[2]   Recycling and Waste Management, Renova AB, Box 156, Gothenburg SE-401 22, Sweden

*   Author to whom correspondence should be addressed; karin.karlfeldt@chalmers.se; Tel.: +46-31-772-21-49; Fax: +46-31-772-56-95.

Academic Editor: Suresh Bhargava
Received: 1 July 2015; Accepted: 14 July 2015; Published: 23 July 2015

Abstract: Excavation followed by landfilling is the most common method for treating soils contaminated by metals. However, as this solution is not sustainable, alternative techniques are required. Chemical soil washing is one such alternative. The aim of this experimental lab-scale study is to develop a remediation and metal recovery method for Cu contaminated sites. The method is based on the washing of soil or ash (combusted soil/bark) with acidic waste liquids followed by electrolytic Cu recovery by means of bioelectrochemical systems (BES). The results demonstrate that a one- or two-step acidic leaching process followed by water washing removes >80 wt. % of the Cu. Copper with 99.7–99.9 wt. % purity was recovered from the acidic leachates using BES. In all experiments, electrical power was generated during the reduction of Cu. This clearly indicates that Cu can also be recovered from dilute solutions. Additionally, the method has the potential to wash co-pollutants such as polycyclic aromatic hydrocarbons (PAHs) and oxy-PAHs.

Keywords: soil washing; Cu; PAH; soil remediation; metal recovery; microbial fuel cell

1. Introduction

Excavation followed by landfilling is the most common method for treating soils contaminated by metals. However, landfilling is not a sustainable management option, as many landfills will close in the coming decades, while new ones are not opened at the same rate. Consequently, alternative treatment methods are necessary. One interesting method is soil washing, which can be divided into physical and chemical washing, although a combination of both is often used. Physical soil washing usually aims at enriching most of the pollutants into a specific soil fraction, see e.g., [1], while chemical soil washing aims to wash out the pollutants. There is thorough lab-scale research on chemical soil washing using a variety of leaching agents, including inorganic acids, bio surfactants and complexing agents such as EDTA or its derivates [2–8]. Apart from cases where water was used as the leaching agent, not many pilot scale experiments have been conducted [9]. A reason might be that when virgin chemicals are used, recirculation or reuse of the leaching agents is essential in order to ensure an efficient and economically viable process. There are examples of recycling and reuse of the commonly used leaching agent EDTA [10,11]. However, as EDTA is non-biodegradable, it is important to ensure that any remaining EDTA does not lead to problems with the solid residues. From that perspective, an acid might be a better option. Probably the best alternative would be a waste liquid for leaching as the need for recycling is lower making the proposed method simpler and potentially less expensive. Acidic wastewater from a flue gas cleaning process has been shown to effectively (>70%) leach copper (Cu) from contaminated soils [12].

In many cases, metal-contaminated sites also contain organic pollutants; especially those at which wood preservation took place using both metal-containing and creosote-based preservers. Creosote contains high concentrations of polycyclic aromatic hydrocarbons (PAHs) that are known to be toxic, mutagenic and carcinogenic. In contaminated sites, the PAHs are degraded to oxygenated PAHs (oxy-PAHs) through chemical oxidation [13], photo oxidation [14] or biological transformation [15]. The oxy-PAHs are of great interest because they are persistent, highly toxic to both humans and the environment and more water-soluble than their corresponding PAHs [16]. When investigating treatment methods for metal-contaminated soils, it is therefore necessary to study the fate of organic pollutants.

An aspect that has not received much attention is the potential to recover the valuable metals released through soil washing processes. Conventionally, Cu is recovered from concentrated liquids (>35 g Cu/L) using large-scale energy-demanding electrolysis [17]. An interesting alternative, which would reduce the amount of electrical energy required during electrolysis, is the use of microbial bioelectrochemical systems (BES) [18] In BES, microorganisms oxidize the organic compounds present in e.g., wastewater and use the anode as an electron acceptor, thereby transforming the chemical energy in the organic compounds into electrical energy. Recent studies on diluted Cu solutions (~1 g/L) demonstrate that the energy consumption is significantly lower with BES compared to traditional electrolysis [19–21]. At certain cathode potentials, electrical energy could even be extracted from the system together with the Cu. As in conventional electrolysis, it is possible to selectively reduce individual metals from a mixture by varying the cathode potential in the system. Modin and co-workers found that high purity Cu (99.9%) could be recovered from a simulated ash leachate containing a mixture of Cu, Cd, Pb and Zn ions (1 g/L) without energy input. In addition, high purity Zn (>99.9%) was recovered from the mixture, although Cd and Pb needed further purification [22]

The aim of this experimental lab-scale study is to develop a remediation and metal recovery method for Cu contaminated sites. The method is based on soil washing using acidic waste liquids and BES to recover Cu. In addition, as contaminated areas often contain both metals and organic pollutants, the behavior of PAHs and oxy-PAHs during the recovery process is studied and evaluated.

2. Experimental Section

In this work, the term "soil" is defined as "the entire mantle of unconsolidated material, whatever its nature or origin", first formulated by G.P. Merrill in his book "Rocks, rock-weathering and soils" published in 1897 and used by other research groups.

2.1. Soil Samples

Soil samples with different characteristics from two sites, Köpmannebro (A) in western Sweden and Björkhult (B) in eastern Sweden, were used in this study. Both sites are highly contaminated with Cu and were previously used for wood preservation by means of $CuSO_4$ according to the Boucherie method. Based on results from chemical analyses in earlier studies on the two sites, representative samples were collected from hotspots with a high metal content. Sample A1 is a bark sample from a depth of 5–80 cm and A2 is a soil sample from the same spot but from a depth of 120–150 cm. Sample B1 is bark from a depth of 5–20 cm and B2 is a soil sample from the same spot at a depth of 40–60 cm. All samples are mixtures of several sub-samples taken from the bark and the soil, respectively.

Pre-Treatment of the Soil Samples

All samples were gently dried at 80 °C until constant weight (*i.e.*, 100% dry solids (DS)). The A1 and B1 samples (*i.e.*, bark) were cut into smaller pieces using an automatic mixer and thereafter most parts of each sample were separately incinerated to ash, while the remaining bark samples were stored for analysis. The incineration processes were performed in two ways. The bark was either incinerated in batches in a laboratory oven at 850 °C for 5 h or in a real full-scale plant (Type-D 200 using the BS Incinerator System) normally employed for the incineration and destruction of e.g., animal cadavers and hospital waste. The bark was continuously incinerated at 830 °C. This temperature was chosen to theoretically prevent vaporization of Cu compounds, thus enriching them in the collected bottom ash. Bark sample A1 was incinerated by both methods and the resulting ash (A1a) was a mixture (50/50%) of the two ashes *i.e.*, A1a$_{lab}$ incinerated in the lab and A1a$_{incin}$ incinerated in the real boiler. This mixture was used in the present study. The B1 bark was only incinerated in the laboratory oven, resulting in ash sample B1a. All samples were stored in airtight containers until use.

2.2. Leaching Experiments

Highly acidic wastewater (pH around 0) produced in the wet flue gas cleaning process during full-scale incineration of municipal solid and industrial waste was used as a leaching agent. Consequently, the wastewater contained metal ions and chlorides from the flue gases. The composition of ions in the process water varies in accordance with incineration conditions and the kind of waste used, but is dominated by Cl$^-$. Representative concentrations of the most common ions are presented in Table S1. Today, acidic water is purified of toxic metal ions and small particles by means of precipitation. Thereafter, the sludge is landfilled, while the clean water is released to the recipient.

2.2.1. Optimizing Leaching Experiments

The influence of the parameters leaching time, liquid-to-solid (L/S) ratio and number of acidic leaching steps was studied to optimize the Cu leaching process. The optimization experiments were carried out on samples A1a and A2. The latter was also tested using significantly longer leaching times of 18 and 24 h. The original bark samples were not leached directly, as previous research revealed a greater leaching efficiency from the corresponding ashes as well as a higher Cu concentration in the final leachates [12]. In each experiment, 0.5–5 g of ash or soil (100% DS) was leached in airtight plastic containers using the acidic process wastewater at L/S ratios of 2, 5 or 10. The leaching times varied between 15 and 90 min, and a reciprocal shaker (SM 25, Edmund Bühler GmbH, Hechingen, Germany) was used at 140 rpm to ensure continuous agitation. After leaching, the solid-liquid-mixture was centrifuged for 15 min at 3000 G (gravity). The supernatant was transferred to a new airtight container before being stored at 4 °C until analysis. The solid residues were either leached in an additional step followed by 15 min of centrifugation at 3000 G, or washed immediately for 5 min with continuous shaking using ultrapure water (18.2 MΩ/cm^2) in L/S ratios of 2 calculated on the basis of the original sample amount. In the initial ash leaching experiments, *i.e.*, one acidic leaching step, an L/S of 5 was used for washing. Finally, the sample-water mixture was centrifuged for 15 min at 3000 G and the solid residues were dried and stored in airtight containers. The leachates were also filtrated before analysis using acid resistant filters with pore size 1.6 μm. All soil samples were studied in triplicate, while the ash samples were generally investigated in duplicate due to shortage of ash.

Larger scale batch leaching experiments were performed based on the parameters identified in the optimization experiments. In the batch leaching experiments, 20–30 g (100% DS) of each sample, *i.e.*, ash (A1a, B1a) and soil (A2, B2), was used.

2.2.2. SS-EN-12457-3 Leaching Tests

The potential natural leaching from the original samples as well as from selected soil and ash residues was studied using a downscaled and somewhat modified SS-EN-12457-3 leaching test procedure, *i.e.*, L/S = 2 for 6 h and L/S = 8 for 18 h [23]. Depending on the amounts available, between 1 and 30 g was used in each experiment. The samples were continuously shaken during leaching in a reciprocal shaker (SM 25, Edmund Bühler GmbH) and then centrifuged to separate the leachates from the solids. The leachates were filtered using acid resistant filters with pore sizes varying from 1–6 μm and stored at 4 °C for further analyses.

2.3. Bioelectrochemical Experiments

The bioelectrochemical reactor consisted of cylindrical anode (2 cm diameter, 6 cm length) and cathode (2 cm diameter, 3 cm length) compartments separated by a 3.1 cm^2 anion exchange membrane (AMI-7001, Membranes International Inc., Ringwood, NJ, USA). The anode was a $5 \times 2 \times 0.2$ cm^3 carbon felt (TMIL, Tsukuba, Japan) and the cathode a 5 cm long, 0.81 mm diameter titanium wire (Sigma Aldrich). A nutrient medium containing 500 mg/L glucose, 114 mg/L NH$_4$Cl, 20 mg/L K$_2$HPO$_4$ and 5.6 g/L NaHCO$_3$ was circulated through the anode compartment from a 1 L glass bottle. At the start of the experiment, the anode was inoculated with sludge from an anaerobic digester. The cathode compartment was fed with either leachate A1a or a salt solution containing 2925 mg/L NaCl, 1179 mg/L CuSO$_4$·5H$_2$O and 1 mL/L 2 M HCl.

The bioelectrochemical reactor was operated for five weeks. During the first four weeks, the reactor was acclimatized with a Cu-containing salt solution in the cathode compartment. In Week 5, Cu recovery from real leachates (A1a) was investigated.

2.4. Leaching and Degradation of PAHs

The A2 soil sample was used for the leaching and degradation test of PAHs and oxy-PAHs. An undried soil sample (~2 kg) was mixed in a mortar, and bigger pieces of bark were retained, but inorganic parts such as stones were excluded. The soil sample was thereafter mixed with ~500 mL ultrapure water for 24 h to form a homogeneous soil/water slurry sample. To ensure a sufficiently high PAH content, the soil sample was spiked with additional PAHs. A sub-sample of 800 g was mixed with 1 mL of PAH-mix 3 (Supelco in 50:50 CH$_2$Cl$_2$:CH$_3$OH, Sigma-Aldrich) containing 18 specific PAHs, each with a concentration ranging from 100 to 1000 μg/mL, together with 2 mL MeOH to enable good mixing and prevent the PAHs from adsorbing to the glass surfaces of the containers. The sample was then stirred for 2 h. Thereafter, the PAH spiked sample was stored in a dark glass bottle in a refrigerator (+4 °C) for eight months. A portion of this sample was sent for external analysis of metals, minerals, PAHs, oxy-PAHs, organic (loss of ignition) and water content. Two other ~200 g portions were leached for 30 min with continuous stirring and L/S = 10 using the acidic process water as described above. The filtrates from the two samples were mixed and sent for US EPA PAH-16 and oxy-PAH-9 analysis after filtration through 0.6 μm glass-fiber filters, while the samples for metal analysis were filtered through a 0.45 μm cellulose acetate filter. It should be noted that these soil samples were not centrifuged and not washed with ultrapure water before chemical analysis.

2.5. Analytical Methods

The total metal content of the original and selected leached bark, soil and ash samples was analyzed using either ASTM D3683, *i.e.*, leaching in closed Teflon bottles heated in a microwave oven using 7 M HCl (A1a and B1a; As, Cd, Cu, Co, Hg, Ni, Pb, Sb, S, Se, Sn and Zn), HNO$_3$/H$_2$C$_2$ (A1, A2, B1, B2; As, Cd, Cu, Co, Hg, Ni, Pb, B, S, Sb, Se and Zn) (not Sb in A1 and B1) or ASTM D3682 *i.e.*, melting in lithium tetra borate followed by dissolution in diluted HCl (Al, Ca, Fe, K, Mg, Na, P, Si, Ti, Ba, Be, Cr, Mn, Mo Nb, Sc. Sr, V, W, Y and Zr). The solutions from the total digestion were analyzed using ICP-MS (inductively coupled plasma mass spectrometry) or ICP-AES (inductively coupled plasma atomic emission spectrometry).

The Cu concentration in the leachates obtained from the optimized Cu leaching experiments was analyzed semi-quantitatively using an ATI UNICAM spectrophotometer with a D2/tungsten lamp. The leachates obtained from the batch leaching experiments (based on the optimized leaching parameters and larger sample volumes) and the acidic process water were analyzed for metals (Al, As, Ba, Be, Pb, Cd, Co, Cu, Cr, Li, Mn, Mo, Ni, Se, Ag, Sr, V and Zn) by means of ICP-MS. The same analytical method was also employed for the BES experiment liquids (As, Pb, Cd, Co, Cu, Cr, Ni, V and Zn). Ion chromatography (Dionex ICS-900, Thermo Scientific) was applied to analyze the K$^+$, Mg^{2+}, Na$^+$, NH$_4^+$, Ca^{2+}, Cl$^-$, NO$_2^-$, NO$_3^-$, PO$_4^{3-}$ and SO$_4^{2-}$ concentrations in the acidic process water. The pH was measured in some of the original samples (pH(H$_2$O)) using a Metrohm SM 702 pH meter. At the end of each leaching experiment, the pH was controlled to be <2 using litmus paper. To measure the influence of the various parameters on metal leaching to guarantee an accurate evaluation, analysis of variance (ANOVA) was conducted [24]. The organic content in the anode compartment of the BES was analyzed by means of a total organic carbon analyzer (TOC-V, Shimadzu, Japan). Cell potentials were logged using a data logger (USB-6008, National Instruments, USA). The amount of Cu recovered on the cathode was quantified by dissolving it in 10 mL HNO$_3$ (50%) and measuring the Cu^{2+} concentration using ICP-AES. The PAHs and oxy-PAHs were solvent extracted and analyzed with GC-MS by a commercial laboratory (ALS, Luleå, Sweden).

3. Results and Discussion

3.1. Characterization of Original Samples

Total Amounts

The total amounts of selected major and minor elements in the original samples (A1, A1a$_{oven}$, A1a$_{incin}$, A2, B1, B1a and B2) are presented in Table 1. For comparison purposes, the Swedish generic guideline values for sensitive and less sensitive land use are also provided [25]. Copper greatly exceeds the guidelines in all samples, while the other elements, with the exception of Ba and Zn, are below the recommended values.

The incineration temperature was chosen so as to avoid vaporization of Cu compounds as discussed above. If no Cu vaporization took place, the amount of Cu/mass unit in the collected bottom ash would have been about 10 and 3 times higher in A1a$_{oven}$ and in B1a, respectively, compared to the original bark samples. The reason for the higher ash fraction in sample B1, *i.e.,* lower Cu content/mass unit, is most likely due to the non-volatile soil particles present in the bark fraction, which was not the case in the bark from site A (Figure 1). Sample A1a$_{incin}$, *i.e.,* the bark sample incinerated in a real incineration plant, resulted in less bottom ash compared to the other samples because of the continuous gas flow during the incineration procedure. In this study no fly ash was collected, in which the volatilized metal compounds are captured. However, in a real incineration situation the fly ash is collected in the flue gas cleaning system and thus interesting metals can also be recovered from that ash fraction, although this was not studied here.

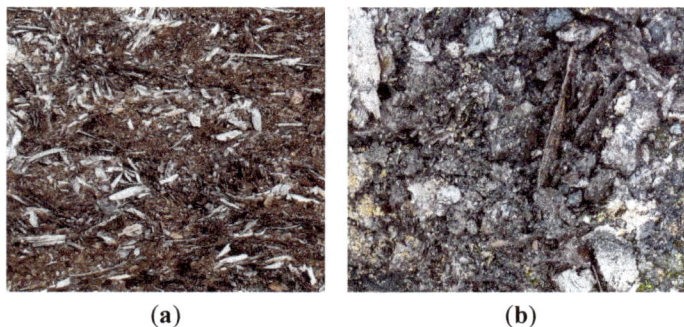

(a)   **(b)**

Figure 1. Bark samples from (**a**) site A (Köpmannebro); and (**b**) site B (Österbybruk).

The concentrations of carcinogenic PAHs have previously been analyzed in selected spots in sites A and B [26,27]. The PAH pollution was heterogeneous at both sites. The highest presence of PAH at site A, 2.3 and 2.1 mg/kg, could be referred to two specific places; where a fire had taken place and where a small turpentine factory was previously located. At site B the highest contamination,

12 mg/kg, could not be related to any specific activity but in several places where soil samples were taken, an ash like material was present, indicating that a fire could have occurred.

Table 1. Average total amounts of selected major (oxide minerals) and minor elements in the original soil samples (A1, A1a$_{oven}$, A1a$_{incin}$, A2, B1, B1a and B2). The Swedish generic guidelines values for sensitive (KM) and less sensitive (MKM) land use are also shown. All amounts are shown in mg/kg DS. Measurement uncertainties vary between 20% and 25%.

Element	A1 (bark)	A1a$_{oven}$ (Bark Ash)	A1a$_{incin}$ (Bark Ash)	A2 (Soil)	B1 (Bark)	B1a (Bark Ash)	B2 (Soil)	Sensitive Land Use [a] (KM)	Less Sensitive Land Use [a] (MKM)
pH(H$_2$O)	4.9 [b]	12.1 [b]	– [c]	5.5	–	–	5.2		
					wt. % DS				
SiO$_2$	1.9	54	8.4	68	26	57	76	* [d]	*
Al2O$_3$	0.7	10	4.6	12	5.3	11	13	*	*
CaO	0.8	4.5	26	1.8	0.5	1.3	0.8	*	*
Fe$_2$O$_3$	0.5	4.0	1.8	2.5	2.0	4.4	1.1	*	*
K$_2$O	0.1	2.7	6.9	2.4	1.6	3.5	4.2	*	*
MgO	0.06	0.7	1.7	0.6	0.1	0.2	0.1	*	*
MnO	0.01	0.07	0.01	0.04	0.01	0.03	0.02	*	*
Na$_2$O	0.08	2.0	7.3	2.6	1.0	2.5	2.7	*	*
P$_2$O$_5$	0.1	0.6	29	0.04	0.2	0.4	0.01	*	*
TiO$_2$	0.02	0.4	0.9	0.5	0.2	0.2	0.2	*	*
					mg/kg DS				
As	2.0	11	<3	0.6	7.8	22	0.3	10	25
Ba	110	930 [e]	430	500	620	230	860	200	300
Be	0.1	1.5	<0.5	1.4	0.7	1.5	1.9	*	*
Cd	0.3	1.3	0.1	0.01	0.1	0.3	0.03	0.5	15
Co	1.2	7.5	2.3	2.2	0.9	2.2	0.3	15	35
Cr	3.9	50	71	42	9.0	24	25	80	150
Cu	11,000	130,000	19,000	1100	15,000	110,000	720	80	200
Hg	0.1	<0.01	<0.01	<0.04	0.4	<0.01	0.1	0.25	2.5
Mo	0.2	2.4	3.1	0.3	0.3	1.4	0.2	40	100
Nb	0.4	8.8	5.4	9.0	7.0	6.4	5.9	*	*
Ni	3.5	27	21	3.1	2.6	7.9	0.3	40	120
Pb	31	360	61	7.9	39	150	3.2	50	400
S	570	4900	2700	76	510	2400	<50	*	*
Sb	–	6.7	8.9	0.2	–	16	0.1	12	30
Sc	0.6	5.9	<1	7.7	1.2	2.9	1.9	*	*
Sn	–	19	33	–	–	19	–	*	*
Sr	30	230	130	190	95	260	200	*	*
V	6.7	38	12	44	7.0	15	9.7	100	200
W	0.6	<50	<50	1.2	0.6	<50	0.7	*	*
Y	2.3	19	2.9	17	6.9	8.0	5.4	*	*
Zn	44	260	1500	14	36	160	3.5	250	500
Zr	5.7	150	13	230	42	95	113	*	*

[a] [25]; [b] [12]; [c] not analyzed; [d] * no guideline values exist; [e] Italic font; above the Swedish guidelines for less sensitive land use.

3.2. Optimization of Leaching Parameters

3.2.1. Ash

In theory, a low L/S-ratio is required in order to obtain the most concentrated leachates possible. Initially, L/S = 2 was used for leaching the ash samples but the ratio turned out to be too low as more or less all the leaching agent was absorbed by the solids, making further leaching impossible. Therefore L/S ratios of 5 or 10 were used for further experiments. The L/S-ratio is more important than the leaching time (Figure 2). The Cu leaching was about twice as high when using L/S = 10 compared to L/S = 5, irrespective of leaching time. This is also confirmed by the ANOVA analysis, where the L/S ratio parameter was $p < 0.05$ (Table S2).

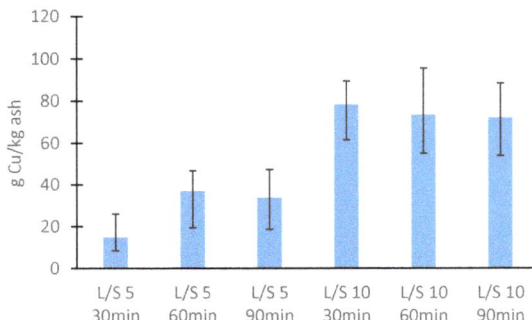

Figure 2. Cu leached from the bark-ash sample A1 with a variation in leaching time (30–90 min) and L/S-ratios (5 and 10). The columns show the average of three replicate tests. The error bars show the maximum and minimum values.

Although Cu release is very efficient when using an L/S ratio of 10, between 1 and 3 wt. % of the Cu initially present in the ash was released in the water washing step following the acidic leaching. This indicates that Cu leaching from the ash residue would exceed the legal limit for depositing it into landfills for hazardous materials (100 mg Cu/kg dry ash), despite the fact that leaching procedures are not totally comparable [28]. Consequently, several leaching steps are required to achieve an acceptable level of Cu leaching and thus a two-step leaching procedure was tested. As expected, most Cu leaching occurred in the first step; on average about 70 wt. % of the Cu, irrespective of leaching time (Figure 3). There was no statistically significant effect of time on Cu leaching (Table S3).

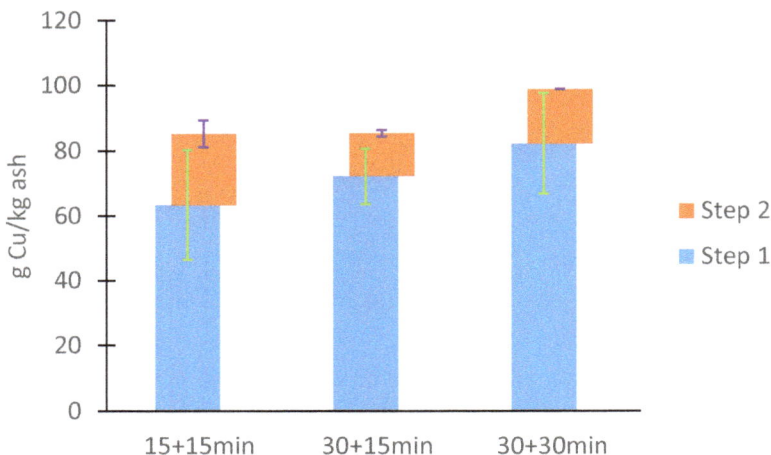

Figure 3. Cu leached from the bark-ash sample A1a using two-step leaching with different leaching times at an L/S ratio of 10. The columns show the average of duplicate tests. The error bars show the maximum and minimum values.

After two 30 min leaching steps and L/S = 10, the Cu release from the ash residue during the water washing step was <100 mg Cu/kg dry ash and thus meets the limit for depositing it into a hazardous waste landfill [28]. Therefore, these leaching parameters were chosen for further investigation, including batch leaching using a larger sample volume.

3.2.2. Soil

For the soil the L/S ratio was also found to be the most important leaching parameter and an L/S = 10 yielded the most efficient Cu leaching irrespective of time (Figure 4 and Table S4). Previous studies have indicated that longer leaching times, *i.e.*, several hours, can increase metal leaching from soils [29,30]. In the present study, the concentration of most metals detected in the leachates (>1 mg/L) after 18 and 24 h was similar to those after shorter periods (Ba, Cu, Mn, Pb and Zn). However, the Al concentration in the leachates was almost twice as high after several hours compared to 30 min leaching. This indicates that the soil matrix to some extent dissolves after several hours leaching, as anorthite was found to be one of the major minerals in soil from site A [12]. In addition, a tendency towards the opposite trend, *i.e.*, lower concentrations in leachates with longer leaching times, was found for Zn and Pb, indicating precipitation of solid compounds or that the metal ions are adsorbed to active sites in the soil matrix. The latter is most likely in the present case, an effect also demonstrated by others [31,32]. Consequently, the soil can act as a cleaning agent for Zn and Pb ions in the process water.

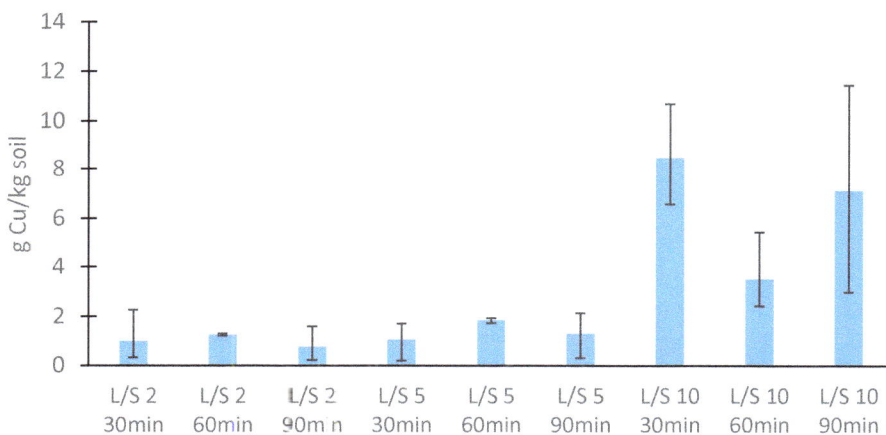

Figure 4. Cu leached from the A2 soil sample at varying leaching times (30–90 min) and L/S-ratios (2, 5 and 10). The columns show the average of three replicate tests. The error bars show the maximum and minimum values.

After one leaching step, <100 mg Cu/kg of dry soil was released in the following washing step, indicating that the soil residues can be deposited in a landfill for hazardous waste after Cu recovery. Consequently, a one-step leaching process with a leaching time of 30 min and an L/S-ratio of 10 was used for the subsequent experiments.

It should be noted that both bark and soil samples were highly heterogeneous. The concentration of Cu differed between locations at the sites and could also differ between subsamples of the collected samples. Therefore, the leached amounts are not directly comparable to the original concentrations shown in Table 1.

3.3. Batch Leaching

Batch leaching experiments were carried out on samples A1a, A2, B1a and B2, in two 30 min steps with an L/S = 10. The concentrations of selected elements in the respective leachates are presented in Table 2. It should be noted that the concentrations include the total Cu present in the leachates, *i.e.*, Cu originally present in the process water, is included (Table S1). Al, Pb and Zn were the only other measured elements present in significant amounts.

Table 2. Concentrations of selected major and minor elements in the final leachates from the ash samples A1a and B1a and from the soil samples A2 and B2. All amounts are in mg/L. Uncertainties in the analyses vary between 25% and 35%. Beryllium, Tl, U and Ag were detected in concentrations of <0.1 mg/L in all leachates.

Element	A1a (Bark Ash) [a]		B1a (Bark Ash) [a]		A2 (Soil) [a]	B2 (Soil) [a]
	Step 1 [b]	Step 2 [c]	Step 1 [b]	Step 2 [c]	Step 1 [b]	Step 1 [b]
	mg/L					
Al	820	340	96	52	150	560
As	0.7	0.3	2.5	0.4	0.2	0.2
Ba	5.2	2.1	1.3	1.5	2.0	1.7
Pb	21	12	11	11	11	9.3
Cd	0.3	0.3	0.3	0.3	0.3	0.3
Co	0.1	<0.1	<0.1	<0.1	<0.1	<0.1
Cu	3400	2400	8400	1500	120	140
Cr	0.8	0.2	0.1	0.1	0.1	0.1
Li	0.6	0.08	0.06	0.04	0.06	0.0
Mn	20	5.0	6.9	4.5	4.9	5.4
Mo	0.2	<0.1	<0.1	<0.1	<0.1	<0.1
Ni	6.6	2.5	9.4	0.7	0.1	0.1
Se	0.1	<0.1	<0.1	<0.1	0.1	<0.1
Sr	13	2.2	4.3	1.1	0.7	0.6
V	0.7	0.1	0.5	<0.1	0.2	0.1
Zn	430	90	70	60	60	50

[a] L/S = 10; [b] 30 min; [c] 30 min.

Cu leaching from B1a ash was significantly higher in the first compared to the second leaching step, while Cu leaching from A1a ash was more evenly distributed between the two steps; 85/15% and 60/40%, respectively (Table 2 and Figure 5a). This is probably due to the higher initial Cu content at site B but the properties of the original bark and the corresponding ash effect, for instance how strongly Cu is bound into the ash matrix, also play a role. It is noticeable that the A1a leaching fraction for the two steps is more evenly distributed in the batch leaching compared to the previous experiments on smaller amounts (Figures 3 and 5). This is probably due to less efficient contact between liquid and solid in larger sample volumes. Although only about 1% of the total Cu for both ashes was leached during the washing steps, it corresponds to >100 mg Cu/kg dry ash (940 mg Cu/kg and 820 mg Cu/kg dry ash for ash A1a and B1a, respectively). Consequently, leaching and washing are less effective when larger amounts of ash and process water are used. The reasons for this could be inefficient mixing or that Cu ions are temporarily adsorbed into the pore water of the particles during centrifugation and released in the following washing step instead of during leaching. In the soils, about 5% of the leached Cu was released in the washing step for both samples (Figure 5b), which corresponds to <100 mg Cu/kg dry soil, *i.e.*, below the legal limit.

As the batch leachates from A1a were more complex than those from B1a due to higher concentrations of most elements, the former were chosen for the electrolysis experiments.

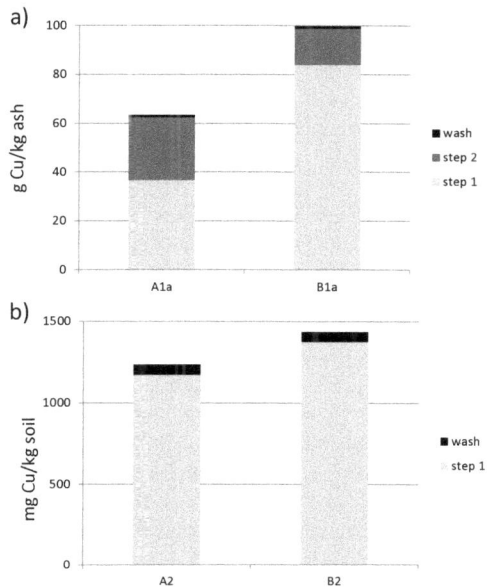

Figure 5. Amounts of Cu leached in steps 1 and 2, each 30 min, L/S = 10, followed by a washing step for (**a**) the A1a and B1a ashes; and (**b**) the A2 and B2 soils. The amounts are shown in (**a**) g Cu/kg dry ash; and (**b**) mg Cu/kg dry soil. Uncertainties in the analyses vary between 25% and 35%.

3.4. Cu Recovery Using BES

The BES was operated as a microbial fuel cell with either a 100 Ω (Ohm) or 1000 Ω resistor connected between the anode and cathode. This means that current flowed spontaneously in the reactor and no electrical energy input was needed to drive Cu reduction. After four weeks of acclimatization with a copper-containing salt solution as a catholyte, four experimental runs were carried out on real leachates. The anode of the microbial fuel cell contained 132–158 mg C/L of glucose during the experimental runs, thus the anode performance was not limited by substrate availability. Instead, the aim was to investigate the purity of the recovered Cu in the microbial fuel cell and the efficiency of Cu extraction from the A1a leachates.

A summary of the results obtained with the microbial fuel cell is presented in Table 3. In addition to Cu, As, Pb, Cd, Co, Cr, Ni, V and Zn were analyzed in the leachate. Cu made up approximately 90% of the total metal content, Zn accounted for 9.4% and Pb for 0.4%. The other metals were present in much lower concentrations. In the metal deposits recovered from the cathode, Cu made up 99.7%–99.9% of the total metal content, Zn 0.1%–0.3% and Pb 0.01%–0.03% except in Run 4 when it constituted 0.2%. This demonstrates that Cu can be selectively extracted from the leachate with only minor contamination from other metals. It is unlikely that Zn will be spontaneously reduced on the cathode in a microbial fuel cell and its presence is likely due to salt deposits or small drops of catholyte remaining on the cathode when it was removed for sampling.

Operation of the microbial fuel cells at lower resistance led to a higher current and more rapid reduction of Cu. At 100 Ω, the maximum current was 141.3 A/m^3 (catholyte volume) and at 1000 Ω 37.5 A/m^3. The currents correspond to Cu recovery rates of 167.5 and 14.0 g Cu/m^3/h, respectively, at 100% cathode efficiency. Based on the charge transfer in the microbial fuel cell, the theoretical amount of Cu reduction can be calculated for each run and compared to the actual removal of Cu from the catholyte as well as the amount recovered on the cathode surface (Table 3). Of the Cu removed from the catholyte solution, 58%–85% was recovered from the cathode surface. The remaining fraction could have been lost during sampling as not all Cu bound strongly to the cathode surface. Of the theoretical Cu reduction based on charge transfer, 73%–116% could be accounted for by the removal of Cu^{2+} ions in the catholyte. A value above 100% could mean that Cu_2O was formed, which only requires one electron per Cu, or precipitation of non-reduced Cu as brochantite ($CuSO_4 \cdot 3Cu(OH)_2$) [20]. A value below 100% could indicate that oxygen on the cathode was also reduced. Of the original amount of Cu present in the catholyte, 20%–23% (at 1000 Ω resistor) and 64%–79% (at 100 Ω) were recovered on the cathode surface. Higher recovery efficiencies could probably be achieved with longer reaction times. Based on the measured current densities, the highest rate of Cu reduction that could have been obtained in this reactor was 167.5 $g/m^3/h$ (normalized to catholyte volume). A higher reduction rate could potentially be achieved by increasing the cathode surface area per unit volume as this reactor only had 14 m^2/m^3. As the reactor was operated as a microbial fuel cell, electrical power was generated during Cu reduction. The power output ranged from 0.09–0.11 kWh/kgCu in Runs 2 and 4 (100 Ω) to 0.32–0.50 kWh/kgCu in Runs 1 and 3 (1000 Ω).

The A1a leachate was a mixture of two leaching steps. From a Cu recovery perspective, it might be more efficient to only recover Cu from the leachate with the highest Cu concentration in order to speed up the process. On the other hand, from a remediation perspective, two leaching steps are needed to release most of the Cu, which means that the less concentrated leachate still has to be taken care of, while failure to recover Cu from both steps implies that valuable Cu is lost. However, this has to be decided on a site-to-site basis, depending on the amount of Cu available and its distribution between leaching steps.

Table 3. Cu recovery from leachate in the microbial fuel cell.

Run	Duration (h)	Resistor (Ω)	Final Cu Conc. [a] (mg/L)	Theoretical Cu Recovery [b] (mg)	Cu removal from Catholyte (mg)	Measured Cu Recovery (mg)	Measured Cu Recovery [c] (%)
1	19	1000	1500	6.4	7.4	4.9	23
2	29	100	420	21	17	13	64
3	22	1000	1500	8.1	7.2	4.2	20
4	72	100	160	27	20	17	79

[a] Final Cu concentration in catholyte. The initial concentration in all runs was 2300 mg/L. This is slightly lower than the Cu concentrations measured in A1a leachate immediately after leaching (Table 2), possibly because of precipitation during storage; [b] Calculated based on charge transfer in BES; [c] Fraction of Cu recovered from the original amount in the leachate.

3.5. Characterization of Leached Soil Samples

3.5.1. Metal Content

The content of major and minor elements in the leaching residues from site A (A1a leached for 30 × 2 min and water washed for 5 min; A2 leached for 30 min) (L/S = 10), was analyzed (Table 4). Please note that the A2 residue was not washed in water after the acidic leaching and prior to metal content analyses. Apart from Cu, most elements in the solids were comparable before and after leaching. Despite the fact that the Cu leaching was efficient (>80%) in both samples, the Cu content in the ash greatly exceeded the MKM limit, which is why it has to be landfilled. However, as the Cu leaching in the EN 12457-3 test exceeded the limit for deposition in hazardous landfill, it cannot be landfilled directly (Figure 6b). The A2 soil residue was in the range of the MKM limits even without water washing and can plausibly be put back into the remediated site (Table 4). However, the Ba and Hg content in the residue exceeded the MKM limits. Barium was high in the original soil and the leaching potential low (Table 1 and Figure 6b). It is probably present as low soluble $BaSO_4$. Mercury was only identified in low concentrations in the original soil samples (Table 1). However, it is probably present in the process water and adsorbed onto the soil particles during leaching, despite the fact that the pH is acidic. Less than 0.5 mg·Hg/L in the process water can account for this phenomenon and as process water is a waste product from flue gas cleaning after waste incineration, it possibly contains such concentrations. If using this kind of leaching agent for soil remediation, it will probably require purification from Hg before utilization. Nevertheless, no water washing steps were used here; by using washing, the Hg content in the residues might fall below the MKM limit.

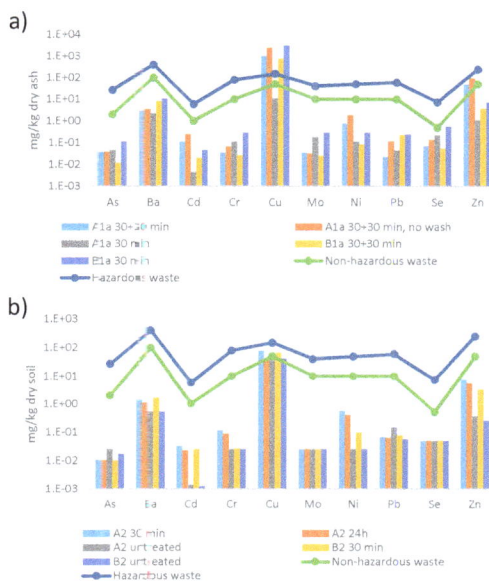

Figure 6. Metal release in the EN-12457-3 leaching test compared to the limit values for acceptance in non-hazardous and hazardous waste landfills: (**a**) treated ash samples A1a and B1a; and (**b**) original and treated soil samples A2 and B2.

The soil matrixes were stable against leaching; <5 wt. % was dissolved during leaching, while the ashes were dissolved to a greater extent. About 10 wt. % of the B1a sample was dissolved after leaching in two steps compared to 50 wt. % of the A1a sample. This difference is probably due to the presence of soil particles in the B1 sample as discussed above (Section 3.1).

3.5.2. Metal Leaching

With the exception of Cu, the leaching of all elements in the original and leached soil samples A2 and B2 was below the limits for acceptance on landfills for non-hazardous waste (Figure 6). Most elements (As, Ba, Cd, Cr, Mo and Se) also met the requirements for acceptance in an inert landfill. However, the leaching of Ba, Cd, Cr, Cu, Ni and Zn was higher from the leached soil compared to the original sample, which is most probably due to the fact that previously encapsulated metal compounds are released when the soil matrix is affected by the acid. Consequently, in this particular case, soil washing to minimize the risk of leaching is not beneficial and the total amount of a specific element is not correlated with the risk of leaching. This was also seen for the ash samples where all elements except Cu and Zn were below the non-hazardous waste limit (As, Ba, Cr and Mo were also below the inert waste limits), irrespective of the higher initial total amounts. There is no general trend as to whether a two-step acid leaching process reduces metal leaching to a greater extent in a subsequent EN-test than a one-step leaching procedure. However, water washing after leaching seems to decrease metal release during the EN-test, even if the washing period is as short as 5 min. This is probably caused by metal ions that were easily adsorbed to the ash particles being released to the first water in contact with these solids. The results also indicate that a higher L/S-ratio during the water washing step might further decrease leaching in the EN-test.

Table 4. Total amounts of selected major (calculated as oxides) and minor elements in the A1a (bark ash) and A2 (soil) residues after leaching experiments. Please note that the A2 residue was not washed in water after the acidic leaching. Uncertainties in the analyses vary between 20% and 25%.

Element	A1a Bark Ash Residue	A2 Soil Residue
	wt. % DS	
SiO_2	54	71
Al_2O_3	12	11
CaO	1.0	1.9
Fe_2O_3	5.6	2.2
K_2O	4.6	2.4
MgO	0.8	0.5
MnO	<0.1	<0.1
Na_2O	1.8	2.6
P_2O_5	1.1	<0.1
TiO_2	8.1	0.6
	mg/kg DS	
As	<6	0.8
Ba	2200	450
Be	1.9	1.3
Cd	0.7	0.1
Co	7.2	1.7
Cr	220	20
Cu	10,000	210
Hg	68	3.6
Mo	4.2	0.4
Nb	34	8.5
Ni	66	3.9
Pb	370	8.4
S	900	220
Sb	64	2.7
Sc	7.2	7.5
Sn	180	2.5
Sr	82	180
V	37	41
W	<50	1.1
Y	10	14
Zn	1000	28
Zr	140	220

3.5.3. PAHs and Oxy-PAHs Content

The concentration of each specific PAH in the spiked soil varied from 0.25 mg/kg (DS) to 4.4 mg/kg (DS) in the unleached sample. In this sample the low-molecular weight PAH-L, *i.e.*, the sum of naphthalene, acenapthylene and acenapthene, was 10 mg/kg (DS), the mean-molecular weight PAH-M, *i.e.*, the sum of fluorene, phenanthrene, anthracene, fluoranthene and pyrene, was 2.0 mg/kg (DS) and the high-molecular weight PAH-H, *i.e.*, the sum of benz[a]anthracene, chrysene, benzo[b]fluoranthene, benzo[k]fluoranthene, benzo[a]pyrene, dibenz[ah]anthracene, benzo[ghi]perylene and indeno[123cd]pyrene, was 2.8 mg/kg (DS), see Figure 7. These concentrations are in the ranges of the generic Swedish guideline values for sensitive and less sensitive land use [33]. Two of the nine oxy-PAHs analyzed were also identified in this unleached sample: 9-fluorenone 0.14 mg/kg (DS), a degradation product of fluorene, and 9,10-anthraquinone 0.15 mg/kg (DS), a degradation product of anthracene, indicating that PAH oxidation may occur during storage of the spiked soil sample. This could occur, e.g., because of aerobic bacterial degradation [15] or biocatalytic activity of fungi in the soil [16]. However, it cannot be ruled out that oxy-PAHs were already present in the original samples. The calculated ratios of the sum of oxy-PAH-9/PAH-16 are 0.04, for 9-fluorenone/fluorene 0.2 and for 9,10-anthraquinone/anthracene 0.6. These ratios are much lower than reported from other sites contaminated by PAHs [16,34].

After acidic washing, the concentration of each specific PAH in the treated soil was lower and varied from 0.19 to 2.5 mg/kg (DS). These concentrations are below the guideline value for less sensitive land use, but not low enough to qualify for sensitive land use [33]. For PAH-L, the concentrations were 1.6 times lower, for PAH-M 1.3 and PAH-L 1.2 times lower, compared to the concentrations before soil washing. In the resulting leachate, the concentration of each specific PAH varied from 0.17 to 16 µg/L. From calculations of the relative compositions of the PAHs in the soil before and after acidic washing, and in the acidic leachates, no signs of PAH degradation or volatilization could be found. If degradation had occurred, the most volatile and water-soluble PAH-L would disappear first causing a decrease in HMW/LMW PAH ratios [35,36]. None of the oxy-PAHs were found after leaching, neither in the residual soil nor in the leachate, indicating that acidic leaching may accelerate the degradation of oxy-PAHs. This may be caused by enhanced leaching of the oxy-PAHs into the water phase, *i.e.*, more exposed for degradation, caused by weathering and degradation of the soil matrix under the strong acidic washing.

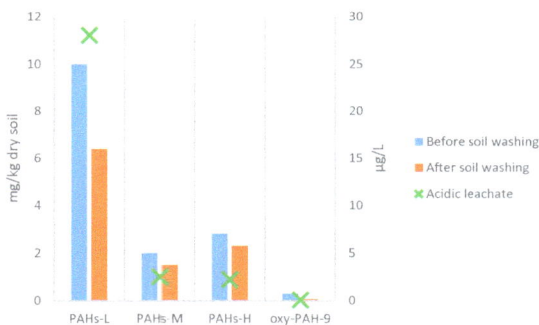

Figure 7. Concentrations of PAHs and oxy-PAHs analyzed in the A1 spiked soil sample before and after acidic washing (mg/kg), and the concentrations in the corresponding acidic leachate (µg/L).

In the leachate water, relatively more water soluble PAH-L occurred, indicating that the transport of the most water soluble PAHs into the acidic water was high. However, this transport to the water phase is expected to be very low and should not be pH dependent, but may follow the humic and fulvic acids released from the soil during mixing with acidic water. In a recent study of the distribution of organic pollutants in dissolved, colloidal and particulate form, relatively high PAH-L concentrations were detected in both the dissolved and the colloidal phases [37]. The results also suggested that dissolved organic carbon (DOC) colloids were carriers of pollutants in the colloidal phase [37,38]. PAHs are also assumed to be sorbed to Fe minerals such as goethite [39]. Iron was released during the acidic leaching (Tables 1 and 4) and in a previous study on site A, Fe minerals were identified in original soil samples but not in the acid leached residues [12]. Organic matter is known to be sorbed to iron oxides in soils [40], and therefore it is expected that the dissolution of iron oxide enhance the solubilization of DOC, and therefore could explain why PAHs are solubilized. The release of Fe minerals followed by an enhanced release of DOC can thus be important factors in the PAH leaching under strong acidic conditions. The leaching and degradation of PAHs in soils from contaminated sites may be improved by addition of a strong oxidant to achieve a Fenton-like oxidation [41], electrochemical degradation [42] or by a combination of the two techniques [43]. The addition of a strong oxidant could also affect the Cu leachability, which is something that should be studied further if this technique is used to remove PAHs.

4. Conclusions

In this study, a lab-scale method to release and recover Cu from contaminated soils has been developed and discussed. Acidic wastewater was used as a leaching agent. In a one- or two-step (depending on soil material) acidic leaching process followed by a water washing step, >80 wt. % of the Cu was leached. Although the Cu leaching efficiency was high, none of the samples are below the guidelines for less sensitive land use (MKM) or the limit values for acceptance in non-hazardous and hazardous waste landfills. Consequently, improvement of the final washing step is needed. Copper (99.7–99.9 wt. % purity) was recovered from the acidic leachates of one of the samples (A1a) using a BES operated as a microbial fuel cell. Electric power was generated, while Cu was reduced and recovered on the cathode surface. Although microbial fuel cells have been previously operated with Cu-containing catholytes, this was the first study to demonstrate recovery of high purity Cu from real soil leachate. According to predictions, there are about 30 tons of Cu in site A. Based on the results of this lab scale study, about 20 tons could potentially be recovered, corresponding to a value of approximately one million SEK (~100,000 euro). A limitation of the proposed method in full scale is the supply of acidic wastewater. Some pre-treatment is necessary in order to reduce the quantity of soil for washing. For example, physical soil washing, e.g., sieving, can concentrate the polluted material into smaller particle sizes, which efficiently reduces the amounts that require acidic washing. This study also revealed that acidic water is a good leaching agent for all PAHs; especially those with low-molecular weight. There are also indications that oxy-PAHs are fully degraded by acidic soil washing. PAH leaching mechanisms are not fully understood and the soil washing method needs to be further developed to increase efficiency.

Acknowledgments: This work was funded by Renova (a Swedish waste and recycling management company), FORMAS (the Swedish Research Council for Environment, Agricultural Sciences and Spatial Planning), the Royal Society for the Arts and Sciences in Gothenburg (KVVS) and the European Commission's career integration grant (project: Bioanode). The authors also wish to express their gratitude to Anna Eriksson, Per Johansson, Olivier Lavergne and Mona Pålsson for their assistance with the laboratory work.

Author Contributions: K.K.F., O.M. and A.M.S. together conceived and designed the experiments; performed the experiments; analyzed the data; and wrote the paper. K.K.F. coordinated the work.

Conflicts of Interest: The authors declare no conflict of interest.

References

1. Sierra, C.; Menendez-Aguadoa, J.M.; Afifb, E.; Carreroc, M.; Gallegoa, J.R. Feasibility study on the use of soil washing to remediate the As-Hg contamination at an ancient mining and metallurgy area. *J. Hazard. Mater.* **2011**, *196*, 93–100. [CrossRef] [PubMed]
2. Bisone, S.; Blais, J.; Drogui, P.; Mercier, G. Toxic metal removal from polluted soil by acid extraction. *Water Air Soil Pollut.* **2012**, *223*, 3739–3755. [CrossRef]
3. Moon, D.H.; Lee, J.R.; Wazne, M.; Park, J.H. Assessment of soil washing for Zn in contaminated soils using various washing solutions. *J. Ind. Eng. Chem.* **2012**, *18*, 822–825. [CrossRef]
4. Sierra, C.; Martínez-Blanco, D.; Blanco, J.A.; Gallego, J.R. Optimization of the concentration operation via magnetic separation: A case study for soil washing. *Chemosphere* **2014**, *107*, 290–296. [CrossRef] [PubMed]
5. Sierra, C.; Martínez, J.; Menéndez-Aguado, J.M.; Afif, E.; Gallego, J.R. High intensity magnetic separation for the clean-up of a site polluted by lead metallurgy. *J. Hazard. Mater.* **2013**, *248–249*, 194–201. [CrossRef] [PubMed]
6. Ghosh, M.; Singh, S.P. A review on phytoremediation of heavy metals and utilization of its byproducts. *Ecol. Environ. Res.* **2005**, *3*, 1–18. [CrossRef]

7.   Camenzuli, D.; Gore, D.B.; Stark, S.C. Immobilisation of metals in contaminated landfill material using orthophosphate and silica amendments: A pilot study. *Int. J. Environ. Pollut. Remediat.* **2015**, *3*, 27–32. [CrossRef]

8.   Chigbo, C.; Batty, L. Chelate-assisted phytoremediation of Cu-pyrene-contaminated soil using Z. Mays. *Water Air Soil Pollut.* **2015**, *226*. [CrossRef]

9.   Travar, I.; Kihl, A.; Kumpiene, J. The release of As, Cr and Cu from contaminated soil stabilized with apc residues under landfill conditions. *J. Environ. Manag.* **2015**, *151*, 1–10. [CrossRef] [PubMed]

10. Voglar, D.; Lestan, D. Chelant soil-washing technology for metal-contaminated soil. *Environ. Technol.* **2014**, *35*, 1389–1400. [CrossRef] [PubMed]

11. Pociecha, M.; Lestan, D. Washing of metal contaminated soil with edta and process water recyling. *J. Hazard. Mater.* **2012**, *235–236*, 334–337. [CrossRef] [PubMed]

12. Karlfeldt Fedje, K.; Yillin, L.; Strömvall, A.M. Remediation of metal polluted hotspot areas through enhanced soil washing—Evaluation of leaching methods. *J. Environ. Manag.* **2013**, *128*, 489–496.

13. Achugasim, O.; Ojinnaka, C.; Osuji, L. Assessment of ability of three chemcial oxidants to remove hydrocarbons from soils polluted by bonny light crude oil. *Eur. Chem. Bull.* **2013**, *2*, 226–230.

14. Xu, C.B.; Dong, D.B.; Meng, X.L.; Su, X.; Zheng, X.; Li, Y.Y. Photolysis of polycyclic aromatic hydrocarbons on soil surfaces under uv irradiation. *J. Environ. Sci. China* **2013**, *25*, 569–575. [CrossRef]

15. Baboshin, M.A.; Golovleva, L.A. Aerobic bacterial degradation of polycyclic aromatic hydrocarbons (PAHs) and its kinetic aspects. *Microbiology* **2012**, *81*, 639–650. [CrossRef]

16. Lundstedt, S.; White, P.A.; Lemieux, C.L.; Lynes, K.D.; Lambert, L.B.; Oberg, L.; Haglund, P.; Tysklind, M. Sources, fate, and toxic hazards of oxygenated polycyclic aromatic hydrocarbons (PAHs) at PAH-contaminated sites. *Ambio* **2007**, *36*, 475–485. [CrossRef]

17. Jenkins, J.; Dodge, P.; Morenci, A. *Solvent Extraction Transfer of Cu from Leach Solution to Electrolyte*; Pergamon: New York, NY, USA, 2002; pp. 307–325

18. Nancharaiah, Y.V.; Venkata Mohan, S.; Lens, P.N. Metals removal and recovery in bioelectrochemical systems: A review. *Bioresour. Technol.* **2015**. [CrossRef] [PubMed]

19. Modin, O.; Fedje, K.K. Combined Wastewater Treatment and Recovery of Copper from Ash Leachate. In Proceedings of the IWA World Water Congress & Exhibition, Busan, Korea, 16–31 September 2012.

20. Tao, H.C.; Liang, M.; Li, W.; Zhang, L.J.; Ni, J.R.; Wu, W.M. Removal of copper from aqueous solution by electrodeposition in cathode chamber of microbial fuel cell. *J. Hazard. Mater.* **2011**, *189*, 186–192. [CrossRef] [PubMed]

21. Ter Heijne, A.; Liu, F.; Weijden, R.V.D.; Weijma, J.; Buisman, C.J.N.; Hamelers, H.V.M. Copper recovery combined with electricity production in a microbial fuel cell. *Environ. Sci. Technol.* **2010**, *44*, 4376–4381. [CrossRef] [PubMed]

22. Modin, O.; Wang, X.; Wu, X.; Rauch, S.; Fedje, K.K. Bioelectrochemical recovery of Cu, Pb, Cd, and Zn from dilute solutions. *J. Hazard. Mater.* **2012**, *235–236*, 291–297. [CrossRef] [PubMed]

23. Swedish Standard Institute. SS-EN-12457-3 characterization of waste—Leaching—Compliance test for leaching of granular waste materials and sludges. In *Part 3: Two Stage Batch Test at a Liquid to Solid Ratio of 2 L/kg and 8 L/kg for Materials with High Soild Content and with Particle Size Below 4 mm (witout or with Size Reduction)*; Swedish Standard Institute: Stockholm, Sweden, 2003.

24. Montgomery, D.C. *Design and Analysis of Experiments*, 7th ed.; John Wiley & Sons Pte Led (Asia): Hongkong, China, 2009.

25. SEPA. *Riktvärden för Förorenad Mark Modellbeskrivning och Vägledning Rapport 5976 (Guidelines for Contaminated Soil Model Description and Guidance Report 5976)*; SEPA: Stockholm, Sweden, 2009.

26. Springfors, A. *Kartläggning av Föroreningssituationen i Mark, Grundvatten och Sediment, Förenklad Riskbedömning, Översiktlig Åtgärdsutredning Samt Riskbedömning för Föreslagna Efterbehandlingsåtgärder för Långön, Köpmannebo*; ÅF-IPK AB: Göteborg, Sweden, 2001.

27. Florell, S.E.; Stubdrup, O.P.; Arnér, M. *Rapport, Televerket Björkhult, Huvudstudie (Report, Televerket Björkhult)*; WSP: Linköping, Sweden, 2008.

28. Naturvårdsverkets föreskrifter om deponering, kriterier och förfaranden för mottagning av avfall vid anläggningar för deponering av avfall. Available online: http://www.naturvardsverket.se/Stod-i-miljoarbetet/Rattsinformation/Foreskrifter-allmanna-rad/NFS/2004/NFS-200410---Mottagning-av-avfall-for-deponering/ (accessed on 17 July 2015).

29.   Yip, T.C.M.; Tsang, D.C.W.; Ng, K.T.W.; Lo, I.M.C. Empirical modeling of heavy metal extraction by edds from singel-metal and mulit-metal contaminated soils. *Chemosphere* **2008**, *74*, 301–307. [CrossRef] [PubMed]
30.   Udovic, M.; Lestan, D. Edta and hcl leaching of calcareous and acidic soils polluted with potentially toxic metals: Remediation efficiency and soil impact. *Chemosphere* **2012**, *88*, 718–724. [CrossRef] [PubMed]
31.   Karlfeldt, K.; Steenari, B.M. Assessment of metal mobility in msw incineration ashes using water as the reagent. *Fuel* **2007**, *86*, 1983–1993. [CrossRef]
32.   Kalmykova, Y.; Strömvall, A.M.; Steenari, B.M. Alternative materials for removal of heavy metals and petroleum hydrocarbons from contaminated leachates. *Environ. Technol.* **2008**, *29*, 111–122. [CrossRef] [PubMed]
33.   Swedish Environmental Protection Agency. Generic Guideline Values for Contaminated Soils. Availble online: http://www.swedishepa.se/en/In-English/Start/Operations-with-impact-on-the-environment/ Remediation-of-contaminated-areas/Classification-and-risk-assessment/Generic-guideline-values-for-contaminated-soils/ (accessed on 29 May 2012).
34.   Bandowe, B.A.M.; Wilcke, W. Analysis of polycyclic aromatic hydrocarbons and their oxygen-containing derivatives and metabolites in soils. *J. Environ. Qual.* **2010**, *39*, 1349–1358. [CrossRef] [PubMed]
35.   Jonsson, S.; Persson, Y.; Frankki, S.; van Bavel, B.; Lundstedt, S.; Haglund, P.; Tysklind, M. Degradation of polycyclic aromatic hydrocarbons (PAHs) in contaminated soils by Fenton's reagent: A multivariate evaluation of the importance of soil characteristics and pah properties. *J. Hazard. Mater.* **2007**, *149*, 86–96. [CrossRef] [PubMed]
36.   Mao, D.; Lookman, R.; Van de Weghe, H.; Weltens, R.; Vanermen, G.; De Brucker, N.; Diels, L. Combining HPLC-GCXGC, GCXGC/TOF-MS, and selected ecotoxicity assays for detailed monitoring of petroleum hydrocarbon degradation in soil and leaching water. *Environ. Sci. Technol.* **2009**, *43*, 7651–7657. [CrossRef] [PubMed]
37.   Kalmykova, Y.; Bjoerklund, K.; Stroemvall, A.-M.; Blom, L. Partitioning of polycyclic aromatic hydrocarbons, alkylphenols, bisphenol A and phthalates in landfill leachates and stormwater. *Water Res.* **2013**, *47*, 1317–1328. [CrossRef] [PubMed]
38.   Nielsen, K.; Kalmykova, Y.; Strömvall, A.M.; Baun, A.; Eriksson, E. Particle phase distribution of polycyclic aromatic hydrocarbons in stormwater—Using humic acid and iron nano-sized colloids as test particles. *Sci. Total Environ.* **2015**, *532*, 103–111. [CrossRef] [PubMed]
39.   Tunega, D.; Gerzabek, M.H.; Haberhauer, G.; Totsche, K.U.; Lischka, H. Model study on sorption of polycyclic aromatic hydrocarbons to goethite. *J. Colloid Interface Sci.* **2009**, *330*, 244–249. [CrossRef] [PubMed]
40.   Oren, A.; Chefetz, B. Successive sorption-desorption cycles of dissolved organic matter in mineral soil matrices. *Geoderma* **2012**, *189–190*, 108–115. [CrossRef]
41.   Yap, L.C.; Gan, S.; Ng, H.K. Fenton based remediation of polycyclic aromatic hydrocarbons-contaminated soils. *Chemosphere* **2011**, *83*, 1414–1430. [CrossRef] [PubMed]
42.   Tran, L.H.; Drogui, P.; Mercier, G.; Blais, J.F. Electrochemcial degradation of polycyclic aromatic hydrocarbons in creostoe solution using ruthenium oxid on titanium expanded mesh anod. *J. Hazard. Mater.* **2009**, *164*, 1118–1129. [CrossRef] [PubMed]
43.   Zheng, X.J.; Blais, J.F.; Mercier, G.; Bergeron, M.; Drogui, P. PAH removal from spiked municipal wastewater sewage sludge using biological, chemical and electrochemical treatments. *Chemosphere* **2007**, *68*, 1143–1152. [CrossRef] [PubMed]

metals

MDPI

Article

Heavy Metal Behavior in Lichen-Mine Waste Interactions at an Abandoned Mine Site in Southwest Japan

Yuri Sueoka [1,*], Masayuki Sakakibara [1] and Koichiro Sera [2]

[1] Graduate School of Science and Engineering, Ehime University, Bunkyo-cho 2-5, Matsuyama, Ehime 790-8577, Japan; sakakibara.masayuki.mb@ehime-u.ac.jp

[2] Cyclotron Research Center, Iwate Medical University, Tomegamori 348-58, Takizawa, Iwate 020-0603, Japan; ksera@iwate-med.ac.jp

* Author to whom correspondence should be addressed; sueoka@sci.ehime-u.ac.jp; Tel.: +81-89-927-9649; Fax: +81-89-927-9640.

Academic Editor: Suresh Bhargava

Received: 13 July 2015; Accepted: 28 August 2015; Published: 2 September 2015

Abstract: The lichen, *Stereocaulon exutum* Nylander, occurring in a contaminated abandoned mine site was investigated to clarify (1) the behavior of heavy metals and As during the slag weathering processes mediated by the lichen; and (2) the distribution of these elements in the lichen thallus on slag. The heavy metals and As in the slag are dissolved from their original phases during the weathering process by lichen substances (organic acids) and hypha penetration, in addition to non-biological weathering. The dissolved elements are absorbed into the lichen thallus. Some of these dissolved elements are distributed in the cells of the hyphae. The others are distributed on the surface of the hyphae as formless particles and show lateral distribution inside the cortex of the thallus. The Cu and Zn concentrations in the thalli are positively correlated with the concentrations in the corresponding substrata and a positive intercept in the regression curve obtained using a linear function. These chemical characteristics make this lichen a good biomarker for Cu and Zn contamination of the substrata of the lichen. Therefore, the present study supposes that *Stereocaulon exutum* has a possible practical application in biomonitoring or risk assessment of heavy metal pollution at abandoned mine sites.

Keywords: *Stereocaulon exutum*; biomarker; smelting slag; weathering

---

1. Introduction

Lichen covers more than 6% of the land surface of the earth [1]. Lichen is found in various areas, including extreme environments such as tropical forests, deserts, alpine regions, polar regions, and urban areas, and even in highly polluted areas [2,3]. Therefore, as organisms that indicate the environmental conditions of their habitats, lichens could have broad utility as bioindicators and biomarkers throughout the world.

The interactions of lichens with their substrata have been investigated. In addition to abiotic weathering, the weathering of rocks is accelerated by lichens [4–6]. Biological weathering by lichens occurs through two main processes [7,8]. One is a chemical process involving lichen substances called lichen acids. Lichen acids are organic acids that affect the chemical decomposition of minerals. The other is a physical process of penetration and expansion of lichen hyphae. The mechanical fragmentation by hyphae increases the surface area of the rocks and minerals and accelerates the chemical decomposition.

Lichens have been well studied as a biomonitoring tool for air pollution [9–13]. The lichens absorb mineral nutrients and trace elements, including metals from dry and wet atmospheric deposition, due to the lack of a protective cuticle and a vascular root system [14,15]. Therefore, some lichens are sensitive to airborne pollutants [16]. Some characteristics of lichens that could be used as bioindicators and biomarkers for heavy metal pollution of their substrata have been revealed in recent studies. Several lichens absorb and/or accumulate heavy metals from the substrata [17,18]. Osyczka and Rola (2013) found that the relationships between the Zn and Cd contents in *Cladonia rei* Schaer. thalli and in the host substrate through specific non-linear regression models could be described by a power function [19].

However, the interactions between lichens and the corresponding substrata have not been discussed comprehensively in relation to the behavior of heavy metals during the weathering process and the heavy metal concentrations. Thus, there is insufficient information on the availability of lichens as bioindicators or biomarkers for heavy metal pollution of their substrata. Therefore, this study has clarified (1) the behavior of Cu, Zn, As, and Pb during the slag weathering processes mediated by *Stereocaulon exutum* Nyl., and (2) the distribution of these elements in *S. exutum* thalli growing on slag. Finally, the present study evaluated the possible practical application of the lichen for environmental monitoring in terms of heavy metal and As pollution.

2. Materials and Methods

2.1. Study Area

This study was conducted at an abandoned mine site in Eastern Okayama, Southwest Japan, at an altitude of approximately 250 m above sea level (Figure 1). The site is located in the temperate zone with an average annual temperature of 14.1 °C (the seasonal averages are 2.8 °C in January and 25.8 °C in July) and a total annual precipitation of 1200 mm (the seasonal totals were 22.0 mm in January and 299 mm in August) in 2014 [20].

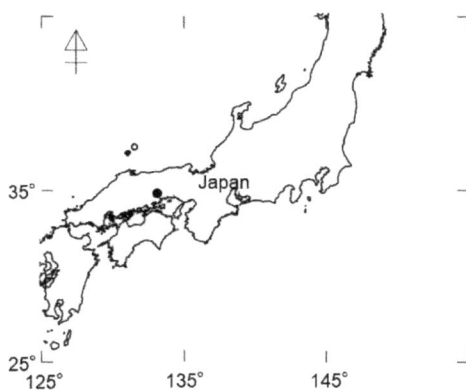

Figure 1. A location map of study area (indicated by a black dot).

Mainly Cu had been smelted at the abandoned mine site more than 50 years ago. The host rocks for the ore deposits are mainly rhyolite, andesite, sandstone, mudstone, and phreatomagmatic breccia [21]. These host rocks contain pyrite and chalcopyrite with quartz and calcite veins.

The surface of the waste dump consists primarily of tailings composed of phreatomagmatic breccia and rhyolite, and fractured slag overlies solidified, coherent slag [21]. The tailing and fractured slag have been kept on-site, in part, by wooden boards. However, these boards are now decaying and are no longer effective [21].

The soil and the stream water that flows through the waste dump are contaminated by Zn and/or As because the slag and tailings leach from the waste dump and accumulate on erosion control dams and sandbanks. The pH of the stream is nearly neutral (7.2–7.8), but the water contains Zn and As in excess of the environmental water quality standards determined by the Ministry of Health, Labor and Welfare, Japan [22,23]. The soil in the waste dump also contains As in excess of Japan's environmental quality standards set by the Ministry of the Environment [22,23].

The slag is one of the potential pollutants in this area and contains high levels of heavy metals [21]. *S. exutum* can grow even on the slag, and it could affect the slag weathering. Therefore, this fruticose lichen was chosen as the object of this study.

2.2. Sampling Methods

The 18 lichens, 12 slags, seven tailings, and four rocks were randomly sampled from the slag and tailing dump, outcrops, and sandbanks at the abandoned mine site from 21 April 2011 to 28 September 2013. The lichen samples were taken with the corresponding substrata and packed into plastic bags that have a zipper. The slag, tailing, and rock samples were packed into plastic bags or put into cartons. The lichen samples for SEM-EDS analysis were stored in cool and dark place. The lichen samples for PIXE analysis were stored in a desiccator after cleaning by hyper-pure water.

2.3. Analytical Methods

The lichens were identified by morphological and biochemical classification using color reactions with the following three solutions: 10% aqueous potassium hydroxide [KOH], saturated aqueous calcium hypochlorite [Ca(OCl)$_2$], and 5% alcoholic *p*-phenylenediamine solutions.

Rocks and slag were sampled at random, and their structures were observed using a scanning electron microscope (SEM). Then, the substrata were investigated by X-ray powder diffraction (XRD) analysis, and a scanning electron microscope equipped with an energy-dispersive X-ray spectrometer (SEM-EDS) was used to identify the primary and subsidiary phases and to clarify the behavior of heavy metals and As during the slag weathering process.

Eight pulverized slags and four pulverized rocks were analyzed by XRD to identify their mineral components. XRD analyses were performed on an Ultima IV (Rigaku, Tokyo, Japan) spectrometer housed at Ehime University (Ehime, Japan) using Cu K$_\alpha$ (λ = 1.54056 Å) radiation. The diffraction patterns for non-constant azimuth analysis were collected using an accelerating voltage of 40 kV, a specimen current of 40 mA, an analytical speed of 2°/min, and an analytical range of 5°–70°.

To investigate the distribution of potentially toxic elements in the slag, the slag and lichen samples were imaged and analyzed by SEM-EDS on a JSM-6510LV (JEOL, Tokyo, Japan) and X-Max 50 (Oxford Instruments, Tokyo, Japan) detector with INCA software (Oxford Instruments, Tokyo, Japan) at Ehime University. The slag samples were embedded in resin and prepared as polished thin sections that were then carbon-coated. The glass-phase content in the slag was estimated by counting the intersection points of a 200 μm mesh overlaid on back-scattered electron images. The SEM was operated with an accelerating voltage of 15 kV and a beam current of 0.8 nA. A counting time of 50 s was used for quantitative analysis, and count times of >1 h were required for element mapping. Enstatite (MgSiO$_3$), K-feldspar (KAlSi$_3$O$_8$), and anorthite (CaAl$_2$Si$_2$O$_8$) (Japan Electron Optics Laboratory Mineral Standard Samples for Electron Probe Micro Analyzer, Tokyo, Japan) were analyzed to confirm the analytical precision and accuracy. The following standards were used: NaAlSi$_2$O$_6$ for Na, Mg$_2$SiO$_4$ for Mg, Al$_2$O$_3$ for Al, CaSiO$_3$ for Si, FeS$_2$ for S, KBr for K, Ti for Ti, Cr for Cr, Mn for Mn, FeS$_2$ for Fe in matte drops, Fe$_2$O$_3$ for Fe in magnetite, Fe$_2$SiO$_4$ for Fe in fayalite and glass, Co for Co, Cu for Cu, ZnO-glass for Zn, InAs for As, Ag for Ag, and Sn for Sn. The Co standard was also used to optimize the quantification of the analyses. The L characteristic X-ray lines were used for the Cu, Zn, As, Ag, and Sn quantification, and the K lines were used for other elements.

The heavy metal concentrations of the substrata, including seven slags, eight tailings, and seven host rocks, were determined by energy dispersive X-ray fluorescence (ED-XRF) spectrometry using an Epsilon5 instrument (PANalytical, Almelo, The Netherlands) at Ehime University. Pressed powder pellet samples were used for the XRF analysis. The samples were powdered to a grain size of <1 μm in a tungsten carbide vibrating sample mill (Sample Mill model TI-100, HEIKO, Tokyo, Japan) and an agate mortar at Ehime University. $C_{13}H_{14}O_6$ was used as a binder (20 wt. % in the pressed powder pellets) and homogenized with the powdered samples by shaking for 1 h. The XRF analyses were conducted at an excitation voltage of 100 kV. The following standards were used as secondary targets for irradiation with the relevant excited X-rays for the individual elements: Al for Na and Mg; Ti for Al–Sc; Ge for Ti–Ga; Mo for Ge–Y, Tl–Ra, and Ac–U; and Al_2O_3 for other elements, including lanthanides. The X-ray source was a Gd anode. A counting time of 300 s was repeated three times on a Ge detector for each secondary target, and elemental concentrations were determined by the fundamental parameter method. Geological standards, including JSd-1, JSd-2, JSd-3, JSl-1, Jlk-1, and JMn-1 (Geological Survey of Japan Referenced Materials) [24,25], were analyzed to confirm the analytical precision (relative standard deviations are 1.86% Fe_2O_3, 11.3% Cu, 3.09% Zn, 10.9% As, 45.5% Sn, and 53.2% Pb).

The heavy metal concentrations of the lichens were determined by particle-induced X-ray emission (PIXE) performed at the Nishina Memorial Cyclotron Center established by the Japan Radioisotope Association. The analytical conditions followed Sera *et al.* (1992) [26]. A small cyclotron provides a 2.9 MeV-proton beam on the target after passing through a beam collimator of graphite. The maximum beam intensity on the target is approximately 40 nA for a beam spot diameter of 2 mm and 80 nA for a diameter of 6 mm. Elements from Na to U are detected by two ORTEC Si (Li) detectors. The elements heavier than Ca are detected by the first detector, which has a 0.025 mm-thick Be window and a 6 mm active diameter, with X-rays with an energy resolution of 154 eV at 5.9 keV and a 300- to 500-μm-thick Mylar absorber inserted between the target and the detector. The other low atomic number elements are detected by the second detector, which has a 0.008 mm Be window and a 4 mm active diameter, a resolution of 157 eV, and a small graphite aperture without an absorber. The upper 1–5 mm portion of the lichen thallus was cut out from the substrata to remove the effects of the trapped substratum fragments by the hypha structure and rinsed with ultrapure water. The rinsed samples were dried at 80 °C for 24 h in an oven. The dried samples (30 mg) were digested with 1 mL of HNO_3 and heated in a microwave (150 W). The analytical accuracy and precision of the analyses were verified using the NIES CRM No. 1 environmental sample.

2.4. Statistical Methods

The distributional normality of the Cu, Zn, As, and Pb concentrations in *S. exutum* and the corresponding substrata was verified by the Shapiro-Wilk normality test. The correlation of the scatter plots was verified by Spearman's rank-correlation coefficient. All statistical analyses were performed with EZR (Saitama Medical Center, Jichi Medical University, Saitama, Japan), a graphical user interface for R (The R Foundation for Statistical Computing, Vienna, Austria). More precisely, it is a modified version of R Commander designed to add statistical functions frequently used in biostatistics [27].

3. Results

3.1. Microstructure of the Hyphae of Stereocaulon exutum Thallus and the Effects of the Lichen on Slag Weathering

The hyphae of the thallus form loosely-complicated structures and adhere to each other near the interface with the slag. The hyphae that compose the medulla of the thallus form lines in the direction of growth. Hyphae that compose the cortex of the upper thallus form complicated structures, and these different hyphae partially adhere to each other.

Stereocaulon exutum has both physical and chemical effects on the slag during the weathering process. The surface of the weathered slag under the lichen is penetrated and fractured by the lichen hyphae during the physical process (Figure 2a–c). The hyphae penetrate and expand into the crack of the slag and along the mineral interfaces. Etch-pits occur on the weathered willemite and matte drops under the lichen during the chemical process (Figure 2e).

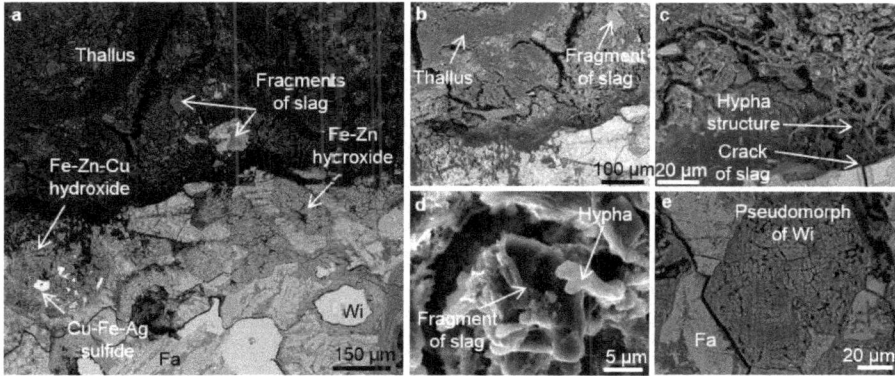

Figure 2. Back-scattered electron images of (**a**) lichen-slag interface; (**b**) slag fragments caught by hyphae structure; (**c**) hyphae penetrating crack of slag; (**e**) etch-pits of pseudomorph of willemite; and (**d**) a secondary electron image of slag fragments caught by hyphae structure. W-Wi: weathered willemite; Wi: willemite; Fa: fayalite.

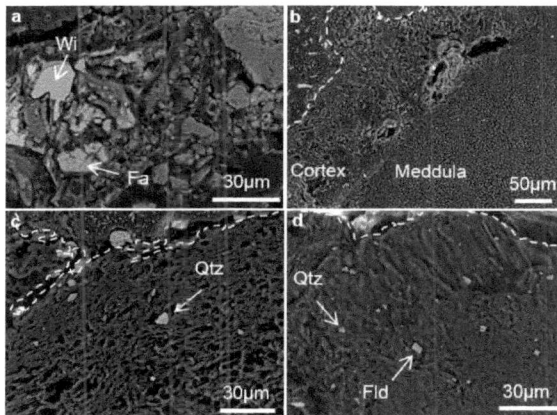

Figure 3. Back-scattered electron images of thallus containing fragments. Distances from substrata increase from (**a**) to (**d**) every 2 mm. (**a**) is the section where 100 μm high from slag; (**b**) and (**c**) are the section of thallus; (**d**) is the section of apothecium. The broken lines indicate thallus-resin interfaces. Wi: willemite; Fa: fayalite; Qtz: quartz; Fld: feldspar.

The slag fragments in the thallus show differences in the distance from the substrata. The lower portion of the thallus contains fragments of the fractured slag (Figure 2b,d). The amounts of the slag fragments in the upper portion (more than 1 mm from the substratum) are less comparable with the portion near the interface between the lichen and the slag. The fragments trapped in the upper portion consist of mainly quartz and plagioclase (Figure 3).

3.2. Heavy Metal Distribution in the Weathered Slag

The slag is composed mainly of willemite, fayalite, and/or magnetite with a silicate glassy matrix and contains matte drops, which are Cu-, Zn-, Pb-metals, -alloys, and -sulfides. Willemite is the main host of Zn, and fayalite is the main host of Fe. Matte drops are the main hosts of heavy metals, S, and As.

The heavy metals and As are dissolved from the original phases during the weathering process. The outermost portion in the weathered zone comprises mainly clay minerals. The willemite and matte drops are converted to Fe-hydroxides during the weathering process. The heavy metal (except for Fe and As) concentrations in the weathered phases are lower than in the original phases.

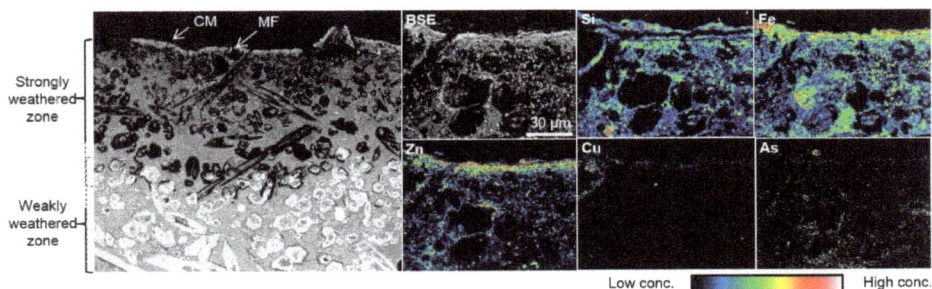

Figure 4. Back-scattered electron images and the elemental maps of weathered slag obtained by SEM-EDS. MF: Metal film-like structure; CM: Clay minerals.

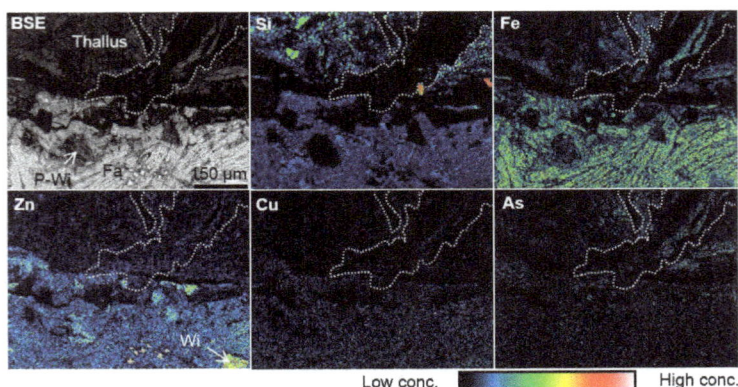

Figure 5. A back-scattered electron image and the elemental maps of a lichen-slag interface obtained by SEM-EDS. The broken lines indicate thallus. Fa: fayalite; P-Wi: pseudomorph of willemite.

Heavy metals and As in the willemite-rich slag are partially concentrated inside the clay minerals and form a film-like structure (Figure 4). However, during the lichen-mediated weathering process, the heavy metals and As do not form the structure under the lichen thallus but rather tend to be concentrated in the lichen thalli (Figure 5).

3.3. Elemental Distribution in the S. exutum Thallus

The thallus contains heavy metals and As. Fe and As are especially contained on the surface of the hyphae in the thallus (Figure 6). Formless particles occur on this portion. The hyphae contain Fe, Cu, Zn, and As in the cells of their rhizome (Figure 7). Several elements, including Fe, Cu, Zn, and As, are contained inside the cortex of the thallus and show lateral distribution (Figure 8). Formless particles occur in this portion. The concentrations of these elements in the phyllocladium were not found.

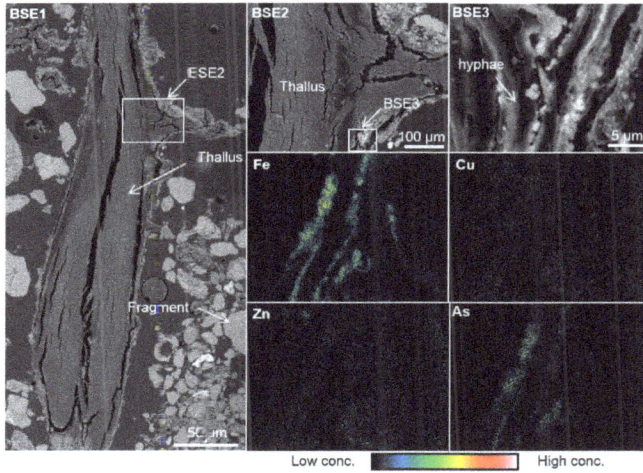

Figure 6. Back-scattered electron images and the elemental maps of hyphae inside the cortex of the thallus. The white box in BSE2 indicates the area of BSE3 and the mapping analysis area.

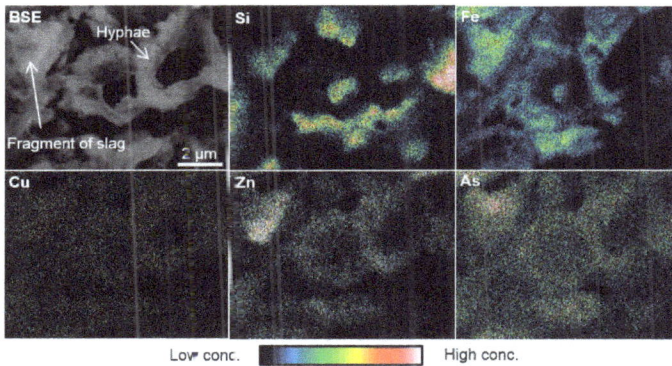

Figure 7. A back-scattered electron image and the elemental maps of the lichen hyphae.

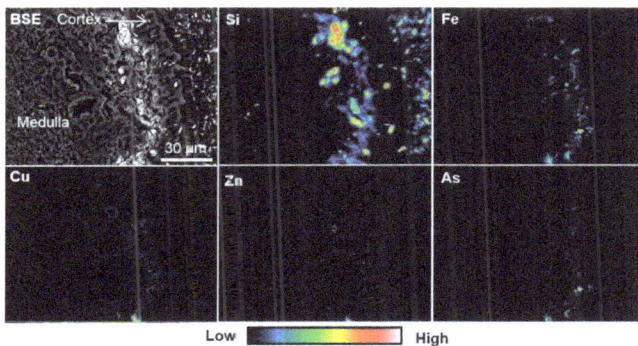

Figure 8. A back-scattered electron image and the elemental maps of the lichen thallus.

3.4. Heavy Metals and As Concentrations of Lichens and the Corresponding Substrata

The heavy metal and As concentrations of *Stereocaulon exutum* on slag and rocks, including tailings, gravel, and the host rock, are 33.6–1250 mg/kg-DW Zn, 6.51–923 mg/kg-DW Cu, 12.0–147 mg/kg-DW Pb, and ND-438 mg/kg-DW As. The heavy metal and As concentrations of the substrata are as follows: slag contains 20.6%–3.27%-DW Zn, 1.62%–0.66%-DW Cu, ND–758 mg/kg-DW Pb, and 31.3–561 mg/kg-DW As; the tailings contain 640–101 mg/kg-DW Zn, 154–34.2 mg/kg-DW Cu, 12.7–1280 mg/kg-DW Pb, and ND–23.0 mg/kg-DW As; the outcrop rocks contain 79.1–1940 mg/kg-DW Zn, 96.6–461 mg/kg-DW Cu, 8.08–97.1 mg/kg-DW Pb, and 1.12–490 mg/kg-DW As.

On determination of the correlations between heavy metal concentrations in lichens and the corresponding substrata, the distributional normality was verified by the Shapiro-Wilk normality test (Table 1). As no significant normal distribution of the scattered plots was shown in corresponding concentrations of Cu, Zn, As, and Pb in the lichen and the corresponding substrata (Table 1, Figure 9), the correlation was verified by Spearman's rank-correlation coefficient. The results of the test show that the Cu and Zn concentrations in *S. exutum* thalli exhibit positive correlations with those in the corresponding substrata under the 5% significance level but not As and Pb concentrations (Table 2, Figure 10). The formulas of the regression curves have positive X-intercepts (Figure 10).

Table 1. *p*-values of heavy metal concentrations in lichens and the corresponding substrata obtained by Shapiro-Wilk normality test.

Elements	*p*-Values	
	S. exutum	Substrata
Cu	1.23×10^{-3}	8.26×10^{-5}
Zn	3.48×10^{-5}	4.39×10^{-6}
As	1.16×10^{-2}	0.109
Pb	0.207	2.07×10^{-5}
log.Cu	0.888	4.31×10^{-3}
log.Zn	5.52×10^{-3}	8.17×10^{-4}
log.As	1.92×10^{-3}	0.437
log.Pb	2.24×10^{-4}	0.802

Table 2. Spearman's correlation rho and *p*-values of the scattered plots of Cu, Zn, As, and Pb concentrations in the lichens and the corresponding substrata. $n = 18$.

Elements	r_s	*p*-Values
Cu	0.618	7.43×10^{-3}
Zn	0.829	1.28×10^{-5}
As	0.421	0.119
Pb	0.336	0.188

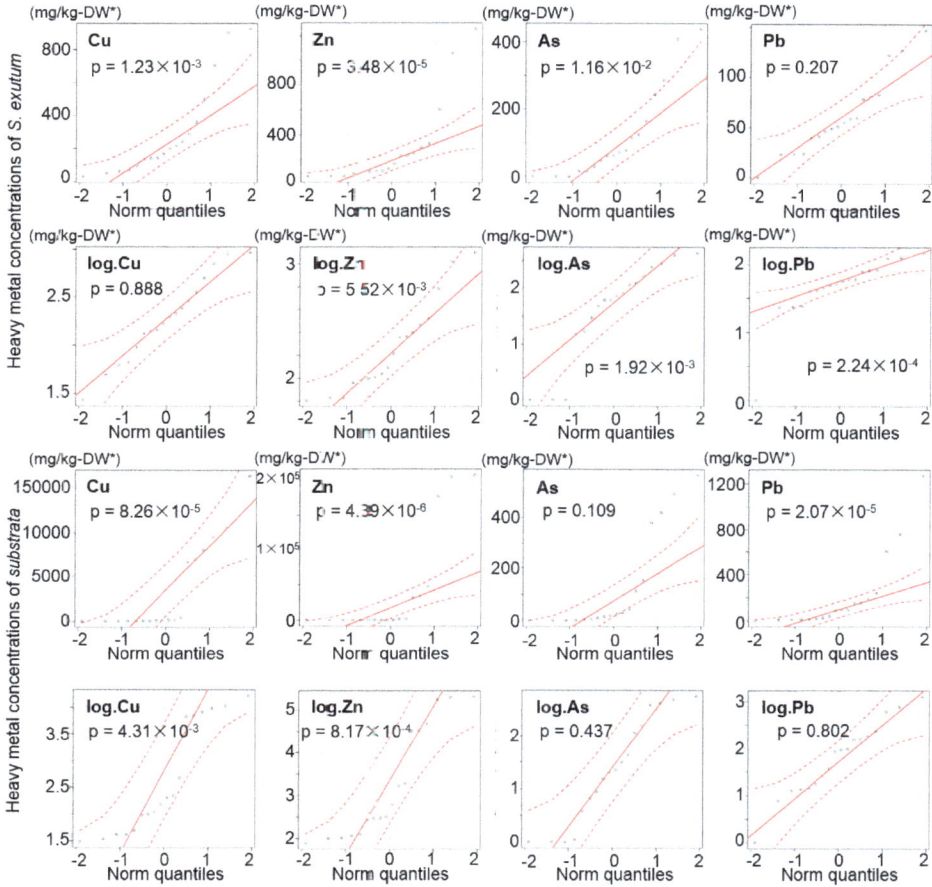

Figure 9. QQ plots of heavy metal concentrations of lichens and the corresponding substrata and the logarithmic data. The *p*-values were obtained by Shapiro-Wilk normality test. *n* = 18.

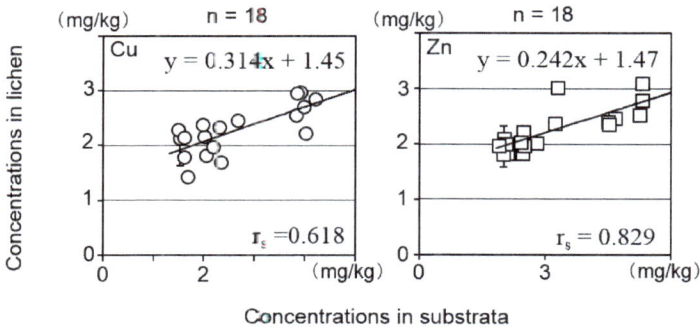

Figure 10. Scatter diagrams show Cu and Zn concentration of *S. exutum* and the corresponding substrata with linear functions after logarithmic transformation. Note: significance level = 5%. *n* = sample number.

4. Discussion

4.1. Effects of S. exutum on Heavy Metals and As Behavior in the Slag Weathering Process

In addition to abiological weathering, *Stereocaulon exutum* affects and accelerates the weathering of the original phases, and the effect can be attributed to both physical and chemical processes. The physical effects are caused by hyphal penetration and expansion, and the chemical effects are caused by lichen substances.

Fragmentation and disaggregation of the lithic surface below the lichens have been well known and investigated in previous studies [4,5]. Various types of rocks, including calcareous rock, siliceous rock, and sedimentary rock, can be weathered physically by the hyphal penetration [7,8,28–31]. The slag that is composed of mainly silicate phases is also fractured by the hyphae even in the coherent portion.

The fragments of the substrata of lichens are incorporated into the thallus and coated by extracellular polymers and secondary minerals [4,5]. The fragments of the weathered slag are observed mainly in the lower 1–5 mm portion of the thallus, closer to the substratum. However, the trapped fragments in the upper than 5 mm portion of the thallus consist of mainly quartz and feldspar. These results indicate that the fragments are not transported to the upper portion of the thallus. Thus, the origin of the fragments trapped in the upper thallus could not be the substrata but atmospheric deposition. The trapped slag fragments consist of the Fe hydroxides containing Zn as weathered secondary minerals. However, no extracellular polymers were found in this study.

In addition to penetration of hyphae, substrata are mechanically disrupted by expansion, contraction, freezing, and thawing of the lichen thallus [5]. The hyphae penetrate into cracks of the willemite and matte drops in the slag and along the interface of these phases. The lichen thallus has an abundant water storage capacity [32]. Therefore, the slag is physically disaggregated through wetting and drying of the hyphae, and chemical weathering could be accelerated by the water supply.

Lichens affect not only physical weathering but also chemical weathering mostly by lichen compounds [6]. The main chemical dissolution processes of minerals by lichens are generated by respiratory CO_2, the excretion of oxalic acid, and the chemical action of lichen compounds [33]. Mineral etching is a frequent feature of weathering at the rock-lichen interface [6]. The etch-pits in the weathered willemite and matte drops are evidence of the chemical weathering of these phases in this study. The excretion of various organic compounds can effectively dissolve the original phases and chelate metallic cations [4–6]. Several lichen compounds such as depsides and depsidones are slightly soluble and have the ability to chelate in conjunction with the presence of polar groups such as OH, CHO, and COOH [34]. Although the dissolution and decomposition of rock-forming minerals by lichen acids has been demonstrated in laboratory experiments [35,36], the dissolution and decomposition processes have not been conclusively proven in previous field studies [37,38].

S. exutum contains the depsidone (lobaric acid) in the thallus and could affect the dissolution and decomposition of slag phases. Although the present study also provides no scientific evidence to clarify the processes, the results show the evidence of heavy metal and As dissolution from the original phases by acids. Willemite and matte drops, which are the main hosts of heavy metals and As, are transformed to Fe hydroxide during the weathering process. The heavy metal and As concentrations of weathered phases in the weathered portion of the slag are lower than the heavy metal and As concentrations in the unweathered portion [21]. The lower heavy metal and As concentrations in the weathered phases and the occurrence of etch-pits in the pseudomorphs of willemite and matte drops in only the lichen-mediated weathered portion are evidence of the effect of lichens on decomposing of slag phases and dissolving of heavy metals from the original phases. Although this study could not clarify whether the *S. exutum* or other microbes are mostly effective in dissolving and decomposing slag phases by excretion of acids, the results indicate the lichen affects biochemical weathering of smelting slag in terms of heavy metal and As dissolution from the original phases for the first time.

4.2. The Accumulation and the Behavior of Heavy Metals in the Lichen Thallus

The heavy metal uptake ability and the accumulation capacity of lichens has been well demonstrated in various previous studies [19,39]. Several lichens, such as *Cladonia* spp. and *Stereocaulon* spp. that grow near Cu smelters and Cu-polluted sites contain high levels of heavy metals [40–42]. *S. exutum* growing on the tailing and slag dump at this study site also contains high concentrations of Cu, Zn, As, and Pb.

The heavy metals accumulated by *S. exutum* through the slag weathering process are contained in the lichen thallus. The kinetics and thermodynamics of ion exchange, such as binding constants and charge balances, have been well demonstrated in previous studies [43]. Cation uptake occurs extracellularly in lichens [44]. Nieboer and Richardson (1980) supposed that the affinity of ions for the exchange sites varies in an order corresponding to monovalent Class A < divalent Class A < borderline divalent < divalent Class B [45]. They categorized ions as follows: Class A ions are composed of alkaline metals, alkaline-earth metals, Al, Y, lanthanoid, and actinoid; borderline ions are composed of Sc, Ti, V, Cr, Mn, Fe, Co, Ni, Cu^{2+}, Zn, Ga, As, Cd, In, Sn, Sb, and Pb^{2+}; and Class B ions are composed of Cu^{1+}, Rh, Pd, Ag, Ir, Pt, Au, Hg, Tl, Pb^{4+}, and Bi [45]. The distribution of the cations has been estimated as follows four fractions into an intercellular and surface fraction, ion exchange site fraction, intracellular fraction, and residual fraction [46].

The SEM observations and elemental mapping analysis of the lichen thalli clarified at least two distributions of the absorbed elements in this study. Some of the elements are distributed throughout cells or the exchange sites of the cell wall of the hyphae in the thallus and show no trend. The others are distributed on the surface of the hyphae as unformed materials and show lateral distribution inside the cortex. As the elements do not concentrate in the phyllocladium, the fungal component positively absorbs the elements rather than the algal partner. Although this study could not clarify the form of the ions that are absorbed in the lichen thallus, the elements in the cell may indicate the results of absorption of the ions into the cytoplasm from external solutions or through positively and negatively charged anionic binding sites. The distribution of Fe and As on the surface of the hyphae as unformed materials could show the results of elemental precipitation or making compounds during the evaporation of external solutions. The elements in the external solutions may be affected by lichen substances during the processes of precipitation and chemical combination. As the carboxyl group can easily form complexes with Cu and Zn [47], these ions could make some compounds with lichen substances or some enzymes that are excreted by *S. exutum*. While the Cu and Zn could be affected by the lichen substances, Fe and As could be affected by inorganic precipitation and adsorption. The Fe and As in the thallus tend to be more concentrated on the surface of the hyphae of the thallus as formless particles in the study samples. As arsenic is easily adsorbed by Fe hydroxides [48], it is possible that the As and Fe absorbed by lichens may deposit on the surface of the hypha structure as Fe hydroxides.

The SEM observations and PIXE analysis indicate that the upper thallus contains heavy metals and As not as slag fragments but rather as ions and/or compounds. The hyphae of *S. exutum* are composed of filamentous fungi with a diameter of generally 1–10 μm. The filamentous fungi of ascomycete commonly have a septum with a stoma in the hyphae, and protoplasmic streaming occurs between the cells of the fungi [49,50]. Therefore, the dissolved elements from the weathered slag phases are taken into the cells of the lichen thallus and are transported to the upper portion.

While biominerals such as moolooite found by Purvis (1984) have not been found in this study [51], formless particles rich in Fe and As have been found on the surface of the hyphae in the thallus. In addition, Fe, Cu, Zn, and As are distributed throughout the cells of the hyphae. These results support the possibility of different mechanisms of Fe and As absorption and accumulation by lichens, compared with previous studies.

4.3. The Correlation of Cu and Zn Concentrations in the Lichen and Substrata

The relationship of heavy metal concentrations between lichens and the corresponding substrata has been demonstrated in previous studies [19,41]. Copper concentrations of *Cladonia subconistea* and *C. humilis* growing on Cu-hyperaccumulator moss *Scopelophila cataractae* at a Cu-polluted site are much higher than the copper concentrations of control samples growing on soil [41]. Osyczka and Rola (2013) [19] demonstrated the relationship of Zn and Cd concentrations between *C. rei* thalli and the host substrata using specific non-linear regression models described by a power function. In this study, the concentrations of Cu and Zn in lichens after logarithmic transformation show a positive correlation with their concentrations in the corresponding substrata that are obtained by linear regression models described by a linear function as well as the result of Osyczka and Rola (2013) [19], while the lichen species and the type of their substrata are different. The regression model clarified an impressive ability of *S. exutum* to accumulate Cu and Zn because the model showed the positive intercept. Furthermore, the relationship between the concentrations in the lichen thalli and the corresponding substrata shows no peak in the scatter diagram, which also supports the characterization of *S. exutum* as a Cu and Zn accumulator.

The dominant source of heavy metals should be estimated on the discussion of the relationships of the heavy metal content between lichens and the corresponding substrata. The fragments of the weathered slag have not been transported to the upper portion, and the analyzed portion is the portion higher than 1–5 mm from the substratum. Although the fragments that are contained in the analyzed portion are composed of mainly quartz and feldspar as material from atmospheric depositions, the Cu and Zn concentrations of the lichen show a positive correlation with the concentrations of their corresponding substrata. Therefore, the Cu and Zn concentrations in the lichen reflect the Cu and Zn contents absorbed by the lichen from the corresponding substratum and not the fragments of the substratum. Accordingly, the *S. exutum* appears to absorb mainly the heavy metals from the substrata below the lichen thallus at the study site.

5. Conclusions

Stereocaulon exutum affects slag weathering in biological processes by lichen substances (organic acids) and hyphae penetration and could accelerate non-biological weathering. Heavy metal and As in the slag are dissolved from the silicate, sulfide, and metal phases during the weathering process. The heavy metals form metal film-like structures in the non-lichen-mediated weathering portion of the slag but not in the lichen-mediated weathered portion. The elements are absorbed into the lichen and contained within the thallus. Some of the elements are distributed in the cells of the hyphae. The others are distributed on the surface of the hyphae as formless particles and show lateral distribution inside the cortex of the thallus.

S. exutum shows other interesting results related to the chemical properties. The Cu and Zn concentrations in the thalli are positively correlated with the concentrations in the corresponding substrata and the positive intercept of the regression curve obtained by a linear function. These chemical characteristics make this lichen a good biomarker for Cu and Zn contamination of the substrata of the lichen, making lichen-solid interactions very important and interesting to the field of environmental studies. Therefore, the present study supposes that *S. exutum* has a possible practical application to biomonitoring or risk assessment of heavy metal pollution at abandoned mine sites.

Acknowledgments: We are grateful to A. Kawamata (Ehime prefectural science museum, Japan) for help with identifications of lichens. The authors are grateful to H. Ohfuji (Ehime University, Japan) for help with analysis by SEM-EDS. We thank H. Akamatsu (Ehime University, Japan) for help with making thin sections. We thank the two anonymous referees for their perceptive comments and recommendations. This study was partially supported by JSPS KAKENHI Grant Number 15J04570 and grant from Fukada Geological Institute (Fukada Grant-in-Aid, 2013).

Author Contributions: Y. Sueoka mainly performed the present study and wrote the paper. M. Sakakibara provided advice and recommendations on analytical methods and polishing of the paper. K. Sera performed the PIXE analysis.

Conflicts of Interest: The authors declare no conflict of interest.

References

1. Purvis, O.W. Chapter Three: Lichens and industrial pollution. In *Ecology of Industrial Pollution*; Batty, L.C., Hallberg, K.B., Eds.; Cambridge University Press: Cambridge, UK, 2010; pp. 41–69.
2. Edwards, H.G.M.; Moody, C.D.; Villar, S.E.J.; Wynn-Williams, D.D. Raman spectroscopic detection of key biomarkers of cyanobacteria and lichen symbiosis in extreme Antarctic habitats: Evaluation for Mars Lander missions. *Icarus* **2005**, *174*, 560–571. [CrossRef]
3. Bačkor, M.; Loppi, S. Interactions of lichens with heavy metals. *Biol. Plant.* **2009**, *53*, 214–222. [CrossRef]
4. Adamo, P.; Violante, P. Weathering of rocks and neogenesis of minerals associated with lichen activity. *Appl. Clay Sci.* **2000**, *16*, 229–256. [CrossRef]
5. Chen, J.; Blume, H.P.; Beyer, L. Weathering of rocks induced by lichen colonization—A review. *Catena* **2000**, *39*, 121–146. [CrossRef]
6. Wilson, M.J.; Jones, D. Lichen weathering of minerals: Implications for pedogenesis. *Geol. Soc. Lond. Spec. Publ.* **2012**, *11*, 5–12. [CrossRef]
7. Banfield, J.F.; Baker, W.W.; Welch, S.A.. Taunton, A. Biological impact on mineral dissolution: Application of the lichen model to understanding mineral weathering in the rhizosphere. *Proc. Natl. Acad. Sci. USA* **1999**. *96*, 3404–3411. [CrossRef] [PubMed]
8. Favero-Longo, S.E.; Gazzano, C.; Girlanda, M.; Castelli, D.; Tretiach, M.; Baiocchi, C.; Piervittori, R. Physical and chemical deterioration of silicate and carbonate rocks by meristematic microcolonial fungi and endolithic lichens (Chaetothyriomycetidae). *Geomicrobiol. J.* **2011**, *28*, 732–744. [CrossRef]
9. Conti, M.E.; Cecchetti, G. Biological monitoring: Lichens as bioindicators of air pollution assessment—A review. *Environ. Pollut.* **2001**, *114*, 471–492. [CrossRef]
10. Golubev, A.V.; Golubeva, V.N.; Krylov, N.G.; Kuznetsova, V.F.; Mavrin, S.V.; Aleinikov, A.Y.; Hoppes, W.G.; Surano, K.A. On monitoring anthropogenic airborne uranium concentrations and 235U/238U isotopic ratio by Lichen—Bio-indicator technique. *J. Environ. Radioact.* **2005**, *84*, 333–342. [CrossRef] [PubMed]
11. Ohmura, Y.; Kawachi, M.; Ohtara, K.; Sugiyama, K. Long-term monitoring of *Parmotrema tinctrumand* qualitative changes of air pollution in Shimizu Ward, Shizuoka City, Japan. *J. Jpn. Soc. Atmos. Environ.* **2008**, *43*, 47–54.
12. Pinho, P.; Augusto, S.; Martins-Loução, M.A.; Pereira, M.J.; Soarea, A.; Máguas, C.; Branquinho, C. Causes of change in nitrophytic and oligotrophic lichen species in a Mediterranean climate: Impact of land cover and atmospheric pollutants. *Environ. Pollut* **2008**, *154*, 380–389. [CrossRef] [PubMed]
13. Guidotti, M.; Stella, D.; Dominici, C.; Blasi, G.; Owczarek, M.; Vitali, M.; Protano, C. Monitoring of traffic-related pollution in a province of central Italy with translocated lichen *Pseudovernia furfuracea*. *Bulll. Environ. Contam. Toxicol.* **2009**, *83*, 852–858. [CrossRef] [PubMed]
14. Čeburnis, D.; Steinnes, E. Conifer needles as biomonitors of atmospheric heavy metal deposition: Comparison with mosses and precipitation, role of the canopy. *Atmos. Environ.* **2000**, *34*, 4265–4271. [CrossRef]
15. Loppi, S.; Pirintsos, S.A. Epiphytic lichens as sentinels for heavy metal pollution at forest ecosystems (central Italy). *Environ. Pollut.* **2003**, *121*, 327–332. [CrossRef]
16. Nash, T.H. Sensitivity of lichens to sulfur dioxide. *Am. Biol. Lichenol. Soc.* **2012**, *76*, 333–339.
17. Purvis, O.W.; Pawlik-Skowronska, B.; Cressey, G.; Jones, G.C.; Kearsley, A.; Spratt, J. Mineral phases and element composition of the copper hyperaccumulator lichens *Lecanora polytropa*. *Mineral. Mag.* **2008**, *72*, 607–616. [CrossRef]
18. Purvis, O.W.; Bennett, J.P.; Spratt, J. Copper localization, elemental content, and thallus colour in the copper hyperaccumulator lichen *Lecanora. sierrae* from California. *Lichenologist* **2011**, *43*, 165–173. [CrossRef]
19. Osyczka, P.; Rola, K. Response of the lichen *Cladonia. rei* Schaer. To strong heavy metal contamination of the substrate. *Environ. Sci. Pollut. Res. Int.* **2013**, *20*, 5076–5084. [CrossRef] [PubMed]

20. Homepage of the Japan Meteorological Agency. Available online: http://www.data.jma.go.jp/gmd/risk/obsdl/index.php (accessed on 30 July 2015).

21. Sueoka, Y.; Sakakibara, M. Primary phases and natural weathering of smelting slag at an abandoned mine site in southwest Japan. *Minerals* **2013**, *3*, 412–426. [CrossRef]

22. Homepage of the Ministry of the environment, Japan. Available online: https://www.env.go.jp/kijun/dojou.html (accessed on 31 August 2015).

23. Sakakibara, M.; Ohmori, Y.; Ha, N.T.H.; Sano, S.; Sera, K. Phytoremediation of heavy metal-contaminated water and sediment by *Eleocharis. acicularis*. *Clean Soil Air Water* **2011**, *39*, 735–741. [CrossRef]

24. Imai, N.; Terashima, S.; Itoh, S.; Ando, A. Compilation of analytical data on nine GSJ geochemical reference samples, "Sedimentary Rock Series". *Geostand. Newsl.* **1996**, *20*, 165–216. [CrossRef]

25. Imai, N.; Terashima, S.; Itoh, S.; Ando, A. Compilation of analytical data for five GSJ geochemical reference samples: The "Instrumental Analysis Series". *Geostand. Newsl.* **1998**, *23*, 223–250. [CrossRef]

26. Sera, K.; Yanagisawa, T.; Tsunoda, H.; Futatsugawa, S.; Hatakeyama, S.; Saitoh, S.; Suzuki, S.; Orihara, H. Bio-PIXE at the Takizawa Facility (Bio-PIXE with a baby cyclotron). *Int. J. PIXE* **1992**, *2*, 325–330. [CrossRef]

27. Kanda, Y. Investigation of the freely available easy-to-use software "EZR" for medical statistics. *Bone Marrow Transplant.* **2013**, *48*, 452–458. [CrossRef] [PubMed]

28. Friedmann, E.I. Endolithic microorganisms in the Antarctic cold desert. *Science* **1982**, *215*, 1045–1053. [CrossRef] [PubMed]

29. Sanders, W.B.; Ascaso, C.; Wierzchos, J. Physical interactions of two rhizomorph-forming lichens with their rock substrate. *Bot. Acta* **1994**, *107*, 432–439. [CrossRef]

30. Wierzchos, J.; Ascaso, C. Application of back-scattered electron imaging to the study of the lichen-rock interface. *J. Microsc.* **1994**, *175*, 54–59. [CrossRef]

31. Wierzchos, J.; Ascaso, C. Morphological and chemical features of bioweathered granitic biotite induced by lichen activity. *Clays Clay Miner.* **1996**, *44*, 652–657. [CrossRef]

32. Creveld, M.C. *Epilithic Lichen Communities in the Alpine Zone of Southern Norway*; J. Gantner Verlag KG: Vaduz, Liechtenstein, 1981; pp. 48–55.

33. Syers, J.K.; Iskandar, I.K. Pedogenetic significance of lichens. In *The Lichens*; Ahmadjian, V., Hale, M.E., Eds.; Academic Press: New York, NY, USA, 1973; pp. 225–248.

34. Iskandar, I.K.; Syers, J.K. Solubility of lichen compounds in water: Pedogenetic implications. *Lichenologist* **1971**, *5*, 45–50. [CrossRef]

35. Iskandar, I.K.; Syers, J.K. Metal-complex formation by lichen compounds. *J. Soil Sci.* **1972**, *23*, 255–265. [CrossRef]

36. Ascaso, C.; Galvan, J. Studies on the pedogenetic action of lichen acids. *Pedobiologia* **1976**, *16*, 321–331.

37. Jones, D.; Wilson, M.J. Chemical activity of lichens on mineral surfaces—A review. *Int. Biodeterior.* **1985**, *21*, 99–104.

38. Wilson, M.J. Interactions between lichens and rocks. *Cryptogam. Bot.* **1995**, *5*, 299–305.

39. Goyal, R.; Seaward, M.R. Metal uptake in terricolous lichens: Metal localization within the thallus. *New Phytol.* **1981**, *89*, 631–645. [CrossRef]

40. Nieboer, E.; Ahmed, H.M.; Puckett, K.J.; Richardson, D.H.S. Heavy metal content of lichens in relation to distance from a nickel smelter in Sudbury, Ontario. *Lichenologist* **1972**, *5*, 292–304. [CrossRef]

41. Nakajima, H.; Fujimoto, K.; Yoshitani, A.; Yamamoto, Y.; Sakurai, H.; Itoh, K. Effect of copper stress on cup lichens *Cladonia. humilis* and *C. subconistea* growing on copper-hyperaccumulating moss *Scopelophila. cataractae*. *Ecotoxicol. Environ. Saf.* **2012**, *84*, 341–346. [CrossRef] [PubMed]

42. Nakajima, H.; Yamamoto, Y.; Yoshitani, A.; Itoh, K. Effect of metal stress on photosynthetic pigments in the Cu-hyperaccumulating lichens *Cladonia. humilis* and *Stereocaulon. japonicum* growing in Cu-polluted sites in Japan. *Ecotoxicol. Environ. Saf.* **2013**, *97*, 154–159. [CrossRef] [PubMed]

43. Nieboer, E.; Richardoson, D.H.S. Lichens as monitors of atmospheric depositions. In *Atmospheric Pollutants in Natural Waters*; Eisenreich, S.J., Ed.; Ann Arbor Science Publisher: Ann Arbor, MI, USA, 1981; pp. 339–388.

44. Nieboer, E.; Richardoson, D.H.S.; Tomassini, F.D. Mineral uptake and release by lichens: An review. *Bryologist* **1978**, *81*, 226–246. [CrossRef]

45. Nieboer, E.; Richardoson, D.H.S. The replacement of the nondescript term "heavy metals" by a biologically and chemically significant classification of mental ions. *Environ. Pollut.* **1980**, *1*, 3–26. [CrossRef]

46. Brown, D.H.; Beckett, R.P. Uptake and effect of cations on lichen metabolism. *Lichenologist* **1984**, *16*, 173–188. [CrossRef]
47. Shimada, T.; Shimizu, T.; Uehara, N. Aggregation property of thermo-sensitive copolymers having polyamino groups and carboxyl groups. *Bunseki Kagaku* **2002**, *51*, 689–695. [CrossRef]
48. Sato, T.; Fukushi, K.; Yonada, T. Environmental behavior and management of hazardous inorganic anions in nature. *J. MMIJ* **2007**, *123*, 132–144. [CrossRef]
49. Yaguchi, T. Classification and examination of fungi. *Jpn. J. Food Microbiol.* **2010**, *27*, 47–55. [CrossRef]
50. Fahselt, D.; Madazia, S. Scanning electron microscopy of invasive fungi in lichens. *Bryologist* **2012**, *104*, 24–39. [CrossRef]
51. Purvis, O.W. The occurrence of copper oxalate in lichens grown on copper sulphide-bearing rocks in Scandinavia. *Lichenologist* **1984**, *16*, 197–204. [CrossRef]

![metals](metals logo) ![MDPI](MDPI logo)

Article

Decontamination of Uranium-Contaminated Soil Sand Using Supercritical CO$_2$ with a TBP–HNO$_3$ Complex

Kwangheon Park *, Wonyoung Jung and Jihye Park

Department of Nuclear Engineering, Kyung Hee University, Yongin 446-701, Korea;
wonyoung1987@gmail.com (W.J.); qkr0126@khu.ac.kr (J.P.)

* Author to whom correspondence should be addressed; kpark@khu.ac.kr; Tel.: +82-31-201-2917;
  Fax: +82-31-202-2410.

Academic Editors: Suresh Bhargava and Rahul Ram
Received: 31 July 2015; Accepted: 18 September 2015; Published: 25 September 2015

Abstract: An environmentally friendly decontamination process for uranium-contaminated soil sand is proposed. The process uses supercritical CO$_2$ as the cleaning solvent and a TBP–HNO$_3$ complex as the reagent. Four types of samples (sea sand and coarse, medium, and fine soil sand) were artificially contaminated with uranium. The effects of the amount of the reagent, sand type, and elapsed time after the preparation of the samples on decontamination were examined. The extraction ratios of uranium in all of the four types of sand samples were very high when the time that elapsed after preparation was less than a few days. The extraction ratio of uranium decreased in the soil sand with a higher surface area as the elapsed time increased, indicating the possible formation of chemisorbed uranium on the surface of the samples. The solvent of supercritical CO$_2$ seemed to be very effective in the decontamination of soil sand. However, the extraction of chemisorbed uranium in soil sand may need additional processes, such as the application of mechanical vibration and the addition of bond-breaking reagents.

Keywords: supercritical CO$_2$; decontamination; uranium; soil sand; TBP–HNO$_3$ complex

---

1. Introduction

Nuclear energy is environmentally friendly because of its high energy density, with no release of greenhouse gases. However, nuclear energy by fission generates radioactive fission products, such as Cs-137, Xe-133, I-131, and Sr-90. Worldwide, there have been two major nuclear accidents: Chernobyl in 1986 and Fukushima in 2011. These major nuclear accidents showed that radioactive products that leak from a plant could contaminate the soil in a large area near the nuclear power plant [1,2]. Soil can also be contaminated by natural radioactive isotopes, such as uranium. Uranium-contaminated soil is common around uranium production, ore, and nuclear fuel production facilities. The pollution of these sites by radioactive isotopes is directly harmful to human health and disrupts soil functions that support terrestrial ecosystems [3]. Cleaning the soil contaminated by radioactive materials requires huge efforts, as the amount of soil needed for decontamination is generally immense due to the large contaminated area that has to be cleaned. Moreover, a large amount of secondary aqueous waste is generated during decontamination, and this waste has to be decontaminated again later. Hence, conventional soil-cleaning methods do not seem to be effective for cleaning soil because they generate secondary wastes [4,5].

Plutonium uranium redox extraction (PUREX), a solvent-extraction process, can be used for the extraction of uranium and plutonium from spent nuclear fuels [6]. Tributylphosphate (TBP) is a key extractant for uranium and plutonium in the PUREX process, and it dissolves well in organic solvents, such as kerosene. According to a recent report, TBP dissolved easily in supercritical carbon dioxide (CO_2) [7]. Supercritical CO_2 has good qualities as a solvent, combining the high solubility of a liquid with the fast reaction speed of a gas. CO_2 has great potential as a supercritical solvent because it becomes supercritical relatively easily at 31 °C and 73.8 bar. It is easy to recycle CO_2 because the phase, in addition to the temperature, can be easily controlled by a depressurization/compression process [8]. CO_2 is a nonpolar organic solvent, which shows low solubility in polar substances, such as inorganic matter, and in ionic substances, such as metal ions. To develop an effective uranium decontamination process using supercritical CO_2, an extraction reagent (e.g., TBP) that combines with uranium metal must be used [9]. TBP and nitric acid form a complex that dissolves well in supercritical CO_2 [10]. A technology to directly reprocess spent fuels using a TBP-nitric acid compound has been proposed [11], and a reprocessing method based on CO_2, referred to as Super-Direx, has been developed. The potential of supercritical CO_2 technology in the extraction of actinide-based metal ions has also been studied [12,13].

In this study, an environmentally friendly decontamination method to clean uranium-contaminated soil is proposed. The method uses a complex of TBP and nitric acid as the extracting agent in supercritical CO_2. CO_2, the solvent in the cleaning process, can be easily recycled by changing its pressure. In addition, the TBP and nitric acid are reusable after the removal of the uranium from the TBP-nitric acid complex using conventional technology applied in the PUREX process. Hence, no secondary waste is generated. The experimental decontamination method is explained, and the results are discussed to the conclusion to see the feasibility of the supercritical CO_2 decontamination method.

2. Experiments

2.1. Preparation of the Reagent for Extraction and the Extraction Conditions

Uranium dioxide in its solid state can be dissolved directly in supercritical CO_2 with a TBP–HNO_3 complex [14–16]. To prepare the reagent (*i.e.*, TBP–HNO_3 complex) for extraction, anhydrous TBP and 70% HNO_3 were fully mixed in a 1:1 ratio in a beaker using a rotating magnetic stirring bar. The mixture was then separated by centrifuging for 40 min. The ratio of HNO_3:TBP:H_2O in the TBP–HNO_3 complex obtained after separation was 1:1.8:0.4 [13]. The solubility of the TBP–HNO_3 and TBP–U–NO_3 complexes in the supercritical CO_2 was observed through a view cell. The solubility of TBP–HNO_3 was measured at 40–60 °C, and the solubility of TBP–U–NO_3 was determined at 50 °C, as illustrated in Figure 1 [13]. As shown in the figure, these complexes can form a single phase with supercritical CO_2 in a wide range of concentrations. At 50 °C, both complexes fully dissolved in supercritical CO_2 when the applied pressure was higher than 17 MPa. In this study, the following conditions were used for the extraction of uranium from contaminated soil sand: a pressure of 20 MPa and temperature of 40 °C.

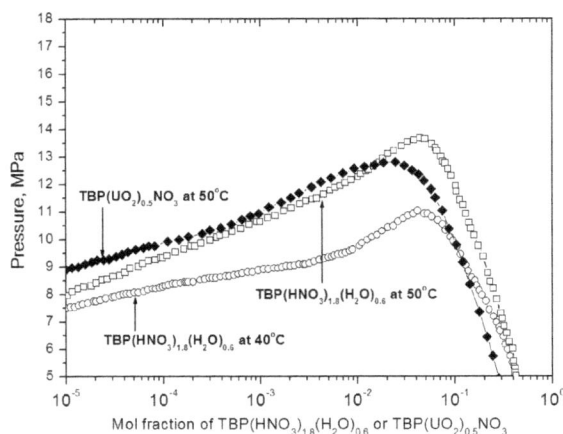

Figure 1. Solubility curves for the transition pressures from two-phase to single-phase TBP·(UO$_2$)$_{0.5}$·NO$_3$ and TBP·(HNO$_3$)$_{1.8}$·(H$_2$O)$_{0.6}$ in supercritical CO$_2$ [13].

2.2. Specimen Preparation

Soil is made up of many components, such as minerals, organic matter, gases, liquids, and countless organisms [17]. In this study, we focused only on the effect of the soil grains (*i.e.*, sand). Two types of sand were prepared: sea sand and soil sand collected from a hill in the campus of Kyung Hee University (located south of Seoul, Korea). The sea sand was pure grade and purchased from JUNSEI Chemical Co., Tokyo, Japan. The soil sand for the experiment was collected at a depth of 50 cm from the ground surface. The soil was immersed in a 30% peroxide solution for about 3 h to remove organic matter. The soil was then cleaned in 6 M nitric acid for 3 h to remove metallic impurities on the surface. After rinsing the soil in hot water (about 80 °C), it was dried in a vacuum glass container for one week. The dried soil was then screened to obtain soil sand. The soil sand was grouped into four sizes according to the diameter of the grains: larger than 1, 0.5–1.0, 0.2–0.5, and less than 0.2 mm. Particles in the sand larger than 1 mm were not considered. In the experiments, the sand was classified as follows: (1) sea sand (0.5–1.0 mm); (2) coarse soil sand (0.5–1.0 mm); (3) medium soil sand (0.2–0.5 mm); and (4) fine soil sand (less than 0.2 mm). The details of the soil sample preparation have been described previously [4].

A nitric acid solution containing uranium was prepared by dissolving a small amount of uranium oxide in the solution. The sand samples (50 µg-U per 1 g of sample) were placed in beakers containing the solution. The beaker was then placed in an ultrasonic cleaner for 1 h to evenly mix the uranium ions in the sand. The beakers containing the sand and solution were placed in a vacuum oven and dried at 90 °C for about 24 h. After full evaporation of the solution, the dried sand was mixed again with a spatula to ensure that the uranium was uniformly distributed in the sand. The sand samples artificially contaminated by uranium are shown in Figure 2. In each experiment, 1 g of the contaminated sand specimen was used.

Figure 2. Sand samples artificially contaminated with uranium. (**a**) Sea sand; (**b**) coarse soil sand; (**c**) medium soil sand; and (**d**) fine soil sand.

2.3. Experimental Process

The experimental setup for the uranium extraction from sand using supercritical CO_2 is shown in Figure 3. The experimental apparatus was composed of a syringe pump, mixing cell, and reaction cell. The mixing cell contained a magnetic bar, which mixed both the TBP–HNO_3 complex and the CO_2 inside. The reaction cell was a stainless tube, inside which the contaminated sand sample was located. The syringe pump supplied CO_2 at a pressure of 20 MPa from the tank to the system. The CO_2 was pumped into the mixing cell and combined with the TBP–HNO_3 complex inside the cell for 30 min. The mixture was then transferred to the reaction cell containing the sand sample. Uranium was extracted from the sand sample in the reaction cell. This dynamic extraction process lasted for more than 30 min. The flow rate of the CO_2 from the syringe pump was about 3 mL (liquid CO_2) per min. The reacted TBP–HNO_3 complex that dissolved in the CO_2 was collected in a beaker after depressurization outside the system. Both the mixing cell and the reaction cell were inside a water bath (*i.e.*, a reaction zone), the temperature of which was maintained at 40 °C. The effects of each of three variables on the extraction of uranium were determined: the amount of the reagent, sand type (or size), and duration of the postcontamination period (one hour, one day, one week, one month, three months, and four years). Table 1 shows the experimental conditions in the uranium-extraction experiments.

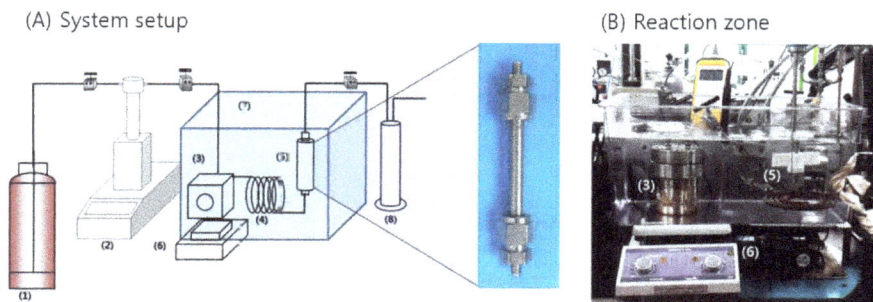

Figure 3. Experiment setup and the reaction zone for uranium extraction from the contaminated sand: (1) CO_2 cylinder; (2) syringe pump; (3) mixing cell; (4) preheating cell; (5) specimen; (6) mixer; (7) thermostat; and (8) collector.

Table 1. Conditions for the extraction experiments.

Variables	Specimen	Description
Amount of the reagent used	Sea sand	0.1–15 mL
Sand size (or type)	Sea sand	1.0 mm (reagent amount: 0.5 mL)
	Soil sand	Coarse (0.5–1.0 mm), medium (0.2–0.5 mm), fine (less than 0.2 mm) (reagent amount: 0.5 mL)
Elapsed time after sample preparation	Sea sand	1 h, 1 day, 1 week, 1 month, 4 years (reagent amount: 0.5 mL)

2.4. Analysis of the Extracted Fraction of Uranium

The extracted fraction of uranium from the sand samples was obtained by comparing the amount of uranium in the sample before and after the experiment using microwave digestion and Inductively Coupled Plasma (ICP) measurements. After the experiment, 1 g of each sand sample was placed in a Teflon XP-1500 container (MARS5 Digestion Microwave System, CEM Co., Matthews, NC, USA) with 20 mL of 35% nitric acid. A reference specimen (*i.e.*, a sample not exposed to the cleaning process) containing the same amount of nitric acid was placed in another XP-1500 container. Both containers were heated slowly by microwaves from room temperature to 180 °C for 15 min. The maximum pressure inside the XP-1500 container was set as 2 MPa. The temperature was maintained at 180 °C for another 15 min. After cooling the containers, the nitric acid was extracted from each container, and the amount of uranium in each solution was analyzed using ICP (Leeman Labs, Lowell, MA, USA). The fraction of uranium extracted after the experiment was obtained by comparing the uranium concentration between the two nitric acids. The extracted fraction of uranium or the extraction ratio, f_U), was estimated as:

$$f_U = \frac{C_{bef} - C_{aft}}{C_{bef}} \tag{1}$$

where C_{bef} and C_{aft} are the concentration of uranium in the solution before and after the experiment, respectively.

The concentration measured from the ICP analysis is known to have an uncertainty of 5%–10%. Based on this uncertainty, the uncertainty of the extraction ratio may be 5%–20%.

3. Results and Discussion

3.1. The Effect of the Amount of Extraction Regent Used

Sea sand was used in these experiments, and the amount of the extraction reagent used was 0.1–15 mL. Figure 4 shows the results. The extraction ratio of uranium increased with the amount of extraction reagent used. The extraction ratio of uranium exceeded 90% when the amount of reagent was 0.4 mL or higher. As the weight of the sample was 1 g in the experiment, the total amount of uranium in the reaction cell was about 50 μg. The extraction reaction of the uranium by the reagent, the TBP–HNO$_3$ complex, can be expressed as follows:

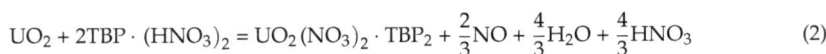

$$UO_2 + 2TBP \cdot (HNO_3)_2 = UO_2(NO_3)_2 \cdot TBP_2 + \frac{2}{3}NO + \frac{4}{3}H_2O + \frac{4}{3}HNO_3 \tag{2}$$

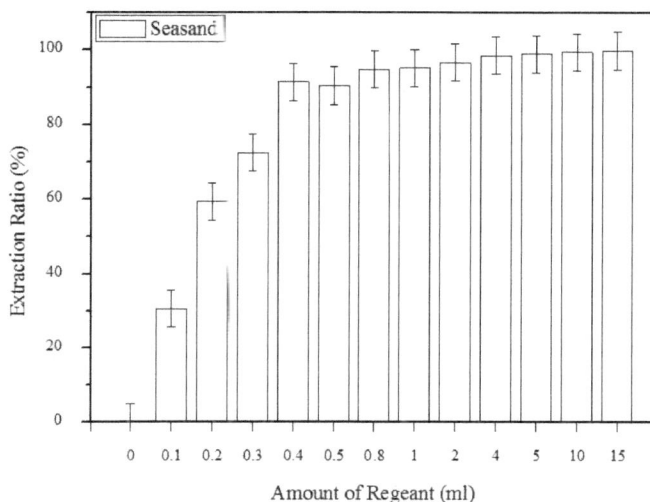

Figure 4. Extracted fraction of uranium as a function of the amount of extraction reagent used (specimen: sea sand with 50 µg-U/g, uncertainty applied: 5%).

Based on the above reaction, the required amount of TBP–HNO_3 complex to extract 1 mol of uranium (238.029 g) would be 2 mol (284.66 g). Hence, the weight of the extraction reagent required to extract 50 µg uranium would be 165 µg. Assuming that the density of the extraction reagent was 0.98, about 0.17 mL of the reagent would be needed theoretically to achieve uranium extraction. In this experimental setup, the process turns out to need the extraction reagent about three times more than the theoretical amount for full extraction of uranium. We selected 0.5 mL of the extraction reagent as a reference, which was the minimum amount for full extraction of uranium from the sea sand specimen in this experiment system.

3.2. The Effects of the Sand Type and Elapsed Time after Sample Preparation

Four types of samples (sea sand and coarse, medium, and fine soil sand) were used for the experiment. Some of the samples were four years old. The sample preparation process of these four-year-old samples was almost identical to that of the current samples, except that the soil was collected in fields surrounding Daejun City, which is located in the central part of the Republic of Korea (specifically, the Korea Atomic Energy Research Institute) [4]. The amount of the reagent used in the experiment was 0.5 mL. The results, extraction ratios with respect to sand type, and elapsed time after the sample preparation are shown in Figure 5. In all four types of samples, when the samples had been prepared only a few days earlier, the extraction ratios of uranium were more than 94%. As the elapsed time increased, the extraction ratios of the four sand types decreased, but the level of decrease differed according to the sample type. The uranium extraction ratios of the sea sand samples were high (higher than 90%), even those of the four-year-old samples. The extraction ratios of the coarse and medium soil sand that were several months old were also good. After four years, the extraction ratio dropped to 70%. However, in the case of the fine soil sand, the reduction in the extraction ratio of uranium started as early as one week later after the sample preparation, and the extraction ratio dropped to less than 50%. In summary, the extraction ratio of the sea sand was good, regardless of the time that had elapsed after the sample preparation, and the extraction ratios of the coarse and medium soil sand were reasonably good (approximately 90% after a few months and 70% after four years). However, the extraction ratio of the fine soil sand decreased rapidly (below 60%) after one week.

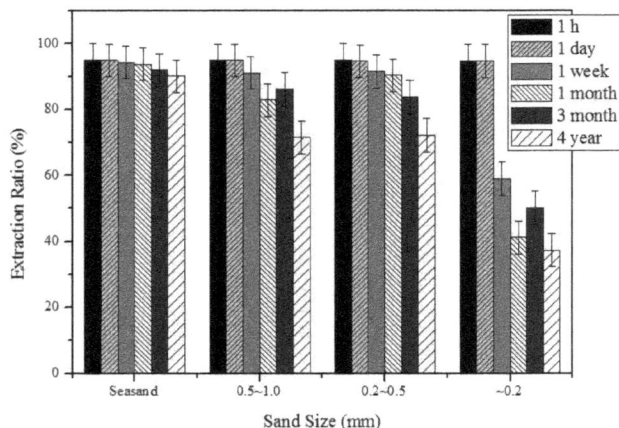

Figure 5. Extracted ratio changes according to soil type and the elapsed time after sample preparation (uncertainty applied: 5%).

In the present study, the reduction in the extraction ratio of uranium ions in the soil sand may have originated from two mechanisms: the trapping of uranium ions in cracks and pores and chemisorption of uranium ions on the surface. The microstructures of the sea sand and soil sand were observed by SEM (Figure 6) (Stereoscan 440, Leica, Cambridge, UK). The surface of the sea sand was relatively smooth and glassy, whereas that of the soil sand was rough and contained cracks and small pores. If uranium ions are adsorbed to the surfaces inside cracks or pores, they may not be easy to remove or extract. In the case of chemisorption, the extraction of uranium may also be difficult. Chemically pure sea sand is mainly silica, which rarely reacts chemically with uranium ions. An SEM-EDX analysis of the sand indicated the existence of various impurities, including Fe, Mg, K, and Al (Figure 7). These impurities may take different structures on the surface, and they can provide sites for chemical reactions with uranium ions. Initially, uranium ions in soil are mainly physically adsorbed on the surface of the soil. Over time, the uranium ions and impurities on the surface of the soil start to chemically react with each other (chemisorption). Once the uranium ions are chemically bonded with the impurities on the soil surface, it becomes more difficult to extract uranium from the surface, and a stronger extraction process is required.

The results shown in Figure 5 indicate that the extraction ratios of all the four types of soil sand were very high when the elapsed time was less than one day. Based on these results, it seems more difficult to extract chemisorbed uranium ions on the surface than trapped uranium ions in cracks and pores. Supercritical CO_2 has a very good ability to penetrate solid gaps, such as cracks and open pores. Although soil sand has a large surface area, with cracks and pores, supercritical CO_2 can carry the extraction reagent to cracks and pores by penetrating fine gaps in the sample. However, as time progresses, physically adsorbed uranium ions transform to a chemically bonded state, with impurities on the surface. Once uranium ions enter a chemisorbed state, they are not easy to extract. Some soil, such as fine soil sand, which has a greater higher surface area, has more chemical bonding sites for uranium ions. In such soil samples, the extraction ratio decreases with the elapsed time since the preparation of the samples. The extraction of chemisorbed uranium may require mechanical energy, such as ultrasonic vibrations, and additional chemical reagents that prevent chemical bonding.

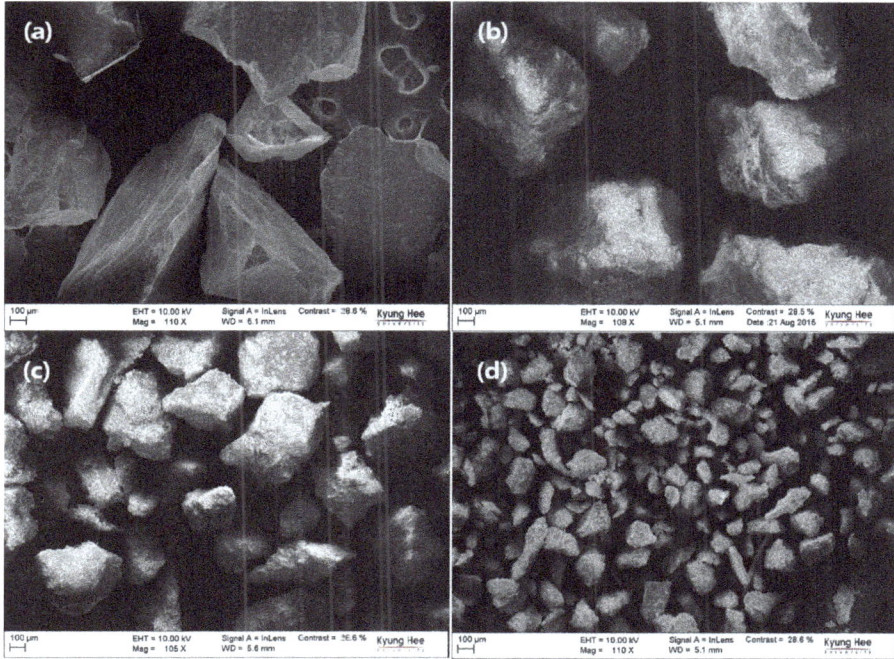

Figure 6. SEM images of the four types of soil samples before the cleaning process (four-year-old samples). (**a**) Sea sand; (**b**) coarse soil sand; (**c**) medium soil sand; (**d**) fine soil sand.

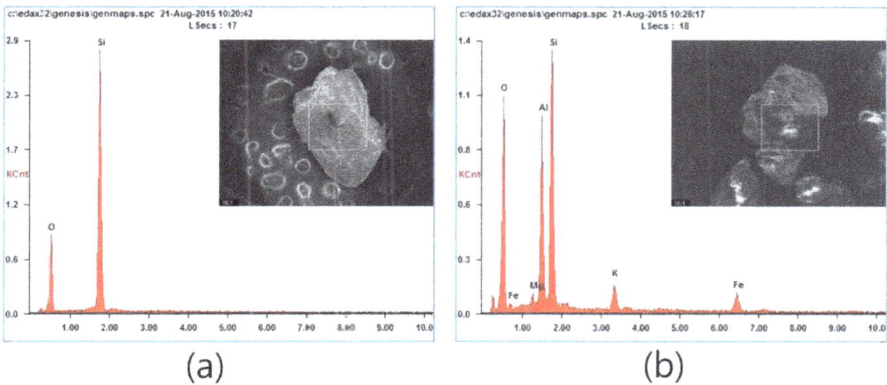

Figure 7. EDX analyses of the sea sand and fine soil sand before the cleaning process (four-year-old samples). (**a**) Sea sand and (**b**) fine soil sand.

4. Conclusions

Sand fractions sorted from soil contaminated with uranium were cleaned by supercritical CO_2 with a TBP–HNO_3 complex. Four types of sand (sea sand, and coarse, medium, and fine soil sand) were used, and experiments were performed to examine the effects of the amount of reagent, sand type, and elapsed time after sample preparation on the uranium extraction.

In this experimental setup, about three times more extraction reagent was needed than the theoretical amount for full extraction of uranium from the sea sand specimens. Only if the elapsed time after the sample preparation was less than a day were the extraction ratios of uranium very high in all the four types of sand. This indicates that supercritical CO_2 can easily pump the extraction reagent into cracks and pores by penetrating through the fine gaps in the sample. Over time, physically adsorbed uranium ions transform to a chemically bonded state, with impurities on the surface. Once uranium ions enter a chemisorbed state, they are not easy to extract. Thus, the extraction ratio of the fine soil sand, which has the largest surface area of the samples, decreased the most in the present study as the elapsed time since the sample preparation increased.

The process suggested in this study is environmentally friendly. CO_2, the solvent in the cleaning process, can be easily recycled, and TBP and nitric acid can be reused. Thus, no secondary waste is generated during the uranium-decontamination process. The solvent of supercritical CO_2 seems very effective in the decontamination of soil. However, the extraction of chemisorbed uranium in soil may require additional processes, such as the application of mechanical vibration and the addition of bond-breaking reagents.

Acknowledgments: This research was supported by a grant from the National Research Foundation of Korea (NRF), funded by the Korean government (MSIP: Ministry of Science, ICT and Future planning). The project number is 2012M2B2B1055502.

Author Contributions: Kwangheon Park interpreted the results and drafted the paper. Wonyoung Jung built the experimental setup, prepared the specimens and produced the results. Jihye Park analyzed the specimens.

Conflicts of Interest: The authors declare no conflict of interest.

References

1. Steinhauser, G.; Brandl, A.; Johnson, T.E. Comparison of the Chernobyl and Fukushima Nuclear Accidents: A review of the environmental impacts. *Sci. Total Environ.* **2014**, *470–471*, 800–817. [CrossRef] [PubMed]
2. Calmon, P.; Gonze, M.A.; Mourlon, C. Modeling the Early-Phase Redistribution of Radiocesium Fallouts in an Evergreen Coniferous Forest after Chernobyl and Fukushima Accidents. *Sci. Total Environ.* **2015**, *529*, 30–39. [CrossRef] [PubMed]
3. Certini, G.; Scalenghe, R.; Woods, W.I. The Impact of Warfare on the Soil Environment. *Earth Sci. Rev.* **2013**, *127*, 1–15. [CrossRef]
4. Park, K.; Lee, J.; Sung, J. Metal Extraction from the Artificially Contaminated Soil Using Supercritical CO_2 with Mixed Ligands. *Chemosphere* **2013**, *91*, 616–622. [CrossRef] [PubMed]
5. Park, K.; Jinhyun, S.; Koh, M.; Kim, H.; Kim, H. Decontamination of Radioactive Contaminants Using Liquid and Supercritical CO_2. In *Radioactive Waste*; Rahman, R.A., Ed.; InTech: Rijeka, Croatia, 2012; pp. 219–238.
6. Benedict, M.; Pigford, T.H.; Levi, H.W. *Nuclear Chemical Engineering*, 2nd ed.; McGraw-Hill Book Company: New York, NY, USA, 1981.
7. Lin, Y.; Smart, N.G.; Wai, C.M. Supercritical Fluid Extraction of Uranium and Thorium from Nitric Acid Solutions with Organophosphorus Reagents. *Environ. Sci. Technol.* **1995**, *29*, 2706–2708. [CrossRef] [PubMed]
8. McHardy, J.; Sawan, S.P. *Supercritical Fluid Cleaning: Fundamentals, Technology and Applications*; Noyes: Westwood, NJ, USA, 1998.
9. Park, K.H.; Kim, H.W.; Kim, H.D.; Koh, M.S.; Ryu, J.D.; Kim, Y.E.; Lee, B.S.; Park, H.T. Application of CO_2 Technology in Nuclear Decontamination. *Han'guk Pyomyon Konghak Hoechi* **2001**, *34*, 62–67.
10. Carrott, M.J.; Waller, B.E.; Smart, N.G.; Wai, C.M. High solubility of $UO_2(NO_3)_2 \cdot 2TBP$ complex in supercritical CO_2. *Chem. Commun.* **1998**. [CrossRef]
11. Shimada, T.; Ogumo, S.; Ishihara, N.; Kosaka, Y.; Mori, Y. A Study on the Technique of Spent Fuel Reprocessing with Supercritical Fluid Direct Extraction Method (super-DIREX method). *J. Nucl. Sci. Technol.* **2002**, *39*, 757–760. [CrossRef]
12. Erkey, C. Supercritical Carbon Dioxide Extraction of Metals from Aqueous Solutions: A Review. *J. Supercrit. Fluids* **2000**, *17*, 259–287. [CrossRef]
13. Enokida, Y.; Yamamoto, I. Vapor-liquid Equilibrium of $UO_2(NO_3)_2 \cdot 2TBP$ and Supercritical Carbon Dioxide Mixture. *J. Nucl. Sci. Technol.* **2002**, *9*, 270–273. [CrossRef]

14. Tamika, O.; Meguro, Y.; Enokida, Y.; Yamamoto, I.; Yoshida, Z. Dissolution Behavior of Uranium Oxides with Supercritical CO_2 Using HNO_3–TBP Complex as a Reactant. *J. Nucl. Sci. Technol.* **2001**, *38*, 1097–1102. [CrossRef]

15. Meguro, Y.; Iso, S.; Yoshida, Z.; Tomicka, O.; Enokida, Y.; Yamamoto, I. Decontamination of uranium Oxides from Solid Wastes by Supercritical CO_2 Fluid Leaching Method Using HNO_3–TBP Complex as a Reactant. *J. Supercrit. Fluids* **2004**, *31*, 141–147. [CrossRef]

16. Sawada, K.; Uruga, K.; Koyama, T.; Shimada, T.; Mori, Y.; Enokida, Y.; Yamanoto, I. Stoichiometric relation for Extraction of Uranium from UO_2 Powder Using TBP Complex with HNO_3 and H_2O in Supercritical CO_2. *J. Nucl. Sci. Technol.* **2005**, *42*, 301–304 [CrossRef]

17. Buol, S.W.; Southard, R.J.; Graham, R.C.; McDaniel, P.A. *Soil Genesis and Classification*, 6th ed.; Wiley-Blackwell: Hoboken, NJ, USA, 2011.

metals

MDPI

Article

Removal of Zn from Contaminated Sediment by FeCl$_3$ in HCl Solution

Sang-hun Lee [1], Ohhyeok Kwon [2], Kyoungkeun Yoo [2],* and Richard Diaz Alorro [3]

[1]   Gas & Mining Plant Division, Samsung Construction & Technology, Seoul 137-956, Korea;
      leeshty747@gmail.com
[2]   Department of Energy & Resources Engineering, Korea Maritime and Ocean University (KMOU),
      Busan 606-791, Korea; koh1338@gmail.com
[3]   Department of Mining Engineering and Metallurgical Engineering, Western Australian School of Mines,
      Curtin University, Kalgoorlie, WA 6430, Australia; richard.alorro@curtin.edu.au
*   Author to whom correspondence should be addressed; kyoo@kmou.ac.kr; Tel.: +82-51-410-4686;
     Fax: +82-51-403-4680.

Academic Editors: Suresh Bhargava, Mark Pownceby and Rahul Ram
Received: 28 July 2015; Accepted: 5 October 2015; Published: 8 October 2015

Abstract: Harbor sediments contaminated with ZnS concentrate were treated by ferric chloride in HCl solution to remove Zn. The sediments were evaluated using Tessier's sequential extraction method to determine the different metal phase associations of Zn. Leaching tests were performed to investigate the effects of experimental factors, such as agitation speed, ferric ion concentration, temperature, and pulp density, on the removal of Zn. The sequential extraction procedure revealed that about 17.7% of Zn in the sediment was associated with soluble carbonate and oxide phases. The results of the leaching tests indicated that higher ferric concentration and temperature increased the leaching efficiencies significantly, while the agitation speed has a negligible effect on the removal of Zn. The removal ratio increased to more than 99% within 120 min of treatment at 1 kmol·m^{-3} HCl solution with 1 kmol·m^{-3} Fe^{3+}, 10% pulp density, and 400 rpm at 90 °C. The dissolution kinetics of Zn were discussed by comparing the two shrinking core models. It was determined that the kinetic data followed the diffusion controlled model well compared to the surface chemical reaction model. The activation energies were calculated to be 76.9 kJ/mol, 69.6 kJ/mol, and 58.5 kJ/mol for 0.25 kmol·m^{-3}, 0.5 kmol·m^{-3}, and 1 kmol·m^{-3} Fe^{3+}, respectively.

Keywords: harbor sediments; ferric chloride; zinc sulfide; metal contamination

1. Introduction

Operational or incidental spillage of materials during unloading activities of metal concentrates has caused contamination problems in some harbor areas in Korea. This has led to increasing concerns regarding metal-contaminated soils and sediments and their impacts on the environment. Various treatment methods have been proposed for remediating contaminated sites, such as gravity separation, phytoremediation, thermal processing, solidification/stabilization, electrokinetic remediation, and soil washing [1,2]. Gravity separation techniques have been considered to be cost-effective and environmentally friendly primarily because of the chemical-free process, but their effectiveness and applicability are rather difficult in the treatment of fine concentrates with particle sizes less than 75 μm [2]. Since fine particles have high specific surface area, chemical treatment methods such as leaching generally are more suitable than physical treatment processes.

Zinc sulfide (sphalerite, ZnS) concentrates are normally subjected to roasting as a pre-treatment method before leaching to enhance the dissolution of Zn. After leaching, the released Zn ions are recovered as metallic zinc in the electrowinning process [3]. The emission of SO$_2$ during roasting is a serious environmental problem and is governed by strict environmental regulations. Due to this, the hydrometallurgical route has gained recognition as an alternative process to recover Zn from concentrates [3]. The leaching of sphalerite has been investigated using various leaching media, such as ferric sulphate and chloride with hydrogen peroxide or oxygen [3], ferric sulphate [4], hydrochloric acid with ferric chloride [5], and sulfuric acid with nitric acid [6]. These studies suggested the use of either ferric ions, H_2O_2, O_2, or HNO_3 for ZnS oxidation using mineral grade sphalerite or concentrates. Only a few studies on the hydrometallurgical removal of Zn from ZnS-contaminated sediments with ferric ions are available in literature.

Harbor sediments contaminated with ZnS are exposed to natural environmental conditions and may undergo weathering over time and may change to a different chemical form. Metals accumulate in sediments from both natural and anthropogenic sources, thus making it difficult to identify and determine their origin. It is therefore important to evaluate the individual fractions of the metals not only to understand their actual and potential environmental effects but also to aid in the process design. However, metal concentrations in contaminated sediments are usually below the detection limit of the X-ray diffractometer (XRD), making their measurements difficult. To overcome this limitation, sequential extraction methods can be used [7,8]. One of the most widely used methods is the Tessier sequential extraction method which consists of five extraction steps: "bound to exchangeable", "bound to carbonate", "bound to oxide", "bound to sulfide or organic matter", and residue. A study conducted by Park *et al.* on the leaching of heavy metals (Cu, Zn, Pb) from contaminated soil using citric acid revealed that the metals in carbonate and oxide phases are easy to dissolve [2].

In the present study, the sediments contaminated with ZnS concentrate were evaluated using Tessier's sequential extraction method. Ferric chloride leaching tests were performed to investigate the effects of leaching factors such as agitation speed, ferric chloride concentration, temperature, and pulp density on Zn removal. The dissolution kinetics were discussed and interpreted by comparing the two shrinking core models.

2. Experiments

2.1. Materials

Sediments contaminated with ZnS were collected from a harbor site that has been used for the unloading of Zn concentrate. The samples were washed three times with distilled water and then dried overnight at 105 °C. The particle size distribution of the sediment as shown in Figure 1 reveals that 48.5% of the sample has a particle size <75 μm. This size fraction was used in the leaching tests. The zinc content of the sample was determined to be 5287 mg/kg by aqua-regia digestion. All chemicals used in this study were of reagent grade.

2.2. Leaching Procedures and Analytical Methods

The leaching tests were performed in a 500 dm^3 three-necked Pyrex glass reactor using a heating mantle to maintain the desired temperature to within ±2 °C. This reactor was fitted with a stirrer and a reflux condenser to prevent solution loss at high temperatures. Leach solutions were prepared with $FeCl_3$ concentrations of between 0.25 kmol·m^{-3} and 1 kmol·m^{-3} in 1 kmol·m^{-3} HCl solution. For every batch of leaching test, 200 cm^3 of leach solution was introduced into the reactor. Once the solution reached thermal equilibrium (40–90 °C), 20–60 g of the fine sediment powder was added and agitated at 200–600 rpm. During these tests, a 2 cm^3 aliquot solution was withdrawn periodically over the desired time interval (5–240 min) for analysis.

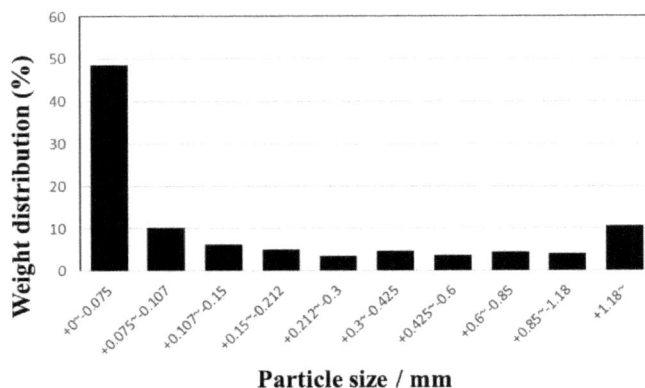

Figure 1. Particle size distribution of the contaminated sediment.

The sequential extraction method used in this study was originally developed by Tessier *et al.* (1979) [7] and later modified by Li *et al.* (1995) [8]. The extraction method used 1 g of dried sample and was performed under the following conditions and reagents: (1) 8 cm^3 of 1 kmol·m^{-3} MgCl$_2$ (adjusted to pH 7) for 1 h at room temperature in the shaking bath; (2) 8 cm^3 of 1 kmol·m^{-3} CH$_3$COONa (adjusted to pH 5) for 5 h at room temperature in the shaking bath; (3) 20 cm^3 of 0.4 kmol·m^{-3} NH$_2$OH·HCl (in 25% CH$_3$COOH) for 6 h at 96 °C in a heating block; (4) 3 cm^3 of 0.02 kmol·m^{-3} HNO$_3$ and 5 cm^3 of 30% H$_2$O$_2$ (adjusted to pH 2) for 2 h at 85 °C in a heating block, followed by 3 cm^3 of 30% H$_2$O$_2$ (adjusted to pH 2) for 3 h at 85 °C in a heating block, and, finally, 5 cm^3 of 3.2 kmol·m^{-3} CH$_3$COONH$_4$ (in 20% HNO$_3$) and about 4 cm^3 deionized water for 30 min at room temperature in the shaking bath; (5) 20 cm^3 of aqua regia for 20 min. All solid/liquid separations were done by centrifugation for 30 min at 10,000 rpm (1065× *g*), designated amounts of supernatant were taken and then diluted with 5% HNO$_3$ solution. The concentration of Zn was determined with AA-7000 atomic absorption spectrophotometer (Shimadzu Scientific Instrument, Ltd., Kyoto, Japan).

3. Results and Discussion

Zinc concentrate is usually imported to Korea in the form of ZnS (sphalerite) and due to operational or incidental spillage during unloading, this concentrate has become the source of Zn contamination in harbor sediments. The chemical association of Zn in this sediment was examined using Tessier's sequential extraction method to understand its characteristics, through which it was found that 7.9% and 9.8% of the Zn was in the "bound to carbonate" and "bound to oxide" phases, respectively. The rest of the Zn content was found to be associated with the "bound to sulfide" phase. This result indicates that 17.7% of Zn is bound to the weathered forms, such as carbonates and oxides, which can be dissolved easily in HCl solution.

Leaching tests using the contaminated sediments were carried out at agitation speeds in the range of 200–600 rpm to examine the effect of liquid film boundary diffusion surrounding the solid particles on the leaching efficiency at the following conditions: 1 kmol·m^{-3} HCl with 1 kmol·m^{-3} Fe^{3+}, 10% pulp density, and 50 °C. Figure 2 shows that the leaching efficiencies of Zn are independent of the agitation speeds. Therefore, in all subsequent leaching tests, a working agitation speed of 400 rpm was selected to ensure effective particle suspension in the solution. It was also observed that the leaching efficiency of Zn increased rapidly to more than 20% within 5 min and then gradually increased afterwards. This can be attributed to the initial dissolution of Zn present in the "bound to carbonate" and "bound to oxide" phases.

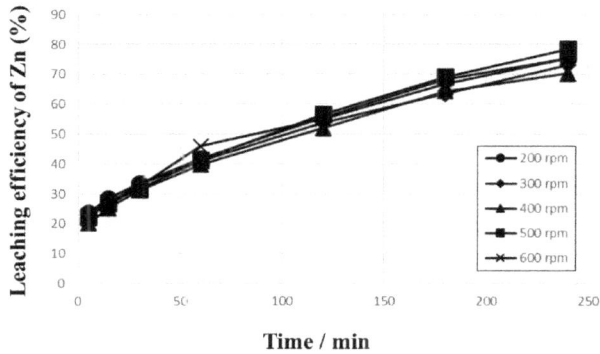

Figure 2. The effects of agitation speed on the leaching efficiency of Zn at 1 kmol·m^{-3} HCl with 1 kmol·m^{-3} Fe^{3+}, 10% pulp density, and 50 °C.

Figure 3 shows the effect of Fe^{3+} concentration on the dissolution of Zn from the contaminated sediment at 1 kmol·m^{-3} HCl solution, 10% pulp density, 0.25–1 kmol·m^{-3} Fe^{3+}, 50 °C, and 400 rpm. The leaching efficiencies of Zn increased gradually with leaching time and improved significantly with higher Fe^{3+} concentrations. A ferric ion was used as an oxidant in this study, and the oxidation of ZnS was promoted by the ferric ion according to the following equation [9]:

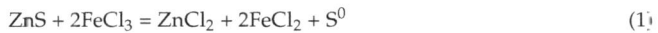

$$ZnS + 2FeCl_3 = ZnCl_2 + 2FeCl_2 + S^0 \tag{1}$$

The sediments contain 5287 mg/kg of Zn, which is equivalent to 8 mol·m^{-3} in a 10% pulp density solution. However, although 250 mol·m^{-3} Fe^{3+} is already almost 15 times more than the stoichiometric requirement of 16 mol·m^{-3} Fe^{3+} to oxidize 8 mol·m^{-3} of ZnS, the leaching efficiency of Zn was only 41.4% at 240 min as shown in Figure 3.

Figure 4 shows the effect of temperature on the leaching efficiency of Zn at 1 kmol·m^{-3} HCl solution, 10% pulp density, 1 kmol·m^{-3} Fe^{3+}, 40–90 °C, and 400 rpm. Higher temperatures yield higher Zn leaching efficiency and it increased to more than 99% within 120 min and 180 min at 90 °C and 80 °C, respectively. These results also show that the Zn content decreased to less than 200 mg/kg, which is the environmental standard for Zn in Korea, and indicate that Zn could be removed successfully. The effect of pulp density on the leaching of Zn from the sediments was investigated under the following conditions: 1 kmol·m^{-3} HCl, 1 kmol·m^{-3} Fe^{3+}, 10%–30% pulp density, 50 °C temperature, and 400 rpm agitation speed, but the results showed no apparent differences in leaching efficiency (data not shown).

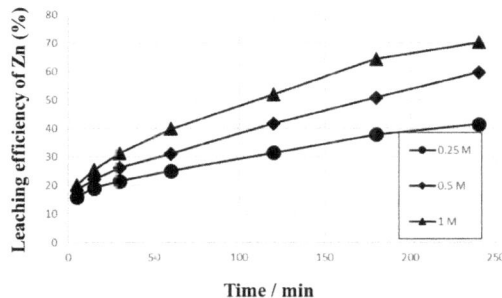

Figure 3. The effects of Fe^{3+} concentration on the leaching efficiency of Zn at 1 kmol·m^{-3} HCl, 10% pulp density, 50 °C, and 400 rpm.

Studies on the dissolution kinetics of ZnS were reported previously and investigations were carried out using pure sphalerite crystals [9,10], low-grade calcareous sphalerite [11], and ZnS concentrate [4,5]. Activation energy values were calculated using the shrinking core model and the Dickson and Heal model [4,5,9–11] and were determined to be 27.5 kJ/mol in sulfuric acid [4] and 42–49.2 kJ/mol in hydrochloric acid [5,9–11], respectively. The shrinking core model could be expressed by [5,12]:

$$1 - (1 - x)^{\frac{1}{3}} = k_r t \tag{2}$$

$$1 - (\frac{2}{3})x - (1 - x)^{\frac{2}{3}} = k_d t \tag{3}$$

Equation (2) represents the shrinking core model based on surface chemical reaction (reaction-controlled model), while Equation 3 is based on diffusion reaction (diffusion-controlled model). The value for x in the equations represents the fraction reacted and is obtained from the leaching efficiencies, while k_r and k_d are rate constants.

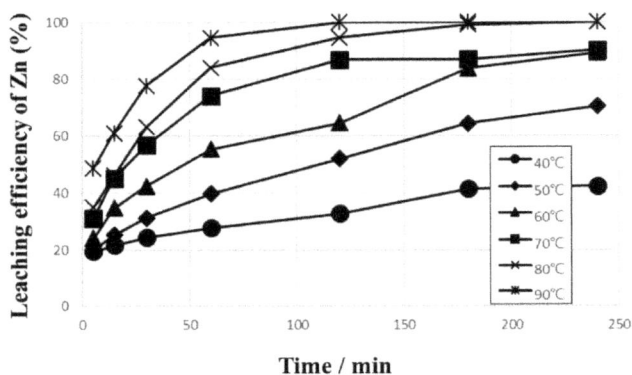

Figure 4. The effects of temperature on the leaching efficiency of Zn at 1 kmol·m^{-3} HCl with 1 kmol·m^{-3} Fe^{3+}, 10% pulp density, and 400 rpm.

The leaching efficiency of Zn increased to more than 90% at 80 °C and 90 °C within 120 min, so the leaching data within 60 min were fitted to the models as shown in Figure 5 (reaction-controlled model) and Figure 6 (diffusion-controlled model), respectively. Based on the calculations, the correlation coefficients ($R2$) for the diffusion-controlled model are higher compared to that of the reaction-controlled model. Previous studies on ZnS leaching with FeCl$_3$ reported that the leaching behavior of Zn followed the reaction-controlled model [5,9,10] while others proposed that the reaction was controlled by both the reaction- and diffusion-controlled model [4,11]. In the current study, the kinetic model parameters indicated that the diffusion-controlled model is more suitable to explain the kinetic behavior of Zn removal. This result can be attributed to the low Zn content of the test sample (about 0.5%) as compared to the 48%–67% Zn contents of the samples used in previous studies [4,5,9–11].

Rate constants for different temperatures were calculated from Figure 6, and the Arrhenius plots of ln$K$ *vs.* T^{-1} for leaching data are shown in Figure 7. From the Arrhenius plots, the activation energies of 76.9 kJ/mol, 69.6 kJ/mol, and 58.5 kJ/mol were calculated for the sphalerite dissolution in 0.25 kmol·m^{-3}, 0.5 kmol·m^{-3}, and 1 kmol·m^{-3} Fe^{3+}, respectively. The results indicate that the reaction is controlled by the diffusion of Fe^{3+} ions.

This study confirmed that the removal of Zn could be achieved by ferric chloride leaching in HCl solution, and the Zn content decreased to less than 200 mg/kg, satisfying the Korean environmental standard. During leaching, ferrous (Fe^{2+}) ion is generated and the Fe^{2+} ion could be oxidized electrochemically or biologically into Fe^{3+}, which can be reused as an oxidant in the process. Further study will be required to optimize the re-oxidation.

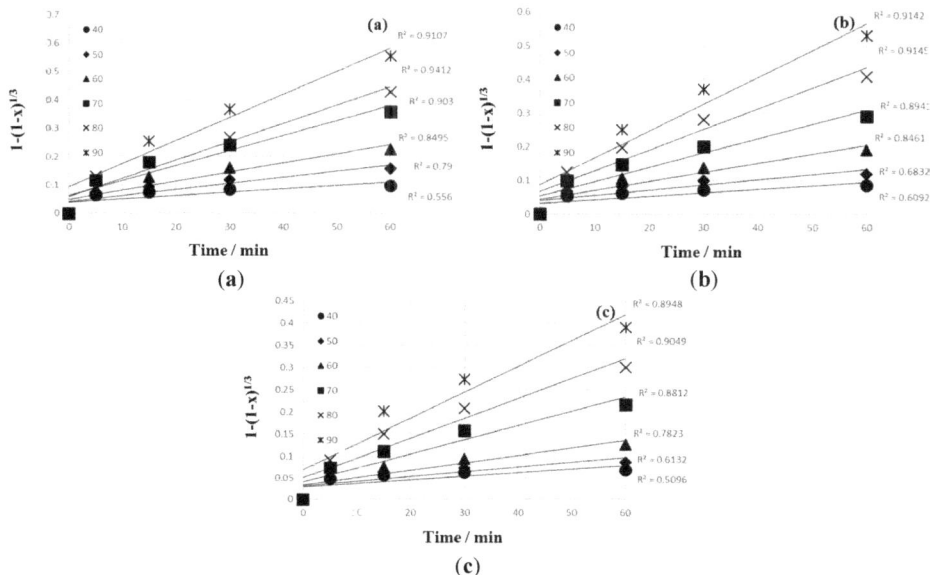

Figure 5. Plot of $1 - (1 - x)^{1/3}$ *vs.* T for different temperatures ([Fe^{3+}]: (**a**) 1.0 M; (**b**) 0.5 M; (**c**) 0.25 M; 10% pulp density; 400 rpm agitation speed).

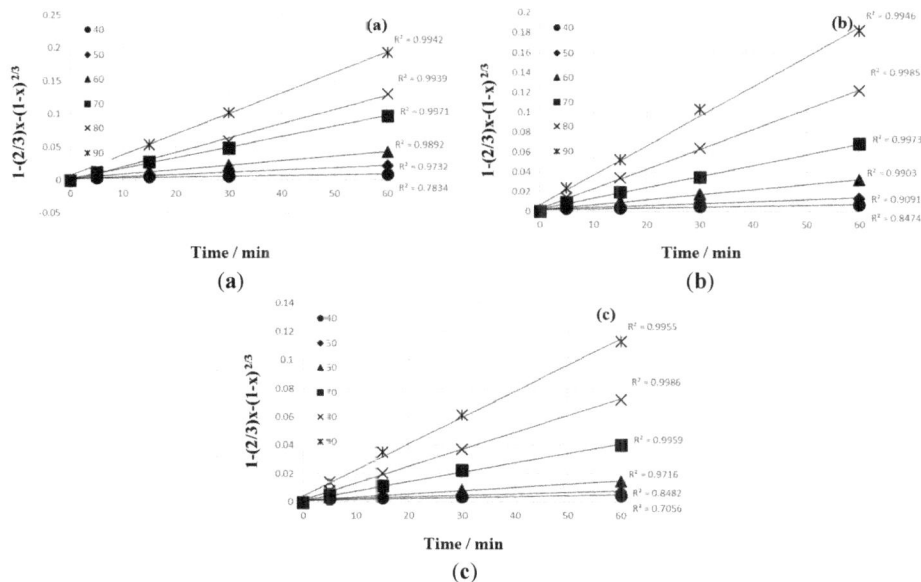

Figure 6. Plot of $1 - 2/3x - (1 - x)^{2/3}$ *vs.* T for different temperatures ([Fe^{3+}]: (**a**) 1.0 M; (**b**) 0.5 M; (**c**) 0.25 M; 10% pulp density; 400 rpm agitation speed).

275

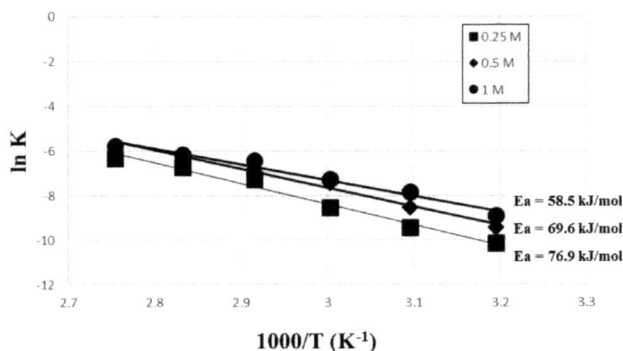

Figure 7. Arrhenius plots. (1 kmol m^{-3} HCl; 0.25 kmol m^{-3}–1 kmol m^{-3} Fe^{3+}; 400 rpm agitation speed; 10% pulp density; 40 °C–90 °C temperature).

4. Conclusions

The leaching behavior of Zn from sediment contaminated with ZnS concentrate using Fe^{3+} ions in HCl solution was investigated. The chemical associations of Zn in the sediment were also evaluated using Tessier's sequential extraction method. By fitting the leaching data to two shrinking core models, the following conclusions were made:

1. The removal of Zn was achieved by ferric chloride leaching in HCl solution. The removal ratio increased to more than 99% within 120 min under the following leaching conditions; 1 kmol·m^{-3} Fe^{3+}; 1 kmol·m^{-3} HCl; 400 rpm agitation speed; 90 °C leaching temperature; 10% pulp density. More importantly, the results obtained meet the Korean environmental standards for Zn of 200 mg/kg.
2. About 17.7% of Zn in the sediment was associated with the soluble carbonate and oxide phases as determined by the sequential extraction procedure. The leaching efficiency of Zn increased quite rapidly in the first 5 min and then gradually in the succeeding treatment time. The rapid increase can be attributed to the dissolution of Zn in carbonate and oxide fractions. These results confirm the importance of evaluating the different metal phase associations through sequential extraction to understand the leaching behavior.
3. The kinetic data indicated that the leaching behavior followed the diffusion-controlled model well rather than the surface chemical reaction model. As ferric ion concentration increases, the calculated activation energy decreases. The results also suggest that higher ferric ion concentration and temperature could enhance the leaching efficiency of Zn.

Acknowledgments: The authors express appreciation for the support of the "Leaders Industry-University Cooperation" Project, funded by the Ministry of Education, Science & Technology (MEST) and the National Research Foundation of Korea (NRF).

Author Contributions: Lee and Yoo wrote the paper and contributed to all activities. Kwon conducted leaching tests and analyses. Alorro contributed to interpretation and discussion of the kinetics data.

Conflicts of Interest: The authors declare no conflict of interest.

References

1. Furukawa, M.; Tokunaga, S. Extraction of heavy metals from a contaminated soil using citrate enhancing extraction by pH control and ultrasound application. *J. Environ. Sci. Health* **2004**, *39*, 627–638. [CrossRef]
2. Park, H.; Jung, K.; Alorro, R.D.; Yoo, K. Leaching behavior of copper, zinc and lead from contaminated soil with citric acid. *Mater. Trans.* **2013**, *54*, 1220–1223. [CrossRef]

3. Santos, S.M.C.; Machado, R.M.; Correia, M.J.N.; Reis, M.T.A.; Ismael, M.R.C.; Carvalho, J.M.R. Ferric sulhate/chloride leaching of zinc and minor elements from a sphalerite concentrate. *Miner. Eng.* **2010**, *23*, 606–615. [CrossRef]

4. Souza, A.D.; Pina, P.S.; Leão, V.A.; Silva, C.A.; Siqueira, P.F. The leaching kinetics of a zinc sulphide concentrate in acid ferric sulphate. *Hydrometallurgy* **2007**, *89*, 72–81. [CrossRef]

5. Aydogan, S.; Aras, A.; Canbazoglu, M. Dissolution kinetics of sphalerite in acidic ferric chloride leaching. *Chem. Eng. J.* **2005**, *114*, 67–72.

6. Peng, P.; Xie, H.; Lu, L. Leaching of a sphalerite concentrate with H_2SO_4–HNO_3 solutions in the presence of C_2Cl_4. *Hydrometallurgy* **2005**, *80*, 265–271. [CrossRef]

7. Tessier, A.; Campbell, P.G.C.; Bisson. M. Sequential extraction procedure for the speciation of particulate trace metals. *Anal. Chem.* **1979**, *51*, 844–850. [CrossRef]

8. Li, X.; Coles, B.J.; Ramsey, M.H.; Thornton, I. Sequential extraction of soils for multielement analysis by ICP-AES. *Chem. Geol.* **1995**, *124*, 109–123. [CrossRef]

9. Dutrizac, J.E.; Macdonald, R.J.C. The dissolution of sphalerite in ferric chloride solutions. *Metall. Trans. B* **1978**, *9B*, 543–551. [CrossRef]

10. Al-Harahsheh, M.; Kingman, S. The influence of microwaves on the leaching of sphalerite in ferric chloride. *Chem. Eng. Prog.* **2008**, *47*, 1246–1251. [CrossRef]

11. Dehghan, R.; Noaparast, M.; Kolahdoozan, M. Leaching and kinetic modelling of low-grade calcareous sphalerite in acidic ferric chloride solution. *Hydrometallurgy* **2009**, *96*, 275–282. [CrossRef]

12. Levenspiel, O. *Chemical Reaction Engineering*; John Wiley & Sons: New York, NY, USA, 1999; pp. 566–588.

![metals logo] **metals**

Article

Use of Nanoscale Zero-Valent Iron (NZVI) Particles for Chemical Denitrification under Different Operating Conditions

Alessio Siciliano

Department of Environmental and Chemical Engineering, University of Calabria, P. Bucci, 87036 Arcavacata di Rende (CS), Italy, alessio.siciliano@unical.it; Tel./Fax: +39-09-84496537

Academic Editor: Suresh Bhargava
Received: 9 July 2015; Accepted: 18 August 2015; Published: 21 August 2015

Abstract: The nitrate pollution of waters and groundwaters is an important environmental and health concern. An interesting method to remove the oxidized forms of nitrogen from waters and wastewaters is chemical denitrification by means of metallic iron (Fe^0). Particularly advantageous is the use of nanoscopic zero-valent iron particles due to the elevated surface area, which allows reaching extremely high reaction rates. In the present paper, the efficiency of nitrate reduction by means of nanoscopic Fe^0 has been investigated under several operating conditions. The iron nanoparticles were synthesized by the chemical reduction of ferric ions with sodium borohydride. The effects of Fe^0 dosage, initial $N–NO_3^-$ concentration and pH on chemical denitrification were identified. In particular, the results of the tests carried out showed that it is possible to reach an almost complete nitrate reduction in treating solutions with a nitrate nitrogen concentration higher than 50 mg/L. Moreover, the process performance was satisfactory also under uncontrolled pH. By means of the trends detected during the experiments, the kinetic-type reaction was identified. Furthermore, a relation between the kinetic constant and the process parameters was defined.

Keywords: denitrification; kinetic analysis; nanoscopic iron; nitrate

1. Introduction

Nitrate pollution is a serious environmental problem, because it causes eutrophication phenomena and may lead to diseases when drinking water sources are contaminated. The sources of nitrogen compounds include nitrogenous fertilizers, animal manure, municipal and industrial wastewaters. Many physicochemical and biological processes, such as ion exchange, reverse osmosis and biological denitrification, have been used to remove nitrate from waters and wastewaters. However, ion exchange and reverse osmosis may be expensive, because they require frequent regeneration of the media and generate secondary waste [1]. Biological denitrification, the most widely-used method, produces excessive biomass and soluble microbial products that require further expensive treatment for use in drinking water supplies. Moreover, the microbial processes are generally slow and sometimes incomplete compared to chemical processes [1]. The $N–NO_3^-$ chemical reduction to ammonia by means of microscopic zero-valent iron has been widely studied, because this metal is abundant, inexpensive and readily available and because its reduction process requires little maintenance [1–12]. The performance of Fe^0 treatment is dependent on the pH value in the aqueous system. Generally, rapid reduction of nitrate by Fe^0 occurred at pH $\leq$ 4, which is probably related to the enhanced iron corrosion in acid solution. Under environmental conditions, the iron corrosion products, such as iron oxides, oxide hydroxides or salts, which are usually coated on the Fe^0 surface, could slow down the nitrate reduction reaction. Thus, the process requires acidic conditions to achieve the best performances. Several compounds, such as H_2SO_4, HCl, acetic acid and buffered agents, were used to speed up the rate of nitrate removal [3]. However, these methods are not convenient in practical applications, because the costs are significantly increased and because the release into waters of undesired compounds, such as sulfate, chloride *etc.*, occurs. Moreover, in the nitrate removal by Fe^0, a high transfer into the solution of soluble iron generally occurs. Therefore, in order to recovery these ions, further treatment is generally needed. In some works, the efficiency of the treatment was improved by the addition to the solution of a sufficient amount of Fe^{2+}, because, in this way, it is possible to locally transform the corrosion products into reactive compounds, so as to facilitate nitrate reduction [4]. Furthermore, in the presence of Fe^{2+}, dissolved oxygen does not seem to interfere with rapid nitrate reduction by Fe^0 [4]. The adaptation of nano-scale zero-valent iron (NZVI) for the decontamination of water and wastewaters has several advantages compared to micro-scale ZVI. Reducing the particle size of granular Fe^0 materials (mm) to 10–100 nm, it increases the surface area and, thus, the reaction rates. Moreover, the nano-Fe^0 can be applied in the treatment process in slurry form, which facilitates the contact with the contaminant. These factors allow operating with lower dosages and obtaining a high and stable reactivity. Furthermore, the recovery of NZVI particles after the treatment could be easy accomplished. In fact, it is a common behavior that nanoparticles would aggregate in clusters in aqueous solution unless they are dispersed by a surfactant. Therefore, synthesized nanoscale iron particles can be easily separated from liquid by filtration or other physical separation processes [5]. In the present paper, the results of an experimental activity conducted to investigate the efficiency of nanoscale zero-valent iron (NZVI) particles for chemical denitrification of nitrate nitrogen in aqueous solutions, under different operating conditions, are reported. Nanosized iron was synthesized by chemical reduction of ferric ions with sodium borohydride. Several batch experiments were carried out to evaluate the influence of nitrate concentrations, iron amounts and pH on the process performance. In particular, contrary to the common zero-valent iron applications, the applicability of NZVI particles without pH setting was also verified. Furthermore, a detailed analysis was conducted to identify the type of reaction kinetics.

2. Experimental Section

2.1. Preparation of Nanosized Iron

In this study nanosized iron was synthesized by chemical reduction of ferric ions with sodium borohydride [1,5] according to the following reaction:

$$4Fe^{3+} + 3BH_4^- + 9H_2O \rightarrow 4Fe^0 + 3H_2BO_3^- + 12H^+ + 6H_2 \tag{1}$$

In a typical synthesis procedure, 33.77 g of $FeCl_3 \cdot 6H_2O$ were dissolved in 250 mL of bi-distilled water (0.5 M of Fe^{3+}). The resulting solution was transferred to a glass beaker of 500-mL capacity. Then, 9.45 g of $NaBH_4$ powder were dissolved in 250 mL bi-distilled water (1 M of B^{5-}), and this was added incrementally to the $FeCl_3 \cdot 5H_2O$ solution by means of a peristaltic pump with a flowrate of about 10 mL/min. The mixture was mechanically stirred at 250 rpm during the $NaBH_4$ feeding. The mixing was stopped 15 min after the entire sodium borohydride solution was added. No surfactant was added to keep the iron particles dispersed. The mixture was centrifuged for 30 min at 4000 rpm, and the recovered solid was washed twice with bi-distilled water. The prepared wet NZVI was used immediately for batch experiments.

2.2. Nitrate Reduction by NZVI

The experiments were carried out in order to investigate the influence of NVZI, nitrate concentrations and pH on process performance. A total of 36 tests were conducted using wet iron amounts of 1 g/L, 2 g/L, 3 g/L and 5 g/L for the treatment of solutions with initial nitrate concentrations (N_i) of about 50 mgN/L, 70 mgN/L and 95 mgN/L, at conditions of pH 3, 5 and uncontrolled. Batch tests for nitrate reduction were conducted in 0.5-L glass beakers. The reactor was filled with 300 mL of nitrate standard solutions (prepared using KNO_3 salt), then wet NZVI particles were added to the reaction solution. The dispersion was continuously mixed with a mechanical stirrer at 350 rpm for 60 min. During the tests conducted with the pH setting, HCl (1 N) was added to hold the established pH value. The reactions were conducted at atmospheric conditions at a temperature of about 20 °C. No attempts were made to exclude oxygen during the experiments. Samples of 5 mL were periodically taken and immediately analyzed, after a filtration at 0.45 μm, with respect to nitrogen compounds ($N-NO_3^-$, $N-NO_2^-$ and $N-NH_4^+$) and dissolved iron ions.

2.3. Analytical Methods and Presentation of Results

The BET surface area of nanoparticles was determined by Micromeritics (Thermo Sorptomatic 1990, Waltham, MA, USA). XRD (Philips PW 1710, Amsterdam, The Netherlands) was used to characterize the iron corrosion products. During the nitrate removal tests, temperature, pH and dissolved oxygen were measured by a multiparametric analyzer (Hanna HI 98196, Padova, Italy); $N-NO_3^-$, $N-NO_2^-$, $N-NH_4^+$ and iron ions by colorimetric methods using a UV spectrophotometer (Thermo Genesys10uv, Waltham, MA, USA) [13]. Each measurement was carried out three times, and the mean value was considered. The results of the efficiencies reported were representative of the actual removal or production of the compounds. Thus, the values were not affected by dilution because of reactant additions in the various processes.

3. Results and Discussion

3.1. Nitrate Abatement by Fe^0 Nanoparticles

The iron particles synthesized in this study were determined to have a specific surface area of 39.36 m^2/g, in line with the values reported in the literature [1,5]. The results of the experiments conducted by means of the prepared nanoparticles are reported in Table 1 and Figures 1–4. For each pH condition, growing nitrate reductions in response to the increase of Fe^0 mass were detected. In particular, in treating solutions with an initial $N-NO_3^-$ concentration of about 50 mg/L, at pH = 3 and with the lower amount of nanoparticles, an abatement close to 56% was obtained, reaching a residual concentration of about 23 mg/L (Figure 1a). This yield rapidly increased up to an almost complete nitrate abatement using a concentration of nanoparticles equal to 2 g/L. (Figure 1a).

Smaller performances were obtained in the tests conducted with higher nitrate concentrations (Table 1). Indeed, the curves plotted in Figure 1b,c show slower decreasing trends in the tests carried out with solutions characterized by initial concentrations of 70 mg/L and 95 mg/L. This is also noticeable from the specific reaction rates (k) identified by interpolating the experimental results (see Section 3.2). By comparing the k values for the two last sets of tests (Table 1), just a moderate reduction was observed. Only with the greatest Fe^0 quantity a notable gap between the kinetic constant values (k) was detected (Table 1). Anyhow, using the Fe^0 amount of 1 g/L, the efficiencies were around 32% and 36% in treating 70 mg/L and 95 mg/L respectively; then they progressively increased, reaching a complete nitrate removal with the highest nanoparticles dosage (Table 1, Figure 1b,c).

These results, taking into account the amounts of nitrate and iron used for the experiments, are consistent with those detected by Yang [5], confirming the reduction power of nanoscopic Fe^0 also in treating highly concentrated solutions. Moreover, our tests showed only slight reductions of the process performance with increasing the pH to five (Table 1, Figure 2).

In particular, the yields obtained during the tests conducted with the lower nitrate concentration were analogous to that obtained at pH = 3. Indeed, although a general reduction occurred in response to pH increase, the reactions rates were such that they ensured particularly high abatements in treating solutions with 50 mg/L of nitrate nitrogen (Figure 2a). Furthermore, compared to those observed at pH = 3, even higher abatements were detected with the initial nitrate amount of 70 mg/L (Table 1, Figure 2b). Instead, during the set of experiments carried out with 95 mg/L, lower performances were detected using Fe^0 amounts of 3 g/L and 5 g/L. In fact, yields higher than 90% were obtained at pH = 3, while a maximum yield of about 70% was observed at pH = 5, reaching a final concentration of about

20 mg/L (Figure 2c). Thus, the above results indicate that, by increasing the process pH up to five, only a meaningful abatement reduction occurred for the highest nitrate concentrations tested in this study. In the experiments conducted without pH control, the initial value was about 5.9, 5.6 and 5.4 for the solutions with 50, 70 and 95 mg/L, respectively. The above initial pH values rapidly increased during the process up to reaching in a few minutes values higher than nine (Figure 3). However, remarkable nitrate reductions were also observed with these process conditions (Table 1). In particular, with the lower nitrate concentration, the yields increased from 42% to about 98% in response to iron dosage enhancement from 1 g/L–5 g/L (Table 1, Figure 4a). Maximum abatements around 87% and 63%, reaching residual concentration of about 9 mg/L and 35 mg/L, were obtained in treating solutions characterized with nitrate amount of 70 mg/L and 95 mg/L, respectively (Figure 4b,c).

Table 1. Nitrate abatement, nitrate removed, ammonium produced, dissolved iron and the specific reaction rate (k) detected for each test.

pH	N_i mg/L	Fe^0 g/L	$N-NO_3^-$ Abatement (%)	$N-NO_3^-$ Removed (mg/L)	$N-NH_4^+$ Produced (mg/L)	Fe^{2+} mg/L	k min^{-1}
3	50	1	56.1	28.2	27.6	198	0.013
		2	99.9	49.2	47.8	491	0.091
		3	99.9	49.6	46.3	705	0.185
		5	99.9	50.8	48.5	1408	0.282
	70	1	36.2	26.9	24.4	197	0.008
		2	74.3	52.9	49.3	605	0.025
		3	84.4	59.2	58.5	803	0.034
		5	99.9	71.7	70.2	1530	0.135
	95	1	32.1	31.2	31.0	198	0.006
		2	60.2	58.6	57.6	491	0.019
		3	92.4	89.6	88.3	705	0.029
		5	99.9	98.5	97.3	1408	0.075
5	50	1	52.0	24.6	23.0	48	0.015
		2	97.1	48.3	47.6	230	0.054
		3	99.9	51.0	48.1	350	0.109
		5	99.9	52.2	48.3	1200	0.209
	70	1	38.3	26.3	23.0	96	0.008
		2	83.0	56.4	55.1	232	0.029
		3	99.9	65.3	62.4	235	0.073
		5	99.9	69.4	69.1	530	0.084
	95	1	28.4	27.5	26.7	32	0.004
		2	61.0	57.9	56.5	126	0.014
		3	69.2	66.4	64.6	168	0.019
		5	73.1	71.2	70.1	370	0.029
uncontrolled	50	1	42.3	20.3	16.0	4.0	0.010
		2	77.1	37.3	33.1	4.2	0.021
		3	80.0	39.2	34.1	4.6	0.022
		5	98.0	44.6	40.5	6.0	0.060
	70	1	32.4	22.0	16.4	5.0	0.007
		2	59.2	42.9	30.1	6.3	0.015
		3	60.5	44.1	34.6	7.0	0.016
		5	87.1	59.2	41.5	6.8	0.034
	95	1	29.5	28.7	20.7	4.0	0.005
		2	45.2	44.0	30.8	5.4	0.010
		3	47.4	46.3	40.3	5.7	0.010
		5	63.0	61.7	43.0	5.7	0.015

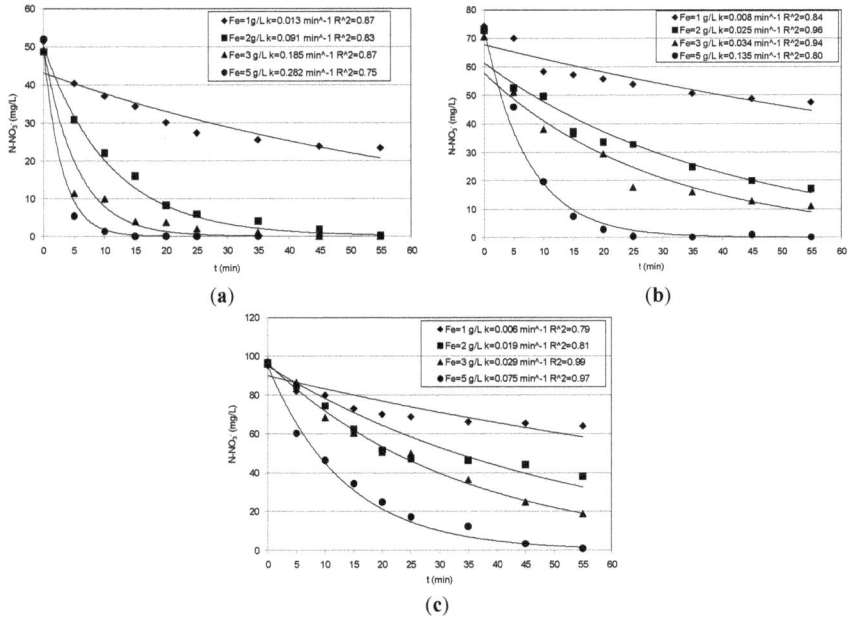

Figure 1. Trends of nitrate concentrations detected during the tests conducted at pH = 3: (a) N_i = 50 mg/L; (b) 70 mg/L; (c) 95 mg/L.

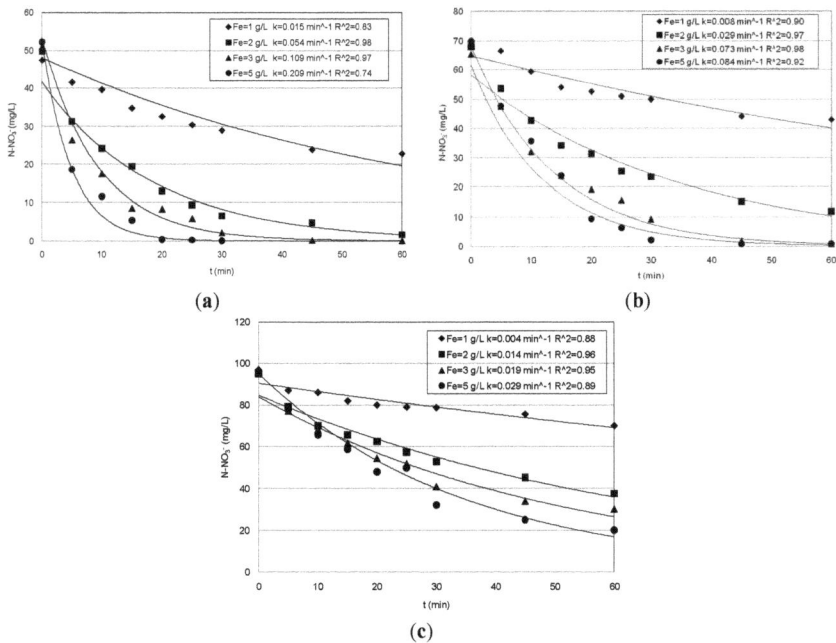

Figure 2. Trends of nitrate concentrations detected during the tests conducted at pH = 5: (a) N_i = 50 mg/L; (b) 70 mg/L; (c) 95 mg/L.

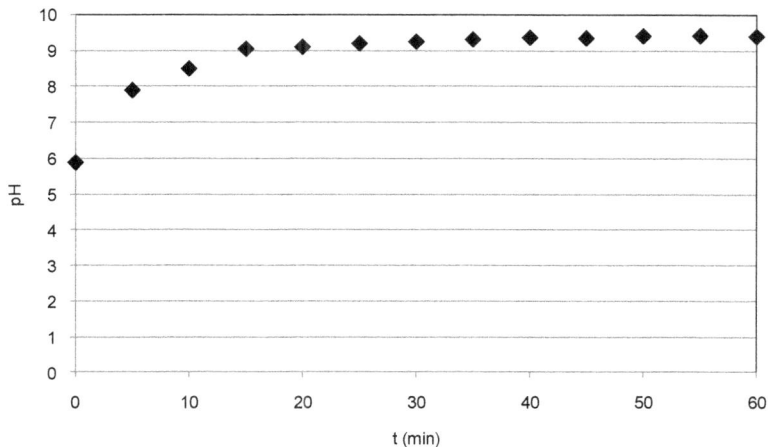

Figure 3. pH trend during the test conducted with Fe^0 = 3 g/L, N_i = 50 mg/L and uncontrolled pH.

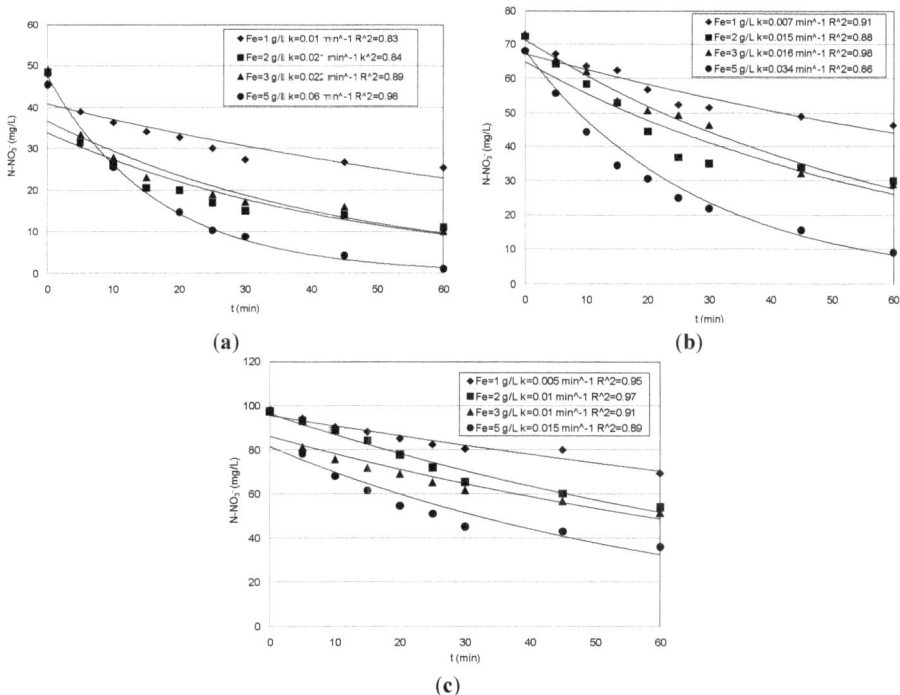

(a)

(b)

(c)

Figure 4. Trends of nitrate concentrations detected during the tests conducted at uncontrolled pH: (a) N_i = 50 mg/L; (b) 70 mg/L; (c) 95 mg/L.

These yields are significantly higher than those reported in other works in which a remarkable loss in process efficiency at a pH value higher than four was observed [5,14]. This is generally attributed to the formation, with the pH increase, of protective layers at the surface of nanosized ZVI that reduce the reactive sites for chemical reduction of nitrate [5]. To increase the process performance under uncontrolled pH, some works have tested Fe/Cu bimetallic nanoscopic materials [1,15]. In this way, however, a transfer in the solution of Cu ions could occur, which can form stable complexes that are difficult to remove, with the ammonia nitrogen [16] produced during the nitrate reduction.

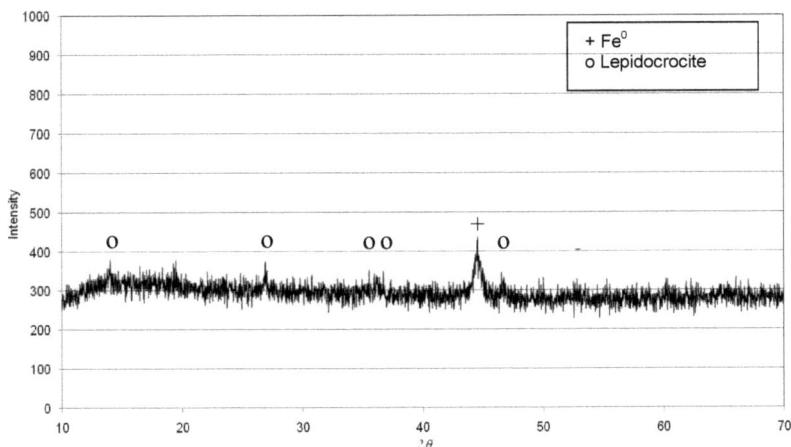

Figure 5. XRD of iron corrosion product after the tests conducted with $Fe^0 = 3$ g/L, $N_i = 50$ mg/L and pH = 3.

Our results, instead, are in agreement with Cheng *et al.* [17], who stated that also iron corrosion products may be responsible for appreciable nitrate reduction. In fact, as shown in Figure 4, in our tests the $N–NO_3^-$ abatement proceeds beyond the time required to reach pH 9 (Figure 3) and, thus, over the corrosion products' formation. Moreover, from the XRD analysis of iron particles recovered after the treatment, appreciable differences between the spectra obtained in the tests carried out at acid conditions and those under uncontrolled pH do not appear (Figures 5 and 6). In fact, in both cases, it seems that the main iron corrosion product could be lepidocrocite (γ-FeOOH) [4]. These statements suggest that the prevailing amount of iron corrosion products is generated during the first minutes of treatment. Lepidocrocite could be initially-formed by O_2 because our tests have been conducted in aerobic environment. Indeed, other works detected, in similar process conditions, mainly lepidocrocite instead of magnetite (Fe_3O_4) or maghemite (γ-Fe_3O_4) [4]. On the contrary, Fe_3O_4 and γ-Fe_3O_4 are commonly observed in experiments conducted in controlled anoxic conditions [4].

The high yields of our treatment could be also attributed to the strong mixing rate that allowed disaggregation of the nanoparticle clusters, enhancing the mass transport of nitrate to the iron surface. Moreover, our tests being conducted under atmospheric conditions, the observed results do not show a meaningful dependence of nitrate reduction from dissolved oxygen. These statements are consistent with the founding of other works conducted both with microscopic and nanoscopic iron particles [5,17]. The experiments of the present study confirmed that the ammonium nitrogen was the major product of nitrate reduction (Table 1). In particular, very low differences between the removed nitrate amounts and the produced ammonia were detected during the experiments conducted at pH values of three and five. The nitrite amounts were always negligible; thus, it is presumable that a slight amount of nitrogen gas has been generated in each test. Lower concentrations of ammonium were detected by performing the process under uncontrolled pH (Table 1). This suggests that part of the produced ammonia has been stripped due to the rapid pH increase beyond nine during the treatment [1]. Without pH control, as expected, also restricted iron ions in the treated solutions were monitored (Table 1). These concentrations were much less than those detected during the other tests, in which the acid conditions promote the iron dissolution. The lower iron and ammonium ions that remain in solutions in the tests conducted in uncontrolled conditions, are clearly a positive aspect. Anyhow, the residual $N-NH_4^+$ makes it necessary to forecast further adequate treatments for its removal. A suitable and profitable technology may be the struvite precipitation process using low expensive reagents that was developed by the author [18,19].

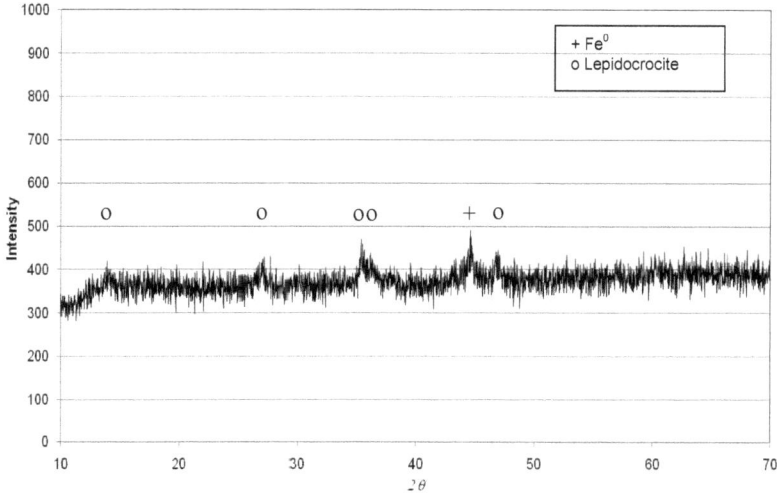

Figure 6. XRD of iron corrosion product after the tests conducted with $Fe^0 = 3$ g/L, $N_i = 50$ mg/L and uncontrolled pH.

3.2. Kinetic Analysis

The results discussed above show how the nitrate reduction through nanoscopic iron is affected by many factors. In order to analyze the influence of these factors, a kinetic analysis of the $N-NO_3^-$ decreasing trends was conducted, which were monitored during the experimental tests. For each operating conditions tested, the $N-NO_3^-$ removal exhibits typical trends of first order kinetics (Figures 1, 2 and 4). Thus, the reduction rate for nitrate can be expressed with the following equation:

$$r_N = \frac{dN}{dt} = -k \cdot N \tag{2}$$

where N is the nitrate concentration and k is the specific reaction rate (first order kinetic constant). By solving the above equation, the following behavior for nitrate concentrations as a function of time can be easily obtained:

$$N = N_i \cdot e^{-k \cdot t} \tag{3}$$

By means of the previous equations, it is possible to accurately simulate the results of the investigation carried out. Indeed, the remarkable agreement between the experimental results and theoretical predictions (Figures 1, 2 and 4) is a good validation of the proposed kinetic model, which, furthermore, is consistent with the statements of other authors [1,5,20]. The abatement curves generally showed faster decreasing trends with the Fe^0 amount enhancement and decreasing reaction rates in response to the growth of initial nitrate concentration and the process pH. However, the kinetic constants did not follow a linear relationship in response to the operating parameters' change. In fact, a concomitant effect of these parameters occurred. By carefully analyzing the detected values of the kinetic constant, it was possible to identify, for the operating conditions tested in this study, a close dependence of the *k* constant on the ratio between the iron amount and the square of initial nitrate concentration ($IN = Fe/N_i^2$) (Figure 7). In particular, the following function type was identified:

$$k = \alpha \cdot IN^\beta \tag{4}$$

where the values of α and β change in response of process pH. The overall reaction rate can be expressed by means of the following relation of the process parameters:

$$r_N = -\alpha(\text{pH}) \left[\frac{Fe}{N_i^2} \right]^{\beta(\text{pH})} N \tag{5}$$

The curves reported in Figure 7 show that, for a given pH, the kinetic constants detected in the various tests follow the same function of the *IN* ratio. Thus, at constant pH, this ratio can be considered the main parameter affecting the process rate. Other works, instead, identified the Fe/N_i ratio as the factor governing the process [5]. Our results clearly underline a greater effect of the nitrate amount to be removed on the reaction rate. From the trends reported in Figure 7, it can be noticed that the kinetic parameters are just lower in the tests conducted at pH 5 with respect to those at pH 3, while higher reductions of *k* values were observed by performing the experiments without pH control. Anyhow, as discussed above, also under uncontrolled pH, the process evolved with reaction rates and was able to ensure satisfactory yields in a reaction time of 60 min.

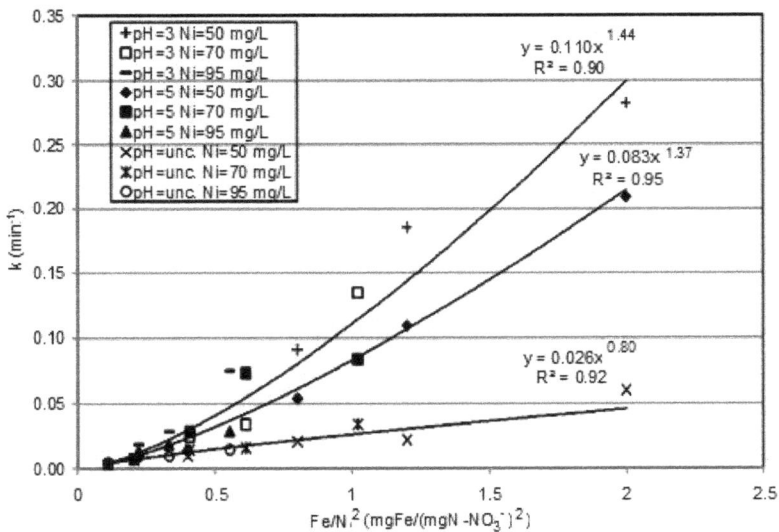

Figure 7. Trends of the *k* constant *vs.* the Fe/N_i^2 ratio.

4. Conclusions

The experiments conducted for this work allowed investigating the effects of operating conditions on nitrate reduction in wastewaters by means of nanoscopic zero-valent iron particles. In particular, the results of the tests carried out showed how the efficiency of the treatment is positively affected by the pH reduction. However, lower increases in process performance were detected by reducing the pH from five to three. Moreover, satisfactory yields were obtained also under uncontrolled pH. In fact, abatements up to 98%, 87% and 63% were reached in 60 min during the treatment of solutions characterized by initial nitrate nitrogen concentrations of 50 mg/L, 70 mg/L and 95 mg/L, respectively.

Metals **2015**, *5*, 1507–1519

The increase of Fe^0 dosage clearly produces an enhancement of the nitrate abatement, while the initial $N-NO_3^-$ amount of wastewaters negatively affected the reduction process. This statement was verified also by the kinetic analysis of trends detected during the conducted experiments. In fact, a first order kinetic-type reaction was identified for each operating condition tested, and the kinetic constant, for a given process pH, resulted in a positive function of the Fe/N_i^2 ratio. The identification of the overall expression of the reaction rate could help to plan treatments for chemical denitrification of wastewaters. The results of the experiments proved, furthermore, that the nitrate reduction efficiently evolved also in aerobic conditions and that the ammonia nitrogen is the main process product. The residual $N-NH_4^+$ amount resulted in being lower in the tests conducted without pH control. In fact the basic conditions that were rapidly reached during the reactions cause ammonia stripping and the precipitation of iron corrosion compounds. The good efficiencies and the low residual ammonium and iron ions make the process under uncontrolled pH preferable. Indeed, in this way, the treatment is easier to manage and less expensive, avoiding the addition of acid. The costs of the process could be further reduced by exploiting unconventional sources, such as iron-rich clay minerals [21,22], for the nanoparticles' generation. Anyhow further experiments are necessary to verify the applicability of the process under field conditions.

Acknowledgments: The author thanks Mr. Giuseppe Bevilacqua for the technical support.

Conflicts of Interest: The authors declare no conflicts of interest

References

1. Hwang, Y.H.; Kim, D.G.; Shin, H.S. Mechanism study of nitrate reduction by nano zero valent iron. *J. Hazard. Mater.* **2011**, *185*, 1513–1521. [CrossRef] [PubMed]
2. Noubactep, C.; Caré, S. On nanoscale metallic iron for groundwater remediation. *J. Hazard. Mater.* **2010**, *182*, 923–927. [CrossRef] [PubMed]
3. Ruangchainikom, C.; Liao, C.H.; Anotai, J.; Lee, M.T. Effects of water characteristics on nitrate reduction by the Fe^0/CO_2 process. *Chemosphere* **2006**, *63*, 335–343. [CrossRef] [PubMed]
4. Huang, Y.H.; Zhang, T.C. Effects of dissolved oxygen on formation of corrosion products and concomitant oxygen and nitrate reduction in zero-valent iron systems with or without aqueous Fe^{2+}. *Water Res.* **2005**, *39*, 1751–1760. [CrossRef] [PubMed]
5. Yang, G.C.C.; Lee, H.L. Chemical reduction of nitrate by nanosized iron: Kinetics and Pathways. *Water Res.* **2005**, *39*, 884–894. [CrossRef] [PubMed]
6. Bhatnagar, A.; Sillanpää, M. A review of emerging adsorbents for nitrate removal from water. *Chem. Eng. J.* **2011**, *168*, 493–504. [CrossRef]
7. Hansen, H.C.B.; Guldberg, S.; Erbs, M.; Koch, C.B. Kinetics of nitrate reduction by green rusts—Effects of interlayer anion and Fe(II):Fe(III) ratio. *Appl. Clay Sci.* **2001**, *18*, 81–91. [CrossRef]
8. Huang, C.P.; Wang, H.W.; Chiu, P.C. Nitrate reduction by metallic iron. *Water Res.* **1998**, *32*, 2257–2264. [CrossRef]
9. Huang, Y.H.; Zhang, T.C. Effects of low pH on nitrate reduction by iron powder. *Water Res.* **2004**, *38*, 2631–2642. [CrossRef] [PubMed]
10. Rodríguez-Maroto, J.M.; García-Herruzo, F.; García-Rubio, A.; Gómez-Lahoz, C.; Vereda-Alonso, C. Kinetics of the chemical reduction of nitrate by zero-valent iron. *Chemosphere* **2009**, *74*, 804–809. [CrossRef] [PubMed]
11. Luo, H.; Jin, S.; Fallgren, P.H.; Colberg, P.J.S.; Johnson, P.A. Prevention of iron passivation and enhancement of nitrate reduction by electron supplementation. *Chem. Eng. J.* **2010**, *160*, 185–189. [CrossRef]
12. Tsai, Y.J.; Chou, F.C.; Cheng, T.C. Coupled acidification and ultrasound with iron enhances nitrate reduction. *J. Hazard. Mater.* **2009**, *163*, 743–747. [CrossRef] [PubMed]
13. American Public Health Association. *Standard Methods for the Examination of Water and Wastewater*, 20th ed.; APHA: Washington DC, USA, 1998.
14. Kassaee, M.Z.; Motamedi, E.; Mikhak, A.; Rahnemaie, R. Nitrate removal from water using iron nanoparticles produced by arc discharge vs. Reduction. *Chem. Eng. J.* **2011**, *166*, 490–495. [CrossRef]

15. Liou, Y.H.; Lo, S.L.; Lin, C.J.; Kuan, W.H.; Weng, S.C. Chemical reduction of an unbuffered nitrate solution using catalyzed and uncatalyzed nanoscale iron particles. *J. Hazard. Mater.* **2005**, *127*, 102–110. [CrossRef] [PubMed]

16. De Rosa, S.; Siciliano, A. A catalytic oxidation process of olive oil mill wastewaters using hydrogen peroxide and copper. *Desalination Water Treat.* **2010**, *23*, 187–193. [CrossRef]

17. Cheng, F.; Muftikian, R.; Fernando, Q.; Korte, N. Reduction of nitrate to ammonia by zero-valent iron. *Chemosphere* **1997**, *35*, 2689–2695. [CrossRef]

18. Siciliano, A.; de Rosa, S. Recovery of ammonia in digestates of calf manure through a struvite precipitation process using unconventional reagents. *Environ. Technol.* **2014**, *35*, 841–850. [CrossRef] [PubMed]

19. Siciliano, A.; Ruggiero, C.; de Rosa, S. A new integrated treatment for the reduction of organic and nitrogen loads in methanogenic landfill leachates. *Process Saf. Environ. Prot.* **2013**, *91*, 311–320. [CrossRef]

20. Choe, S.; Liljestrand, H.M.; Khim, J. Nitrate reduction by zero-valent iron under different pH regimes. *Appl. Geochem.* **2004**, *19*, 335–342. [CrossRef]

21. Stucki, J.W.; Su, K.; Pentrakova, L.; Pentrak, M. Methods for handling redox-sensitive smectite dispersions. *Clay Miner.* **2014**, *49*, 359–377. [CrossRef]

22. Pentrak, M.; Pentrakova, L.; Stucki, J.W. Iron and manganese reduction-oxidation. In *Methods in Biogeochemistry of Wetlands*; DeLaune, R.D., Reddy, K.R., Richardson, C.J., Megonigal, J.P., Eds.; Soil Science Society of America: Madison, WI, USA, 2013; pp. 701–721.

metals

MDPI

Article

Palladium(II) Recovery from Hydrochloric Acid Solutions by *N,N'*-Dimethyl-*N,N'*-Dibutylthiodiglycolamide

Ana Paula Paiva [1,*], Mário E. Martins [1,†] and Osvaldo Ortet [1,2,†]

[1]   Centro de Química e Bioquímica, Faculdade de Ciências, Universidade de Lisboa, Lisboa 1749-016, Portugal; mario.mem@live.com.pt (M.E.M.); oaortet@fc.ul.pt (O.O.)

[2]   Departamento de Ciência e Tecnologia, Universidade de Cabo Verde, 379C, Praia 279, Santiago Island, Cape Verde

*   Author to whom correspondence should be addressed; appaiva@fc.ul.pt; Tel.: +351-217-500-953; Fax: +351-217-500-088.

†   These authors contributed equally to this work.

Academic Editors: Suresh Bhargava, Mark Pownceby and Rahul Ram
Received: 27 October 2015; Accepted: 30 November 2015; Published: 8 December 2015

Abstract: *N,N'*-dimethyl-*N,N'*-dibutylthiodiglycolamide (DMDBTDGA) has been synthesized, characterized, and is investigated in this work as a potential liquid-liquid extractant for palladium(II), platinum(IV), and rhodium(III) from hydrochloric acid solutions. Pd(II) is the only ion which is efficiently removed by DMDBTDGA in toluene from 1.5 M to 4.5 M HCl, but it is not extracted from 7.5 M HCl. Pd(II) stripping is quantitatively achieved by an acidic thiourea solution. Pd(II) extraction kinetics are highly favored (2–5 min). Distribution data points to a DMDBTDGA:Pd(II) species with a 1:1 molar ratio. Pd(II) can selectively be recovered by DMDBTDGA from 4.0 M HCl complex mixtures containing equivalent concentrations of Pt(IV) and Rh(III). When five-fold Fe(III) and Al(III) concentrations are present, only Pt(IV) in the presence of Fe(III), and Fe(III) itself, are extensively co-extracted together with Pd(II). However, Fe(III) can easily be eliminated through an intermediate scrubbing step with water.

Keywords: Palladium(II); hydrochloric acid; liquid-liquid extraction; thiodiglycolamide derivative

1. Introduction

Platinum, palladium, and rhodium, three elements included in the class of the platinum-group metals (PGMs), are rather scarce in the earth, with their core mineral resources concentrated in South Africa and Russia [1]. The preservation of day-to-day life in the Western world would be threatened if the supply of PGMs was to rely on their mines only, since PGMs are currently used in several technological applications, and with a tendency for further development. The exclusive catalytic properties of PGMs place these elements in the forefront of the fabrication of automobile catalysts (to drop the harmful emissions of the motor engines) and also in the chemical industry [2]. Therefore, recycling practices are nowadays determinant for balancing the overall supply of PGMs worldwide [3].

Spent automobile catalytic converters are the main secondary source of PGMs recycled by the metallurgical industry, through the use of pyro and/or hydrometallurgical methods [4]. The direct hydrometallurgical leaching of PGMs from spent catalysts involves hydrochloric acid alone, or with different oxidizing agents [5], with the liquid-liquid extraction (solvent extraction, SX) [6] and the ion exchange [7] techniques being the most common approaches to separate, concentrate, and purify solutions containing PGMs. The composition of the leaching media coming from the treatment of secondary sources is rather different from the original leaching solutions of the PGMs mines [6];

therefore, SX research for improving the performance of well-known and/or commercial extractants [8, 9] or for developing new molecules [10,11] to process the typical leaching media of secondary raw materials is currently very active.

Within the novel molecules created to improve the selective and efficient recovery of PGMs from chloride solutions, various sorts of amide derivatives deserve a special mention. For the recovery of platinum(IV), several malonamides, e.g., [11,12], and thiodiglycolamides [10,13] have proven their adequacy; sulfide-containing monoamides [14], thiodiglycolamides [15,16], thioamides [17], and a dithiodiglycolamide [18] have also been proposed for palladium(II) extraction. Recently, amide-containing tertiary amine compounds have been reported as suitable for rhodium(III) recovery [19].

Following the promising results found for *N,N'*-dimethyl-*N,N'*-dicyclohexylthiodiglycolamide (DMDCHTDGA) toward Pd(II) extraction [13,20,21], this article reports some preliminary research regarding the synthesis and use of another member of the same family, *N,N'*-dimethyl-*N,N'*-dibutylthiodiglycolamide (DMDBTDGA), as a SX agent for Pd(II) (Figure 1). This work is an attempt to better understand the role played by different alkyl substituents (butyl *vs.* cyclohexyl) when *N,N'*-thiodiglycolamide derivatives are used as Pd(II) extractants, in addition to the methyl groups placed at the nitrogen atoms. Accordingly, a step-by-step comparison of the performances of DMDBTDGA and DMDCHTDGA is carried out, highlighting the similarities and main differences observed. Generally, it can be considered that the replacement of the cyclohexyl groups of DMDCHTDGA by the butyl substituents in DMDBTDGA worsens the ability of this latter compound to efficiently and selectively recover Pd(II) when compared with the performance shown by DMDCHTDGA [13,20]. Accordingly, a tentative rationalization of the potential main factors contributing to the differences found between DMDBTDGA and DMDCHTDGA in Pd(II) extraction is presented and discussed.

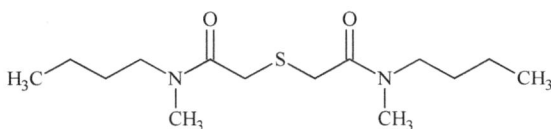

Figure 1. Structure of *N,N'*-dimethyl-*N,N'*-dibutylthiodiglycolamide (DMDBTDGA).

2. Experimental Section

2.1. Synthesis and Characterization

The FTIR spectrum was recorded in a Satellite Mattson spectrophotometer (Thermo Mattson, Madison, WI, USA) sing NaCl cells. The ^{1}H and ^{13}C NMR spectra (400 MHz for ^{1}H and 100 MHz for ^{13}C nuclei) were acquired on a Bruker Avance 400 (Bruker Corporation, Billerica, MA, USA) using deuterated chloroform (Sigma-Aldrich, St Louis, MO, USA, 99.8%) as solvent. For the GC-MS analysis an Agilent 6890 gas chromatograph (Agilent Technologies, Santa Clara, CA, USA), equipped with a non-polar capillary column Hewlett-Packard HP5-MS (Hewlett-Packard, Palo Alto, CA, USA) and coupled to a mass spectrometer Agilent 5973 (electronic impact of 70 eV, Agilent Technologies), was used. The temperature program consisted of an initial heating at 50 °C for 1 min, a raise of 15 °C/min until 180 °C, and another sequential increase of 25 °C/min until 280 °C, remaining at this latter temperature for 3 min. The temperature of the injector was fixed at 280 °C. The preparation of the samples was accomplished by dilution of a small portion of the compound in 200 μL of dichloromethane, and encapsulating the solution in a small vial prior to its analysis.

The synthesis of *N,N'*-dimethyl-*N,N'*-dibutylthiodiglycolamide (DMDBTDGA) was carried out similarly to the one previously reported for *N,N'*-dimethyl-*N,N'*-dicyclohexylthiodiglycolamide (DMDCHTDGA) [13]. A yellowish viscous residue was obtained, with a global yield of 41%.

FTIR (NaCl cells, cm^{-1}): 1643 (C=O).

^{1}H NMR (CDCl$_3$, δ in ppm): 3.47–3.50 (4H, multiplet, C$\underline{H}_2$–S), 3.28–3.38 (4H, multiplet, N–C$\underline{H}_2$), 3.03 (3H, singlet, N–C$\underline{H}_3$), 2.92 (3H, singlet, N–C$\underline{H}_3$), 1.47–1.61 (4H, multiplet, N–CH$_2$–C$\underline{H}_2$), 1.26–1.36 (4H, sextuplet, CH$_3$–C$\underline{H}_2$), 0.90–0.96 (6H, multiplet, C$\underline{H}_3$–CH$_2$).

^{13}C NMR (CDCl$_3$, δ in ppm): 13.83 ($\underline{C}$H$_3$–CH$_2$), 19.94, 19.97 (CH$_3$–$\underline{C}$H$_2$), 29.27, 30.60 (CH$_3$–CH$_2$–$\underline{C}$H$_2$), 33.53, 33.64, 34.09, 34.25 ($\underline{C}$H$_2$–S), 33.65, 35.71 (N–$\underline{C}$H$_3$), 47.87, 50.33 (N–$\underline{C}$H$_2$), 168.42 ($\underline{C}$=O).

GC-MS: retention time = 13.70 min, 91% abundance; m/z = 288 (molecular ion, [M]$^+$), 114 (base peak, [CH$_3$(CH$_2$)$_3$NCH$_3$CO]$^+$), 87 [CH$_3$(CH$_2$)$_3$NCH$_3$ + H]$^+$ and 57 ([CH$_3$(CH$_2$)$_3$]$^+$).

2.2. Solvent Extraction Experiments

Feed aqueous phases containing about 9.4 × 10^{-4} M Pd(II), 5.1 × 10^{-4} M Pt(IV), or 9.7 × 10^{-4} M Rh(III) (about 100 mg·L^{-1} each) were prepared from atomic absorption spectroscopy standards (1003 ± 4 mg·L^{-1} Pd(II) in 5% HCl, 1000 ± 4 mg·L^{-1} Pt(IV) in 5% HCl, 1001 ± 6 mg·L^{-1} Rh(III) in 5% HCl, all provided by Fluka, a subsidiary company of Sigma-Aldrich), in the required volumes of hydrochloric acid (~37%, p.a., Fisher Chemicals, Fisher Scientific Unipessoal, Lisboa, Portugal). Organic phases with 0.02 M DMDBTDGA were settled in toluene (Sigma-Aldrich, ≥99.3%). For the determination of the stoichiometry of the DMDBTDGA:Pd(II) species, organic phases with DMDBTDGA concentrations varying between 1.6 × 10^{-3} to 2.0 × 10^{-2} M were considered. A 0.1 M thiourea solution (99%, Sigma-Aldrich, in 1.0 M HCl (~37%, p.a., Fisher Chemicals) was systematically used as Pd(II) stripping media [13,17].

For the selectivity experiments, 4.0 M HCl solutions containing 9.4 × 10^{-4} M Pd(II) and 5.1 × 10^{-4} M Pt(IV) were prepared, as well as similar ones with the additional presence of 9.7 × 10^{-4} M Rh(III) (about 100 mg·L^{-1} of each metal ion). Quaternary and five-metal ion solutions were further arranged, containing the three PGMs with 9.0 × 10^{-3} M Fe(III) (prepared from iron(III) chloride hexahydrate, 99%, Riedel-de Haën, a subsidiary company of Sigma-Aldrich) and/or 1.9 × 10^{-2} M Al(III) (prepared from aluminum(III) chloride, >98%, Merck, Merck & Co., Inc., Kenilworth, NJ, USA). The Fe(III) and Al(III) contents in the aqueous solutions were of about 500 mg·L^{-1}.

Extraction and stripping experiments were generally performed by stirring equal volumes of aqueous and organic phases (A/O = 1) at room temperature (25 ± 2 °C) and adopting a rotation speed between 900 and 1000 rpm. The preliminary experiments carried out to evaluate the extraction affinity of DMDBTDGA toward Pd(II), Pt(IV) and Rh(III) in different HCl concentrations were performed considering a contact time of 30 min. After the investigation of the kinetics associated with Pd(II) extraction, the adopted agitation time for the subsequent experiments was 5 min. The period of shaking time considered for the stripping assays was always 30 min.

After separation of the two phases, the aqueous and organic extracts were always filtrated to minimize mutual entrainments. In the selectivity experiments, the assays including the presence of Fe(III) involved an additional scrubbing step of the loaded organic phase with distilled water.

The determination of the metal contents in single ion aqueous phases, before and after extraction, was performed by flame atomic absorption spectrophotometry (AAS, novAA 350, Analytik Jena, Jena, Germany). For the analysis of the multi-metallic aqueous solutions, inductively coupled plasma-atomic emission spectrometry (ICP-AES, model Ultima from Horiba Jobin-Yvon S.A.S., Longjumeau, France) was used instead. Metal ion concentrations in organic phases were calculated by mass balance. At least two replicates were considered for all the experiments, and the analysis of the aqueous solutions before and after extraction was systematically made in triplicate. The uncertainties determined for the extraction and stripping results are in the range of ±5% to ±8%.

3. Results and Discussion

3.1. Palladium(II), Platinum(IV) and Rhodium(III) Extraction, and Palladium(II) Stripping

In order to appraise the extraction characteristics of DMDBTDGA toward Pd(II), Pt(IV), and Rh(III), several experiments were programmed to screen the effect of hydrochloric acid concentration on their distribution ratios (D). Accordingly, equal volumes of aqueous solutions containing about 9.4×10^{-4} M Pd(II), 5.1×10^{-4} M Pt(IV), or 9.7×10^{-4} M Rh(III) (in varying HCl concentrations between 1.0 M and 8.0 M) and organic phases with 0.02 M DMDBTDGA in toluene were put in contact for 30 min. The extraction results obtained are displayed in Figure 2, in which a plot showing the dependence of the log D values of Pd(II), Pt(IV), and Rh(III) on HCl concentrations can be observed.

Figure 2 illustrates that DMDBTDGA does not exhibit any appreciable capacity to extract Pt(IV) and Rh(III). In fact, the maximum log D values found for Pt(IV) and Rh(III) were −0.12 and −0.75, respectively, both at 7.0 M HCl. Inversely, Pd(II) extraction is rather efficient between 1.5 M (log D = 1.17) and 4.5 M HCl (log D = 1.01), but a decreasing tendency is clearly visible for higher HCl concentrations. At 7.5 M HCl there is no Pd(II) extraction.

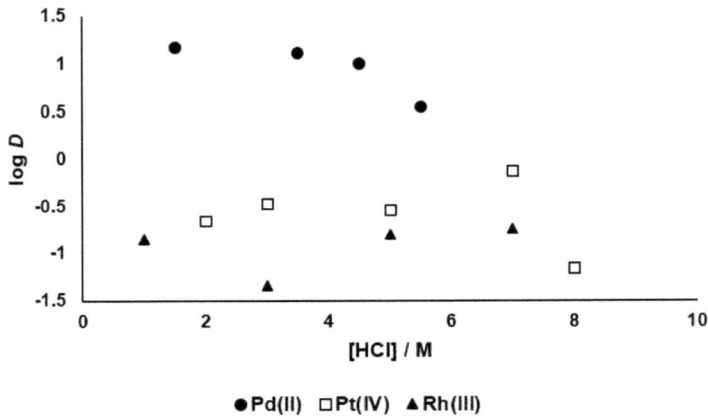

Figure 2. Variation of the log D values of Pd(II), Pt(IV), and Rh(III) with the HCl concentration (9.4×10^{-4} M Pd(II), 5.1×10^{-4} M Pt(IV), or 9.7×10^{-4} M Rh(III) in 1.0 M to 8.0 M HCl; 0.02 M DMDBTDGA in toluene; A/O = 1, room temperature, 900–1000 rpm, 30 min). Standard deviations: ±5%.

A direct comparison of the Pd(II) extraction behavior shown by DMDBTDGA can be established with DMDCHTDGA [20], since the adopted experimental conditions are similar. The two extraction profiles are depicted in Figure 3. It is rather visible that both extractants reduce their extraction ability of Pd(II) as the HCl concentration in the aqueous phases increases, but DMDCHTDGA is much more efficient than DMDBTDGA. It is likely that the diluent (toluene) may play a determinant role in what concerns the dependence on the HCl concentration of both extraction systems [20], since Pd(II) was quantitatively extracted regardless of the HCl content (until 8.0 M HCl) by 0.05 M DMDCHTDGA in 1,2-dichloroethane [13].

Furthermore, the more HCl-concentrated equilibrium aqueous phases (from 5.5 M HCl onwards) resulting from contact with the DMDBTDGA toluene solutions showed a very visible stronger yellow tonality than the initial feeds, which could denote the loss of the organic compound to the aqueous phases. This assumption was confirmed after extraction of those aqueous solutions with toluene, the analysis of the residues showing the presence of DMDBTDGA. Hence, at least when toluene is used as the diluent, it is likely that the decrease of Pd(II) extraction efficiency on the enhancement of HCl concentration may partially be explained by the dissolution of DMDBTDGA in the more acidic HCl solutions, in addition to other phenomena of interfacial nature [20]. The DMDCHTDGA system involves a more hydrophobic extractant, having four extra carbon atoms when compared with DMDBTDGA; hence, a stronger yellowish tone in the equilibrium aqueous phase after contact with DMDCHTDGA was only visible for the 8.0 M HCl aqueous solution.

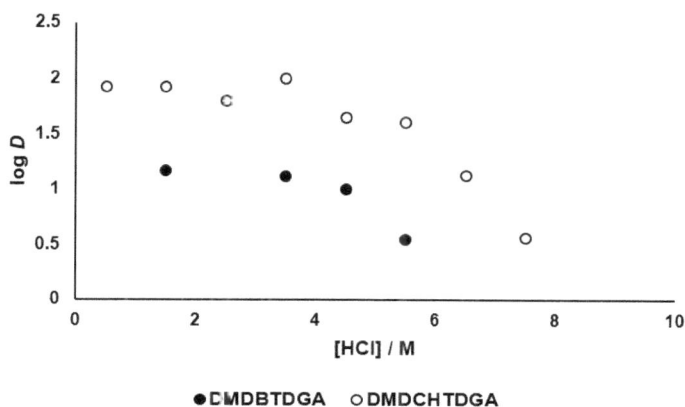

Figure 3. Variation of the log D values of Pd(II) with the HCl concentration by DMDBTDGA and DMDCHTDGA (9.4×10^{-4} M Pd(II) in 0.5 M to 7.5 M HCl; 0.02 M DMDBTDGA or DMDCHTDGA* in toluene; A/O = 1, room temperature, 900–1000 rpm, 30 min). Standard deviations: ±5% (*: data collected from [20]).

A few other thiodiglycolamide derivatives have been previously investigated to extract Pd(II), for instance N,N'-dimethyl-N,N'-diphenylthiodiglycolamide (DMDPHTDGA) in chloroform [22], and N,N,N',N'-tetraoctylthiodiglycolamide (TODTDGA), in 80 vol. % n-dodecane-20 vol. % 2-ethylhexanol [15], with Pd(II) being totally extracted from HCl solutions until 8.0 M HCl. A completely different behavior has been found for N,N'-dimethyl-N,N'-didecylthiodiglycolamide (MDTDGA), dissolved in 80 vol. % n-dodecane-20 vol. % 2-ethylhexanol, as D values of about 1–2 were achieved for 1.0–2.0 M HCl, of about 5 for 4.0 M HCl, 60 for 5.0 M, and >100 for 6.0 M and 7.0 M HCl [16]. Therefore, features such as the length and/or combination of substituent groups, and diluents as well, cause diverse Pd(II) extraction profiles when HCl concentration in the aqueous chloride media vary.

Nonetheless, the adequacy of DMDBTDGA in toluene to extract Pd(II) from HCl solutions until 4.5 M HCl is unquestionable, and therefore, further research on this extractive system is rather justifiable.

In order to confirm that Pd(II) stripping from DMDBTDGA is similarly efficient as when DMDCHTDGA is used [13,20], the Pd(II) loaded organic phases coming from the contact with the HCl solutions until 4.5 M HCl were equilibrated with equal volumes of 0.1 M thiourea in 1.0 M HCl. Pd(II) was always quantitatively stripped to the thiourea aqueous phases, with these results opening good perspectives for the utilization of this solvent system for selective Pd(II) recovery from complex metallic solutions.

3.2. Kinetics of Palladium(II) Extraction

This set of experiments was carried out using a feed aqueous solution containing 9.4×10^{-4} M Pd(II) in 4.5 M HCl, and involving a 0.02 M DMDBTDGA in toluene as the organic phase. The time of contact for both phases varied between 2 and 60 min, and all the other parameters were kept constant.

It was found that the extraction kinetics for these systems is very favorable, as can be seen in Figure 4a 2 min contact results in the highest log D value obtained, with an apparent tendency to decrease for longer contact periods. A higher uncertainty of the results (±8%) may partially justify this unexpected trend. However, the pattern achieved also suggests the possible occurrence of a more complex phenomenon: if the direct kinetic constant is larger than the one corresponding to the forward extraction reaction, this means that Pd(II) would be extensively extracted at very short times, and after a few minutes, when the forward reaction takes place, D values would decrease until equilibrium is reached. Further investigation is obviously needed to support such an assumption.

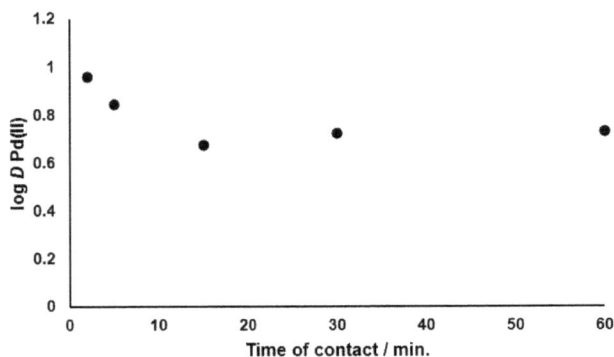

Figure 4. Variation of the log D values of Pd(II) with the contact time (9.4×10^{-4} M Pd(II) in 4.5 M HCl; 0.02 M DMDBTDGA in toluene; A/O = 1, room temperature, 900–1000 rpm). Standard deviations: ±8%.

Nevertheless, the period of time necessary for DMDBTDGA to efficiently extract Pd(II) is quite short, being in general agreement with the ones previously reported for other thiodiglycolamide derivatives, e.g., less than 5 min for DMDPHTDGA [22], 10 min for MDTDGA [16], and 15 min for both TODTDGA [15] and DMDCHTDGA [20].

Accordingly, a contact time of 5 min was adopted for all the subsequent experiments with DMDBTDGA.

3.3. Effect of DMDBTDGA Concentration on Palladium(II) Extraction

The evaluation of the effect of the extractant concentration on the D values of Pd(II) provides information about the stoichiometry of the extractant:Pd(II) species, contributing to the establishment of the Pd(II) extraction reactions. These sets of experiments were carried out under the usual extraction conditions and involved a 9.4×10^{-4} M Pd(II) in 4.5 M HCl aqueous solution, which was contacted with DMDBTDGA organic phases with concentrations varying between 1.6×10^{-3} to 2.0×10^{-2} M.

The log-log plot between the Pd(II) D values and the initial DMDBTDGA concentrations indicates a slope value of about 1.2, as shown in Figure 5. However, as the DMDBTDGA concentrations are not in a sufficiently large excess when compared with that of Pd(II), the DMDBTDGA-Pd(II) species concentrations cannot be neglected or, in other words, the initial DMDBTDGA concentration should not be considered as similar to the DMDBTDGA concentrations at equilibrium. Accordingly, an iterative method previously developed in our group to overcome these situations was applied [23].

In this method, calculations for the free extractant concentrations are carried out upon consideration of different stoichiometric molar ratios between the extractant and the metal ion. When plotting the log-log plots involving the equilibrium ligand concentrations, the coincidence between the slope value obtained and the estimated one is sought [23]. For the DMDBTDGA system, slopes were coincident for a 1.10 value (Figure 5). Hence, this result suggests that species having a 1:1 DMDBTDGA:Pd(II) stoichiometry should exist in the organic solutions.

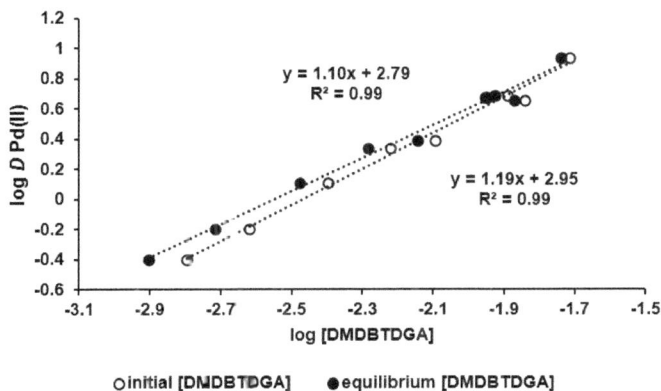

O initial [DMDBTDGA]   ● equilibrium [DMDBTDGA]

Figure 5. Dependence of the Pd(II) log D values on the log of the initial and equilibrium DMDBTDGA concentrations (9.4×10^{-4} M Pd(II) in 4.5 M HCl; 1.6×10^{-3} to 2.0×10^{-2} M DMDBTDGA in toluene; A/O = 1, room temperature, 900–1000 rpm). Standard deviations: ±5%.

Knowing that the predominant Pd(II) species for 1 M and higher HCl concentrations is $[PdCl_4]^{2-}$ [24–26], the general data on the Pd(II) extraction reactions involving different amine and amide derivatives indicates the occurrence of two different mechanisms: the direct extraction of $[PdCl_4]^{2-}$ by a positively charged extractant, resulting in an outer-sphere electrostatic attraction, and/or the replacement of chloride anions from the complexation sphere of $[PdCl_4]^{2-}$ by ligand molecules, due to the liability nature of this chlorocomplex, giving rise to inner-sphere complexes in the organic phase [24]. Previous studies with other thiodiglycolamide derivatives, e.g., MDTDGA [16] and TODTDGA [15], point to inner-sphere complexation reactions when Pd(II) is extracted from 3.0 M HCl (in spite of their completely different profile when the influence of HCl in the feed aqueous solution is taken into account) whereas the metal ion seems to be extracted by a mixing of outer-sphere ion pairs and inner-sphere complexation by DMDPHTDGA [22].

Further research is needed to determine how is Pd(II) extracted by DMDBTDGA from 4.5 M HCl, although the fast extraction rate of the metal ion in this system may eventually suggest the existence of ion-pair extraction reactions [22,27]. Assuming this is the case, a proposal for the Pd(II) species extracted by DMDBTDGA would be the outer-sphere electrostatic attraction between the $[PdCl_4]^{2-}$ species and the protonated extractant on its two amide sites, in a similar scheme as the one suggested for the DMDCHTDGA-Pt(IV) species [13].

3.4. Selectivity for Palladium(II) Extraction

The selectivity shown for Pd(II) extraction by DMDBTDGA is a relevant key point whose evaluation is rather pertinent. On one hand, the separation of Pd(II) from the other two PGMs can be envisaged; subsequently, its isolation from base metals such as Al(III) and Fe(III), often appearing in the leaching solutions of end-of-life anthropogenic sources, is again interesting. Therefore, focusing on the use of HCl media with a more practical interest, 4.0 M HCl solutions with two—Pd(II) and Pt(IV)—and three PGMs—Pd(II), Pt(IV) and Rh(III)—were considered, all with concentrations of about 100 mg·L^{-1}.

The overall data obtained for the DMDBTDGA extractive performance toward the three PGMs are illustrated in Table 1, shown by the separation factors (SF) of Pd(II) in relation to each of the other metal ions (calculated as SF = D_{Pd}/D_{metal}). Hence, the higher the SF, the more selective for Pd(II) recovery DMDBTDGA is. As similar tests were previously carried out with DMDCHTDGA [20], the results obtained with this compound are also included in Table 1 for a direct comparison.

Table 1. Selectivity of DMDBTDGA for extraction of two and three PGM solutions (solutions 1 and 2, respectively), evaluated through the correspondent separation factor values (SF). Standard deviations: ±5%.

[HCl] = 4 M	Solution 1	Solution 2	
	Pt(IV)	Pt(IV)	Rh(III)
DMDBTDGA	261	356	~2880
DMDCHTDGA *	354	289	~2000

* Data collected from [20].

For the binary aqueous media—solution 1—it can be observed that the $SF_{Pd/Pt}$ achieved for DMDBTDGA is good, and a similar situation is extended for the three PGMs solution. Rh(III) is practically not extracted. The selective performance shown by DMDBTDGA for Pd(II) recovery when Pt(IV), and Pt(IV) and Rh(III), are present in the 4.0 M HCl solution is comparable to the data previously found for DMDCHTDGA [20]. Pd(II) was efficiently stripped by 0.1 M thiourea in 1.0 M HCl from the DMDBTDGA loaded organic phases—84 to 86 mg·L^{-1}—with a maximum Pt(IV) contamination of 7 mg·L^{-1}.

Other experiments were subsequently carried out involving four metals in the feed aqueous phases, namely the three PGMs with either 500 mg·L^{-1} Al(III) (solution 3) or Fe(III) (solution 4), and finally five-metal aqueous solutions were also considered, containing the three PGMs, Al(III) and Fe(III) (solution 5), all with similar concentrations as described above. The overall results are depicted in Table 2.

Table 2. Selectivity of DMDBTDGA for extraction of three PGMs with Al(III) or Fe(III) (solutions 3 and 4, respectively) and three PGMs with Al(III) and Fe(III) (solution 5), evaluated through the correspondent SF values. Standard deviations: ±5%.

[HCl] = 4 M	Solution 3			Solution 4			Solution 5			
	Pt(IV)	Rh(III)	Al(III)	Pt(IV)	Rh(III)	Fe(III)	Pt(IV)	Rh(III)	Al(III)	Fe(III)
DMDBTDGA	271	759	673	5	66	26	5	126	169	26
DMDCHTDGA *	314	~2180	~7300	20	996	30	32	~1650	~2260	38

* Data collected from [20].

In the presence of Al(III), the $SF_{Pd/Pt}$ by DMDBTDGA is comparable to the ones previously achieved, whereas Rh(III) is more extracted than before. Al(III) is not extracted extensively. As expected, due to what previously occurred with DMDCHTDGA [20], Fe(III) is widely extracted by DMDBTDGA, and Pt(IV) withdrawal markedly increases as well. In spite of that, the Pd(II) distribution ratio is high (D = 20). The general selective behavior toward Pd(II) found earlier for DMDCHTDGA [20] is better than that determined for DMDBTDGA.

The results displayed in Table 2 for the extraction of the five-metal aqueous phase (solution 5) show that Rh(III) and Al(III) are now generally more extracted by DMDBTDGA, whereas the SF values for the pairs Pd/Pt and Pd/Fe are the same as those obtained for solution 4. The D value for Pd(II) was about 11. These overall results are worse than the ones achieved for DMDCHTDGA [20], also included in Table 2.

Although extensively extracted, Fe(III) can efficiently be scrubbed from the loaded organic media with water [20]. Hence, focusing on the treatment of solution 5, 90% of Fe(III) from the loaded organic phase transferred to water, accompanied by a maximum of 2 mg·L^{-1} Pd(II). Pd(II) was efficiently stripped by 0.1 M thiourea in 1.0 M HCl from the loaded organic phase—86 mg·L^{-1}—with residual contaminations of Pt(IV) and Fe(III) less than 1 mg·L^{-1}. Neither water nor the thiourea solution were able to remove the Pt(IV) content from the loaded DMDBTDGA solution. An efficient stripping agent to recover Pt(IV) has not still been found to date, but its discovery would be crucial to take profit of the extensive co-extraction of Pt(IV) with Pd(II), as the two PGMs could be further separated by selective stripping [13]. The extraction of Pt(IV) together with Pd(II) was not so effective when DMDCHTDGA was used, as can be seen in Table 2 [20].

In summary, the selectivity patterns observed for DMDBTDGA toward Pd(II) are relatively good; among the metal ions tested, only Pt(IV) in the presence of Fe(III), and Fe(III) itself, are extensively extracted together with Pd(II). Fe(III) is easily removed through an intermediate scrubbing step with water. Nevertheless, the selective behavior for Pd(II) recovery shown by the other thiodiglycolamide derivative previously investigated—DMDCHTDGA [20]—is generally much better than that of DMDBTDGA, particularly if Al(III) and Fe(III), and not only Pt(IV) and Rh(III), are present in the feed 4.0 M HCl aqueous phase. These selectivity differences can be rationalized by resorting to the role the substituent groups in the thiodiglycolamide skeleton are likely to play; the butyl groups are apparently too short to favor a more specific selectivity trend, and probably to avoid significant losses of the extractant to the more acidic HCl phases as well. The cyclohexyl substituents, on the other hand, proved to be more adequate to achieve better extraction efficiency and selectivity profiles for Pd(II) recovery by DMDCHTDGA [20].

The evaluation of the performance shown by MDTDGA toward Pd(II) extraction has already been carried out [16], but under quite different experimental conditions than those adopted in this work. A direct comparison between the extraction systems involving DMDBTDGA and DMDCHTDGA with MDTDGA is not possible at present, but the screening of the extraction behavior for Pd(II) depicted by a similar thiodiglycolamide derivative with octyl or decyl groups, in addition to the two methyl substituents, would surely be worthwhile.

4. Conclusions

N,N'-dimethyl-*N,N'*-dibutylthiodiglycolamide (DMDBTDGA) has been synthesized to test its practical usefulness as a SX reagent to recover PGMs from concentrated HCl media. DMDBTDGA in toluene is efficient for Pd(II) extraction until 4.5 M HCl, with a gradual decrease afterwards. Pd(II) in the loaded organic phases is quantitatively stripped by a 0.1 M thiourea in 1 M HCl solution. The extraction kinetics are very favorable (5 min maximum), and the distribution data points out to an extractant: Pd(II) ratio of 1:1.

Pd(II) is selectively recovered from 4.0 M HCl solutions containing the three PGMs under study. However, the selectivity patterns become worse when excess concentrations of Al(III) and Fe(III) co-exist in the HCl solution: Pd(II) distribution ratios are not critically affected, but Pt(IV) is much more extracted than before, together with Fe(III). The Fe(III) interference can nonetheless be eliminated by an intermediate scrubbing step of the loaded organic phases with water. Al(III), one of the main contaminants in real leaching solutions of the hydrometallurgical treatment of secondary raw materials, is not extensively extracted.

The selectivity figures obtained for DMDBTDGA, although reasonable, are poorer than those achieved for DMDCHTDGA [20], another thiodiglycolamide derivative tested under similar conditions.

Acknowledgments: The financial support for the investigation reported in this article has been kindly provided by Portuguese national funds through "FCT—Fundação para a Ciência e a Tecnologia" (Lisbon, Portugal) under the projects with references UID/MULTI/00612/2013 and PTDC/QUI-QUI/109970/2009, including the PhD grant offered to Osvaldo Ortet (SFRH/BD/78289/2011). The GC-MS analysis performed by José Manuel Nogueira and collaborators is gratefully acknowledged.

Author Contributions: Mário E. Martins and Osvaldo Ortet carried out the synthesis and characterization of DMDBTDGA, as well as the SX experiments. Ana Paula Paiva supervised all the research and wrote the manuscript.

Conflicts of Interest: The authors declare no conflict of interest.

References

1. Report on Critical Raw Materials for the EU—European Commission, May 2014. Available online: http://www.google.pt/url?sa=t&rct=j&q=&esrc=s&source=web&cd=1&cad=rja&uact=8&ved= 0CB4QFjAAahUKEwj0q5HDpdHIAhVCURoKHZwgCBQ&url=http%3A%2F%2Fec.europa.eu% 2FDocsRoom%2Fdocuments%2F10010%2Fattachments%2F1%2Ftranslations%2Fen%2Frenditions% 2Fnative&usg=AFQjCNEF6SkHHErPeB2gXS_dxBBFtlw_7A&sig2=qkh5AeKsO4cyRpuIZtVrWQ (accessed on 20 October 2015).

2. Johnson Matthey, Precious Metals Management. Available online: http://www.platinum.matthey.com/ about-pgm/applications (accessed on 20 October 2015).

3. Hagelüken, C. Recycling the platinum group metals: A European perspective. *Platinum Met. Rev.* **2012**, *56*, 29–35. [CrossRef]

4. Crundwell, F.K.; Moats, M.S.; Ramachandran, V.; Robinson, T.G.; Davenport, W.G. *Extractive Metallurgy of Nickel, Cobalt and Platinum-Group Metals*, 1st ed.; Elsevier: Amsterdam, The Netherlands, 2011; pp. 537–549.

5. Nogueira, C.A.; Paiva, A.P.; Oliveira, P.C.; Costa, M.C.; Costa, A.M.R. Oxidative leaching process with cupric ion in hydrochloric acid media for recovery of Pd and Rh from spent catalytic converters. *J. Hazard. Mater.* **2014**, *278*, 82–90. [CrossRef] [PubMed]

6. Cox, M. Solvent extraction in hydrometallurgy. In *Solvent Extraction Principles and Practice*, 2nd ed.; Rydberg, J., Cox, M., Musikas, C., Choppin, G.R., Eds.; Marcel Dekker: New York, NY, USA, 2004; pp. 455–505.

7. Nikoloski, A.N.; Ang, K.-L. Review of the application of ion exchange resins for the recovery of platinum-group metals from hydrochloric acid solutions. *Miner. Process. Extr. Metall. Rev.* **2014**, *35*, 369–389. [CrossRef]

8. Lee, J.Y.; Raju, B.; Kumar, B.N.; Kumar, J.R.; Park, H.K.; Reddy, B.R. Solvent extraction separation and recovery of palladium and platinum from chloride leach liquors of spent automobile catalyst. *Sep. Purif. Technol.* **2010**, *73*, 213–218. [CrossRef]

9. Reddy, B.R.; Raju, B.; Lee, J.Y.; Park, H.K. Process for the separation and recovery of palladium and platinum from spent automobile catalyst leach liquor using LIX 84I and Alamine 336. *J. Hazard. Mater.* **2010**, *180*, 253–258. [CrossRef] [PubMed]

10. Narita, H.; Tanaka, M.; Morisaku, K.; Tamura, K. Extraction of platinum(IV) in hydrochloric acid solution using diglycolamide and thiodiglycolamide. *Solvent Extr. Res. Dev. Jpn.* **2006**, *13*, 101–106.

11. Costa, M.C.; Assunção, A.; Costa, A.M.R.; Nogueira, C.; Paiva, A.P. Liquid-liquid extraction of platinum from chloride media by *N,N′*-dimethyl-*N,N′*-dicyclohexyltetradecylmalonamide. *Solvent Extr. Ion Exch.* **2013**, *31*, 12–23. [CrossRef]

12. Malik, P.; Paiva, A.P. Solvent extraction studies for platinum recovery from chloride media by a *N,N′*-tetrasubstituted malonamide derivative. *Solvent Extr. Ion Exch.* **2009**, *27*, 36–49. [CrossRef]

13. Paiva, A.P.; Carvalho, G.I.; Costa, M.C.; Costa, A.M.R.; Nogueira, C. Recovery of platinum and palladium from chloride solutions by a thiodiglycolamide derivative. *Solvent Extr. Ion Exch.* **2014**, *32*, 78–94. [CrossRef]

14. Narita, H.; Morisaku, K.; Tamura, K.; Tanaka, M.; Shiwaku, H.; Okamoto, Y.; Suzuki, S.; Yaita, T. Extraction properties of palladium(II) in HCl solution with sulfide containing monoamide compounds. *Ind. Eng. Chem. Res.* **2014**, *53*, 3636–3640. [CrossRef]

15. Narita, H.; Tanaka, M.; Morisaku, K. Palladium extraction with *N,N,N′,N′*-tetra-*n*-octyl-thiodiglycolamide. *Miner. Eng.* **2008**, *21*, 483–488. [CrossRef]

16. Huang, Y.; Li, N.; Li, Y.; Wu, J.; Li, S.; Chen, S.; Zhu, L. Extraction of precious metals with a new amide extractant. *Adv. Mater. Res.* **2014**, *878*, 399–405. [CrossRef]

17. Ortet, O.; Paiva, A.P. Development of tertiary thioamide derivatives to recover palladium(II) from simulated complex chloride solutions. *Hydrometallurgy* **2015**, *151*, 33–41. [CrossRef]

18. Das, A.; Ruhela, R.; Singh, A.K.; Hubli, R.C. Evaluation of novel ligand dithiodiglycolamide (DTDGA) for separation and recovery of palladium from simulated spent catalyst dissolver solution. *Sep. Purif. Technol.* **2014**, *125*, 151–155. [CrossRef]

19. Narita, H.; Morisaku, K.; Tanaka, M. Highly efficient extraction of rhodium(III) from hydrochloric acid solution with amide-containing tertiary amine compounds. *Solvent Extr. Ion Exch.* **2015**, *33*, 407–417. [CrossRef]

298

20. Ortet, O.; Paiva, A.P. Liquid-liquid extraction of palladium(II) from chloride media by *N,N'*-dimethyl-*N,N'*-dicyclohexylthiodiglycolamide. *Sep. Purif. Technol.* **2015**, *156*, 363–368. [CrossRef]

21. Paiva, A.P.; Carvalho, G.I.; Costa. M.C.; Costa, A.M.R.; Nogueira, C.A. Recovery of Palladium(II) from a Spent Automobile Catalyst Leaching Solution. In Proceedings of the 3rd International Conference on Wastes: Solutions, Treatments and Opportunities, Viana do Castelo, Portugal, 14–16 September 2015; Vilarinho, C., Castro, F., Carvalho, J., Russo, M. Araújo, J., Eds.; CVR (Centro para a Valorização de Resíduos): Guimarães, Portugal, 2015; pp. 61–63.

22. Narita, H.; Tanaka, M.; Morisaku, K. Extraction Properties of Platinum Group Metals with Diamide Compounds. In Proceedings of the 17th International Solvent Extraction Conference (ISEC 2005), Beijing, China, 19–23 September 2005; Conference Proceeding Editorial Department: Beijing, China, 2005; pp. 227–232.

23. Ribeiro, L.C.; Santos, M.S.; Paiva, A.P. Apparent molar volumes of *N,N*-disubstituted monoamides: A convenient tool to interpret iron(III) extraction profiles from hydrochloric acid solutions. *Solvent Extr. Ion Exch.* **2013**, *31*, 281–296. [CrossRef]

24. Cleare, M.J.; Charlesworth, P.; Bryson, D.J. Solvent extraction in platinum group metal processing. *J. Chem. Technol. Biotechnol.* **1979**, *29*, 210–224. [CrossRef]

25. Levitin, G.; Schmuckler, G. Solvent extraction of rhodium chloride from aqueous solutions and its separation from palladium and platinum. *React. Funct. Polym.* **2003**, *54*, 149–154. [CrossRef]

26. Bernardis, F.L.; Grant, R.A.; Sherrington, D.C. A review of methods of separation of the platinum-group metals through their chloro-complexes. *React. Funct. Polym.* **2005**, *65*, 205–217. [CrossRef]

27. Preston, J.S.; du Preez, A.C. Solvent extraction of platinum-group metals from hydrochloric acid solutions by dialkyl sulphoxides. *Solvent Extr. Ion Exch.* **2002**, *20*, 359–374. [CrossRef]

Article

Copper and Cyanide Extraction with Emulsion Liquid Membrane with LIX 7950 as the Mobile Carrier: Part 1, Emulsion Stability

Diankun Lu [1], Yongfeng Chang [1], Wei Wang [1], Feng Xie [1],*, Edouard Asselin [2] and David Dreisinger [2]

[1] School of Material and Metallurgical Engineering, Northeaestern University, 3-11 Wenhua Road, Shenyang 110819, China; ludk@smm.neu.edu.cn (D.L.); changyf@smm.neu.edu.cn (Y.C.); wangwei@smm.neu.edu.cn (W.W.)

[2] Department of Materials Engineering, University of British Columbia, 309-6350 Stores Road, Vancouver, BC V6T 1Z4, Canada; edouard.asselin@ubc.ca (E.A.); david.dreisinger@ubc.ca (D.D.)

* Author to whom correspondence should be addressed; xief@smm.neu.edu.cn; Tel.: +86-24-8367-2298; Fax: +86-24-8368-7750.

Academic Editors: Suresh Bhargava, Mark Pownceby and Rahul Ram
Received: 19 August 2015; Accepted: 14 October 2015; Published: 4 November 2015

Abstract: The potential use of emulsion liquid membranes (ELMs) with LIX 7950 as the mobile carrier to remove heavy metals from waste cyanide solutions has been proposed. Relatively stable ELMs with reasonable leakage and swelling can be formed under suitable mixing time and speed during emulsification. The concentration of LIX 7950 and Span 80 in the membrane phase, KOH in the internal phase and the volume ratio of membrane to internal phases are also critical to ELM formation. The efficiency of copper and cyanide removal from dilute cyanide solution by ELMs is related to ELM stability to some extent. More than 90% copper and cyanide can be removed from dilute cyanide solutions by ELMs formed under suitable experimental conditions.

Keywords: emulsion liquid membrane; LIX 7950; stability; copper cyanides

1. Introduction

Waste cyanide solutions are generated industrially on a large scale in gold and silver extraction from ores, electroplating, coal cooking and organic chemical formulation. Heavy metals such as copper zinc and iron may also occur in these solutions, e.g., those waste effluents arising from the gold cyanidation process [1]. Due to the highly toxicity of cyanide and heavy metals to humans and aquatic organisms, strict environmental regulations on the management of waste cyanide solutions have been established. Environmental constraints controlling the discharge of cyanide from gold plants are being tightened by local governments worldwide [2,3]. A considerable number of methods have been developed for recovering valuable metals and cyanide from waste cyanide solutions. Acidification, volatilization, and recovery (AVR) and some modifications have been long practiced in some gold operations, but the high consumption and cost of reagents has significantly limited their application [4–6]. The indirect recovery of metals and cyanide by activated carbon and ion-exchange resin has been proposed. However, the low adsorption capability of carbon and the high cost of resin have severely hampered their wide application in practice [7–10]. The use of solvent extraction process to recover copper from cyanide solution has been suggested. The extraction and stripping of metal cyanide complexes is believed to occur via an ion-exchange mechanism [11–15]. These investigations prove that copper can be effectively extracted from alkaline cyanide solutions by proposed solvent extraction systems.

Extraction processes using emulsion liquid membrane (ELM) have also received significant attention due to their potential application in treating industrial liquid wastes [16–20]. Use of ELMs for hydrometallurgical recovery of heavy metals from waste industrial solutions has been reported by many investigators and some processes have been successfully commercialized to remove various metals from wastewater in industry [21–27]. The potential use of liquid membranes to extract precious metals from alkaline cyanide solutions was also proposed [28–32]. In ELM processes, simultaneous purification/concentration of the target solute can be realized through combining extraction and stripping into one stage, which results in the high interfacial area for mass transfer, the ability to remove/concentrate selectively or collectively, and the requirement of only small quantities of organic solvent. It is believed that the large specific interfacial area associated with ELMs result in higher permeation rates and the transport is governed by kinetic rather than equilibrium parameters which may enable the achievement of higher metal concentrations in fewer separation stages than solvent extraction [33–35]. Consequently, there is a substantial saving in the volume of mixers and settlers for extraction and stripping which are generally carried out in conventional solvent extraction circuits. However, although the technology has been successfully practiced for removing some heavy metals from waste solutions, its operating efficiency is highly dependent on the emulsion instability which commonly includes membrane leakage, coalescence, and emulsion swelling. In order to obtain relatively stable emulsion liquid membranes, its formulation design including selection of carrier, strip agent, surfactant, diluents, and preparation method is critical [36].

The process of recovering copper and cyanide from waste alkaline cyanide solutions through ELMs using the guanidine extractant LIX 7950 as the mobile carrier has been suggested in this research. The factors which may potentially influence the stability of ELMs and the removal of copper and cyanide by ELMs including membrane formula, mixing speed, and phase ratio (volume ratio of membrane to internal phases) have been examined with the purpose to determine the feasibility of extracting copper and cyanide from waste cyanide solutions by ELMs.

2. Experimental Section

2.1. Materials

LIX 7950, a trialkylguanidine extractant, was used as supplied by the manufacturer (Cognis, Mississauga, ON, Canada). The non-ionic surfactant commercially known as Span 80 (Sinopharm Chemical Reagent, Shanghai, China) was used for stabilizing the emulsion. The diluent was a commercial kerosene which was a complex mixture of aliphatic origin containing about 15%–20% w/w aromatics. A small amount of liquid paraffin was used as the additive to facilitate formation of ELMs. Some 1-dodecanol was added as the modifier to facilitate dissolution of LIX 7950 in the diluent. Synthetic copper cyanide solutions were made up from CuCN and NaCN. All chemicals were of reagent grade and doubly distilled water was used throughout.

2.2. Procedure

To prepare ELMs, designated volume of LIX 7950, 1-dodecanol, Span 80, liquid paraffin, and kerosene were emulsified in a 250 mL beaker through a mechanical mixer with an overhead disk-propeller. A portion of KOH solution used as the internal stripping solution was added dropwise to the stirred membrane solution till obtaining the desired volume ratio of internal to membrane phases. The mixed solution was stirred continuously to obtain stable ELMs.

For stability and extraction tests, designated volume of ELMs prepared through the described procedure was added to a 500 mL beaker filled with the feed solution (100 mL). Mixing was provided with a mechanical stirrer with four Teflon-coated turbine-type impellers. All tests were conducted at the indoor temperature of 20 ± 1 °C (except defined otherwise). After the desired extraction time, the emulsions were separated from the mixture through a 125 mL separator and the obtained aqueous solution was filtrated with Whatman No. 3 filter paper to remove any entrained organic before analysis.

The emulsions were recovered and subsequently broken into its constituent organic and aqueous phases through a high speed centrifuge. Samples of organic phase were filtered with 1 PS Phase Separation paper and samples of aqueous phase were filtered with Whatman No. 3 filter paper to remove any entrained organic phase before analysis.

2.3. Analysis and Calculation

The solution pH was measured with a pH meter using a combined glass electrode. Metals in aqueous samples were analyzed by atomic absorption spectrophotometry (AAS) and Ion coupled Plasma (ICP-MS, PerkinElmer, Shanghai, China) and their content in the organic phase was calculated by mass balance. Total cyanide content in the aqueous solution was determined by the standard distillation method [37]. The morphology of ELMs is determined by a metallographic microscope. To determine the leakage of ELMs, a portion of the feed solution (without K^+) was mixed with equal volume of ELMs with K^+ in the internal stripping solution; after mixing for the designated time, ELMs were separated from the external aqueous phase (the raffinate) and was further de-emulsified thoroughly through heating and centrifuge. The volume of the raffinate, the membrane phase and the internal phase and the concentration of tracer K^+ in each aqueous phase were measured. The apparent leakage (breakage) of ELMs is calculated by the following equation:

$$\delta = \frac{C_{KR} \times V_R}{C_{Ki} \times V_i} \times 100\% \tag{1}$$

where

δ: the apparent leakage (breakage) of ELMs, %;
C_{KR}: the molar concentration of K^+ in the raffinate, mol/L;
C_{Ki}: the molar concentration of K^+ in the initial internal phase, mol/L;
V_R: the volume of the raffinate, L;
V_i: the volume of the initial internal phase, L;

The apparent swelling of ELMs is simply defined as following:

$$\eta = \frac{V'_E/V_R - V_E/V_f}{V_E/V_f} \times 100\% \tag{2}$$

where

V_E: the apparent volume of emulsions before contacting the feed, L;
V'_E: the apparent volume of emulsions after extraction, L;
V_f: the initial volume of feed, L;

The removal (or extraction) of the target species from feed by ELMs is calculated through the following equation:

$$E = \frac{C_f \times V_f - C_R \times V_R}{C_f \times V_f} \times 100\% \tag{3}$$

where

E: the removal (or extraction) of the target species from the feed solution by ELMs, %;
C_f: the molar concentration of the target species in the feed, mol/L;
C_R: the molar concentration of the target species in the raffinate, mol/L;

3. Results and Discussion

3.1. Influence of Mixing for ELM Formation

In order to ensure a potential transport of copper cyanides through ELMs, a relatively stable ELM with reasonable leakage and swelling has to be constructed. Preliminary investigations indicated the input energy for the preparation emulsions played an important role in forming such meta-stable ELMs. When the stirring speed of the mixer is fixed at 4500 RPM, images of ELMs formed under different emulsification time are shown in Figure 1. At the emulsification time of 5 min, the formed ELMs have a relatively large size. They tend to decrease with increasing stirring time, resulting in more uniform droplets of ELMs, e.g., more uniform ELMs are produced after 15 min of stirring than 5 min of stirring. However, when the emulsification time exceeds 25 min, more ELMs start to distort to connect with others to form a semi-continuous phase. It is believed that small droplet tends to have better breaking resistant and rapid extraction, but will be very difficult to de-emulsify by mechanical methods. In addition, too many of them are packed into each organic globule and, consequently, the liquid membrane becomes too thin and ruptures easily. However, large droplets may cause poor stability and extraction efficiency due to the low surface-to-volume ratio. Thus, emulsions with the droplet diameter in the range 0.3–10 μm (preferably 0.8–3 μm) have been suggested considering reasonable ELM stability [16,18].

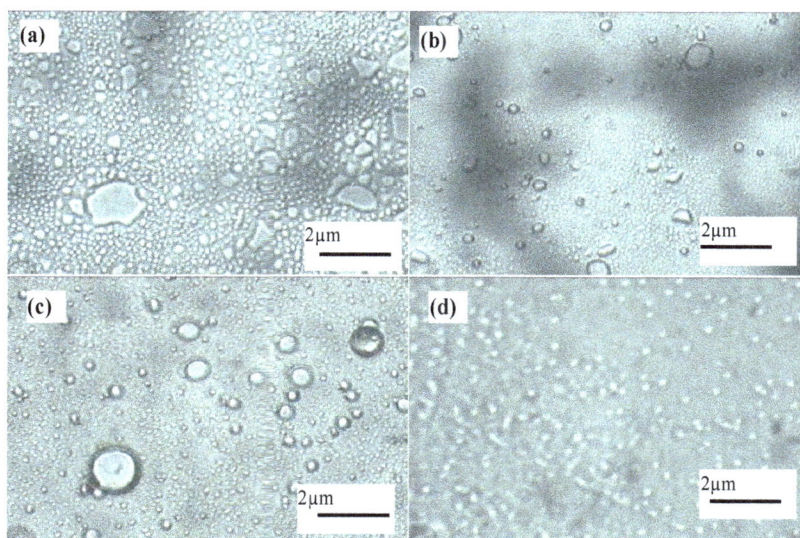

Figure 1. Photographic of ELM droplets formed under different stirring times. (Membrane phase: 6% v/v LIX 7950, 2% v/v 1-dodecanol, 5% v/v Span 80, 1% v/v paraffin, balanced with kerosene; Internal stripping solution: 0.1 mol/L KOH; Volume ratio of the membrane phase to the internal phase: 1:1; Mixing speed: 4500 RPM; Mixing time: (**a**) 5 min; (**b**) 15 min; (**c**) 25 min; (**d**) 35 min).

The effect of emulsification time on ELM stability and removal of copper and cyanide from the external solution by ELMs is shown in Figure 2. Under the experimental conditions, a high leakage (>40%) was observed when the emulsification time is 5 min. This is probably because the short stirring time for emulsification results in relatively large but unstable ELM droplets which tend to break up easily when contacting with the external solution. Accordingly, a negative value for ELM swelling (data not shown in Figure 2) was obtained at emulsification time of 5 min due to high breakage of ELMs (partial internal solution has released to the external phase). The ELM leakage decreases to 8% when the emulsification time is 10 min and maintains at 5% when the emulsification is further increased. Relatively small ELM swelling ($\eta < 2\%$) was observed when the stirring time for emulsification varied from 10–35 min. Copper removal from the external phase by ELMs is only about 50% at the emulsification time of 5 min. This is probably due to high leakage of the resulted ELMs. It increases up to 90% at the emulsification time of 10 min and remains stable when the emulsification time increases from 10–35 min, indicating effective ELMs have formed under these conditions. Under experimental conditions, cyanide extraction with ELMs shows virtually the same values as copper extraction. Thermodynamic calculation indicates that copper mainly occurs as $Cu(CN)_3^{2-}$ in the feed. The analysis of the internal phase after extraction showed the extracted species also mainly occur as $Cu(CN)_3^{2-}$. Since LIX 7950 is a strong organic basic extractant, the extraction of $Cu(CN)_3^{2-}$ through ELMs formed with LIX 7950 as the mobile carrier is thus an anion exchange process. The extraction chemistry of copper cyanides by resulted ELMs can be expressed as following:

At the interface of external and membrane phases:

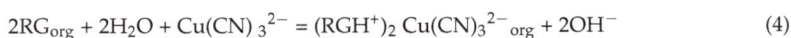

$$2RG_{org} + 2H_2O + Cu(CN)_3^{2-} = (RGH^+)_2\, Cu(CN)_3^{2-}{}_{org} + 2OH^- \tag{4}$$

At the interface of membrane and internal phases:

$$(RGH^+)_2\, Cu(CN)_3^{2-}{}_{org} + 2OH^- = 2RG_{org} + 2H_2O + Cu(CN)_3^{2-} \tag{5}$$

where RG denotes the guanidine extractant LIX 7950 and RGH^+ is its protonated form. The effect of mixing speed for ELM formation on leakage and swelling of ELMs is shown in Figure 3. The mixing speed exhibits an insignificant effect on the ELM swelling ($\eta < 2\%$) when it varies from 2500–7500 RPM. However, a leakage of 12% was observed when the mixer speed for ELM formation was maintained at 2500 RPM, which probably was due to high breakage of unstable ELM droplets formed under low input energy for emulsification. Correspondingly, the removal of copper and cyanide from the external phase by ELMs is only about 45% in this case. Copper and cyanide removal increases to 90% when the mixing speed for ELM formation increases to 3500 RPM. Further increase of the mixing speed (up to 7500 RPM) shows insignificant effect on copper and cyanide removal by ELMs.

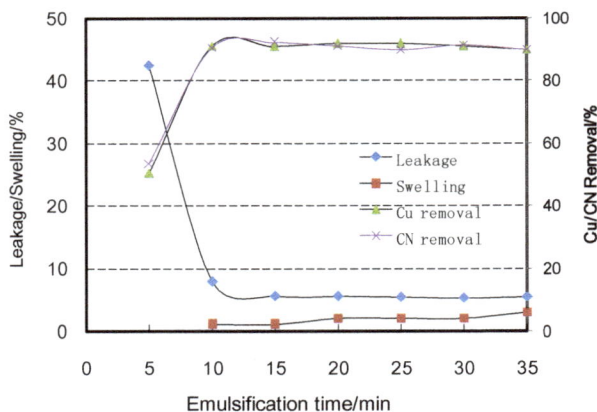

Figure 2. Effect of emulsification time on ELM stability and copper and cyanide removal. (For ELMs formation, membrane phase: 6% v/v LIX 7950, 2% v/v 1-dodecanol, 5% v/v Span 80, 1% v/v liquid paraffin, balanced with kerosene; Internal solution: 0.05 mol/L KOH; Volume ratio of membrane to internal phases: 1:1; Mixing speed: 4500 RPM. The external aqueous phase: dilute copper cyanide solution, $[Cu]_T$ = 19.3 mg/L, $[CN]_T$ = 23.6 mg/L, pH 9.5. Mixing of ELMs and the external solution: 15 min at 300 RPM).

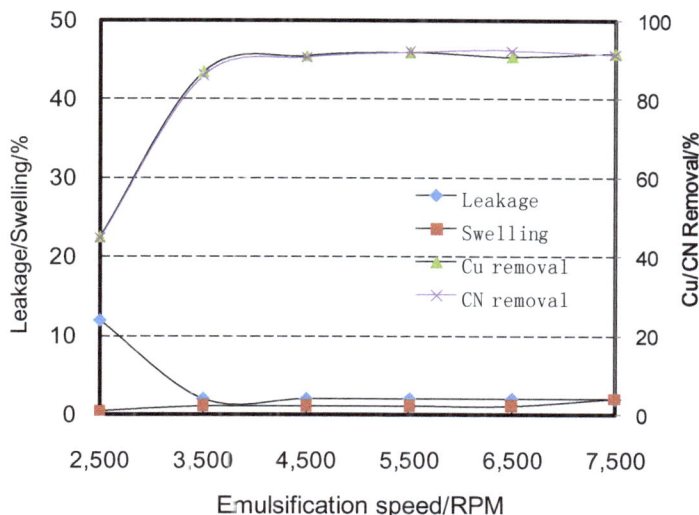

Figure 3. Effect of emulsification speed on ELM stability and copper and cyanide removal. (For ELMs formation, membrane phase: 6% v/v LIX 7950, 2% v/v 1-dodecanol, 5% v/v Span 80, 1% v/v liquid paraffin, balanced with kerosene; internal phase: 0.05 mol/L KOH; Volume ratio of membrane to internal phases: 1:1; Mixing time: 15 min. The external aqueous phase: dilute copper cyanide solution, $[Cu]_T$ = 19.3 mg/L, $[CN]_T$ = 23.6 mg/L, pH 9.5. Mixing of ELMs and the external phase: 15 min at 300 RPM).

3.2. Influence of KOH in the Internal Phase

The mobile carrier LIX 7950, a tri-alkylguanidine, is a strong organic base having an intermediate basicity between that of primary amines and quaternary ammonium salts (pKa is approximately 12). It is capable of being protonated to form a guanidinium cation at the operating pH of cyanidation (usually 10–11). This guanidinium cation can form an ion-pair with metal cyanides such as $Cu(CN)_3^{2-}$, resulting in metal extraction from the cyanide solution. By increasing the basicity of the aqueous phase, the guanidinium cation is converted to the neutral guanidine functionality. The neutral guanidine functionality no longer forms an ion-pair with metal cyanides, causing metal cyanides to be stripped from the organic phase [13]. Therefore, in order to extract metal cyanide species from alkaline cyanide solutions, the ideal internal phase of ELMs would be a solution with pH higher than 12. ELM leakage and swelling with different KOH concentration in the internal solution are shown in Figure 4. Insignificant emulsion swelling (η < 1%) was observed when KOH concentration in the internal phase varied from 0.05–1 mol/L, indicating the increase of ionic strength in the internal phase exhibits an insignificant effect on the change of volume ratio of ELMs to the external aqueous solution under experimental conditions. However, the ELM leakage increases significantly when the KOH concentration in the internal phase is higher than 0.25 mol/L. It was believed that high concentration of KOH might significantly increase the osmotic pressure difference between internal and external phases and thus result in emulsion leaking. It was also reported that the potential reaction between KOH and the surfactant Span 80 might result in reduced properties of the surfactant and consequently lead to an emulsion destabilization [36]. The removal of copper and cyanide from the external solution by ELMs exhibits insignificant change when KOH concentration varies from 0.05–0.5 mol/L (Figure 4), indicating 0.05 mol/L KOH in the internal phase is good enough for stripping copper cyanides from the membrane phase under experimental conditions. High KOH concentration (>0.5 mol/L) in the internal phase results in a slight decrease of copper and cyanide removal probably due to increasing breakage of ELMs.

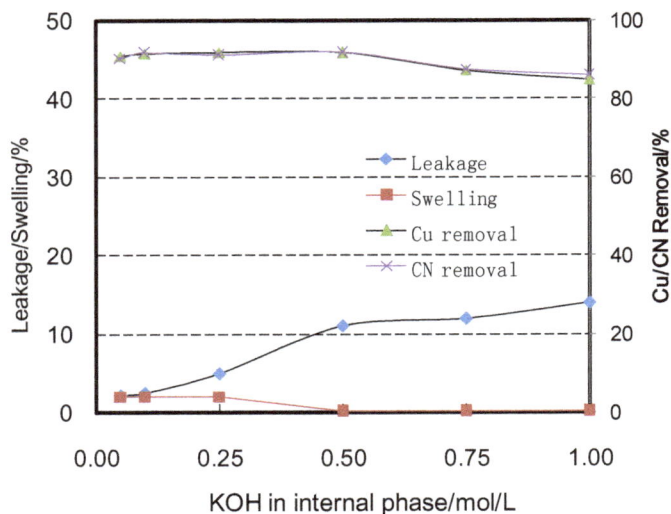

Figure 4. Effect of KOH on ELM stability and copper and cyanide removal. (For ELMs formation, membrane phase: 6% v/v LIX 7950, 2% v/v 1-dodecanol, 5% v/v Span 80, 1% v/v liquid paraffin, balanced with kerosene; Volume ratio of membrane to internal phases: 1:1; Mixing for emulsification: 15 min at 4500 RPM. The external aqueous phase: dilute copper cyanide solution, $[Cu]_T$ = 19.3 mg/L, $[CN]_T$ = 23.6 mg/L, pH 9.5. Mixing of ELMs and the external phase: 15 min at 300 RPM).

3.3. Influence of Extractant and Surfactant

A great number of studies indicated that both extractant and surfactant played important roles in the ELM formation [16,36]. The mobile carrier, LIX 7950, acts as a "shuttle" to carry the target species through ELMs in this case. Effects of concentration of LIX 7950 and Span 80 in the membrane phase on ELM stability and copper and cyanide removal are shown in Figures 5 and 6, respectively. Under the experimental conditions, both ELM leakage and swelling slightly decrease when the concentration of LIX 7950 in the membrane phase increases from 1% v/v–6% v/v. Both ELM leakage and swelling are lower than 2% when LIX 7950 concentration varies from 6% v/v–10% v/v. Correspondingly, the removal of copper and cyanide from the external phase with ELMs increases with an increase of LIX 7950 concentration when it varies from 1% v/v–6% v/v. Further increase of LIX 7950 concentration exhibits an insignificant effect on copper and cyanide removal from the external phase by ELMs. A number of studies has given evidence that both extractant and surfactant concentration would potentially affect the emulsion stability. On one hand, increasing extractant concentration in the membrane phase may potentially facilitate the transport of target species from the external phase to the membrane phase. On the other hand, a high content of extractant in the membrane phase may result in increased viscosity which may affect the dispersion behavior of ELMs and cause a decline in interfacial area, and, consequently, lead to a slow extraction rate of target species. Besides, using less extractant is always preferred in practice from an economic point view since, in most cases, extractant is usually the most expensive agent among membrane components [36].

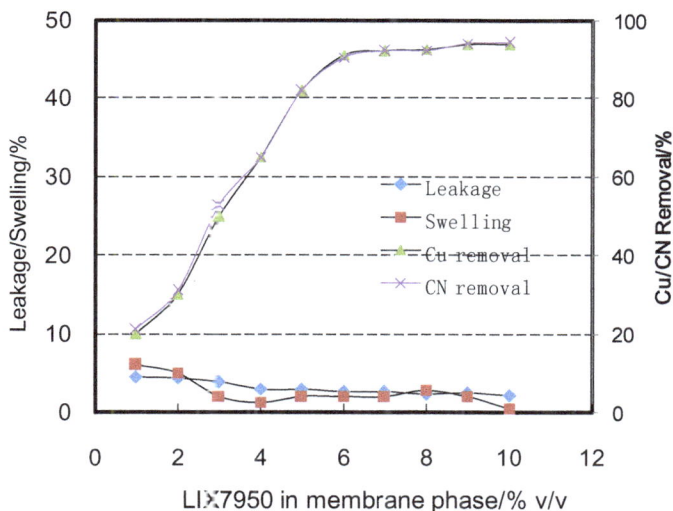

Figure 5. Effect of LIX 7950 concentration on ELM stability and copper and cyanide removal. (For ELMs formation, membrane phase: 2% v/v 1-dodecanol, 5% v/v Span 80, 1% v/v liquid paraffin, balanced with kerosene; Internal phase: 0.05 mol/L KOH solution. Volume ratio of the membrane phase to the internal phase: 1:1; Mixing for emulsification: 15 min at 4500 RPM. The external aqueous phase: dilute copper cyanide solution, $[Cu]_T = 19.3$ mg/L , $[CN]_T = 23.6$ mg/L, pH 9.5. Mixing of ELMs and the external phase: 15 min at 300 RPM)

For the surfactant (Span 80) aspect, under the experimental conditions, high ELM leakage was observed when the content of Span 80 in the membrane phase was low, e.g., more than 20% of leakage with 3% v/v Span 80. Accordingly, ELM swelling exhibits little change when the Span 80 concentration varies from 1% v/v–9% v/v. Test results indicate that stable ELMs can be only formed when the concentration of Span 80 is higher than 5% under the experimental conditions. The removal of copper and cyanide from the external phase with ELMs is relatively low (only 55%) when the concentration of Span 80 in the membrane phase is 3% v/v. Copper and cyanide removal increases to about 91% when the content of Span 80 in the membrane phase is 4% v/v and remains stable when it varies from 4% v/v–6% v/v. Further increase of Span 80 in the membrane phase leads to a decrease of copper and cyanide extraction. Many studies propose that surfactant can reduce the interfacial tension between oil and water by adsorbing at the liquid–liquid interface; thus, maintaining the emulsion stability and the transport of target species through the membrane [16,34–36]. However, it should be noted that, at a high concentration, surfactant may form emulsions that are stable with a higher emulsion viscosity. Consequently, decreasing copper and cyanide extraction with ELMs was observed at a high content of Span 80 in the membrane phase.

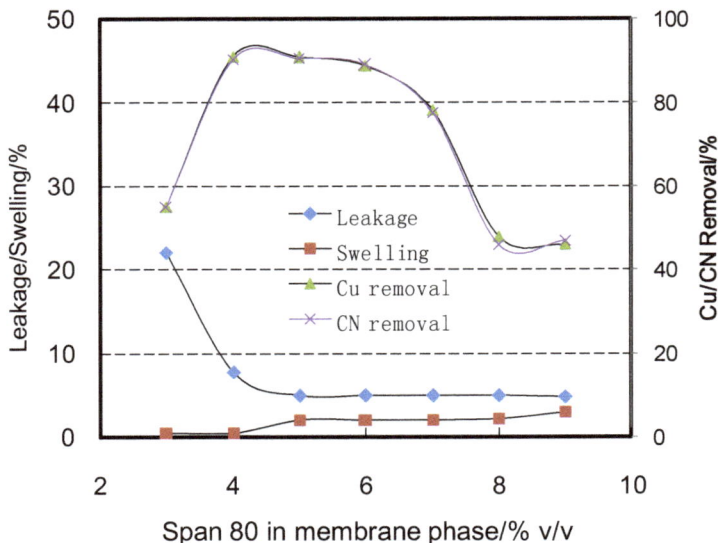

Figure 6. Effect of Span 80 concentration on ELM stability and copper and cyanide removal. (For ELMs formation, membrane phase: 2% *v/v* 1-dodecanol, 1% *v/v* liquid paraffin, balanced with kerosene; Internal phase: 0.05 mol/L KOH solution. Volume ratio of the membrane phase to the internal phase: 1:1; Mixing for emulsification: 15 min at 4500 RPM. The external aqueous phase: dilute copper cyanide solution, $[Cu]_T$ = 19.3 mg/L, $[CN]_T$ = 23.6 mg/L, pH 9.5. Mixing of ELMs and the external phase: 15 min at 300 RPM).

3.4. Influence of Volume Ratio of Membrane to Internal Phases

The influence of volume ratio of membrane phase to internal phase (V_m/V_{int}) on ELM stability was also examined (Table 1). It was found that at low V_m/V_{int}, ELMs exhibited higher leakage, for example, a leakage of 12.5% when V_m/V_{int} was maintained at 1:3. Relatively lower leakage was observed when V_m/V_{int} varied from 2:1 to 1:2. The swelling of ELMs was also insignificant in this range of V_m/V_{int}. These results are generally in accordance with the literature. It was believed that increasing the volume of internal solution made ELMs unstable due to an increase of the size of internal droplets, resulting in decreasing interfacial contact area between the emulsion and the external phase. Other authors reported that the thickness of film in droplets thinned off when the volume of the stripping phase increased [34,36]. For higher V_{int}/V_m, the volume of membrane phase is not enough for enclosing all the stripping solution, thus resulting in higher leakage [36]. ELM leakage and swelling significantly increased with increasing V_m/V_{int} when it was higher than 3:1. Though, usually lower V_{int}/V_m leads to a thicker and more stable membrane phase, it was found that too high V_m/V_{int} also resulted in unstable ELMs (high leakage and swelling). Wang and Zhang [38] reported that V_{int}/V_m might affect the surfactant concentration at the interface of membrane/aqueous phases and in the bulk membrane phase, and the entrainment and osmotic swelling might increase decreasing V_{int}/V_m, which resulted in unstable ELMs. Correspondingly, when V_{int}/V_m varies from 1:2 to 1:1, more than 90% of copper and cyanide removal with ELMs was achieved and lower copper and cyanide removal was observed at both higher and lower V_{int}/V_m.

Table 1. Effect of volume ratio of membrane to internal phases on ELM stability and copper and cyanide removal. (For ELMs formation, membrane phase: 2% v/v 1-dodecanol, 1% v/v liquid paraffin, balanced with kerosene; Internal phase: 0.05 mol/L KOH solution. Mixing for emulsification: 15 min at 4500 RPM. The external aqueous phase: dilute copper cyanide solution, $[Cu]_T$ = 19.3 mg/L, $[CN]_T$ = 23.6 mg/L, pH 9.5. Mixing of ELMs and the external phase: 15 min at 300 RPM).

V_m/V_{int}	Leakage, %	Swelling, %	Cu_{ext}, mg/L	CN_{ext}, mg/L	Cu Removal, %	CN Removal, %
1:3	12.5	1	4.3	5.1	78.0	78.2
1:2	2.5	2	0.9	1.2	95.5	94.9
1:1	3	2	1.7	2.2	91.0	90.8
2:1	4	8	10.8	13.0	44.3	45.0
3:1	15	10	11.7	13.9	39.3	41.2
4:1	18	12	11.3	13.3	41.3	43.8
5:1	28	18	10.8	13.2	44.0	44.0

4. Conclusions

The use of emulsion liquid membranes with LIX 7950 as the mobile carrier to recover metals and cyanide from waste cyanide solutions has been proposed. The concentration of LIX 7950 and Span 80 in the membrane phase, KOH in the internal phase, volume ratio of internal to membrane phases, and mixing during emulsification are critical factors for forming stable ELMs. Relatively stable ELMs with reasonable leakage and swelling can be formed under suitable formulas of membrane components. The extraction of copper and cyanide can be usually obtained when ELMs with relatively low leakage and swelling are formed. However, excessive Span 80 in the membrane phase may result in low copper and cyanide removal from dilute cyanide solutions with ELMs.

Acknowledgments: The authors thank Yan Fu for his generous help on the project. The project was financially supported by Fundamental Research Funds for Central Universities of China (No. N130202001).

Author Contributions: Dianku Lu wrote and edited the manuscript. Yongfeng Chang and Wei Wang helped with emulsion liquid membrane preparation and data collection. Feng Xie supervised experimental work and data analysis. Edouard Asselin and David Dreisinger helped with data analysis and interpretation.

Conflicts of Interest: The authors declare no conflict of interest.

References

1. Marsden, J.; House, I. *Chemistry of Gold Extraction*, 2nd ed.; Society for Mining, Metallurgy, and Exploration: Englewood, CO, USA, 2006; pp. 15–45.
2. DeVries, F. Brief overview of the Baia Mare Dam Breach. In *Cyanide: Social, Industrial and Economic Aspects*; Young, A.A., Twidwell, L.G., Anderson, C.G., Eds.; TMS: Warrendale, PA, USA, 2001; pp. 11–14.
3. Fleming, C.A. Cyanide Recovery. In *Advances in Gold Ore Processing*; Adams, M.D., Ed.; Elsevier: Amsterdam, The Netherlands, 2005; pp. 703–727.
4. Jay, W.H. Copper Cyanide Recovery System. In *Cyanide: Social, Industrial and Economic Aspects*; Young, A.A., Twidwell, L.G., Anderson, C.G., Eds.; TMS: Warrendale, PA, USA, 2001; pp. 317–340.
5. Botz, M.M.; Mudder, T.I.; Akcil, A.U. Cyanide Treatment: Physical, Chemical and Biological Process. In *Advances in Gold Ore Processing*; Adams, M.D., Ed.; Elsevier: Amsterdam, The Netherlands, 2005; pp. 672–702.
6. Barter, J.; Lane, G.; Mitchell, D.; Kelson, R.; Dunne, R.; Trang, C.; Dreisinger, D. Cyanide Management by SART. In *Cyanide: Social, Industrial and Economic Aspects*; Young, A.A., Twidwell, L.G., Anderson, C.G., Eds.; TMS: Warrendale, PA, USA, 2001; pp. 549–562.
7. Adams, M.D. Removal of cyanide from solution using activated carbon. *Miner. Eng.* **1994**, *7*, 1165–1177. [CrossRef]
8. Goldblatt, E. Recovery of cyanide from waste cyanide solution by ion exchange. *Ind. Eng. Chem.* **1956**, *48*, 2107–2114. [CrossRef]

9. Le Vier, K.M.; Fitzpatrick, T.A.; Brunk, K.A.; Ellett, W.N. AuGMENT Technologies: An update. In Proceedings of the Randol Gold Forum, Monterey, CA, USA, 18–21 May 1997; pp. 135–137.
10. Leão, V.A.; Ciminelli, V.S.T.; Costa, R.S. Cyanide recycling using strong-base ion exchange resins. *JOM* **1998**, *50*, 66–69. [CrossRef]
11. Davis, M.R.; MacKenzie, M.W.; Sole, K.C.; Virnig, M.J. A proposed solvent extraction route for the treatment of copper cyanide solutions produced in leaching of gold ores. In Proceedings of the Alta Copper Hydrometallurgy Forum, Brisbane, Australia, 20–21 October 1998; pp. 42–56.
12. Xie, F.; Dreisinger, D. Copper solvent extraction from waster cyanide solution by a solvent mixture of a quaternary amine and nonylphenol. In Proceedings of the ISEC 2008, Tucson, AZ, USA, 15–19 September 2008; Moyer, B.A., Ed.; CIM: Tuscon, AZ, USA, 2008; pp. 107–112.
13. Xie, F.; Dreisinger, D. Recovery of copper cyanide from waste cyanide solution by LIX 7950. *Miner. Eng.* **2009**, *22*, 190–195. [CrossRef]
14. Riveros, P.A. Studies on the solvent extraction of gold from cyanide media. *Hydrometallurgy* **1990**, *24*, 135–156. [CrossRef]
15. Flynn, C.M.; Sandra, L.M. *Cyanide Chemistry—Precious Metals Processing and Waste Treatment*; Information Circular (United States Bureau of Mines) 9429; United States Department of Interior: Washington, DC, USA, 1995; pp. 1–28.
16. Bodzek, M. Membrane Technologies for the Removal of Micropollutants in Water Treatment. In *Advances in Membrane Technologies for Water Treatment: Materials, Processes and Applications*; Basile, A., Cassano, A., Rastogi, N.K., Eds.; Elsevier: Cambridge, UK, 2015; pp. 465–517.
17. Li, N.N.; Chan, R.P.; Naden, D.; Lai, R.W.M. Liquid membrane processes for copper extraction. *Hydrometallurgy* **1983**, *9*, 277–305. [CrossRef]
18. Li, Q.M.; Liu, Q.; Wei, X. Separation study of mercury through an emulsion liquid membrane. *Talanta* **1996**, *43*, 1837–1842. [CrossRef]
19. Kumbasar, R. Selective extraction of cobalt from strong acidic solutions containing cobalt and nickel through emulsion liquid membrane using TIOA as carrier. *J. Ind. Eng. Chem.* **2012**, *18*, 2076–2082. [CrossRef]
20. Pabby, A.K.; Sastre, A.M. State-of-the-art review on hollow fibre contactor technology and membrane-based extraction processes. *J. Membr. Sci.* **2013**, *430*, 263–303. [CrossRef]
21. García, M.G.; Acosta, A.O.; Marchese, J. Emulsion liquid membrane pertraction of Cr(III) from aqueous solutions using PC-88A as carrier. *Desalination* **2013**, *318*, 88–96. [CrossRef]
22. Martin, T.P.; Davies, G.A. The extraction of copper from dilute aqueous solutions using a liquid membrane process. *Hydrometallurgy* **1977**, *2*, 315–334. [CrossRef]
23. Kitagawa, T.; Nishikawa, Y.; Frankenfeld, J.; Li, N.N. Waste water treatment by liquid membrane process. *Environ. Sci. Technol.* **1977**, *11*, 602–605. [CrossRef]
24. Chakraborty, M.; Bhattacharya, C.; Datta, S. Emulsion Liquid Membranes: Definitions and Classification, Theories, Module Design, Applications New Directions and Perspectives. In *Liquid Membranes: Principles and Applications in Chemical Separations and Wastewater Treatment*; Kislik, V.S., Ed.; Elsevier: Amsterdam, The Netherlands, 2010; pp. 141–199.
25. Marr, R.J.; Draxler, J. Emulsion Liquid Membrane Application. In *Membrane Handbook*; Ho, W.S., Sirkar, K.K., Eds.; Chapman & Hall: New York, NY, USA, 1992; pp. 701–717.
26. Chiha, M.; Hamdaoui, O.; Ahmedchekkat, F.; Petrier, C. Study on ultrasonically assisted emulsification and recovery of copper(II) from wastewater using an emulsion liquid membrane process. *Ultrason. Sonochem.* **2010**, *17*, 318–325. [CrossRef] [PubMed]
27. Kumbasar, R.A. Selective separation of chromium(VI) from acidic solutions containing various metal ions through emulsion liquid membrane using trioctylamine as extractant. *Sep. Purif. Technol.* **2008**, *64*, 56–62. [CrossRef]
28. Aydiner, C.; Kobya, M.; Demirbas, E. Cyanide ions transport from aqueous solutions by using quaternary ammonium salts through bulk liquid membranes. *Desalination* **2005**, *180*, 139–150. [CrossRef]
29. Alguacil, F.J.; Alonso, M.; Sastre, A.M. Facilitated supported liquid membrane transport of gold(I) and gold(III) using Cyanex 921. *J. Membr. Sci.* **2005**, *252*, 237–244. [CrossRef]
30. Alguacil, F.J.; Alonso, M. Transport of Au(CN)$_2$$^-$ across a supported liquid membrane using mixtures of amine Primene JMT and phosphine oxide Cyanex 923. *Hydrometallurgy* **2004**, *74*, 157–163. [CrossRef]

31. Pabby, A.K.; Haddad, R.; Alguacil, F.J.; Sastre, A.M. Improved kinetics-based gold cyanide extraction with mixture of LIX79 + TOPO utilizing hollow fiber membrane contactors. *Chem. Eng. J.* **2004**, *100*, 11–22.
32. Kumar, A.; Haddad, R.; Benzal, G.; Ninou, R.; Sastre, A.M. Use of modified membrane carrier system for recovery of gold cyanide from alkaline cyanide media using hollow fiber supported liquid membranes: Feasibility studies and mass transfer modeling. *J. Membr. Sci.* **2000**, *174*, 17–30. [CrossRef]
33. León, G.; Guzmán, M.A. Kinetic study of the effect of carrier and stripping agent concentrations on the facilitated transport of cobalt through bulk liquid membranes. *Desalination* **2005**, *184*, 79–87. [CrossRef]
34. Uddin, M.S.; Kathiresan, M. Extraction of metal ions by emulsion liquid membrane using bi-functional surfactant: Equilibrium and kinetic studies. *Sep. Purif. Technol.* **2000**, *19*, 3–9. [CrossRef]
35. Singh, R.; Mehta, R.R.; Kumar, V. Simultaneous removal of copper, nickel and zinc metal ions using bulk liquid membrane system. *Desalination* **2011**, *272*, 170–173. [CrossRef]
36. Ahmada, A.L.; Kusumastuti, A.; Dereka, C.J.C.; Ooi, B.S. Emulsion liquid membrane for heavy metal removal: An overview on emulsion stabilization and destabilization. *Chem. Eng. J.* **2011**, *171*, 870–882. [CrossRef]
37. ASTM. Standard Test Methods for Cyanides in Water. In *D2036-09, Annual Book of ASTM Standards 2015*; ASTM International: West Conshohocken, PA, USA, 2015; pp. 1–20.
38. Wan, Y.; Zhang, X. Swelling determination of W/O/W emulsion liquid membranes. *J. Membr. Sci.* **2002**, *196*, 185–201. [CrossRef]

![metals](metals logo) MDPI

Article

Separation and Recycling for Rare Earth Elements by Homogeneous Liquid-Liquid Extraction (HoLLE) Using a pH-Responsive Fluorine-Based Surfactant

Shotaro Saito [1], Osamu Ohno [1], Shukuro Igarashi [1,*], Takeshi Kato [2] and Hitoshi Yamaguchi [3]

[1] Department of Biomolecular Functional Engineering, College of Engineering, Ibaraki University, 4-12-1 Nakanarusawa-cho, Hitachi, Ibaraki 316-8511, Japan; 14nd103r@vc.ibaraki.ac.jp (S.S.); osamu.ohno.31@vc.ibaraki.ac.jp (O.O.)

[2] Industrial Technology Institute of Ibaraki Prefecture, 3781-1 Nagaoka, Ibaraki-machi, Ibaraki-gun, Ibaraki 311-3195, Japan; katou@kougise.pref.ibaraki.jp

[3] Research Center for Strategic Materials, National Institute for Material Science, 1-2-1 Sengen, Tsukuba, Ibaraki 305-0047, Japan; yamaguchi.hitoshi@nims.go.jp

* Author to whom correspondence should be addressed; shukuro.igarashi.3@vc.ibaraki.ac.jp; Tel.: +81-294-38-5059; Fax: +81-294-38-5078.

Academic Editor: Suresh Bhargava
Received: 27 July 2015; Accepted: 25 August 2015; Published: 27 August 2015

Abstract: A selective separation and recycling system for metal ions was developed by homogeneous liquid-liquid extraction (HoLLE) using a fluorosurfactant. Sixty-two different elemental ions (e.g., Ag, Al, As, Au, B, Ba, Be, Bi, Ca, Cd, Ce, Co, Cr, Cu, Dy, Er, Eu, Fe, Ga, Gd, Ge, Hf, Hg, Ho, In, Ir, La, Lu, Mg, Mn, Mo, Nb, Nd, Ni, Os, P, Pb, Pd, Pr, Pt, Re, Rh, Ru, Sb, Sc, Se, Si, Sm, Sn, Sr, Ta, Tb, Te, Ti, Tl, Tm, V, W, Y, Yb, Zn, and Zr) were examined. By changing pH from a neutral or alkaline solution (pH $\geq$ 6.5) to that of an acidic solution (pH < 4.0), gallium, zirconium, palladium, silver, platinum, and rare earth elements were extracted at >90% efficiency into a sedimented Zonyl FSA® ($CF_3(CF_2)_n(CH_2)_2S(CH_2)_2COOH$, n = 6–8) liquid phase. Moreover, all rare earth elements were obtained with superior extraction and stripping percentages. In the recycling of rare earth elements, the sedimented phase was maintained using a filter along with a mixed solution of THF and 1 M sodium hydroxide aqueous solution. The Zonyl FSA® was filtrated and the rare earth elements were recovered on the filter as a hydroxide. Furthermore, the filtrated Zonyl FSA was reusable by conditioning the subject pH.

Keywords: homogeneous liquid-liquid extraction; HoLLE; phase separation phenomenon; fluorosurfactant; ion association; rare earth elements; separation and recycling; hydrometallurgy

1. Introduction

Many kinds of rare earth elements such as neodymium, europium, and gadolinium have been recently used in the development of new materials. However, the production of scandium, yttrium, and lanthanoids, collectively referred to as rare earth elements, is not yet widespread across the world. Hence, the potential production of these materials via waste recycling processes has become increasingly important; however, the costs and efficiencies of these recycling processes pose inherent problems. The mutual metal separation method is generally classified into wet and dry methods. Here, the wet method has been employed for reasons such as separation accuracy and energy savings. This typical method utilizes a solvent extraction process which uses solvents that are immiscible with each other. In addition to this process, various separating agents and methods for metal separation have also been recently developed. These include solvent extraction using ionic liquids [1,2], phase separation extraction using stimuli-responsive polymers [3–6], and solid-phase extraction using the adsorption effect of biological materials [7–9]. Among these, it is especially noteworthy that in 1988, Igarashi and Yotsuyanagi discovered the pH-dependent phase separation phenomenon, resulting from the addition of acid to a perfluorooctanoic acid (PFOA) aqueous solution containing trace amounts of acetone [10]. An oily, small secondary phase is resultantly produced, with the target substance ultimately being extracted from therein. Furthermore, the formation of a stable liquid phase which consists of PFOA$^-$ and quaternary salt (Q$^+$ such as tetrabutylammonium ion (TBA$^+$)) was subsequently reported in the following year (1989) [11,12]. The produced liquid phase is stable in water and air; hence, it is known to be the same type as the early ionic liquid reported in 1992 [13] because the liquid is produced at room temperature by ionic associations of the subject cationic and anionic organic compounds (e.g., melting point; -2.3 °C in PFOA$^-$/TBA$^+$ [14]). This formation reaction of an ionic liquid is applied at the same time as the homogeneous liquid-liquid extraction method (HoLLE, type I in Figure 1). Many research papers have been published from the viewpoint of microextraction methods for trace components [15–20]. Hence, HoLLE, a separation method that uses these characteristics effectively, was conceived. In HoLLE, because an aqueous phase and an organic phase are in a homogeneous state at the starting point of extraction, a mechanically vigorous agitation for increasing the contact interface between the aqueous and organic phases in the solvent extraction process is not necessary. This separation method is the extraction to the water-immiscible sedimented phase for the target substance by providing stimuli such as changes in pH, temperature, and light. Some separation and concentration methods for metal chelates using surfactants such as PFOA and slight amounts of water-miscible organic solvents (such as acetone, tetrahydrofuran, *etc.*) have been reported [15–17,19]. In addition, a concentration method for rare earth elements using phosphoric acid and di(2-ethylhexyl)ester (D2EHPA) as an extracting agent in PFOA$^-$/TBA$^+$ has also been reported [21].

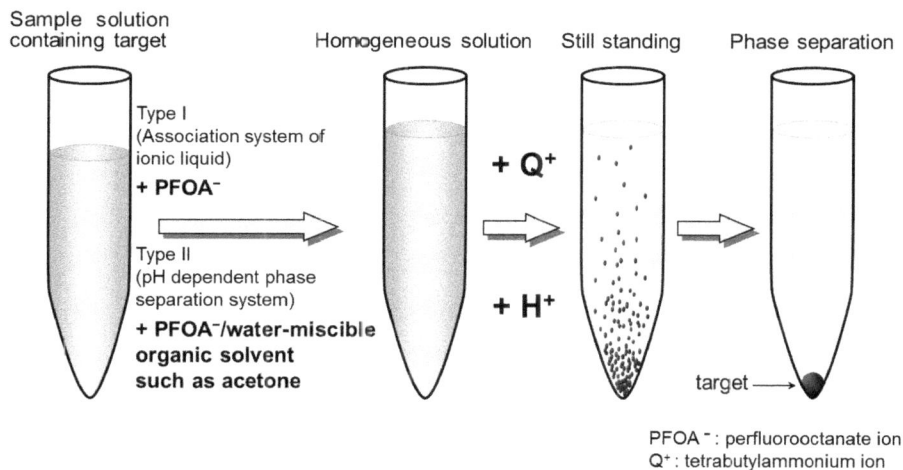

Figure 1. Scheme of homogeneous liquid-liquid extraction method (HoLLE).

In this study, Zonyl FSA®, which is an alternative substance of PFOA$^-$, was used as a phase separation agent in HoLLE. Zonyl FSA® is a 3-[2-(perfluoroalkyl)ethylthio]propanoic acid (CF$_3$(CF$_2$)$_n$(CH$_2$)$_2$S(CH$_2$)$_2$COOH, $n = 6$–8), with an acid dissociation constant (pK$_a$) of approximately 6 [22]. Therefore, this compound displays weak acidity because a spacer of some methylene groups is introduced between the fluorocarbon chain and the hydrophilic group, as compared to the strongly acidic PFOA. Applications of the separation and concentration of chlorophyll-a [22], silver complex [23], copper complex [24], indium complexes [25], and aluminum/titanium with 1,10-phenanthroline complexes [26] have been reported by HoLLE using this phase-separating agent.

In this paper, the extraction for 62 kinds of elements were examined and a high-efficiency separation and recycling system of rare earth elements using Zonyl FSA® as a phase-separating agent based on HoLLE, without the use of a chelating reagent, is described.

2. Experimental Section

2.1. Reagents

Zonyl FSA® was purchased from DuPont (Wilmington, DE, USA). Sixty-two types of elemental ions (Ag, Al, As, Au, B, Ba, Be, Bi, Ca, Cd, Ce, Co, Cr, Cu, Dy, Er, Eu, Fe, Ga, Gd, Ge, Hf, Hg, Ho, In, Ir, La, Lu, Mg, Mn, Mo, Nb, Nd, Ni, Os, P, Pb, Pd, Pr, Pt, Re, Rh, Ru, Sb, Sc, Se, Si, Sm, Sn, Sr, Ta, Tb, Te, Ti, Tl, Tm, V, W, Y, Yb, Zn, and Zr) were selected and their predetermined concentrations were prepared by diluting 1000-ppm standard solutions (received from Kanto Chemicals, Tokyo, Japan) with distilled water. Concretely, Ag(I), Al(III), Ba(II), Be(II), Bi(III), Ca(II), Cd(II), Ce(III), Co(II), Cr(VI), Cu(II), Dy(III), Er(III), Eu(III), Fe(II), Ga(III), Gd(III), Hg(II), Ho(III), In(III), La(III), Lu(III), Mg(II), Mn(II), Nd(III), Ni(II), Pb(II), Pd(II), Pr(III), Sc(III), Se(IV), Sm(III), Sr(II), Tb(III), Tl(I), Tm(III), V(V), Y(III), Yb(III), Zn(II), and Zr(IV) were 0.01−1 M of nitric acid aqueous solution. As(III), Au(III), Ir(IV), Mo(VI), Os(IV), Pt(II), Rh(III), Ru(III), Sb(III), Sn(II), and Te(IV) were 0.4−6 M of hydrochloric acid aqueous solution. Hf(IV), Nb(V), and Ta(V) were 1–3 M of hydrofluoric acid aqueous solution. Ge(IV) and Si(IV) were 0.2–0.4 M potassium hydroxide solution. Ti(IV) was 2 M sulfuric acid aqueous solution. Each of B(III), P(V), Re(VII), and W(VI) were aqueous solution of oxo acid salts containing some ammonium ion. All element solutions were diluted with distilled water to 100 fold. Tetrahydrofuran (THF), acetic acid, sodium acetate, sodium hydroxide, acetone, and all other reagents used were of commercially available analytical grade unless otherwise specified. The utilized filter (for all cases) was a Merck Millipore hydrophilic PTFE filter JAWP02500 (Omnipore, pore size: 1.0 µm, Billerica, MA, USA).

2.2. Apparatuses

The following apparatuses were used: pH meter = Model F-51, manufactured by Horiba Ltd. (Kyoto, Japan); centrifugal separator = Model LC-100, manufactured by TOMY SEIKO Co., Ltd. (Tokyo, Japan); vacuum pump = Model MDA-015, manufactured by ULVAC (Kanagawa, Japan); ICP optical emission spectrometer = Model iCAP6300, manufactured by Thermo Fisher Scientific (Waltham, MA, USA).

2.3. Experimental Procedure

2.3.1. Extraction Characteristics of Elements by HoLLE with Zonyl FSA®

The following were added into a glass vial: 0.5 mL of elemental mixed solution (each element concentration: 10 ppm), 1.0 mL of THF, and 1.0 mL of Zonyl FSA®. Next, 7.5 mL of acetic acid/sodium acetate buffer solution (pH = 4) were added to bring the total mixture volume up to 10 mL. The mixture was then centrifuged for 10 min at 2000 rpm to separate the sedimented (Zonyl FSA®) phase and the supernatant (aqueous) phase. Each rendered post-centrifugal phase volume was $V_{aq.}$ = 9.95 mL and $V_{Zonyl\ FSA^{(r)}}$ = 50 µL. The concentration of elemental ions in the aqueous phase was determined by ICP-OES.

The extraction percentage (*E* %) of elemental ions was calculated as

$$E\ \% = 100\ (1 - C_{aq.}/C_{Total}) \tag{1}$$

where C_{Total} and $C_{aq.}$ represent the total concentration of the added element and the concentration of the element in the aqueous phase after separation, respectively. Of note, the error range of each element's extraction percentage is ±5% or less.

2.3.2. The Recovery of Elements from the Zonyl FSA® Phase and the Redissolution of the Zonyl FSA® Phase

A water-immiscible Zonyl FSA® phase has retained the state of a stable liquid phase in the aqueous solution, and the elemental ions were recovered by filtration. In the recovery of targets, rare earth elements (Sc, Y, and lanthanoids) were well extracted by the experiment described in Section 2.3.1, and they were thus selected as targets. A Zonyl FSA® phase (approximately 50 µL) was produced after rare earth elemental extraction, which held on the hydrophilic PTFE filter. Water-immiscible Zonyl FSA® was filtered by adding 1 mL of sodium hydroxide and THF solution, which were mixed at a volume ratio of 1:1. The elements were trapped on the filter as hydroxides, the recovery percentage of elemental ions in the filtrate phase was determined by ICP-OES. Furthermore, the rare earth elements were recycled again by the same procedure outlined in Section 2.3.1 using the filtrated Zonyl FSA®.

3. Results and Discussion

3.1. The Extraction of THF within the Phase Separation Phenomenon

As solubilizing agents for Zonyl FSA®, acetone and THF are examined. THF was selected from the viewpoint of its concentration time and operability, because it was already reported that the sedimented volume and its associated viscosity increase in the case of using acetone [22]. The relationship of the volume percentage of THF in the sedimented volume showed linear behavior in the range of 10–35 vol. % for the THF volumes shown in Figure 2 below. In the case of below 10 vol. % THF, the Zonyl FSA® phase precipitated fine solid particles. Moreover, Zonyl FSA® did not form a sedimented phase in the case of THF exceeding 50 vol. %. Therefore, THF volume in this system was ultimately deemed appropriate for 10 vol. %.

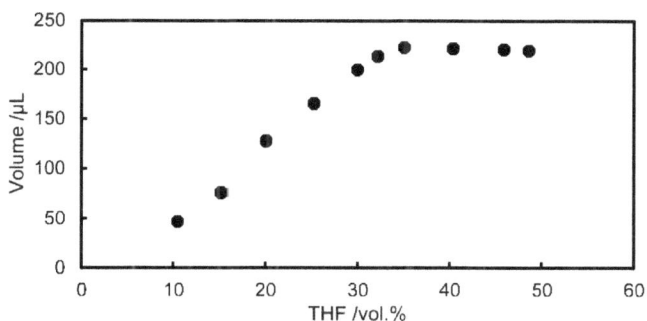

Figure 2. Relationship between the volume percentage of THF and the volume of sedimented phase. (Zonyl FSA®) = 10 vol. %; pH = 4.

3.2. Screening of Elements

The experimental procedure was performed according to Section 2.3.1. The percentage of extracted element ions in the sedimented phase is shown in Figure 3. Here, some element ions in the sample solution are estimated to exist as different type species such as a free metal ion, an anion complex such as gold chloride complex or platinum chloride complex, and an oxo anion species such as boric acid or phosphoric acid. Gallium, zirconium, palladium, silver, platinum, and rare earth elements (except for praseodymium) were obtained with a high extraction percentage; however, gold, indium, and germanium were obtained with a very low extraction percentage. With respect to rare earth elements, most rare earth elements are stable in the trivalent state, and they do not produce a hydroxide unless the pH range is of weak basicity to neutrality, e.g., since the solubility product constant (K_{sp}) of lanthanum is 1.0×10^{-19} [27], it does not precipitate as a hydroxide during the extraction procedure (pH = 4.0). On the other hand, only cerium was obtained with a notably low extraction percentage. This is because cerium is thought to exist in trivalent and tetravalent states in an aqueous solution with a pH of approximately 4. Based on this reason, tetravalent cerium is precipitated as a hydroxide in an acidic condition and, thus, is still susceptible to hydrolysis [28,29]. Therefore, the extraction into the sedimented Zonyl FSA® phase can present inherent challenges because of the fact that tetravalent cerium has usually already hydrolyzed at the time of extraction in a pH equal to 4.

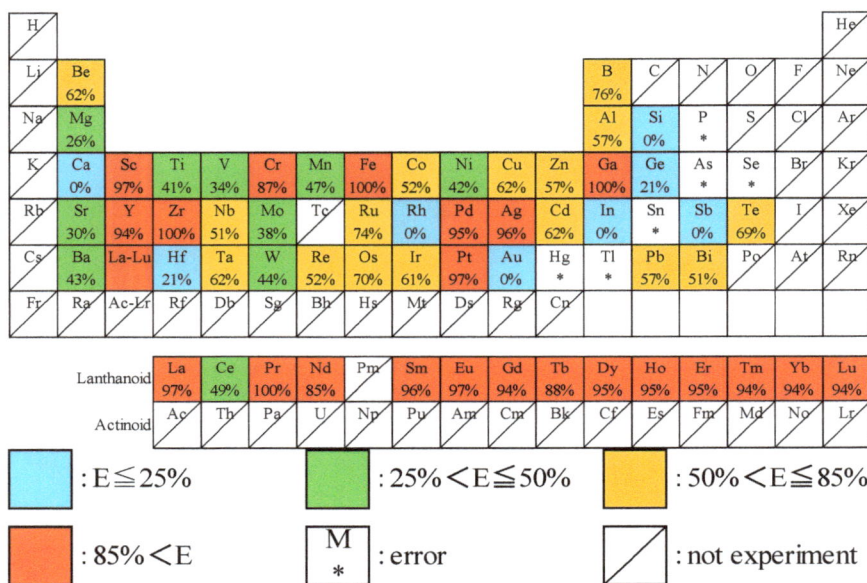

Periodic table extraction percentages:

- Be 62%; B 76%
- Mg 26%; Al 57%; Si 0%
- Ca 0%; Sc 97%; Ti 41%; V 34%; Cr 87%; Mn 47%; Fe 100%; Co 52%; Ni 42%; Cu 62%; Zn 57%; Ga 100%; Ge 21%; As *; Se *
- Sr 30%; Y 94%; Zr 100%; Nb 51%; Mo 38%; Tc; Ru 74%; Rh 0%; Pd 95%; Ag 96%; Cd 62%; In 0%; Sn *; Sb 0%; Te 69%
- Ba 43%; La-Lu; Hf 21%; Ta 62%; W 44%; Re 52%; Os 70%; Ir 61%; Pt 97%; Au 0%; Hg *; Tl *; Pb 57%; Bi 51%

Lanthanoid: La 97%; Ce 49%; Pr 100%; Nd 85%; Pm; Sm 96%; Eu 97%; Gd 94%; Tb 88%; Dy 95%; Ho 95%; Er 95%; Tm 94%; Yb 94%; Lu 94%

Actinoid: Ac; Th; Pa; U; Np; Pu; Am; Cm; Bk; Cf; Es; Fm; Md; No; Lr

Legend:
- : $E \leq 25\%$
- : $25\% < E \leq 50\%$
- : $50\% < E \leq 85\%$
- : $85\% < E$
- M * : error
- : not experiment

Figure 3. Summary of the extraction percentages for elemental ions with Zonyl FSA®.

3.3. Stripping of Rare Earth Elements

The extraction behavior of elements into the Zonyl FSA® liquid phase was examined by focusing on rare earth elements. Filtration operation by the filter was evaluated for the sedimented phase containing rare earth elements. The sedimented phase was held/maintained by a microsyringe on a Teflon filter (Pore size: 1 µm). The rare earth elements were collected on the filters as hydroxides by a 1:1 mixed solvent of 1 mL THF and 1 M sodium hydroxide aqueous solution, whereas the Zonyl FSA® was dissolved in the procedure. That rare earth elements had been stripped was confirmed by measuring Zonyl FSA® aqueous solution that was dissolved and filtrated using ICP-OES (Table 1).

Table 1. Stripping percentages of rare earth elements from the Zonyl FSA® phase.

Metal	Sc	Y	La	Ce	Pr	Nd	Sm	Eu	Gd	Tb	Dy	Ho	Er	Tm	Yb	Lu
Recovery %	81	92	88	90	91	88	91	92	92	91	92	92	84	92	92	92

3.4. Reuse of Spent Zonyl FSA®

The subject rare earth elements could be separated from the sedimented Zonyl FSA® phase by filtration, and the sedimented liquid phase similar to what existed before extraction was formed by adding a buffer solution (pH = 4) in the filtrate containing the Zonyl FSA®. Based on this phenomenon, the stripping experiments for rare earth elements were performed again using the spent Zonyl FSA®. The related findings are shown below in Table 2. Part of the heavy elements has a higher extraction percentage in the second extraction compared to the first extraction. Because part of the Zonyl FSA was lost during the recycling procedure, the component balance was changed in each fluorinated carbon chain number (6–8) of Zonyl FSA®. Thereby, the extraction selectivity for heavy elements was relatively increased. Zonyl FSA® was found to be reusable for the extraction and stripping of rare earth elements.

Table 2. Comparison of extraction percentages in first and second extraction times.

Metal	Sc	Y	La	Ce	Pr	Nd	Sm	Eu	Gd	Tb	Dy	Ho	Er	Tm	Yb	Lu
1st Extraction %	97	94	97	49	100	85	96	97	94	88	95	95	95	94	94	94
2nd Extraction %	80	100	93	53	100	84	100	100	100	94	100	100	94	100	100	100

3.5. Overview of Separation and Recycling System of Rare Earth Elements Using Zonyl FSA®

Extracting rare earth elements by Zonyl FSA® is a plausible recovery technique using a simple filtration operation, and the spent Zonyl FSA® can be reused. Therefore, a separation and recycling system for rare earth elements using Zonyl FSA® was constructed, with the subject scheme shown below in Figure 4. In this system, the Zonyl FSA® and THF were added to metal solutions containing rare earth elements (depicted in the figure), with the solution resultantly forming a homogeneous state by gentle shaking. In addition, a weak acid buffer solution is introduced into the solution to extract the rare earth elements, and the Zonyl FSA® is subsequently aggregated by centrifugation for a period. The sedimented Zonyl FSA® phase is ultimately retained on the filter. Then, only Zonyl FSA® is filtrated and the rare earth elements are recovered on the filter as hydroxides. In contrast, Zonyl FSA® solution is obtained as a filtrate to around a neutral pH (pH $\geq$ 6.5). Accordingly, it is capable as a second recycling product to be used again for separation and recovery, as demonstrated below in the figure.

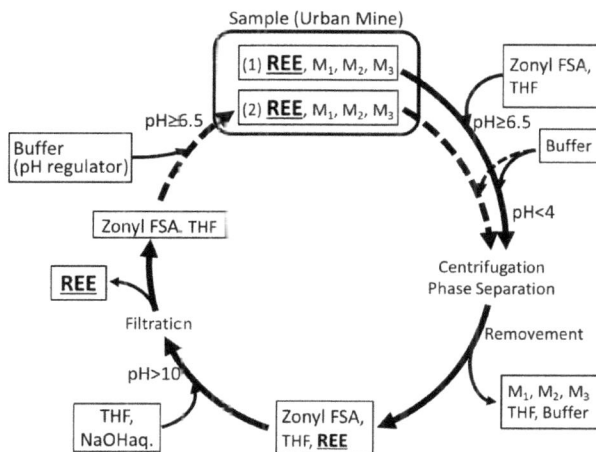

Figure 4. Separation and recycling system for rare earth elements using Zonyl FSA®. REE: rare earth elements; M_X: coexistent metals.

4. Conclusions

A homogeneous liquid-liquid extraction method (HoLLE) for metal elements using a Zonyl FSA® fluorosurfactant instead of a chelating agent was proposed. As a result, metal elements such as palladium, platinum, and rare earth elements were extracted at a very high percentage. This method is highly capable of extracting and stripping (with high efficiency) rare earth elements by HoLLE via the use of Zonyl FSA®. This method is expected to be a preprocessing technique for separating and recycling rare earth elements in electrical appliance components. Moreover, the developed process is capable of both the recycling of rare earth elements and the reuse of Zonyl FSA® as an extraction agent.

Acknowledgments: The authors thank Toshinori Fukazawa for assistance with the screening experiment of metals. This work was supported by JSPS KAKENHI Grant Numbers 15H02847. The authors would like to thank Enago (www.enago.jp) for the English language review.

Author Contributions: S.S. and S.I. conceived the experiments, and wrote the paper; S.S. designed experiments with contributions from S.I., O.O., T.K. and H.Y.; H.Y. was measured of metals by ICP-OES; S.S. performed experiments and interpretation.

Conflicts of Interest: The authors declare no conflict of interest.

References

1. Yang, J.; Kubota, F.; Baba, Y.; Kamiya, N.; Goto, M. One step effective separation of platinum and palladium in an acidic chloride solution by using undiluted ionic liquids. *Solvent Extr. Res. Dev. Jpn.* **2014**, *21*, 129–135. [CrossRef]
2. Yang, J.; Kubota, F.; Baba, Y.; Kamiya, N.; Goto, M. Separation of precious metals by using undiluted ionic liquids. *Solvent Extr. Res. Dev. Jpn.* **2014**, *21*, 89–94. [CrossRef]
3. Saitoh, T.; Sugiura, Y.; Asano, K.; Hiraide, M. Chitosan-conjugated thermos-responsive polymer for the rapid removal of phenol in water. *React. Funct. Polym.* **2009**, *69*, 792–796. [CrossRef]
4. Kato, T.; Igarashi, S.; Ohno, O.; Watanabe, Y.; Murakami, K.; Takemori, T.; Yamaguchi, H.; Ando, R. Separation and recovery properties of rare earth elements using a pH-sensitive polymer having benzoic acid substituent group. *Bunseki Kagaku* **2012**, *61*, 235–242. [CrossRef]
5. Igarashi, S.; Saito, S.; Kato, T.; Okano, G.; Yamaguchi, H. Development of separation and recovery system of gold and rare earth elements using stimuli-responsive polymers—G-MOVE system and La-VEBA system. *J. Surf. Finish. Soc. Jpn.* **2012**, *63*, 630–632. [CrossRef]
6. Saito, S.; Igarashi, S.; Yamaguchi, H. Selective collection characteristics and separation/recovery method of gold(III), silver(I) and platinum(II) with pH-sensitive polymer under addition of L-ascorbic acid. *Bunseki Kagaku* **2014**, *63*, 791–795. [CrossRef]
7. Inoue, K.; Ohto, K.; Yoshizuka, K.; Shinbaru, R.; Baba, Y.; Kina, K. Adsorption behavior of metal ions on some carboxymethylated chitosans. *Bunseki Kagaku* **1993**, *42*, 725–731. [CrossRef]
8. Parajuli, D.; Kawakita, H.; Inoue, K.; Ohto, K.; Kajiyam, K. Persimmon peel gel for the selective recovery of gold. *Hydrometallurgy* **2007**, *87*, 133–139. [CrossRef]
9. Gurung, M.; Adhikari, B.B.; Kawakita, H.; Ohto, K.; Inoue, K. Recovery of Au(III) by using low cost adsorbent prepared from persimmon tannin extract. *Chem. Eng. J.* **2011**, *174*, 556–563. [CrossRef]
10. Igarashi, S.; Yotsuyanagi, T. A novel homogeneous liquid-liquid Extraction by pH dependent phase separation with fluorocarbon ionic surfactant. In Proceedings of the Symposium on Solvent Extraction 1988, Tokyo, Japan, 5–7 December 1988; pp. 175–180.
11. Igarashi, S.; Yotsuyanagi, T. Temperature dependent phase transformation in the homogeneous liquid-liquid extraction with fluorocarbon surfactant. In Proceedings of the Symposium on Solvent Extraction 1989, Sendai, Japan, 9–11 November 1989; pp. 51–54.
12. Igarashi, S.; Yotsuyanagi, T. New homogeneous liquid-liquid extraction by phase separation and transformation with fluorocarbon surfactant and quaternary ammonium salt. In Solvent extraction 1990, Proceedings of the International Solvent Extraction Conference (ISEC '90), Kyoto, Japan, 16–21 July 1992; Sekine, T., Ed.; Elsevier: New York, NY, USA, 1992; pp. 1725–1730.
13. Wilkes, J.S.; Zaworotko, M.J. Air and water stable 1-ethyl-3-methylimidazolium based ionic liquids. *J. Chem. Soc. Chem. Commun.* **1992**, *13*, 965–967. [CrossRef]

14. Yamaguchi, H. Did the study of ionic liquids begin from analytical chemistry? *Bunseki* **2007**, *11*, 608–609.

15. Igarashi, S.; Yotsuyanagi, T. Homogeneous liquid-liquid extraction by pH dependent phase separation with a fluorocarbon ionic surfactant and its application to the preconcentration of porphyrin compounds. *Mikrochimica Acta* **1992**, *106*, 37–44. [CrossRef]

16. Oshite, S.; Igarashi, S. Homogeneous liquid-liquid extraction using perfluorooctanesulfonic acid and calcium and its application to the separation and recovery vitamin B_{12}. *J. Chem. Technol. Biotechnol.* **1999**, *74*, 1183–1187. [CrossRef]

17. Oshite, S.; Furukawa, M.; Igarashi, S. Homogeneous liquid-liquid extraction method for the selective spectrofluorimetric determination of trace amoutns of tryptophan. *Analyst* **2001**, *126*, 703–706. [CrossRef] [PubMed]

18. Takahashi, A.; Ueki, Y.; Igarashi, S. Highly efficient homogeneous liquid-liquid extraction of rare earth metal ions from perfluorocarboxylate surfactant solutions using ion-pair phase separation systems. *Solvent Extr. Res. Dev. Jpn.* **2001**, *8*, 235–240.

19. Hoogerstraete, T.V.; Ohghena, B.; Binnemans, K. Homogeneous liquid-liquid extraction of rare earths with the betaine-batainium bis(trifluoromethylsulfonyl)imide ionic liquid system. *Int. J. Mol. Sci.* **2013**, *14*, 21353–21377. [CrossRef] [PubMed]

20. Takagai, Y.; Igarashi, S. Homogeneous liquid-liquid extraction method as simple and powerful preconcentration for capillary gas chromatography and capillary electrophoresis. *Am. Lab.* **2002**, *34*, 29–30.

21. Fuchimukai, J.; Yamaguchi, H.; Meguro, Y.; Kubota, T.; Igarashi, S. Highly efficient homogeneous liquid-liquid extraction of lanthanoid ons in a strong acidic solution. *Solvent Extr. Res. Dev. Jpn.* **2006**, *13*, 139–146.

22. Sudo, T.; Igarashi, S. Homogeneous liquid-liquid extraction method for spectrofluorimetric determination of chlorophyll a. *Talanta* **1996**, *43*, 233–237. [CrossRef]

23. Ghiasvand, A.R.; Moradi, F.; Sharghi, H.; Hasaninejad, A.R. Determination of silver(I) by electrothermal-AAS in a microdroplet formed from a homogeneous liquid-liquid extraction system using tetraspirocyclohexylcalix[4]pyrroles. *Anal. Sci.* **2005**, *21*, 387–390. [CrossRef] [PubMed]

24. Ghiasvand, A.R.; Shadabi, S.; Kakanejadifard, A.; Khajehkoolaki, A. Synthesis of a new α-dioxime derivative and its application for selective homogeneous liquid-liquid extraction of Cu(II) into a microdroplet followed by direct GFAAS determination. *Bull. Korean Chem. Soc.* **2005**, *26*, 781–785.

25. Kato, T.; Igarashi, S.; Ishiwatari, Y.; Furukawa, M.; Yamaguchi, H. Separation and concentration of indium from a liquid crystal display via homogeneous liquid-liquid extraction. *Hydrometallurgy* **2013**, *137*, 148–155. [CrossRef]

26. Yamaguchi, H.; Itoh, S.; Igarashi, S.; Kobayashi, T. Homogeneous liquid-liquid extraction of metal-1,10-phenanthoroline chelates in a weak acidic solution. *Bunseki Kagaku* **2005**, *54*, 227–230. [CrossRef]

27. Sinha, S.P. *Complexes of the Rare Earths*; Pergamon Press Ltd.: Oxford, UK, 1966; p. 26.

28. Suzuki, Y. *Kidorui no Hanashi [The Story of Rare Earth Elements]*; Shokabo Publishing Co., Ltd.: Chiyoda, Japan, 1998; pp. 66–72.

29. Burgess, J. *Metal Ions in Solution*; Ellis Horwood Ltd.: New York, NY, USA, 1978; pp. 266–267.